T0309731

Biophilic Design

Biophilic Design

The Theory, Science, and Practice of Bringing Buildings to Life

EDITED BY:

Stephen R. Kellert

Judith H. Heerwagen

Martin L. Mador

John Wiley & Sons, Inc.

This book is printed on acid-free paper. ♾

Published by John Wiley & Sons, Inc., Hoboken, New Jersey
Published simultaneously in Canada

Library of Congress Cataloging-in-Publication Data:

Biophilic design : the theory, science, and practice of bringing buildings to life / edited by Stephen R. Kellert, Judith H. Heerwagen, Martin L. Mador.
p. cm.
Includes bibliographical references and index.
ISBN 978-0-470-16334-4 (cloth)
1. Architecture—Environmental aspects 2. Architecture—Human factors. I. Kellert, Stephen R. II. Heerwagen, Judith H., 1944– III. Mador, Martin L., 1949–
NA2542.35.E44 2008
720'.47—dc22

2007023228

Printed in the United States of America

SKY10062297_121123

Contents

Preface vii
Stephen R. Kellert and Judith H. Heerwagen

Acknowledgments xi

Prologue: In Retrospect xiii
Hillary Brown

PART I The Theory of Biophilic Design 1

Chapter **1** Dimensions, Elements, and Attributes of Biophilic Design 3
Stephen R. Kellert

Chapter **2** The Nature of Human Nature 21
Edward O. Wilson

Chapter **3** A Good Place to Settle: Biomimicry, Biophilia, and the Return of Nature's Inspiration to Architecture 27
Janine Benyus

Chapter **4** Water, Biophilic Design, and the Built Environment 43
Martin L. Mador

Chapter **5** Neuroscience, the Natural Environment, and Building Design 59
Nikos A. Salingaros and Kenneth G. Masden II

PART II The Science and Benefits of Biophilic Design 85

Chapter **6** Biophilic Theory and Research for Healthcare Design 87
Roger S. Ulrich

Chapter **7** Nature Contact and Human Health: Building the Evidence Base 107
Howard Frumkin

Chapter **8** Where Windows Become Doors 119
Vivian Loftness with Megan Snyder

Chapter **9** Restorative Environmental Design: What, When, Where, and for Whom? 133
Terry Hartig, Tina Bringslimark, and Grete Grindal Patil

Chapter **10** Healthy Planet, Healthy Children: Designing Nature into the Daily Spaces of Childhood 153
Robin C. Moore and Clare Cooper Marcus

Chapter **11** Children and the Success of Biophilic Design 205
Richard Louv

Chapter **12** The Extinction of Natural Experience in the Built Environment 213
David Orr and Robert Michael Pyle

Part III The Practice of Biophilic Design 225

Chapter **13** Biophilia and Sensory Aesthetics 227
Judith H. Heerwagen and Bert Gregory

Chapter **14** Evolving an Environmental Aesthetic 243
Stephen Kieran

Chapter **15** The Picture Window: The Problem of Viewing Nature Through Glass 253
Kent Bloomer

Chapter **16** Biophilic Architectural Space 263
Grant Hildebrand

Chapter **17** Toward Biophilic Cities: Strategies for Integrating Nature into Urban Design 277
Timothy Beatley

Chapter **18** Green Urbanism: Developing Restorative Urban Biophilia 297
Jonathan F. P. Rose

Chapter **19** The Greening of the Brain 307
Pliny Fisk III

Chapter **20** Bringing Buildings to Life 313
Tom Bender

Chapter **21** Biophilia in Practice: Buildings That Connect People with Nature 325
Alex Wilson

Chapter **22** Transforming Building Practices Through Biophilic Design 335
Jenifer Seal Cramer and William Dee Browning

Chapter **23** Reflections on Implementing Biophilic Design 347
Bob Berkebile and Bob Fox, with Alice Hartley

Contributors 357

Image Credits 365

Index 371

Preface

Stephen R. Kellert and Judith H. Heerwagen

This book immodestly aspires to help mend the prevailing breach existing in our society between the modern built environment and the human need for contact with the natural world. In this regard, the chapters in this volume focus on the theory, science, and practice of what we call *biophilic design*, an innovative approach that emphasizes the necessity of maintaining, enhancing, and restoring the beneficial experience of nature in the built environment. Although we present biophilic design as an innovation today, ironically, it was the way buildings were designed for much of human history. Integration with the natural environment; use of local materials, themes and patterns of nature in building artifacts; connection to culture and heritage; and more were all tools and methods used by builders, artisans, and designers to create structures still among the most functional, beautiful, and enduring in the world.

The authors in this book represent widely diverse disciplines, including architects, natural scientists, social scientists, health professionals, developers, practitioners, and others who offer an original and timely vision of how we can achieve not just a sustainable but also a more satisfying and fulfilling modern society in harmony with nature. Collectively, they articulate a paradigm shift in how we design and build with nature in mind. Still, biophilic design is not about greening our buildings or simply increasing their aesthetic appeal through inserting trees and shrubs. Much more, it is about humanity's place in nature, and the natural world's place in human society, a space where mutuality, respect, and enriching relation can and should exist at all levels and emerge as the norm rather than the exception.

Biophilic design at any scale from buildings to cities begins with a simple question: How does the built environment affect the natural environment, and how will nature affect human experience and aspiration? Most of all, how can we achieve sustained and reciprocal benefits between the two?

The idea of biophilic design arises from the increasing recognition that the human mind and body evolved in a sensorially rich world, one that continues to be critical to people's health, productivity, emotional, intellectual, and even spiritual well-being. The emergence during the modern age of large-scale agriculture, industry, artificial fabrication, engineering, electronics, and the city represents but a tiny fraction of our species' evolutionary history. Humanity evolved in adaptive response to natural conditions and stimuli, such as sunlight, weather, water, plants, animals, landscapes, and habitats, which continue to be essential contexts for human maturation, functional development, and ultimately survival.

Unfortunately, modern technical and engineering accomplishments have fostered the belief that humans can transcend their natural and genetic heritage. This presumption has encouraged a view of humanity as having escaped the dictates of natural systems, with human progress and civilization measured by its capacity for fundamentally altering and transforming the natural world. This dangerous illusion has given rise to an architectural practice that encourages overexploitation, environmental degradation, and separation of people from natural systems and processes. The dominant paradigm of design and development of the modern built environment has become one of unsustainable energy and resource consumption, extensive air and water pollution, widespread atmospheric and climate alteration, excessive waste generation, unhealthy indoor environmental conditions, increasing alienation from nature, and growing "placelessness." One of the volume's authors, David Orr (1999:212–213), described this lamentable condition in this way:

> Most [modern] buildings reflect no understanding of ecology or ecological processes. Most tell its users that knowing where they are is unimportant.

> Most tell its users that energy is cheap and abundant and can be squandered. Most are provisioned with materials and water and dispose of their wastes in ways that tell its occupants that we are not part of the larger web of life. Most resonate with no part of our biology, evolutionary experience, or aesthetic sensibilities.

Recognition of the necessity to change this self-defeating paradigm has led to significant efforts at minimizing and mitigating the adverse environmental and human health impacts of modern development. These efforts have resulted in the growth of the sustainable or green design movement, dramatically illustrated by the extraordinary rise of the U.S. Green Building Council's LEED certification and rating system. While commendable and necessary, these efforts will ultimately be insufficient to achieving the long-term goal of a sustainable, healthy, and well-functioning society.

The basic deficiency of current sustainable design is a narrow focus on avoiding harmful environmental impacts, or what we call *low environmental impact design*. Low environmental impact design, while fundamental and essential, fails to address the equally critical needs of diminishing human separation from nature, enhancing positive contact with environmental processes, and building within a culturally and ecologically relevant context, all basic to human health, productivity, and well-being. These latter objectives are the essence of biophilic design. True and lasting sustainability must combine both low environmental impact and biophilic design, the result being what is called *restorative environmental design* (Kellert 2005). This book, in effect, contends that biophilic design has been until now the largely missing link in current sustainable design. The various chapters attempt to redress this imbalance.

The notion of biophilic design derives from the concept of *biophilia*, the idea that humans possess a biological inclination to affiliate with natural systems and processes instrumental in their health and productivity. Originally proposed by the eminent biologist and one of the volume's authors, Edward O. Wilson, biophilia has been eloquently described by Wilson in this way (1984:35): "To explore and affiliate with life is a deep and complicated process in mental development. To an extent still undervalued . . ., our existence depends on this propensity, our spirit is woven from it, hope rises on its currents." The idea of biophilia is elucidated elsewhere (Wilson 1984, Kellert and Wilson 1993, Kellert 1997), and described in chapters in this volume by Kellert and E. O. Wilson.

Biophilic design is the expression of the inherent human need to affiliate with nature in the design of the built environment. The basic premise of biophilic design is that the positive experience of natural systems and processes in our buildings and constructed landscapes remains critical to human performance and well-being. Various chapters in the volume cite growing scientific evidence to corroborate this assumption in studies of health care, the workplace, childhood development, community functioning, and more. More generally, the authors offer insight and understanding regarding the theory, science, and practice of biophilic design.

Part I of the book focuses on a conceptual understanding of biophilia and biophilic design. Chapters by Kellert, E. O. Wilson, Benyus, Mador, and Salingaros and Masden offer various biological and cultural understandings of the human need to affiliate with natural systems, and how this inclination can be achieved through design of the built environment. The authors address the neglect of the human-nature connection in modern architecture and construction, a condition the eminent architectural historian Vincent Scully described in this way (1991:11): "The relationship of man-made structures to the natural world . . . has been neglected by architecture. . . . There are many reasons for this. Foremost among them . . . is the blindness of the contemporary urban world to everything that is not itself, to nature most of all."

A major cause for this blindness has been the lack of empirical evidence revealing the illogical and self-defeating consequences of designing in adversarial relation to the natural environment. Part II of the book provides much of this needed evidentiary material, particularly the many health and productivity benefits of biophilic design, as well as the harmful consequences of impeding and degrading human contact with natural systems and processes. Chapters by Ulrich, Frumkin, Loftness, and Hartig and colleagues delineate a range of health, physical, emotional, and intellectual advantages of building and landscape designs that facilitate the positive experience of nature. Additional chapters by Moore and Marcus, Louv, and Pyle and Orr describe the importance of nature in childhood maturation, how to foster this connection through the design of residential and educational settings, and the deleterious and potentially disastrous consequences of doing otherwise.

Part III focuses on the practical challenge of implementing biophilic design, most particularly how to transform conventional and prevailing sustainable design practice. Chapters by Heerwagen and Gregory, Kieran, Bloomer, Hildebrand, Fisk, and Bender provide insight and guidance regarding the architectural expression of biophilic design, focusing largely on the building and site scale. Additional chapters by Beatley and Rose emphasize how to foster the human-nature connection at the neighborhood, community, and urban scales, even what Beatley ambitiously calls the creation of "biophilic cities." The challenge of transforming the process of design and development essential to implementing biophilic design is addressed in chapters by Alex Wilson, Cramer and Browning, and Fox and Berkebile.

We believe this volume will greatly advance our notions of sustainable, biophilic, and restorative environmental design. Still, our efforts remain a work in progress, with much more to learn about the elusive expression of the inherent human need to affiliate with nature in the design and construction of our buildings, landscapes, communities, neighborhoods, and cities.

References

Kellert, S. 1997. *Kinship to Mastery: Biophilia in Human Evolution and Development*. Washington, DC: Island Press.

Kellert, S. 2005. *Building for Life: Understanding and Designing the Human-Nature Connection*. Washington, DC: Island Press.

Kellert, S., and E.O. Wilson, eds. 1993. *The Biophilia Hypothesis*. Washington, DC: Island Press.

Orr, D. 1999. "Architecture as Pedagogy." In *Reshaping the Built Environment*, edited by C. Kibert. Washington, DC: Island Press.

Scully, V. 1991. *Architecture: The Natural and the Manmade*. New York: St. Martin's Press.

Wilson, E. O. 1984. *Biophilia: The Human Bond with Other Species*. Cambridge, MA: Harvard University Press.

Acknowledgments

This timely and, we hope, highly relevant book emerged from a three-day meeting in a beautiful retreat setting known as "Whispering Pines" in rural Rhode Island in May 2006. This extraordinary setting and gathering of leading scientists, designers, practitioners, and others was made possible by the support of many generous benefactors. We particularly appreciate the major assistance of the Geraldine R. Dodge Foundation and its visionary president, David Grant. Additional critical support was provided by the William and Flora Hewlett Foundation, the Edward and Dorothy Kempf Fund at Yale University, the Vervane Foundation and especially Josephine Merck, the Hixon Center for Urban Ecology at the Yale School of Forestry and Environmental Studies, and Rev. Albert P. Neilson. Further support for the project was provided by the Henry Luce Foundation.

A number of Yale University students were especially helpful in hosting the symposium and in other vital ways. Particular thanks are due Ben Shepherd. Others who provided critical assistance included Roderick Bates, Christopher Clement, Gwen Emery, Maren Haus, Sasha Novograd, Judy Preston, Chris Rubino, Jill Savery, Ali Senauer, Adrienne Swiatocha, Terry Terhaar, and Christopher Thompson.

Finally, we very much thank our editor at John Wiley, Margaret Cummins, for her considerable confidence and support.

Prologue: In Retrospect

Hillary Brown

During a visit to Turkey more than two decades ago, my companions and I shared pilgrimages to that country's Arcadian ruins, the rock-cut underworld of Cappadocia, and other rewarding sights. At one stop along the Aegean coast, we spent the night seaside at a resort community. With construction detritus everywhere, it was in a graceless stage of formation, its platted but unbuilt streets undoing the modesty of the village. A dozen hotels along the beach elbowed for sea frontage, gleaming glass and concrete towers, each straining to trump the other as more formally prominent, more luxuriously endowed.

In contrast, the entry to our hotel was undistinguished, even obscure, a suggestive breach in a white wall, solid for its several-storied height. Over the threshold, we found ourselves within a long narrow courtyard open to the elements. The sky overhead (experienced as one would an artwork by James Turrell—not as passive observer, but as participant) was an azure slash. At the far end, the sky ballooned above what appeared to be a plaza.

We were seduced down this street that was mostly self-shaded and cooled by a gentle updraft. Trees and plantings dotted the surfaces, muting the noise of our progress. Underfoot, the upended and sea worn cobble paving was punctuated with sandstone slabs at the entries to adjacent spaces, texturing our sound as alternatively smooth or gritty down the length of the corridor.

Overhead, the walls were faced with windows and doors that opened onto balconies hanging out over this narrow street, beaming like so many smiles. Most casements were flung open, others still shuttered against the morning. Quite a few were peopled, elbows on sills, whispering shared delight at awakening in this communal scene.

The building was vocalizing, its diverse din a contemporary rendering of an ancient Mediterranean village. From the far end came soft social sounds—footfalls, a child's exclamation, the soft rise and fall of treble and bass voices. Fountains and laughter stippled the air, while clattering silverware broadcast the locale of a café. From here, just as our ears took in the softness of breaking waves, our nostrils detected and eyes at once confirmed the full expanse of the Aegean. Magnifying our senses while buffering us from everything else, the hotel was channeling the sea.

I remember my sense of gratification as well as curious agitation in taking in this unexpected place, an experience of architectural pleasure that resonated as both new and unfathomably familiar. For the first (and since then, only) time I knew, as I turned to my companions and announced with conviction, that a woman had designed this building. To my friends' astonishment, the hotel manager readily confirmed that yes, in fact, a woman-led practice in Istanbul had won the commission.

For years since, I've given thought to that sharp, almost physiological insight, that instant knowing-in-my-bones that arose from a shared design sensibility. Was it how she closed our eyes and ears to the chaos of this beach community, or how she choreographed our movements to dilate the experience in time, intensifying this sensual introduction to the sea? Perhaps it was her preference for socialized space, invoking a primordial practice of *sharing* exquisite places rather than reserving them for private consumption. In setting itself apart, her retreat, after all, recalled the archetypal Islamic caravansary—that protective, walled compound found at intervals along desert trading routes where travelers together sought refreshment and protection. How compelling was this concept, in contrast to the extravagant resorts next door that claimed visual primacy and exclusivity, ignoring the cultural landscape.

Given an emergent environmental consciousness at the time, I now more fully appreciate this architect's accomplishment. My ecstatic moment responded to an artistry that was inventive yet contextual—and deeply ecological. Her rendering of bioregion and climate expressed the essence of *genius loci*—the spirit of the place. Rather than facing the private rooms seaward, she spurned convention by turning them inward, unfolding the sea to us as singular, shared experience.

Just as she intensified the revelation of place, this architect refurbished our faculties by exploiting the intelligence and detailed richness of the natural world, using local resources metamorphosed by time and human agency. She distilled natural materials to their elegant simplicity and rightness of fit. As with ecological designers today, nature was employed here as intrinsic to our biological being, a voice converging through several senses. Our wayfinding to the sparkling sea was intensified with textural, acoustic, and olfactory clues.

Today, many of us realize that successfully communicating the ethical imperative of the green design movement will depend on innovative and compelling expression. In this building, long ago, I glimpsed just such an aesthetic of persuasion—one fundamentally place-based and participatory, experienced through all the senses. While then, this distinguishing green voice struck me as gender-specific, I recognize it today as a responsiveness by no means exclusive to women.

Unnamed at the time, such design sensibilities have recently coalesced for me around the word *biophilic* and now raise central questions framed by a book on biophilic design. First and not least is the curious significance of its only recent arrival as a legitimate topic for investigation. Why isn't biophilic design—perhaps succinctly defined as a creative process driven by, or predisposed toward, bio-logic, which seeks to protect and enhance our link with the forces and faces of nature—an obvious and inherent organizing principle of all works of architecture?

In exploring the dimensions, theories, benefits and practicalities of biophilic design, these essays undertake a range of inquiries. In each new building endeavor, as we renegotiate the boundary between man and the elements, what kind of transactions should take place at the interface? How does the wall become a filter that admits beneficial, yet excludes stressful, sensations? How should we frame a window to function as lens, to better focus on nature while providing a controlled aperture for light, air exchange, and thermal conditioning? If human well-being, productivity and health at home, work, or school may be conferred by an occupant's access to daylight, views of vegetation and fauna, wind currents, and diurnal and seasonal information, why aren't these outcomes already a paramount consideration in all building endeavors? Why shouldn't these natural rights (entitlements, really) feature prominently in our building codes and permitting processes?

An investigation of this intentional, affirmative connection between man and nature makes a provocative contribution to the case for sustainable design. Biophilic design is an emerging voice in building green—a chorus increasingly voluble. It is one that attends to the vital shades and nuances of how we experience environments built for life. For today, in a world of impending climate change and species loss, this design sensibility, one more intuitively *biologic* in nature, is taking on ever greater social and political urgency.

PART I

The Theory of Biophilic Design

chapter 1

Dimensions, Elements, and Attributes of Biophilic Design

Stephen R. Kellert

Biophilic design is the deliberate attempt to translate an understanding of the inherent human affinity to affiliate with natural systems and processes—known as biophilia (Wilson 1984, Kellert and Wilson 1993)—into the design of the built environment. This relatively straightforward objective is, however, extraordinarily difficult to achieve, given both the limitations of our understanding of the biology of the human inclination to attach value to nature, and the limitations of our ability to transfer this understanding into specific approaches for designing the built environment. This chapter provides some perspective on the notion of biophilia and its importance to human well-being, as well as some specific guidance regarding dimensions, elements, and attributes of biophilic design that planners and developers can employ to achieve this objective in the modern, especially urban, built environment.

BIOPHILIA AND HUMAN WELL-BEING

As noted, biophilia is the inherent human inclination to affiliate with natural systems and processes, especially life and life-like features of the nonhuman environment. This tendency became biologically encoded because it proved instrumental in enhancing human physical, emotional, and intellectual fitness during the long course of human evolution. People's dependence on contact with nature reflects the reality of having evolved in a largely natural, not artificial or constructed, world. In other words, the evolutionary context for the development of the human mind and body was a mainly sensory world dominated by critical environmental features such as light, sound, odor, wind, weather, water, vegetation, animals, and landscapes.

The emergence during the past roughly 5,000 years of large-scale agriculture, fabrication, technology, industrial production, engineering, and the modern city constitutes a small fraction of human history, a period that has not substituted for the benefits of adaptively responding to a largely natural environment. Most of our emotional, problem-solving, critical-thinking, and constructive abilities continue to reflect skills and aptitudes learned in close association with natural systems and processes that remain critical in human health, maturation, and productivity. The assumption that human progress and civilization is measured by our separation from if not transcendence of nature is an erroneous and dangerous illusion. People's physical and mental well-being remains highly contingent on contact with the natural environment, which is a necessity rather than a luxury for achieving lives of fitness and satisfaction even in our modern urban society.

Biophilia is nonetheless a "weak" biological tendency that is reliant on adequate learning, experience, and sociocultural support for it to become functionally robust. As a weak biological tendency, biophilic values can be highly variable and subject to human choice and free will, but the adaptive value of these choices is ultimately bound by biology. Thus, if our biophilic tendencies are insufficiently stimulated and nurtured, they will remain latent, atrophied, and dysfunctional. Humans possess extraordinary capacities for creativity and construction in responding to weak biological tendencies, and this ability constitutes in a sense the "genius" of humanity. Yet, this innovative capacity is a two-edged sword, carrying with it the potential for distinctive individual and cultural expression, as well as the potential for self-defeating expression through either insufficient or exaggerated expression of inherent tendencies. Thus, our creative constructions of the human built environment can be either a positive facilitator or a harmful impediment to the biophilic need for ongoing contact with natural systems and processes.

Looking at biophilic needs as an adaptive product of human biology relevant today rather than as a vestige of a now-irrelevant past, we can argue that the satisfaction of our biophilic urges is related to human health, productivity, and well-being. What is the evidence to support this contention? The data is sparse and diverse, but a growing body of knowledge supports the role of contact with nature in human health and productivity. This topic is extensively discussed elsewhere, such as in chapters in this book by Ulrich, Hartig, Frumkin, and others. Still, the following findings are worth noting (summarized in Kellert 2005):

- Contact with nature has been found to enhance healing and recovery from illness and major surgical procedures, including direct contact (e.g., natural lighting, vegetation), as well as representational and symbolic depictions of nature (e.g., pictures).
- People living in proximity to open spaces report fewer health and social problems, and this has been identified independent of rural and urban residence, level of education, and income. Even the presence of limited amounts of vegetation such as grass and a few trees has been correlated with enhanced coping and adaptive behavior.
- Office settings with natural lighting, natural ventilation, and other environmental features result in improved worker performance, lower stress, and greater motivation.
- Contact with nature has been linked to cognitive functioning on tasks requiring concentration and memory.
- Healthy childhood maturation and development has been correlated with contact with natural features and settings.
- The human brain responds functionally to sensory patterns and cues emanating from the natural environment.
- Communities with higher-quality environments reveal more positive valuations of nature, superior quality of life, greater neighborliness, and a stronger sense of place than communities of lower environmental quality. These findings also occur in poor urban as well as more affluent and suburban neighborhoods.

These studies provide scientific support for the ancient assumption that contact with nature is critical to human functioning, health, and well-being. As the psychiatrist Harold Searles concluded some years ago (1960, 117): "The nonhuman environment, far from being of little or no account to human [health and] personality development, constitutes one of the most basically important ingredients of human existence."

RESTORATIVE ENVIRONMENTAL AND BIOPHILIC DESIGN

Unfortunately, the prevailing approach to design of the modern urban built environment has encouraged the massive transformation and degradation of natural systems and increasing human separation from the natural world. This design paradigm has resulted in unsustainable energy and resource consumption, major biodiversity loss, widespread chemical pollution and contamination, extensive atmospheric degradation and climate change, and human alienation from nature. This result is, however, not an inevitable by-product of modern urban life, but rather a fundamental design flaw. We designed ourselves into this predicament and theoretically can design ourselves out of it, but only by adopting a radically different paradigm for development of the modern built environment that seeks reconciliation if not harmonization with nature.

This new design paradigm is called here "restorative environmental design," an approach that aims at both a low-environmental-impact strategy that minimizes and mitigates adverse impacts on the natural environment, and a positive environmental impact or biophilic design approach that fosters beneficial contact between people and nature in modern buildings and landscapes.

Recognition of how much the modern built environment has degraded and depleted the health and productivity of the natural environment prompted the development of the modern sustainable or green design movement, and years of hard work has started to yield significant change in design and construction practices. Unfortunately, the prevailing approach to sustainable design has almost exclusively focused on the low-environmental-impact objectives of avoiding and minimizing harm to natural systems (e.g., Mendler et al. 2006). While necessary and commendable, this focus is ultimately insufficient, largely ignoring the importance of achieving long-term sustainability of restoring and enhancing people's positive relationship to nature in the built environment, what is called here biophilic design. Low-environmental-impact design results in little net benefit to productivity, health, and well-being. Buildings and landscapes, therefore, will rarely be sustainable over time, lacking significant benefits derived from our ongoing experience of nature. Cutting-edge low-environmental-impact technology inevitably becomes obsolete, and when this occurs, will people be motivated to renew and restore these structures? Sustainability is as much about keeping buildings in existence as it is about constructing new low-impact efficient designs. Without positive benefits and associated attachment to buildings and places, people rarely exercise responsibility or stewardship to keep them in existence over the long run.

Biophilic design is, thus, viewed as the largely missing link in prevailing approaches to sustainable design. Low-environmental-impact and biophilic design must, therefore, work in complementary relation to achieve true and lasting sustainability. The major objectives of low-environmental-impact design have been effectively delineated, focusing on goals such as energy and resource efficiency, sustainable products and materials, safe waste generation and disposal, pollution abatement, biodiversity protection, and indoor environmental quality. Moreover, the detailed specification of design strategies to achieve these goals has been incorporated into certification systems such as the U.S. Green Building Council's LEED rating approach.

In contrast, a detailed understanding of biophilic design remains meager (Kellert 2005, Heerwagen 2001). For the remainder of this chapter, therefore, dimensions, elements, and attributes of biophilic design will be described to partially address this need. The following description identifies two basic dimensions of biophilic design, followed by six biophilic design elements, which in turn are related to some 70 biophilic design attributes. This specification can assist designers and developers in pursuing the practical application of biophilic design in the built environment.

The first basic dimension of biophilic design is an *organic or naturalistic* dimension, defined as shapes and forms in the built environment that directly, indirectly, or symbolically reflect the inherent human affinity for nature. Direct experience refers to relatively unstructured contact with self-sustaining features of the natural environment such as daylight, plants, animals, natural habitats, and ecosystems. Indirect experience involves contact with nature that requires ongoing human input to survive such as a potted plant, water fountain, or aquarium. Symbolic or vicarious experience involves no actual contact with real nature, but rather the representation of the natural world through image, picture, video, metaphor, and more.

The second basic dimension of biophilic design is a *place-based or vernacular* dimension, defined as buildings and landscapes that connect to the culture and ecology of a locality or geographic area. This dimension includes what has been called a sense or, better, spirit of place, underscoring how buildings and landscapes of meaning to people become integral to their individual and collective identities, metaphorically transforming inanimate matter into something that feels lifelike and often sustains life. As René Dubos (1980, 110) argued:

> People want to experience the sensory, emotional, and spiritual satisfactions that can be obtained only from an intimate interplay, indeed from an identification with the places in which [they] live. This interplay and identification generate the spirit of the place. The environment acquires the attributes of a place through the fusion of the natural and human order.

People are rarely sufficiently motivated to act as responsible stewards of the built environment unless they have a strong attachment to the culture and ecology of place. As Wendell Berry (1972, 68) remarked: "Without a complex knowledge of one's place, and without the faithfulness to one's place on which such knowledge depends, it is inevitable that the place will be used carelessly and eventually destroyed." A tendency to affiliate with place reflects the human territorial proclivity developed over evolutionary time that has proven instrumental in securing resources, attaining safety and security, and avoiding risk and danger.

Despite the modern inclination for mobility, most people retain a strong physical and psychological need for calling some place "home." This attachment to territory and place remains a major reason why people assume responsibility and long-term care for sustaining buildings and landscapes. Conversely, lacking a sense of place, humans typically behave with indifference toward the built environment. An erosion of connection to place has unfortunately become a common affliction of modern society—what Edward Relph called "placelessness," and described in the following way (1976, 12):

> If places are indeed a fundamental aspect of existence in the world, if they are sources of security and identity for individuals and for groups of people, then it is important that the means of experiencing, creating, and maintaining significant places are not lost. There are signs that these very means are disappearing and that "placelessness"—the weakening of distinct and diverse experiences and identities of places—is now a dominant force. Such a trend marks a major shift in the geographical bases of existence from a deep association with places to rootlessness.

The two basic dimensions of biophilic design can be related to six biophilic design elements:

- Environmental features
- Natural shapes and forms
- Natural patterns and processes
- Light and space
- Place-based relationships
- Evolved human-nature relationships

These six elements are then revealed in more than 70 biophilic design attributes.

The remainder of this chapter describes these elements and attributes of biophilic design. This description is necessarily brief, due to space limitations, and insufficient. Additionally, this initial formulation will be modified in the future with increasing knowledge, and some of this categorization will inevitably overlap. This classification should, therefore, be viewed as a work in progress. At the end of the chapter, all the design elements and attributes are listed in Table 1.1, and a small number of illustrations are provided.

Environmental Features

The first and most obvious of the biophilic design elements is *environmental features*, involving the use of relatively well-recognized characteristics of the natural world in the built environment. Twelve attributes are identified, including the following:

1. *Color.* Color has long been instrumental in human evolution and survival, enhancing the ability to locate food, resources, and water; identify danger; facilitate visual access; foster mobility; and more. People for good and obvious reasons are attracted to bright flowering colors, rainbows, beautiful sunsets, glistening water, blue skies, and other colorful features of the natural world. Natural colors, such as earth tones, are thus often used to good effect by designers.
2. *Water.* Water is among the most basic human needs and commonly elicits a strong response in people. The famous architectural critic John Ruskin remarked in this regard (Hildebrand 2000, 71)): "As far as I can recollect, without a single exception, every Homeric landscape, intended to be beautiful, is composed of a fountain, a meadow, and a shady grove." Roger Ulrich similarly observed (1993) based on a review of many studies: "Water features constantly elicit especially high levels of liking or preference." The effective use of water as a design feature is complex, well described in the chapter by Mador, and often contingent on such considerations as perceptions of quality, quantity, movement, clarity, and other characteristics.
3. *Air.* People prefer natural ventilation over processed and stagnant air. Important conditions include quality, movement, flow, stimulation of other senses such as feel and smell, and visual appeal despite the seeming invisibility of the atmosphere.
4. *Sunlight.* Daylight is consistently identified as an important and preferred feature by most people in the built environment. The simple use of natural rather than artificial light can improve morale, comfort, and health and productivity. This preference reflects the fact that humans are a largely diurnal animal, heavily reliant on sight for securing resources and avoiding hazard and danger. People depend on visual acuity to satisfy various physical, emotional, and intellectual needs. Additional consideration of the importance of light is addressed in a later section on the more general biophilic design element of light and space.
5. *Plants.* Plants are fundamental to human existence as sources of food, fiber, fodder, and other aspects of sustenance and security. The mere insertion of plants into the built environment can enhance comfort, satisfaction, well-being, and performance.
6. *Animals.* Animals are similarly basic to human existence as sources of food, resources, protection, and companionship, and occasionally as precipitators of fear and danger. Designing animal life into the built environment can be difficult and problematic, although sometimes effective in aviaries, aquaria, and even the presence of free-roaming creatures associated with certain designs like green roofs. Animals in building interiors typically occur in representational rather than literal form, many through the use of ornament, decoration, art, and in stylized and highly metaphorical disguise. The presence of animal forms, nonetheless, often provokes satisfaction, pleasure, stimulation, and emotional interest.
7. *Natural materials.* People generally prefer natural over artificial materials, even when the artificial forms are close or seeming exact copies of natural products. Part of the aversion is likely due to the inability of artificial materials to reveal the organic processes of aging, weathering, and other dynamic features of natural materials, even inorganic forms like stone. The patina of time may provoke an intuitive understanding among some people of the benefits flowing from the movement of nutrients and energies through natural systems.
8. *Views and vistas.* People express a strong and consistent preference for exterior views, especially when the vistas contain natural features and vegetation. These views are often most satisfying when the scale is compatible with human experience—for example, not overly restricted or confined, unfamiliar, or out of scale or proportion (e.g., too large or too high).
9. *Façade greening.* Buildings with vegetative façades, such as ivy walls or green roofs, often provoke interest and satisfaction. This likely reflects the historic benefits associated with organic materials as sources of insulation, camouflaging protection, or even food. Plants on buildings and constructed landscapes can also evoke a powerful vernacular, such as the thatched or vegetative roofs of many cultures.

10. *Geology and landscape.* The compatible connection of buildings to prominent geological features is often an effective design strategy. These structures are sometimes described as rooted or grounded. Frank Lloyd Wright achieved particular success with his Prairie-style architecture in part by creating structures that worked in strong parallel relation to rather than dominating their savanna-type landscape.
11. *Habitats and ecosystems.* Buildings and landscapes that possess a close and compatible relationship to local habitats and ecosystems also tend to be highly effective and preferred. Important ecosystems in this regard are often wetlands, forests, grasslands, and watersheds.
12. *Fire.* Fire in the built environment, while a complicated and difficult design challenge, is often a preferred feature, generally associated with the benefits of heating and cooking. The manipulated experience of fire within building interiors has long been celebrated as a sign of comfort and civilization, providing pleasing qualities of color, warmth, and movement.

Natural Shapes and Forms

The second biophilic design element is *natural shapes and forms*. This element includes representations and simulations of the natural world often found on building façades and within interiors. Eleven attributes are associated with this design element:

1. *Botanical motifs.* The shapes, forms, and patterns of plants and other vegetative matter are a frequent and often important design element of the built environment (Hersey 1999). These representations often mimic or simulate plant forms such as foliage, ferns, cones, shrubs, and bushes, both literally and metaphorically.
2. *Tree and columnar supports.* Trees have also played a vital role in human affairs as sources of food, building material, paper products, heating supply, and other uses. The appearance or simulation of tree-like shapes, especially columnar supports, is a common and often coveted design feature in the built environment. Some of our most appealing structures contain tree forms and shapes that frequently include leaf capitals. When revealed in multiples, they can sometimes suggest a forested setting.
3. *Animal (mainly vertebrate) motifs.* The simulation of animal life is widespread in building interiors and facades, although to a less extent than with plants. The appearance of animal parts is often encountered, such as claws or heads, rather than entire creatures. Animal forms are frequently revealed in highly stylized, fictionalized, and sometime contorted shapes and forms.
4. *Shells and spirals.* Simulations and depictions of invertebrate creatures are widespread design features in the built environment, particularly shell and spiral forms of actual and imagined mollusks. The shapes and forms of bees (and their hives), flies, butterflies, moths, and other insects, as well as spiders (and their webs) and other invertebrates, are also common. Some building designs mimic invertebrate processes, such as the bioclimatic controls of termite mounds, the structural strength of seashells and hives, and the patterns of webs, a subject considered at the end of this section under the topic of "biomimicry," and in the chapter by Benyus.
5. *Egg, oval, and tubular forms.* Egglike and tubular forms are also design elements in some building interiors, facades, and exterior landscapes such as gardens and fountains. These shapes often occur literally and metaphorically, both important expressions of ornament and sometimes for structural purposes.
6. *Arches, vaults, domes.* Arches, vaults, and domes in the built environment resemble or copy forms found in nature, including beehives, nest-like structures, shell forms, and cliffs. These forms can be used for both decorative and functional purposes.
7. *Shapes resisting straight lines and right angles.* Natural shapes and forms are often sinuous, flowing, and adaptive in responding to forces and pressures found in nature. Natural features are thus rarely revealed as straight lines and right angles characteristic of human engineering and manufactured products and structures. The large-scale modern built environment has often been characterized by

standardized and rigid shapes. People nonetheless generally prefer designs that resemble the tendency of organic forms to resist hard mechanical edges, straight lines and angles.

8. *Simulation of natural features.* This attribute reaffirms the tendency to simulate rather than replicate actual natural forms in the built environment. Ornamentation and decoration especially employ imagined forms only vaguely reminiscent of those found in the natural world. These designs are often most successful when they possess a logic that intimates functional features occurring in nature, such as shapes, patterns and processes that suggest structural integrity and adaptive advantage in response to environmental pressures rather than mere superficial decoration.
9. *Biomorphy*. Some interesting architectural forms bear very little resemblance to life forms encountered in nature, yet are clearly viewed as organic. These resemblances to living forms are usually unconscious products of design, sometimes called "biomorphy" (Feuerstein 2002). Powerful examples of biomorphic architecture that provoke observers to impute known animal and plant labels even when the designer did not deliberately create these life-forms include the birdlike shape of Jörn Utzon's Sydney Opera House and the fernlike or less reverently labeled "pregnant whale" of Eero Saarinen's Yale University hockey rink.
10. *Geomorphology.* Some building designs mimic or metaphorically embrace landscape and geology in relative proximity to the structure. This relationship to the ground can lend the appearance of solidity to the built environment, making structures appear integral rather than separate from their geological context.
11. *Biomimicry.* Some successful designs borrow from adaptations functionally found in nature, particularly among other species. Examples include the structural strength and bioclimatic properties of shells, crystals, webs, mounds, and hives, effectively incorporated into the built environment. This tendency has been called "biomimicry" by Janine Benyus, elucidated in her book of this title (Benyus 1997) and connected to biophilic design in a later chapter in this volume. The knowledge of biomimetic properties is growing rapidly and will likely result in a revolution of product development with enormous biophilic design implications.

Natural Patterns and Processes

A third biophilic design element is *natural patterns and processes.* This element emphasizes the incorporation of properties found in nature into the built environment, rather than the representation or simulation of environmental shapes and forms. Fifteen attributes have been identified and are described below, although this complex element is likely to be altered in the future with additional understanding.

1. *Sensory variability.* Human fitness and survival has always required coping with a highly sensuous and variable natural environment, particularly responding to light, sound, touch, smell, and other sensory environmental conditions. Human satisfaction and well-being continue to be reliant on perceiving and responding to sensory variability, especially when this occurs in structured and organized ways within the built environment.
2. *Information richness.* The cognitive richness of the natural world reflects its likely being the most intellectually challenging environment people will ever encounter even in our modern information age. This quality constitutes one of its most beguiling features, and when effectively incorporated into the built environment in actual or metaphorical form can stimulate curiosity, imagination, exploration, discovery, and problem-solving. Most people, therefore, respond positively to buildings and landscapes that possess information richness, variety, texture, and detail that mimic natural patterns when coherently revealed.
3. *Age, change, and the patina of time.* A fundamental feature of the natural world is aging through time, particularly organic forms. This dynamic progression evokes a sense of familiarity and satisfaction among people, despite the eventual occurrence of senescence, death, and decay. A patina of time is characteristic of natural materials, even inorganic ones, and is one reason, as noted above, that artificial

products rarely evoke sustained positive response even when they are exact copies.

4. *Growth and efflorescence.* Growth and development are specific expressions of aging that when found in the built environment typically provoke pleasure and satisfaction. Efflorescence marks the progressive unfolding of a maturational process that when encountered in buildings and landscapes, especially through ornamentation, is often highly appealing (Bloomer 2000). These temporal and transitional attributes often lend a dynamic quasi-living character to the built environment despite its immutable character.
5. *Central focal point.* The navigability of natural landscapes is often enhanced by the presence of a centrally perceived focal point. This point of reference frequently transforms what otherwise is a chaotic setting into an organized one that facilitates passage and way-finding. As the poet Wallace Stevens described (1955): "I placed a jar in Tennessee/ And round it was, upon a hill./ It made a slovenly wilderness/ surround that hill." Many successful buildings and constructed landscapes similarly achieve coherence despite complexity and large scale when a centrally organized reference point has been effectively incorporated.
6. *Patterned wholes.* People respond positively to natural and built environments when variability has been united by integrated and patterned wholes. What may have previously been experienced as inchoate becomes structured in a manner that fosters understanding and often feelings of mastery and control.
7. *Bounded spaces.* Humans have a strong proclivity for bounded spaces. This territorial tendency, over evolutionary time, likely fostered resource exploitation and security. People also value delineated spaces within the built environment, which enhance the recognition of clear and consistent boundaries and place demarcations.
8. *Transitional spaces.* Transitional spaces within and between built and natural environments often foster comfort by providing access from one area to another. Important passageways in the built environment include thresholds, portals, doors, bridges, and fenestration.
9. *Linked series and chains.* Clear physical and temporal movement in both natural and built environments is often facilitated by linked spaces, especially when occurring in connected chains. These relational spaces convey meaning and organization, as well as sometimes a sense of mystery that both stimulates and entices.
10. *Integration of parts to wholes.* People prefer in natural and built environments the feeling that discrete parts comprise an overall whole, particularly when the whole is an emergent property consisting of more than the sum of the individual parts. This integrative quality fosters a feeling of structural integrity, even in complexes of considerable size and detail.
11. *Complementary contrasts.* Meaning and intelligibility, as well as interest and stimulation, in natural and constructed settings often reveal the blending of contrasting features in complementary fashion. This can occur through the compatible rendering of seeming opposites, such as light and dark, high and low, and open and closed.
12. *Dynamic balance and tension.* The dynamic balancing of different and sometimes contrasting forms often fosters a sense of strength and durability in both natural and built environments. This blending of varying forces often produces a quality of creative tension that transforms static forms into organic-like entities.
13. *Fractals.* Elements in nature are rarely if ever exact copies of one another, even among highly related entities. Snowflakes or leaves of a single species or tree may be highly similar but never the same. Orderly variation on a basic pattern is the norm, whether it be thematic diversity based on size, or spatial or temporal scale. Related and similar forms are often called "fractals," and these patterns are found in some of our most successful buildings and landscapes. These structures frequently include repeated but varying patterns of a basic design, such as ornamentation in parallel or closely linked rows that differ slightly from one another.
14. *Hierarchically organized ratios and scales.* Successful natural and built forms often occur in hierarchically connected ways, sometimes arithmetically or geometrically related. This thematic congruence

can facilitate the assimilation of highly complex patterns that otherwise might be experienced as overwhelmingly detailed or even chaotic. Arithmetic and geometric expressions of this tendency in both natural and built settings include the golden proportion and the Fibonacci ratio (Portoghesi 2000).

Light and Space

A fourth biophilic design element is *light and space.* Twelve design attributes of this element follow, seven focusing on qualities of light and five focusing on spatial relationships:

1. *Natural light.* This attribute includes the effects of daylighting as previously described, as well as inclusion of the full color spectrum of natural light. Chapters by Loftness and Frumkin note studies showing that natural light is both physically and psychologically rewarding to people, frequently contributing to their health, productivity, and well-being in the built environment.
2. *Filtered and diffused light.* The benefits of natural light are often enhanced by modulating daylight, particularly by mitigating the effects of glare. Filtered or diffused sunlight can also stimulate observation and feelings of connection by providing a variable and mediated connection between spaces, particularly inside and outside areas such as described in the chapter by Bloomer.
3. *Light and shadow.* The complementary contrast of light and dark spaces can produce significant satisfaction in both buildings and landscapes. The creative manipulation of light and shadow can foster curiosity, mystery, and stimulation. This attribute likely evolutionarily enhanced human movement and the ability to discern objects over long distances, particularly from a protected refuge.
4. *Reflected light.* Lighting designs are frequently enhanced by light reflecting off surfaces such as light-colored walls, ceilings, and reflective bodies like water. Functional benefits include mitigation of glare, enhanced penetration of light into interior spaces, and spying resources at a distance.
5. *Light pools.* People are often drawn into and through interior spaces by the presence of pools of connected light. Light pools can assist movement and way-finding by providing lighted patches across shadowed or obscured areas such as a forest or darkened halls and passageways. Light pools can also foster feelings of security and protection, such as a lighted hearth.
6. *Warm light.* The perception of warmly lit areas, often islands of modulated sunlight surrounded by darker spaces, can enhance the feeling of a nested, secure, and inviting interior.
7. *Light as shape and form.* The manipulation of natural light can create stimulating, dynamic, and sculptural forms. Beyond the aesthetic pleasure, these shapes facilitate mobility, curiosity, imagination, exploration, and discovery.
8. *Spaciousness.* People prefer feelings of openness in natural and built environments, especially when it occurs in complementary relation to sheltered protected refuges at the surrounding edges. Effective designs often include spacious settings in close alliance with smaller spaces, which in contemporary architecture can often be encountered in airports, train stations, and some commercial and educational buildings.
9. *Spatial variability.* Spatial variability fosters emotional and intellectual stimulation. Spatial diversity is often most effective when in complementary relation to organized and united spaces.
10. *Space as shape and form.* Space can be creatively manipulated to convey shapes and forms. This effect can add beauty to the built environment, which stimulates interest, curiosity, exploration, and discovery.
11. *Spatial harmony.* The manipulation of space in the built environment tends to be most effective when it blends light, mass, and scale within a bounded context. This achievement evokes a sense of harmony, which fosters a sense of security and facilitates movement within diverse settings.
12. *Inside-outside spaces.* Appealing interior spaces in the built environment often appear connected to the outside environment. These areas also mark the transition of nature with culture. Important design forms in the built environment that evoke this quality include colonnades, porches, foyers, atriums, and interior gardens.

Place-Based Relationships

A fifth biophilic design element is *place-based relationships*. This element refers to the successful marriage of culture with ecology in a geographical context. The connection of people to places reflects an inherent human need to establish territorial control, which during the long course of our species' evolution facilitated control over resources, attaining safety, and achieving security. Locational familiarity—the yearning for home—remains a deeply held need for most people. Eleven attributes of place-based relationships are described, the last (placelessness) being the antithesis of the others rather than a stand-alone attribute.

1. *Geographic connection to place.* Secure feelings of connection to the geography of an area often foster feelings of familiarity and predictability. This can be achieved by emphasizing prominent geological features associated with the siting, orientation, and views of buildings and landscapes.
2. *Historic connection to place.* Meaningful relation to place often marks the passage of time, which fosters a sense of participation and awareness of an area's culture and collective memory. Buildings and landscapes that elicit this continuity with the past encourage the belief that the present and future are meaningfully linked to the history of a place.
3. *Ecological connection to place.* Places are sustained by an affirmative connection to ecology, particularly prominent ecosystems such as watersheds and dominant biogeographical features (e.g., mountains, deserts, estuaries, rivers, and oceans). The design of the built environment inevitably refashions nature, but this can occur in ways that do not diminish the overall biological productivity (e.g., nutrient flux), biodiversity, and ecological integrity of proximate ecological communities. Humans, like any ecologically transformative organism (e.g., elephants on the savanna, sea otters in a kelp bed), can add as well as subtract value from their natural systems. The design of the built environment can, therefore, aspire to achieve net ecological productivity.
4. *Cultural connection to place.* Cultural connection to place integrates the history, geography, and ecology of an area, becoming an integral component of individual and collective identity. The need for culture is a universal human need, sustained over time by repetition, normative events, and the architectural heritage of a people, particularly its treasured and distinctive vernacular forms.
5. *Indigenous materials.* A positive relation to place is generally enhanced by the utilization of local and indigenous materials. Native resources can provide a vivid and resonant reminder of local culture and environment, as well as require less energy for manufacture and transport.
6. *Landscape orientation.* Buildings and landscapes that compatibly connect to the local environment contribute to a sense of place. These constructions typically emphasize landscape features such as slope, aspect, sunlight, wind direction, and others that take advantage of prevailing biometeorological conditions. This orientation to landscape frequently evokes a sense of being a part of and embedded within local settings, rather than being separated from them.
7. *Landscape features that define building form.* Landscape features can embellish and distinguish building form, particularly prominent geological features, natural objects, and water. The built environment can, therefore, integrate with rather than be isolated from its biophysical context. When this fails to occur, even extraordinary buildings can be perceived as standing apart, perhaps impressive products of human engineering but largely abstract forms divorced from context and barren.
8. *Landscape ecology.* Effective place-based designs reinforce landscape ecology over the long term. This can be achieved through design that considers landscape structure, pattern, and process such as ecological connectivity, biological corridors, resource flows, biodiversity, optimal scale and size, ecological boundaries, and other parameters of functioning natural systems (Dramstad et al. 1996).
9. *Integration of culture and ecology.* The fusion of culture with ecology fosters long-term sustainability. The result marks the point where nature and humanity are positively transformed and mutually enriched by their association. When this occurs, buildings and landscapes often provoke considerable

loyalty, responsibility, and stewardship among the people who reside nearby.

10. *Spirit of place.* The spirit of a place signifies a level of commitment and meaning that people extend to both natural and built environments when they become cherished components of individual and collective identity, more than simply inanimate matter. The spirit of a place metaphorically signifies the built environment having become life-life and serving as the motivational basis for long-term stewardship and responsibility. While not technically alive, these structures and places give rise to and sustain human culture and ecology over time.
11. *Avoiding placelessness.* "Placelessness" is the antithesis of place-based design, to be avoided whenever possible. One of the insidious and damaging effects of much modern architecture has unfortunately been the divorce of design from connection to the culture or ecology of place. This corrosive separation of the built environment from its biocultural context has resulted in the decline of human-nature relationships and environmental sustainability.

Evolved Human-Nature Relationships

The sixth and final biophilic design element is *evolved human-nature relationships*. The term is somewhat misleading, as all the described biophilic design elements presumably reflect biologically based human affinities for the natural environment. The attributes described in this section, however, more specifically focus on fundamental aspects of the inherent human relationship to nature. Twelve attributes are described, the last eight of which are derived from a typology of environmental values developed by the author and described elsewhere (Kellert 1996, 1997):

1. *Prospect and refuge.* Refuge reflects a structure or natural environment's ability to provide a secure and protected setting. In the built environment, this often occurs through the design of comfortable and nurturing building interiors and secreted landscape places. Prospect, on the other hand, emphasizes discerning distant objects, habitats and horizons, evolutionarily instrumental in locating resources, facilitating movement, and identifying sources of danger. Some of our most satisfying buildings and landscapes capture the complementary relation of prospect with refuge (Hildebrand 2000, Appleton 1975).
2. *Order and complexity.* Order is achieved in the built or natural environment by imposing structure and organization. Extreme order often results in repetition, monotony, and boredom. By contrast, complexity reflects the occurrence of detail and variability. Excessive complexity can also be troublesome, making it difficult to assimilate detail and sometimes leading to a sense of chaos. Designs that effectively meld order with complexity tend to be successful, stimulating the desire for variety but in ways that seem controlled and comprehensible.
3. *Curiosity and enticement.* Curiosity reflects the human need for exploration, discovery, mystery, and creativity, all instrumental in problem solving (Kaplan et al. 1998). Enticement fosters curiosity. These complementary tendencies can engage the flywheel of human intellect and imagination. Some of our most effective buildings and landscapes foster curiosity, exploration, and discovery of natural process and diversity.
4. *Change and metamorphosis.* Change is a constant in both natural and human systems, reflected in the processes of growth, maturation, and metamorphosis (Bloomer 2000). Many powerful designs capture this dynamic and developmental quality, where one form or state appears to flow into another in a quasi-evolutionary sequence.
5. *Security and protection.* A fundamental objective of the built environment is ensuring protection from threatening forces in nature. Yet, the most successful designs over the long run never accomplish this need at the expense of other equally legitimate environmental values. Security in the built environment must not excessively insulate or isolate people from the natural world.
6. *Mastery and control.* Buildings and constructed landscapes reflect the human desire for mastery and control over nature. When accomplished with moderation and respect, mastering nature facilitates the satisfactory expression of human ingenuity and

cleverness that fosters self-confidence and self-esteem.

7. *Affection and attachment.* Affection for the natural world has been a critical component in engendering the human capacities for bonding and attachment, important in a largely social creature. Buildings and landscapes that elicit strong emotional affinities for nature are typically recipients of lasting loyalty and commitment.
8. *Attraction and beauty.* The aesthetic attraction to nature is one of the strongest inclinations of the human species. This biologically encoded tendency has been instrumental in fostering the capacities for curiosity, imagination, creativity, exploration, and problem solving. Some of our most successful buildings and landscapes foster an aesthetic appreciation for natural process and form.
9. *Exploration and discovery.* Nature is the most information-rich and intellectually stimulating environment that people ever encounter. Buildings and constructed landscapes that facilitate opportunities for exploration and discovery of natural process elicit considerable interest and appreciation, even when these environmental features are largely revealed in representational ways.
10. *Information and cognition.* Intellectual satisfaction and cognitive prowess can be fostered through designs that emphasize the complexity of natural shapes and forms. This can be achieved through the direct and indirect experience of nature, as well as by the creative use of ornamentation in the built environment that fosters critical thinking and problem solving.
11. *Fear and awe.* It may seem odd to emphasize negative and unwanted feelings such as fear and aversion of nature as components of biophilic design. Yet, protecting ourselves from threatening elements of the natural world has always been a primary objective of the built environment. Fear of nature can also be a motivational basis for designing peril and adventure into the built environment, such as overhanging precipices or proximity to fearsome forces like rushing water. Feelings of awe for the natural world can further combine reverence with fear, and some of our most celebrated structures achieve this effect through extolling majestic natural features that engender an appreciation for powers greater than ourselves.
12. *Reverence and spirituality.* Some of our most cherished buildings similarly affirm the human need for establishing meaningful relation to creation. These designs provoke feelings of transcendence and enduring connection that defy the aloneness of a single person isolated in space and time. Structures that achieve this reverential feeling of connection are also typically sustained generation after generation.

CONCLUSION

Six biophilic design elements and roughly 70 attributes have been described, and are summarily listed in Table 1-1. A small number of illustrations are provided at the chapter's conclusion depicting some of these design features. This categorization is a work in progress, which inevitably will be modified and improved over time.

All design of the built environment, including the biophilic desire to harmonize with nature, reflects what René Dubos called the active "wooing of the earth" (Dubos 1980). This objective, in other words, results in some degree of deliberate refashioning of nature to satisfy human needs, but in ways that celebrate the integrity and utility of the natural world. Thus, human intervention, if practiced with restraint and respect, can avoid arrogance and environmental degradation. With humility and understanding, effective biophilic design can potentially enrich both nature and humanity. As Dubos remarked (1980, 68):

> Wooing of the earth suggests the relationship between humankind and nature [can] be one of respect and love rather than domination. The outcome of this wooing can be rich, satisfying, and lastingly successful if both partners are modified by their association so as to become better adapted to each other.

TABLE 1-1 Elements and Attributes of Biophilic Design

Environmental features	Natural shapes and forms	Natural patterns and processes
Color	Botanical motifs	Sensory variability
Water	Tree and columnar supports	Information richness
Air	Animal (mainly vertebrate) motifs	Age, change, and the patina of time
Sunlight	Shells and spirals	Growth and efflorescence
Plants	Egg, oval, and tubular forms	Central focal point
Animals	Arches, vaults, domes	Patterned wholes
Natural materials	Shapes resisting straight lines and right angles	Bounded spaces
Views and vistas	Simulation of natural features	Transitional spaces
Façade greening	Biomorphy	Linked series and chains
Geology and landscape	Geomorphology	Integration of parts to wholes
Habitats and ecosystems	Biomimicry	Complementary contrasts
Fire		Dynamic balance and tension
		Fractals
		Hierarchically organized ratios and scales
Light and space	**Place-based relationships**	**Evolved human-nature relationships**
Natural light	Geographic connection to place	Prospect and refuge
Filtered and diffused light	Historic connection to place	Order and complexity
Light and shadow	Ecological connection to place	Curiosity and enticement
Reflected light	Cultural connection to place	Change and metamorphosis
Light pools	Indigenous materials	Security and protection
Warm light	Landscape orientation	Mastery and control
Light as shape and form	Landscape features that define building form	Affection and attachment
Spaciousness	Landscape ecology	Attraction and beauty
Spatial variability	Integration of culture and ecology	Exploration and discovery
Space as shape and form	Spirit of place	Information and cognition
Spatial harmony	Avoiding placelessness	Fear and awe
Inside-outside spaces		Reverence and spirituality

1. ENVIRONMENTAL FEATURES

Figure 1-1: Jubilee Campus, University of Nottingham, Nottingham, England. This design by Hopkins Architects effectively incorporates water as a positive experiential and low-impact (e.g., evaporative cooling) element.

Figure 1-2: University of Michigan law quadrangle. The ivy-covered walls provide a pleasing integration of vegetation into the building façade.

2. NATURAL SHAPES AND FORMS

Figure 1-3: Foliated sculpture by Kent Bloomer, Ronald Reagan Airport terminal. This metaphorical representation of nature draws well on instinctual affinities for vegetative forms.

Figure 1-4: Sydney Opera House, Jörn Utzon, architect. This building dramatically juxtaposes bird- and sail-like forms against the waters of Sydney Harbour.

3. NATURAL PATTERNS AND PROCESSES

Figure 1-5: New York City building façade. The shapes in this façade draw on foliated patterns and fractal geometries encountered in nature.

Figure 1-6: Harkness Tower, Yale University. This tower mimics many organic features often encountered in Gothic architecture.

4. LIGHT AND SPACE

Figure 1-7: San Francisco hotel lobby. This lobby combines the sculptural qualities of light with a highly organic space.

Figure 1-8: Genzyme Building, Cambridge, Massachusetts, Behnisch, Behnisch and Partner, architects. This office building innovatively includes light, water, and vegetation in a deep building interior, purportedly resulting in enhanced worker comfort, morale, and productivity.

5. PLACE-BASED RELATIONSHIPS

Figure 1-9: Mixed-used development, Portland, Oregon. The combined residential and commercial uses along a restored and revegetated riverfront has engendered a renewed sense of connection to place.

Figure 1-10: Bastille viaduct or Promenade Plantée, Paris. This elevated linear greenway situated on a former railroad line has stimulated commercial and social activity in this section of Paris.

6. EVOLVED RELATIONSHIPS TO NATURE

Figure 1-11: Fallingwater, Frank Lloyd Wright, architect. The strong appeal of this residence partly reflects its prominent prospect and refuge elements, as well as its connection to the hillside and adjacent stream course.

Figure 1-12: Mont-Saint-Michel, France. The timeless fascination of this structure derives in part from its powerful combination of order and complexity against a dramatic hill and ocean backdrop.

REFERENCES

Appleton, J. 1975. *The Experience of Nature*. London: Wiley.

Benyus, J. 1997. *Biomimicry: Innovation Inspired by Nature*. New York: Murrow.

Berry, W. 1972. *A Continuous Harmony: Essays Cultural and Agricultural*. New York: Harcourt.

Bloomer, K. 2000. *The Nature of Ornament: Rhythm and Metamorphosis in Architecture*. New York: Norton.

Dramstad, W., J. Olson, R. Forman. 1996. *Landscape Ecology Principles in Landscape Architecture and Land-Use Planning*. Washington, DC: Island Press.

Dubos, R. 1980. *Wooing of the Earth*. New York: Scribner.

Feuerstein, G. 2002. *Biomorphic Architecture: Human and Animal Forms in Architecture*. Stuttgart: Axel Menges.

Heerwagen, J., and B. Hase. 2001. "Building Biophilia: Connecting People to Nature." *Environmental Design and Construction*. March–April: 30–34.

Hersey, G. 1999. *The Monumental Impulse: Architecture's Biological Roots*. Cambridge, MA: MIT Press.

Hildebrand, G. 2000. *The Origins of Architectural Pleasure*. Berkeley: University of California Press.

Kaplan, S., R. Kaplan, and R. Ryan. 1998. *With People in Mind: Design and Management of Everyday Life*. Washington, DC: Island Press.

Kellert, S. 1996. *The Value of Life: Biological Diversity and Human Society*. Washington, DC: Island Press.

Kellert, S. 1997. *Kinship to Mastery: Biophilia in Human Evolution and Development*. Washington, DC: Island Press.

Kellert, S. 2005. *Building for Life: Designing and Understanding the Human-Nature Connection*. Washington, DC: Island Press.

Kellert, S., and J. Heerwagen. Forthcoming. "Nature and Healing: The Science, Theory, and Promise of Biophilic Design." In *Sustainable Architecture for Health*, edited by G. Vittori and R. Guenther.Hoboken, NJ: Wiley.

Kellert, S., and E. O. Wilson, eds. 1993. *The Biophilia Hypothesis*. Washington, DC: Island Press.

Mendler, S., W. Odell, M. Lazarus. 2006. *The HOK Guidebook to Sustainable Design*. Hoboken, NJ: Wiley.

Portoghesi, P. 2000. *Nature and Architecture*. Milan: Skira.

Relph, E. 1976. *Place and Placelessness*. London: Pion.

Searles, H. 1960. *The Nonhuman Environment in Normal Development and in Schizophrenia*. New York: International Universities Press.

Stevens, W. 1955. *Collected Poems*. New York: Knopf.

Ulrich, R. 1993. "Biophilia, Biophobia, and Natural Landscapes." In *The Biophilia Hypothesis*, edited by S. Kellert and E. O. Wilson. Washington, DC: Island Press.

Wilson, E. O. 1984. *Biophilia: The Human Bond with Other Species*. Cambridge, MA: Harvard University Press.

chapter 2

The Nature of Human Nature

Edward O. Wilson

The most useful term to capture the unity of knowledge is surely *consilience*. It means the interlocking of cause-and-effect explanations across different disciplines, as for example between physics and chemistry, chemistry and biology, and, more controversially, biology and the social sciences. The word *consilience* was introduced in 1840 by William Whewell, the founder of the modern philosophy of science. It is more serviceable than the words *coherence* or *interconnectedness*, because its rarity of usage since 1840 has preserved its original meaning, whereas *coherence* and *interconnectedness* have acquired many meanings scattered across the different disciplines.

Consilience, defined as cause-and-effect explanation across the disciplines, has plenty of credibility. It is the mother's milk of the natural sciences. Its material understanding of how the world works and its technological spin-off are the foundation of modern civilization. The time has come, I believe, to consider more seriously its relevance to the social sciences and humanities. I will grant immediately that belief in the possibility of consilience beyond the natural sciences and across to the other great branches of learning is not the same as science, at least not yet. It is a metaphysical worldview, and still a minority one at that. The evidence is fragmentary, and in some cases it is still relatively thin.

But I believe also that it is a matter of practical urgency to focus on the unity of knowledge. Let me illustrate that claim with an example. Think of two intersecting lines forming a cross, and picture the four quadrants thus created. Label one quadrant environmental policy, the next ethics, the next biology, and the final one social science. Each of these subjects has its own experts, its own language, rules of evidence, and criteria of validation. Think of each as an island of knowledge in a sea of ignorance.

Now, if we focus on more specific topics within each of the quadrants, we see how general theory translates into the analysis of practical problems. And we understand that in each case we somehow have to learn how to travel from one subject to the next—say, from forest management to ecology to economics to ethics and back around again. In a single discussion, maybe in a sentence or two in the discussion, it is necessary to travel the entire circuit.

Now move through concentric circles toward the intersection of the disciplines. As we approach the intersection, where most real-world problems exist, the circuit becomes more difficult and the process more disorienting and contentious.

The nub of the problem vexing a great deal of human thought is the general belief that a fault line exists between the natural sciences on one side and the humanities and humanistic social sciences on the other—in other words, very roughly, between the scientific and the literary cultures as defined by C. P. Snow in his famous 1959 Rede Lecture. The solution to the problem, I believe, is the recognition that this boundary is not a fault line. It is not a permanent epistemological division, and it is not a Hadrian's Wall, as many would have it, needed to protect high culture from the reductionist barbarians of science. What we are beginning at last to understand is that this line does not exist as a line at all. It is instead a broad domain of poorly understood material phenomena awaiting cooperative exploration from both sides.

During the past 20 years, four borderland disciplines have grown dramatically in the natural sciences, or more precisely in the biological sciences that bridge this intermediate domain. From the biology side, these disciplines include cognitive neuroscience, behavioral genetics, evolutionary biology, and environmental sciences. From the social sciences side, the bridging disciplines include cognitive psychology and biological anthropology. To an increasing degree, cognitive psychology and biological anthropology are becoming consilient with biology; in fact, they are anastomosing with its disciplines in cause-and-effect explanations.

Why is this conjunction among the great branches of learning important? Because it offers the prospect of characterizing human nature with greater objectivity and precision, an exactitude that is the key to human self-understanding. The intuitive grasp of human nature has been the substance of the creative arts. It has been the underpinning of the social sciences and a beckoning mystery to the natural sciences. To grasp human nature objectively, to explore it to its depths scientifically, and to grasp its ramifications, would be to approach if not attain the grail of scholarship, and to fulfill the dreams of the Enlightenment.

Now, rather than let the matter hang in the air just rhetorically, I want to suggest a consilient definition of human nature, and then illustrate it with examples. Human nature is not the genome that prescribes it. Human nature is not the cultural universals (such as the incest taboos and rites of passage) that are the products of human nature. Rather, human nature is the collectivity of epigenetic rules, the inherited regularities of mental development. These rules are the genetic biases in the way our senses perceive the world, the symbolic coding by which we represent the world, the options we open to ourselves, and the responses we find easiest and most rewarding to make. In ways that are beginning to come into focus at the physiological and even in a few cases at the genetic level, the epigenetic rules alter the way we see and linguistically classify color. They cause us to evaluate the aesthetics of artistic design according to elementary abstract shapes and the degree of complexity. They lead us differentially to acquire fears and phobias concerning dangers in the environment (as from snakes and heights), to communicate with certain facial expressions and forms of body language, to bond with infants, to bond conjugally, and so on across a wide range of categories in behavior and thought. Most epigenetic rules are evidently very ancient, dating back millions of years in mammalian ancestry. Others, like the stages of linguistic development, are uniquely human and probably only hundreds of thousands of years old.

As an example of epigenetic rules, consider the instinct to avoid incest. Its key element is the Westermarck effect, named after Edward Westermarck, the Finnish anthropologist who discovered it a century ago. When two people live in close domestic proximity during the first 30 months in the life of either one, both are desensitized to later close sexual attraction and bonding. The Westermarck effect has been very well documented in anthropological studies, although the genetic prescription and neurobiological mechanics underlying it remain to be studied. What makes the human evidence the more convincing is that all of the nonhuman primates whose sexual behavior has been closely studied also display the Westermarck effect. It therefore appears probable that the trait prevailed in the human ancestral line millions of years before the origin of *Homo sapiens*, our present-day species.

The existence of the Westermarck effect runs directly counter to the more widely known Freudian theory of incest avoidance. Freud argued that members of the same family lust for one another, making it necessary for societies to create incest taboos in order to avoid the social damage that would follow if within-family sex were allowed. But the opposite is evidently true. That is, incest taboos arise naturally as products of response mediated by a relatively simple inherited epigenetic rule. The epigenetic rule is the Westermarck effect. The adaptive advantage of the Westermarck effect is, of course, that it reduces inbreeding depression and the production of dead or defective children. That relentless pressure is almost surely how it arose through evolution by natural selection.

In another, wholly different realm, consider the basis of aesthetic judgment. Neurobiological monitoring, in particular measurements of the damping of the alpha wave, during presentations of abstract designs, have shown that the brain is most aroused by patterns in which there is a 20 percent redundancy of elements, or put very roughly, the amount of complexity found in a simple maze, or two turns of a logarithmic spiral, or an asymmetric cross. It may be a coincidence that about the same property is shared by a great deal of the art in friezes, grillwork, colophons, logographs, and flag designs. It crops up again in the glyphs of ancient Egypt and Mesoamerica as well as the pictographs of modern Asian languages. None of this is proof, and the idea needs much more testing, but the universal nature and preponderance of the effect has to be considered very suggestive.

To take the same approach but in another direction, I would also like to mention biophilia, the innate affiliation people seek with other organisms and especially with the natural world, a concept on which I have worked. Studies have shown that, given complete freedom to choose the setting of their homes or offices, people gravitate toward an environment that combines three features, intuitively understood by landscape architects and real estate entrepreneurs: people want to be on a height looking down, they prefer open, savanna-like terrain with scattered trees and copses, and they want to be near a body of water, such as a river or lake, even if all these elements are purely aesthetic and not functional. They will pay enormous prices to have this view.

They look for two other, crosscutting elements: they want both a retreat in which to live and a prospect of fruitful terrain in which to forage, and in the prospect, they like distant, scattered large animals and trees with low, nearly horizontal branches.

In short, if you will allow me now to take a deep breath and then plunge where you may not wish to follow, people want to be in the environments in which our species evolved over millions of years, that is, hidden in a copse or against a rock wall, looking out over savanna and transitional woodland, at acacias and similar dominant trees of the African environment. And why not? Is that such a strange idea? All mobile animal species have a powerful, often highly sophisticated inborn guide for habitat selection. Why not human beings?

And then again, in the biologically important realm of erotic aesthetics, the basis of sexual attraction, there is the matter of preferred female facial beauty open to objective analysis, now under new scrutiny by psychologists. The ideal subjectively preferred in tests is not the exact average, as once thought. It is not the average of faces from the general population, which can be readily blended by computer. It is the average instead of the subset considered most attractive and then blended by computer. The ideal has higher cheekbones than the average, a smaller chin, shorter upper lip, and wider eyes, all relative to the size of the face. The evolutionary biologist might surmise that these traits are the signs of juvenescence still on the faces of the young women, hence relative youth and reproductive potential. If all this seems irrational, ask any middle-aged professor whose second wife is a graduate student.

How much do we know about the innate basis of such aesthetics? Not a lot, and certainly very little about the genetics and neurobiology in particular of the epigenetic rules—not because they have been investigated and then found lacking, not because they are technically daunting, but simply because they have not been studied; only recently have researchers begun to ask the right questions within the borderland disciplines. Bear with me while I move farther into this subject with yet a bit more speculation to suggest the range of opportunity.

In the arts, I believe, we convey emotion with what students of animal behavior call *releasers*, simple stimuli that evoke complex, mostly hardwired responses. The ethological elements may be teased out in sophisticated works. For example, in paintings of many kinds, we see mood conveyed by the preponderance of curving strokes, to convey a sense of relative calm, versus angular strokes to evoke tension and violence. Is the response to these elements part of the hereditary epigenetic rules? I would expect that to be the case. Artists have returned to them faithfully for centuries.

My main point is that genetic evolution and cultural evolution are closely interwoven, and we are only beginning to obtain a glimmer of the nature of this process. We know that cultural evolution is shaped substantially by biology, and that biological evolution of the brain, especially the neocortex, has occurred in a social context. But the principles and the details are the great challenge in the emerging borderland disciplines to which I referred. In my opinion, gene-culture coevolution is the central problem of the social sciences and much of the humanities, and it is one of the great remaining problems of the natural sciences. Solving it is the obvious means by which the branches of learning can be foundationally united. I believe its solution will be one of the important advances of twenty-first-century scholarship.

In closing, let me acknowledge that some critics have said, and will continue to say, that whether the conception is correct or not the program is impossible. The critics argue that the two major gaps to traverse in the borderland between the natural sciences on one side and the social sciences and humanities on the other—that is, genes to brain and brain to culture—are just too wide and complex to master. There exist furthermore, in this view, emergent properties that can never be reduced. Perhaps, the critique continues, they even reflect separate epistemologies.

My answer to radical antireductionism is that, quite to the contrary, the first steps are being taken. Biologists, social scientists, and humanities scholars, by meeting with the borderland disciplines, have begun to discover increasing numbers of epigenetic rules such as the ones I have illustrated and speculated on here. Many more rules and their biological processes, I am confident, will come to light as scholars shift their focus to search for these phenomena explicitly. These rules will next be taken to the levels of brain circuitry and then genomics.

I'm very aware that the conception of a biological foundation of complex social and cultural structures runs against the grain for a lot of scholars. They object that too few such inherited regularities have yet been found to make the case solid, and in any case higher mental processes and cultural evolution are too complex, shifting, and subtle to be encompassed this way. Reduction, they say, rips human thought from its context, it is vivisectional, and it bleeds away the artist's true intended meaning. It melts the Inca gold of the humanities.

But the same was said by the vitalists about the nature of life when the first enzymes and other complex organic molecules were discovered. The same was declared about the physical basis of heredity even as early evidence pointed straight to the relatively simple DNA molecule as the carrier of the genetic code. And most recently, doubts about the accessibility of the physical basis of mind are fading before the successes of sophisticated imaging techniques.

In the history of the natural sciences a common sequence has predictably unfolded as follows: An entry point to a complex system is found by analytic probing. At first one and then more such paradigmatic reductions are accomplished. Examples are multiplied as the whole system opens up and the foundational architecture is laid bare. Finally, when the mystery is at least partly solved, the cause-and-effect explanations seem in retrospect to have been obvious, even inevitable.

The value of the consilience program—or renewal

of the Enlightenment agenda, if you prefer—is that at long last we appear to have acquired the means either to establish the truth of the fundamental unity of knowledge or to discard the idea. I think we are going to establish it. The great branches of learning seem destined to meet this way, and if so, it will be a historic event that happens only once.

I can think of no more important way to apply the naturalistic approach to human behavior than in the design of the places in which we live and work. The evidence is overwhelming that, given a choice, people wish to bring the beauty and harmony of nature within sight. When possible, they like to blend these qualities into the details of their daily existence, because in so doing, they add to their own sense of worth and security. If architecture and design are ever to become science as well as art, it will be through scholarship of the kind exemplified by the contributions to *Biophilic Design*.

chapter 3

A Good Place to Settle: Biomimicry, Biophilia, and the Return of Nature's Inspiration to Architecture

Janine Benyus

In an unscripted moment that happens all over the world, a child tosses a maple seed into the air, clapping with delight as it helicopters to the ground on its perfectly shaped wing. The maple samara plays gravity against a cushion of air, allowing the seeds of the next generation to escape their parent's shade. Like all good design, it never fails to inspire wonder, and, eventually, imitation.

One of the human universals, I would argue, is our appreciation for good design. An object's beauty emanates in part from how well it works, how snugly it fits its function, and how elegantly—with a minimum of effort or extras—it is made. Our delight in the presence of good design is probably millions of years in the making, since the first objects of our admiration were most certainly not in museums or shop windows, but in our native habitats. Everywhere we looked, we saw beautiful forms and systems—designs that made life possible.

Figure 3-1: The samaras, or winged seeds, of this maple are perfectly shaped to catch the wind and spin to a spot beyond their parent's shade.

Our aesthetic and practical appreciation for these wonders grew into a desire to emulate them in art and implement. A stone chisel fashioned after a beaver's tooth or a snowshoe shaped like the hind foot of a hare are early examples of an apprenticeship that continues to this day. We humans, who instinctively took design cues from the natural world, are returning to the practice in a discipline called *biomimicry.*

Biomimicry is the act of learning from nature, borrowing designs and strategies that have worked in place for billions of years. This conscious emulation of life's genius is a natural part of *biophilia,* which E. O. Wilson defines as our innate tendency to focus on life and life-like processes, to affiliate with other life-forms.

I believe it is part of our nature to be drawn to life's mastery and to try, with equal parts of awe and envy, to do what birds and fish and insects can do. The fact that our most beloved buildings contain within them the same ratios we see ubiquitously in the natural world is no surprise. The early Greek practitioners of sacred geometry mimicked nature's proportions in the belief that this universal math would bring us closer to the cosmos (Hale 1994). Vernacular architects, struck by the practical beauty all around them, may have learned mud-daubing from swallows and termites, weaving from birds and spiders, and masonry from caddis flies. As recently as the 1800s, Hispanic settlers of the San Luis Valley of Colorado learned about proper insulation from a burrowing mammal called the Columbian ground squirrel. New to the mountain climate, the settlers didn't know how thick to make their adobe walls to buffer winter and summer temperatures. To this day, adobe walls in the valley are built as thick as the average depth of the squirrel's bedroom chambers.

Figure 3-2: Columbian ground squirrels build snug underground abodes. Early Hispanic settlers of the San Luis Valley in Colorado measured the depth of the squirrel's bedroom chambers to determine how thick to make their adobe walls.

It's only recently that we've turned a blind eye to nature's tutelage, focusing instead on each other's latest fashion. My hunch is that we used to apprentice with nature as naturally as we breathed, and that our ability to learn from our biological elders was one of the ways we ratcheted ourselves to higher evolutionary planes. It makes good evolutionary sense for us to be drawn to other life-forms, not just for their company and their role in provisioning us, but also, I would submit, for their advice.

When I introduce biomimicry to architects, designers, and engineers in workshops, it feels like a remembering of something long lost. As I spread seashells and feathers and bones before them, they fall quickly under a spell, exploring life's designs with the eyes of that child under the maple. I can hear their relief as they return to that familiar source of inspiration—the forests and prairies and seashores—to find ideas for products and processes that are well-adapted to life on earth.

Could it be that we who study nature's adaptations are rewarded with a sense of delight as a way of reinforcing our homing instinct for good design? Might we naturally gravitate toward ancient, evolution-honed designs because they harbor lessons essential to our own survival? A new breed of architect is convinced that it's time to go back to school, and as they take nature as their mentor, buildings are becoming more lifelike than ever.

A FOCUS ON FUNCTION

Biomimicry is not a style of building, nor is it an identifiable design product. It is, rather, a design *process*—a way of seeking solutions—in which the designer defines a challenge functionally (flexibility, strength under tension, wind resistance, sound protection, cooling, warming, etc.), seeks out a local organism or ecosystem that is the champion of that function, and then begins a conversation: How are you doing what I want to do? And how might I emulate your design?

This listening in order to emulate is an inherently biophilic process, one that brings us into a close intimacy with our biological mentors. But the final design that comes from a biomimicry process may or may not look organic or visually resemble the organism from which the lesson came. For instance, a solar cell derived from the way leaves photosynthesize may look nothing like a leaf, but it nevertheless works like a leaf and contains a design concept honed over billions of years (Gratzel 2001).

Figure 3-3a: The principles of photosynthesis have been mimicked in dye-sensitized solar cells. These low-cost cells, made from benign materials, can be fashioned into windows, roof tiles, and façade panels, giving the Commonwealth Scientific & Industrial Research Organization (CSIRO) building in Australia the ability to gather its own energy.

This focus on function points to a key difference between buildings that mimic nature to "look as nature looks" for decorative or symbolic purposes and those that mimic nature to "do as nature does" in order to enhance functional performance. Leaves curling from a Corinthian column, the floral exuberance of Art Nouveau, the winged seed motif of Frank Lloyd Wright's Samara House, and more recently, the bold figures of Gehry's fish and Calatrava's bird are examples of mimicking natural form for aesthetic or symbolic purposes. While these are vital to biophilic design, they are an artistic mimesis, not a biomimetic one. Painting the twisting grain of an old tree onto a column's surface is quite different from actually spiraling the column material as trees do in order to twist rather than break in strong winds. One borrows the likeness of nature, while the other borrows its lessons. Biomimicry, like the twisting column, is decidedly more focused on function.

Today's biomimics are on a quest to create more lifelike buildings—buildings that meet their own needs for energy and water, repair and clean themselves, sense fire or toxins, respond to seasons, bounce back from hurricanes, etc. Architects are attempting to bring nature's wisdom into the way a building works—to animate its very bones and skin and organ systems with practical design tips learned from local organisms. Planners seek to infuse whole neighborhoods with interliving strategies learned from local ecosystems.

Biomimics study nature's design principles in order to be granted life-friendly function, to create an architecture that works the way life works. They believe a building need not look exactly like a tree, but, as Frank Lloyd Wright reminded us, it should work like one.

Figure 3-3b: Roll-to-roll manufacturing processes are used by Konarka to make photosynthesis-mimicking solar cells that are flexible and sensitive to low or even indoor light levels.

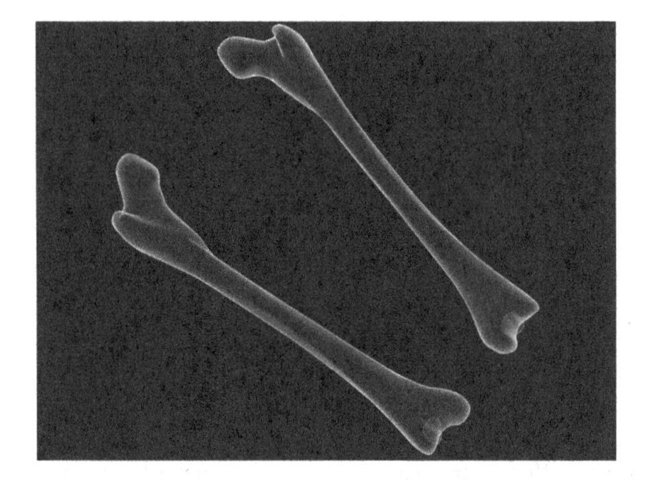

Figures 3-4a and 3-4b: The lower curve of the Eiffel Tower is based on the human femur or thigh bone, where it curves to join the hip. Both femur and tower base must accommodate off-center stresses.

THE BIOMIMETIC BUILDING IS A CHIMERA

While no building today can claim to be completely bio-inspired, a few incorporate features that were inspired by actual organisms. The Eiffel Tower was inspired by the human femur bone, which is expert at handling off-center stresses. The ceiling of the Crystal Palace in London, an engineering marvel of its time, was inspired by the ribbing of the Amazon water lily (Meadows 1999). The membrane structures of Frei Otto (Munich Olympics) use joinery and anchoring mechanisms mimicked from spiders, and suspension wisdom from pneu structures like crab shells (Otto 1967). Buckminster Fuller's domes and the Eden Center owe a debt to radiolarians, and Norman Foster's Swiss Re building was tutored in flow dynamics by the water transport systems of the marine sponge (see Figures 3-6a and 3-6b in color insert). Eugene Tsui's residences incorporate the conical form of barnacles to be granted tornado resistance (Aldersey-Williams and Albert 2003). And there are many more.

Bio-inspired architects need not depend on ideas from just one organism. Instead, they can create a chimera, using ideas from plants to gather sunlight, from mammals to insulate, from fish to streamline, and from microbes to purify. Green building expos already feature several bio-inspired product lines, including leaf-inspired solar cells, mussel-inspired plywood, lotus-inspired building paint that cleans itself with rainwater (see Figures 3-8a, 3-8b, and 3-8c in color insert), butterfly-inspired colorants, and soil-inspired waste treatment systems. More are making their way to commercialization each year. The next section describes how biophilic architects of the future might shop these expos, intent on taking their cue from organisms, using bio-inspired methods to create biophilic design.

BIOPHILIC DESIGN ELEMENTS INSPIRED BY NATURE

The more technological we become, the more we have to remind ourselves of the bodies that we inhabit and the biological communities to which we belong. In

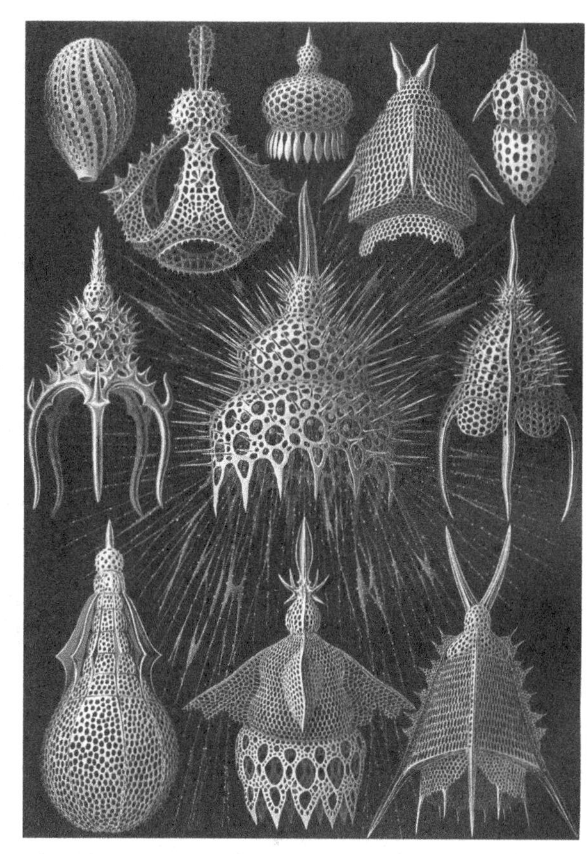

Figures 3-5a and 3-5b: Buckminster Fuller's geodesic domes (like the Epcot Center dome pictured here) incorporate structural principles found in nature at many scales, from the C_{60} molecule to the skeletons of radiolarians.

Figure 3-7a: To tether themselves to surfaces in turbulent tidal flows, blue mussels create tough adhesive filaments called byssus. The adhesive cures underwater and attaches to anything, even Teflon!

Figure 3-7b: The mussel glue recipe has been mimicked and is being used in all the hardwood plywood produced by Columbia Forest Products, North America's largest manufacturer of hardwood plywood and veneer. The waterproof glue in PureBond™ panels replaces current glues that contain urea formaldehyde, a chemical that degrades indoor air quality.

Building for Life, Stephen Kellert says the goal of biophilic design is to "reestablish positive connections between people and nature in the built environment" (Kellert 2005). Some biophilic elements that might bring us back to our animal selves include organically inspired forms; exposure to natural light; natural ventilation; natural patterns of change; natural rhythms and sounds; and the incorporation of indigenous species of plants and animals in and around the building. In practical terms, each of these elements is mediated by some sort of building technology—structural materials, windows, skylights, paints, decor, and garden infrastructure. These are technologies that are also being informed by biomimetic inquiry, so there is the delicious possibility that someday, the very elements that make us feel more at home and connected to nature may be derived from lessons learned in wild habitats. Biophilic buildings might someday not only feature nature and feel like nature; their biophilic elements might be directly inspired by nature. Some of this mirroring is already occurring.

Figure 3-9: The daffodil has a stem shaped like a lemon in cross section, ideal for twisting rather than bending and breaking in a wind. Wind-resistant strategies like these might inspire a strong yet yielding architecture in storm-tossed areas.

Organic Forms and Structures

From a distance, you might not even notice the difference between a biomimetic and conventional building. Come closer, and you will see the signs of a biomimetic quest, in forms and structures that echo nature's wisdom. The building surface may be pleated like a barrel cactus, giving it a self-shading quality. Or the entire building might curve gracefully to allow wind to smoothly flow along its surface, in imitation of a dolphin or whale shape. The roof line—mimicking a rain forest leaf—might swoop into a series of "drip tips" for efficient water harvest. These forms will be more evident in years to come as we learn more about how organisms use shape to minimize turbulent flow, to direct wind or current where they want it (and away from where they don't), and to yield to rather than fight extreme pressure. A light pole may mimic the lemon-shaped cross section of a daffodil in order to spill the wind (Etnier and Vogel 2000). Or a steel column may be filled with a lightweight, porous matrix fashioned after some of nature's strongest cylinders—hedgehog spines and porcupine quills (Karam and Gibson 1994).

D'Arcy Wentworth Thompson, in his epic work *On Growth and Form*, refers to morphology as "a diagram of forces" rendered in flesh by the long evolutionary journey toward accommodation of stresses. Architects climb this adaptive landscape as well, but some are skipping the trial and error and borrowing blueprints already tested through time. The structure of today's biomimetic buildings might benefit from the tutelage of a cantilevering tree branch, the backbone of a bison, the hydraulic tensioning of an earthworm's skin, or the ribs of a dragonfly's wings. Membrane structures might also be lightweighted through the secrets of tensegrity, the way that compressive and tensile stresses are distributed and balanced in a system of beams and cables. Tensegrity, which features so prominently in Buckminster Fuller's iconic domes and Kenneth Snelson's miraculous sculptures, is thought to be found at every scale of biological systems, from the cytoskeleton of our cells through the tendon and bone strutting of our skeletons (Ingber 1998).

One thing that our bones can do that our buildings cannot yet do is to reform themselves in real time in response to stress. Bones beef up with material along lines of stress, removing material where it is not needed. Building scientists would love to mimic this, allowing buildings to get stronger in place. Trees do the same thing; they build reaction wood and spiral grain in response to lines of stress. Claus Mattheck, a biologist who has studied self-strengthening mechanisms in bones and trees, has embedded their design principles in two structural engineering software programs (Mattheck and Tesari 2002). A car proposed for Mercedes-Benz was lightweighted by 40 percent by allowing the software to create a skeleton that featured material only where it needed to be.

Designer Joris Laarman has used the bone-and-tree software to make his "bone chair," an organic-looking piece that gains strength with a bare minimum of material (http://www.jorislaarman.com/bonefurniture.htm). This is where biomimicry finds its nexus with biophilia—where structures derived from natural principles wind up reminding us of the life-forms around and within us. By collaborating with the people who study nature's structures, our efforts to create organic forms need not be "blobitecture"—the vaguely organic forms that Hugh Aldersey-Williams classifies as "Animal by Accident" in his book *Zoomorphic: New Animal Architecture* (Aldersey-Williams and Albert 2003). If biophilic designers continue to cross-pollinate with biomimetic inventors, the organic forms could be both provocative and practical, contributing to a restorative architecture by using less material without sacrificing strength or safety.

Form is also key to how organisms manage extreme forces, and architects in a climate-changed world may want to pay special attention to the champion survivors in hurricane- or flood-prone areas. Perhaps the New Orleans rebuilding teams should study the canopy shape, rooting pattern, and branch morphology of the Seven Sisters live oak, a 1,200-year-old tree near New Orleans, for whom Katrina was just the latest in a long line of wind tunnel tests.

Figures 3-10a, 3-10b, and 3-10c: Engineers at Daimler-Chrysler created a 70-mpg car by mimicking the shape of a coral-reef dweller called boxfish. They lightweighted the skeleton of the car with a software program that uses principles learned from bone growth.

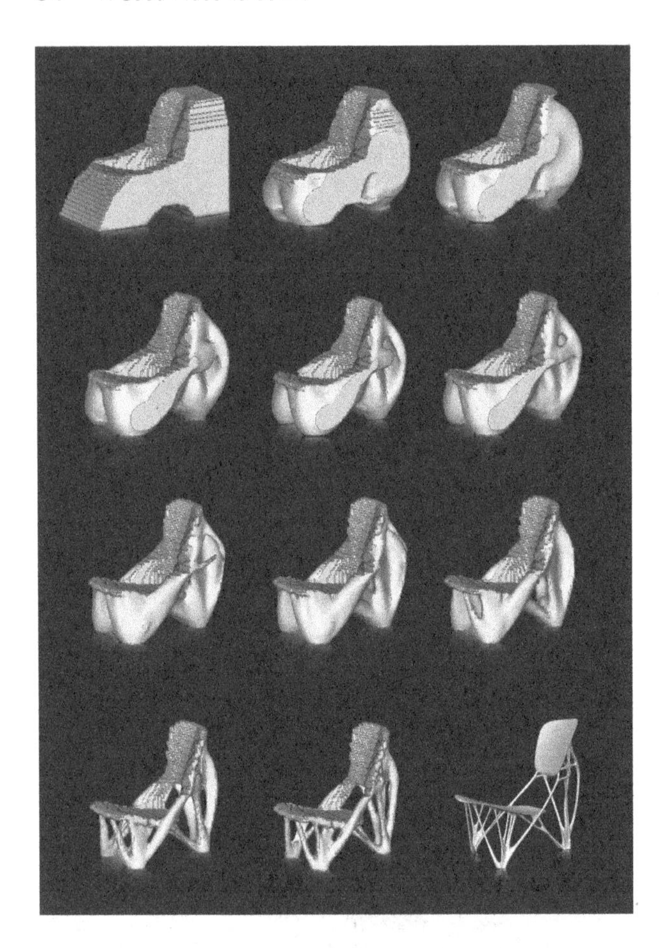

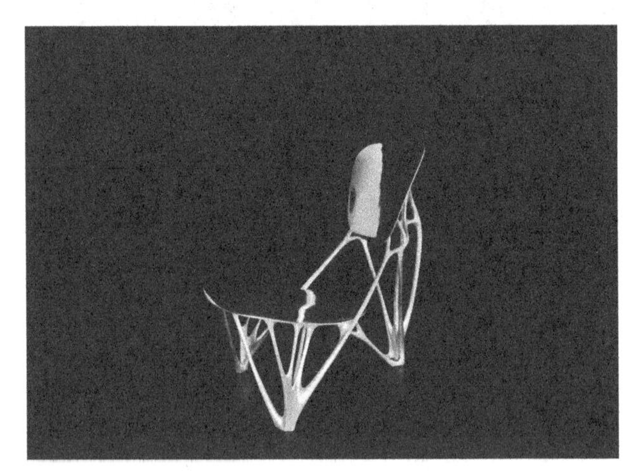

Figures 3-11a and 3-11b: Joris Laarman was able to minimize material in his chair by using software that incorporates the shape-optimizing principles of bone growth. The software reinforced the chair in the areas of highest stress, removing material where it wasn't needed.

Figure 3-12: Well adapted to hurricane-force gales, the Seven Sisters Oak on Lake Pontchartrain had already survived 1,200 years when Hurricane Katrina hit. It leafed out the next spring, just like always, offering powerful design lessons for rebuilders humble enough to listen.

Daylighting

We are not the first organism to want to pipe sunlight into dark recesses. The Venus's flower basket, a sea sponge on the dark ocean floor, has specially shaped filaments that channel light as well as our fiber optics, but with a key difference: they can be tied in a knot without breaking. Alcatel-Lucent scientists would like to mimic their manufacture to create fiber optics that can snake light in and around any bend (Sundar, Yablont, et al. 2003).

Windows and skylights may also be bio-inspired. Rather than being manufactured in kilns offsite, they may crystallize from water solutions right at the building site, using a silicate manufacturing technique that researchers at the University of California-Santa Barbara are learning from diatoms and sponges (Cha, Shimizu, et al. 2000). The windows might also be able to gather the sun's energy, thanks to a technology borrowed from the only true producers on this planet, the photosynthesizers. Unlike photovoltaic (PV) cells, dye-sensitized solar cells (manufactured by Konarka, among others) work the way leaves do, lassoing the sun's energy with dye, even at shallow angles (Gratzel 2005). The flexible films are not yet as efficient as PVs, but are far less toxic, cheaper to produce, and can be sandwiched into windows or glued to walls or roofs. To in-

Figure 3-13: While working at Alcatel-Lucent Labs, Joanna Aizenberg found that the filaments at the base of the sea sponge called Venus's flower basket guide light as well as fiber optics do, but are flexible enough to tie in a knot. Her team hopes to mimic the benign, low-energy manufacturing process.

Figure 3-14: The pillar-like structures on a moth's eye create the perfect antireflective surface, which helps the moth avoid predator-attracting "eye-shine." Autotype's MARAG™ film mimics the pillars to create an antireflective, antiglare surface for displays and solar cells.

crease efficiency, a layer of Autotype's MARAG™, a film that mimics the pillarlike protrusions on the eyes of a night-flying moth, might be added (Parsons 2007). The pillars drink in light and trap it, helping moths avoid the "eye-shine" that attracts bird predators. Since birds are fooled by this antireflective trick in the wild, perhaps MARAG™-coated windows could reduce bird collisions by blocking the reflection of sky in glass.

Finally, the windows might darken or lighten in the same way that cuttlefish change colors. Fuji Xerox has mimicked the skin cells of the cuttlefish in a polymer window material that changes color as the day warms, hinting at how buildings might sense and respond without complicated electronics (Akashi, Tsutsui, et al. 2002).

Natural Ventilation

In his book *The Extended Organism*, termite specialist J. Scott Turner floats a startling proposition: the mound of the termite is actually an extension of its physiology, a giant lung and thermoregulation organ that the colony creates for itself (Turner 2000). The 6- to 18-foot mound visible above the surface has no termites living in it. The true city is underground, including a farm where fungus is cultivated for food. To maintain just the right cultivation temperature, the termites construct tunnels that pipe surface air deep into cool mud chambers. The air rises through the living quarters, drawn upward by a chimney stack and Venturi effect created by the sun-warmed mound above. Meanwhile, the vast network of "bronchial tubes" in the mound circulates ground-level breezes, regulating humidity and exchanging gases in what Turner describes as a "tidal" breath. Amazingly, the ants manage to keep temperatures within one degree of 87°F, despite daily swings of 37°F to 107°F.

When architect Mick Pearce and engineers Ove Arup & Partners wanted to construct a naturally ventilated building in Harare, Zimbabwe, they consulted the local mound-building termites *Macrotermes michaelseni* (Webb 1994). Eastgate, a seven-story, one-block-square office complex that locals call the "anthill," has no air conditioner and uses 35 percent less energy than six conventional buildings in Harare combined. The ventilation system costs one-tenth of a comparable air-conditioned building, and in the first five years, the building saved its owner $3.5 million in energy costs. An atrium, ductwork, and hollow floors mimic the bronchial tubes; concrete floor slabs mimic the mud; and 48 rooftop chimneys let rising air escape. At night, large fans (louder than the wings of termites) flush cool air through the building seven times an hour to chill the hollow floors. By day, smaller fans circulate two changes of air an hour.

Even this is rudimentary compared to the termites, admits Pearce. To move closer to their mentor, the TERMES project led by researchers at Loughborough University is taking an MRI-type scan of actual termite mounds (www.sandkings.co.uk). Using free-form fabrication (a way of "printing" buildings layer by layer with CAD), they hope to introduce termite-inspired channels into walls and cores to help bring fresh air to a desk near you. What is striking about the biomimicry process is how respectfully these architects speak of termites, compared to most builders, for whom termites are pests to be eliminated.

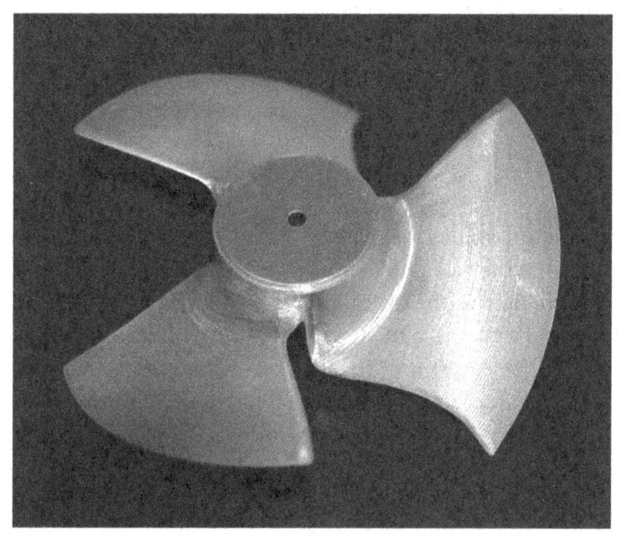

Figures 3-16a and 3-16b: PAX fans use nature's geometry to radically reduce energy and noise. The logarithmic spiral shape of the fan blade is found in seashells, the cochlear of our ears, and even in our skin pores. It's how flow goes.

Natural Sounds and Enough Quiet to Hear Them

In a discussion of scale, Wendell Berry describes a "proper human sound" as one that "allows other sounds to be heard." Perhaps we could hear birdsong if it weren't for one of the main culprits of noise pollution—the incessant whirring of fans in range hoods, computers, air-conditioning units, and more. Fans are everywhere, and the noise that you hear is the sound of friction and electric bills rising.

Biomimic Jay Harman decided to find out how nature accommodates flow and found that the logarithmic spiral (also called the Fibonacci spiral, related to phi and the math of sacred architecture) occurs wherever flow happens—in the spiraling of tidal kelp, the swirl of hurricane clouds, the vortex draining your bath, even the curving channel that lines your skin pores to allow water vapor a friction-free escape. Harman's company, PAX Scientific (www.paxscientific.com), has mimicked nature's path of least resistance to make fans, impellers, and propellers that are astonishingly efficient and quiet. Compared to conventional fans, they reduce energy use by 30 to 80 percent, noise by 75 percent. Imagine the impact this design could have if we retrofitted every fan within hearing as you read this. Lots of fossil fuel saved, and silence enough for a duet between a warbler and your own thoughts.

A Changing Palette of Colors

In our evolutionary past, the forest was a timepiece, with colors changing hour by hour as the sun arched across the sky. Since then, designers have pledged alle-

Figures 3-17a and 3-17b: Peacocks and morpho butterflies create color with structure rather than pigment. Their scales and feathers have intricate layers and matrixes that play with light to create the colors you see.

giance to unchanging dye lots, and the sameness of our interiors no longer makes sense to the deepest part of us. This could change if designers began to spec the many commercially available products (paints, fibers, thin films) that mimic nature's strategy for creating iridescent color without pigments.

It's called structural color, and it's what gives peacocks, hummingbirds, butterflies, and beetles their brilliance. Instead of expensive chemical pigments, these organisms have layered, porous, or ridged structures on their wings, shells, and scales that bend, bounce, diffract, or interfere with ambient light. In organisms with a layered structure, some wavelengths penetrate through the transparent layers, while others bounce back at the layer boundaries, perfectly synced to amplify the color. The color you see depends on the thickness of the layers (refractive index), the space between grooves (diffraction), or the arrangement of bubbles in the material (scattering) (Vukusic and Sambles 2003). Structural color is four times brighter than pigmented color, and it never fades.

Since many of these effects are color-shifting, structural color can also contribute to a more biophilic interior. In a simulated video of a structural color carpet, designer David Oakey (www.davidoakeydesigns.com) shows how a carpet would take on slightly different hues as the sun streamed through various windows during the day. A next step is to find a way to tune the color by moving layers closer or farther apart. In this way, walls or product surfaces could have a dial-a-color feature to accommodate one of the most human of desires—a fresh color palette every season.

Bringing Working Ecosystems Inside

To mimic nature's technologies out of context is not enough. The real lesson of the forest lies not in the individual adaptations, but in the community's magic. Together, in ensemble, the inhabitants of the forest manage to improve the habitat for their collective offspring 10,000 generations from now. There's no need for an Environmental Protection Agency because every action—breathing and breeding, feeding and dying—helps to build soil, clean water, filter air, and cycle nutrients. As a part of everyday living, *life creates conditions conducive to life.*

And life, if we invite it into our buildings, can also create conditions conducive to our life. In addition to their beauty, indoor plants can filter wastes, absorb excess water, mask sounds, and purify air. In fact, these functions are also being performed by ecosystems in our local bioregion. Studying these communities and then mimicking their species composition and structure may grant us their function. For instance, John Todd of John

Figure 3-18: Eco-machines from John Todd Ecological Design, Inc. mimic the patterns of nature's water-purifying ecosystems to clean sewage to pure water.

Todd Ecological Design, Inc. mimics the patterns of local marshes by assembling ecologies of organisms that polish water to a potable state. His eco-machines are connected tanks of creatures from all five kingdoms, and have been installed in everything from individual residences to city-wide systems (Todd, Brown, et al. 2003). He is now installing "restorers"—floating rafts of aquatic organisms that work in concert to heal heavily polluted canals in China.

Biolytix (www.biolytix.com) has created a septic system that mimics a riparian soil community, employing an ecology of local soil organisms to create a humus layer that then acts as the filter. Raw sewage and food waste are treated, and irrigation-quality water is released to landscape plants and lawns through a network of shallow pipes. It features a very compact system that can be networked at a neighborhood scale for one-half the cost of a conventional sewage treatment, and one-tenth its energy use (Cameron 2005).

These examples, which actually use organisms as partners, combine bio-assisted technologies with the biomimetic process. The biomimicry comes in recreating the structural patterns (which species in which spatial arrangements) that have been perfected by native ecosystems over millennia. Mimic the structures of ancient working landscapes, say the biomimics, and you'll be granted function.

Landscaping That Mimics and Restores Landscape Function

Landscape design is one of the areas where biophilic design can make a global difference, especially as the climate changes. A mass movement of plants and animals is already afoot as organisms shift their ranges toward cooler climes. This migration translates into huge ecological disruptions as organisms encounter places (including cites) that are foreign to them. Creating landscaped corridors that allow safe passage for these refugees would be a new kind of "green necklace" for our times, one that would benefit more than just runners, skaters, and bicyclists.

Landscape design also has the opportunity to stitch together some of the fragmentation that has occurred due to uncontrolled urban sprawl. In watersheds where continuous cloaks of vegetation once provided critical services such as water storage and release, air and water purification, nutrient cycling, et cetera, our checkerboard of pavement and lawn doesn't come close to replacing the functions we've lost. For our built landscapes to play their part in the larger whole, we need to bring back the salient features of working watersheds.

That's what Australia decided to do to reverse the soil salinization that has plagued the southeast ever since native systems were displaced by agriculture (Lefroy 1999). The original savanna-type systems, with scattered, deep-rooted trees surrounded by perennials, were keystone regulators of water balance. A government-sponsored "ecosystem mimicry" campaign subsidized landowners willing to resuscitate the land by reintroducing similar structural patterns into residential and commercial landscaping. The beautification project is bringing soil back from a slow death, while bringing residents back in touch with what their land wants to be.

Bio-Inspiration Gardens

When you are surrounded by good design, it seeps into your work. Imagine if you could push back your chair from a stalled project and walk into a garden for inspiration. What if the garden had lotus plants and a mag-

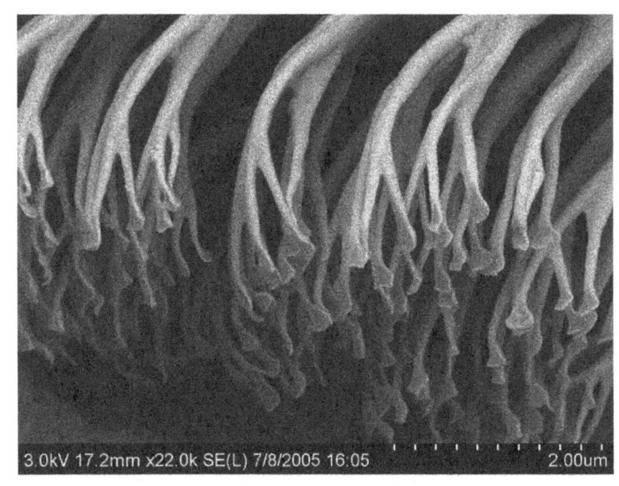

Figures 3-19a and 3-19b: A gecko can stick to walls and ceilings, thanks to the bristles at the base of their specially shaped toes. These setae (magnified by a scanning electron microscope) adhere to the wall with intermolecular forces called van der Waals. BAE has mimicked these in a tape; one square meter would hold the weight of a four-passenger car.

nifying glass so you could watch as a handful of water pearled dirt particles away from the leaf's bumpy surface (the basis for a lotus-effect paint called Lotusan™) (Barthlott and Neinhuis 1997)? What if interpretive signs told you that the feet of the geckos in the terrarium were sticky enough to suspend 280 pounds without toxic glue (Autumn and Peattie 2002)? What if a vine's tendrils suddenly struck you as a phenomenal way to attach items as soon as they touched?

I've seen first hand how immersion in a garden or ecosystem can change what designers design. When I teach in the thrumming jungles of Costa Rica, the projects tend to be system-savvy and complex. When I teach in the company of desert plants in the American Southwest, the designs are all about elegance and making the most of resources. I think it's because we are hardwired to pay attention to habitat cues, and to act as other organisms do—to call as they do, hunt as they do, even build tools like theirs. It is relatively recently that we stopped imitating organisms. What if this keen attention returned and we were encouraged not just to learn *about* organisms, but to learn *from* them?

Bringing designers into nature and nature into buildings reveals the kindred soul of biomimicry and biophilia. Both kinds of matchmaking bring us into respectful contact with organisms, inviting us to admire living beings while dreaming how much more alive our buildings based on their technologies could be.

BRINGING NATURE'S WISDOM BACK INTO THE BUILDING PROCESS

How can we use biomimicry to help our buildings reach a level of not just benign, but benevolent, presence? Applying a suite of products to a building is only the first step toward a biomimetic aesthetic. To build a truly biomimetic building will take a new kind of architectural inquiry that begins with admitting that there is an order of being in the natural world which, as Václav Havel has said, "exceeds all our competence." It will depend on an iterative, deepening conversation with an organism or ecosystem. Learning how an organism keeps itself warm or how it recirculates waste is often the easy part; the difficult bit is to actually emulate that strategy with our own technologies. With emulation comes humbleness, and the desire to find more mentors. In this way, biomimicry leads the practitioner into a renewed relationship with the natural world based on respect, awe, even reverence. The act of asking nature's advice, of valuing nature for its wisdom, bridges the distance that has developed between humans and the rest of nature. In this way, biomimicry is a process of homecoming akin to biophilia, stemming from a desire to not just be with organisms that we admire, but to be more like them.

To mimic the way life works, biomimetic architects and designers must delve into an organism's life to learn how it is meeting its needs, flourishing in its context, and evolving over time. A biomimic consults nature as model, measure, and mentor: What would nature do here? What wouldn't nature do here? Why and why not? A full emulation engages at least three levels of mimicry: form, process, and ecosystem. Form asks, "What is the shape?" Process asks, "How does it perform and how is it made?" And ecosystem mimicry asks, "How does it fit with the whole?"

The best way to bring this kind of inquiry into the building process is to institute new traditions:

1. *A functional survey at every site.* Before you design, survey the organisms that are on the site (or that would be there). How are they caching water; building homes; coping with fire, flood, or wind; communicating, transporting, or restoring themselves? These organisms are the embodied wisdom of living in place, and they can tell you more about site conditions than any text.
2. *Biologist at the design table.* Biologists and ecologists can answer questions such as: How do organisms filter salt out of water? How do they track the sun? How do they dampen vibrations? How do they recycle everything? An amoeba-through-zebra look at how organisms have solved engineering challenges can give designers new (to us) ideas.
3. *A biological filter for all design decisions.* Natural selection works because it has a consistent definition of success—that which enhances the chances of you and your progeny surviving. What about a filter that asks: Is it safe for bodily tissues? Does this action create conditions conducive to life? Is it well adapted to life on earth over the long haul?
4. *A biomimetic innovation credit in building-rating systems (e.g., LEED).* Rewarding good behavior with some sort of advantage that ensures the success of the person practicing that behavior would be a great route to well-adapted buildings. It's how evolution works, after all.
5. *A thanksgiving loop.* Contribute a percentage of your proceeds, or operational savings, to the Innovation for Conservation program to preserve the habitat of the organism that inspired your innovation. It's the least we can do to say thank you.

CAN BIOMIMICRY BRING US BACK HOME?

All this begs the biophilic design question: Will our spirits feel more at home, more connected to nature, in a building designed via a biomimetic process? Surely the builders themselves will feel a greater connection, but what about after the creative spark flies? My sense is that yes, we will recognize the handiwork of nature around us not just in the plantings and the organic forms, but also in the elegance and simplicity of the designs, and in the fact that they work well without our intervention. Walls that need no paint, sensors that require no electricity, exteriors that are cleansed with just rainwater, all of these feel closer to what we experience in fully self-sufficient wild places. When our landscaping finds its own water, fixes its own nitrogen, battles it own pests, and provides a home for rough-legged hawks, we will feel as if our built environments are coming closer to the qualities that we value and innately recognize in healthy habitats. Perhaps we will actually breathe easier, knowing that we've found a good place to settle.

What are biologists doing in this book about buildings? Our lives—those of biologists and ecologists and builders and designers—are beginning to converge because we are having a common dream of who we are as a species. We want to design a world in which we are welcome, no longer estranged from the species that surround us, but, as Mohawk elder Owen Lyons says, "part and parcel of all creation."

There are a thousand cuts in Gaia's side and a thousand ways to stitch them. But are our buildings capable of healing the wounds? Why wouldn't they be? Why is it that we are not surprised that the flanks of Mount St. Helens bloom with lush vegetation that stops the erosion? Or that the coral fish rush in to fertilize the reefs broken by the tsunami, or that the leaves on the sabal palm grow back with more vigor after each hurricane? We expect these restorative acts from natural systems, so why not from ours?

I believe we are as capable as any organism—capable of healing the earth and creating conditions conducive to life. If that is what we ask our buildings to do, if that is the function that we select, then that is what we will turn our prodigious design imaginations to. And we will have plenty of help finding examples of how to do this, right outside our window. It starts as soon as we decide to see the solutions that have been there—scampering, slithering, buzzing, and blooming—all along.

The child infatuated with the maple samara, and Frank Lloyd Wright creating his homage to the winged seed in Samara House, are both examples of our innate adoration for life's designs (see Figures 3-20a and 3-20b in color insert). Somehow Wright kept his child-like worship of natural form, perhaps because he knew that a winged seed represents a miracle worth celebrating. The sweptback membrane is more than adornment; it's what makes it possible for a species' life to continue. And in that matching of perfect form for worthy function, there is a grace that we recognize as beauty.

We know good design when we see it because we grew up immersed, over long evolutionary time, in ecosystems full of competence. We were biomimics and biophilics long before there were names for these affections. As we bring life back into our cities and our homes, and as we take living creatures as our mentors, I am hopeful that the nests that we build with their help will earn natural selection's highest praise. May our dwellings fit their worthy function like a glove—nurturing people and earth, body and soul.

REFERENCES

Akashi, R., H. Tsutsui, et al. 2002. "Polymer Gel Light-Modulation Materials Imitating Pigment Cells." *Advanced Materials* 14(24): 1808–1811.

Aldersey-Williams, H. C. A. V., and M. Albert. 2003. *Zoomorphic: New Animal Architecture*. London: Laurence King Pub. in association with Harper Design International.

Autumn, K., and A. M. Peattie. 2002. "Mechanisms of Adhesion in Geckos." *Integrative and Comparative Biology* 42(6): 1081–1090.

Barthlott, W., and C. Neinhuis. 1997. "Purity of the Sacred Lotus, or Escape from Contamination in Biological Surfaces." *Planta* 202(1): 1–8.

Cameron, D. 2005. "Biolytix: A Low-Cost, High-Performance Sewerage Alternative." *Water* 32(2): 128–129.

Cha, J. N., K. Shimizu, et al. 2000. "Learning from Biological Systems: Novel Routes to Biomimetic Synthesis of Ordered Silica Structures." *Materials Research Society Symposium-Proceedings* 599: 239–248.

Etnier, S. A., and S. Vogel. 2000. "Reorientation of Daffodil (Narcissus: *Amaryllidaceae*) Flowers in Wind: Drag Reduction and Torsional Flexibility." *American Journal of Botany* 87(1): 29–32.

Gratzel, M. 2001. "Molecular Photovoltaics that Mimic Photosynthesis." *Pure and Applied Chemistry* 73(3): 459–467.

Gratzel, M. 2005. "Solar Energy Conversion by Dye-Sensitized Photovoltaic Cells." *Inorganic Chemistry* 44(20): 6841–6851.

Hale, J. 1994. *The Old Way of Seeing*. Boston: Houghton Mifflin.

Ingber, D. E. 1998. "The Architecture of Life." *Scientific American* 278(1): 48–57.

Karam, G. N. and L. J. Gibson. 1994. "Biomimicking of Animal Quills and Plant Stems: Natural Cylindrical Shells with Foam Cores." *Materials Science and Engineering: C* 2(1-2): 113–132.

Kellert, S. R. 2005. *Building for Life : Designing and Understanding the Human-Nature Connection*. Washington, DC: Island Press.

Lefroy, R. 1999. *Agriculture as a Mimic of Natural Ecosystems*. Boston: Dordrecht.

Mattheck, C., and I. Tesari. 2002. "Integrating Biological Optimisation Methods into Engineering Design Process." In *Design and Nature: Comparing Design in Nature with Science and Engineering*, edited by C. A. Brebbia, L.J. Sucharov, and P. Pascolo. Southhampton, UK: WIT Press. 27–36.

Meadows, R. 1999. "Designs from Life." *Zoogoer* 28(4).

Otto, F. T. R. S. F. K. 1967. *Tensile Structures: Design, Structure, and Calculation of Buildings of Cables, Nets, and Membranes* (uniform title: *Zugbeanspruchte Konstruktionen*). English. Cambridge, Mass.: MIT Press.

Parsons, K. 2007. "A Clear View." *Materials World* 15(3): 25–27.
Sundar, V. C., A. D. Yablont, et al. 2003. "Fibre-Optical Features of a Glass Sponge." *Nature* 424(6951): 899.
Todd, J., E. J. G. Brown, et al. 2003. "Ecological Design Applied." *Ecological Engineering* 20(5): 421–440.
Turner, J. S. 2000. *The Extended Organism: The Physiology of Animal-Built Structures*. Cambridge, Mass.: Harvard University Press.
Vukusic, P., and J. R. Sambles. 2003. "Photonic Structures in Biology." *Nature* 424(6950): 852–855.
Webb, R. 1994. "Offices That Breathe Naturally." *New Scientist* no. 1929.

chapter 4

Water, Biophilic Design, and the Built Environment

Martin L. Mador

> *Water, you have no taste, no color, no odor, you cannot be defined, you are relished while ever mysterious. Not necessary to life, but rather life itself, you fill us with a gratification that exceeds the delight of the senses. Of the riches that exist in the world, you are the rarest and also the most delicate: you, water, are a proud divinity.*
>
> ANTOINE DE SAINT-EXUPÉRY, *Wind, Sand and Stars (Terre des Hommes)*

FIRST QUESTIONS

In this chapter, we will consider, in order, several questions. What is water, looking beyond its mere physical characteristics? How do human cultures interact with water? What does water do for us, and how have we treated it in return? How does water fare as a biophilic element? How does our emotional attachment compare with our interest in the animate world? Are there opportunities for water as a biophilic component of the built environment? Ultimately, the answers will help us arrive at a determination of the usefulness of considering water an important element of biophilic design.

WATER 101

Water is in many ways the reticent component of our natural surroundings. Water covers 70 percent of the earth's surface. Water has a ubiquitous presence in the landscape as oceans, rivers, creeks, ponds, marshes, wetlands, springs, ice, vernal pools, brooks, bays, estuaries, and dew. It pervades the atmosphere as clouds, rain, humidity, haze, hail, mist, fog, virga, sleet, snow, rainbows, and in suspension as vapor. Necessary for life, it accompanies every instance of human habitation, as reservoirs, wells, sewage conveyances, storm drains, aquifers, cisterns, and supply piping. It forms a major component of the cellular structure of living organisms. Even the earth, classically regarded as one of the four distinct elements (earth, water, fire, air), is permeated by water as soil moisture, the vadose zone, groundwater, water tables, aquifers, subterranean rivers, and fossil water. In the words of Craig Campbell, "water is the supreme sculptor of our environment" (Campbell 1978).

Water is indeed the unifying element of nature. Because it is mobile, it connects all aspects of the landscape. It defines the landscape by providing edges and boundaries. The watershed, defined as an area of land that shares a common rainfall drainage point, is now regarded as the most appropriate mapping tool for studying almost all environmental issues save air.

Ocean, or salt, water, comprises about 97 percent of all water on earth. Of the remaining 3 percent, most is bound up as frozen water in glaciers and the two icecaps, and in groundwater that does not circulate (fossil water). While salt water provides many services useful to man, such as transportation, food, climate moderation, and waste disposal, its value to a discussion of biophilia is primarily aesthetic. For example, oceanfront property is becoming more and more valued, as the oceanic backdrop adds immeasurably to human satisfaction. For this reason, this chapter will focus primarily on freshwater, the less-than-1-percent of the earth's water that circulates through the hydrologic cycle (rain, ground circulation, evaporation), which is most visible to people and which has significant application to the built environment.

Despite the ubiquitous presence of water surrounding us, when we think of our genetically driven need to connect with the natural world *(biophilia)*, we tend to think of charismatic animal life and dramatic landscapes, of a walk in the woods, the open space of a nearby park, of bird-watching and occasional sightings of backyard raccoons, or visits to the zoo. We are all somehow aware of water, yet we take its presence in our lives for granted.

BIOPHILIA GETS WET

Water enjoys scant conscious appreciation for its necessary and pervasive utilitarian value. Emergence of water of adequate quality and quantity when turning the faucet handle is taken for granted in most developed areas. We are scarcely aware that water is present to serve us. In fact, the only visible presence of water in almost all residences is the standing water in the toilet bowl. Could that thought be what led one author, otherwise a most eloquent spokesman for water's fascinations, to write, "The flush toilet is perhaps man's greatest invention"? (French 1970). Forty years ago, twenty years before E. O. Wilson encapsulated our understanding of the need to connect with nature in the word *biophilia* (Wilson 1984), landscape architect John Ormsbee Simonds wrote, "To some degree we humans still seem to share with our earliest predecessors the urgent and instinctive sense that drew them to the water's edge" (Simonds 1961).

A brief review of the range of biophilic aspects of water would provide a helpful precursor to a discussion of the opportunities of incorporating water in the built environment. Stephen Kellert has composed a typology of nine values or constructs of our attachment to the natural world (Kellert 2005, 1997). This framework provides us with a comprehensive structure for enumerating our many attachments to water, as we briefly consider each:

Dominionistic: the thrill of a whitewater kayaker triumphing over the river; the satisfaction of an engineer designing a hydro dam or installing storm-water piping; a scuba diver successfully completing a trip into a life-threatening environment; the downhill 70-mph run of a skier on snow

Humanistic: the ability of man to form a bond with this natural element, to value its existence, its significance in his sense of place, and its value as a life-giving element

Naturalistic: the thrill of traveling by canoe through a river inaccessible otherwise; hiking through woods along a stream alive with motion and sound

Negativistic: the fear of flooding; drowning; falling through the ice; being swept away by a riptide; threatening weather; waterborne disease

Aesthetic: from a rainbow to deep blue ocean waves to a sunset over the water (sight); crash of the surf, gurgle of a mountain brook, whitewater roar of a swollen river, patter of a light rain, lapping of shoreline waves, drumbeat of a thunderstorm (sound); caress of a morning shower or the initial dip of a toe in the pond (touch); scent of life after a light rain or the salt air of an ocean-side encounter (smell); life-affirming satisfaction of a thirst-quenching drink (taste)

Moralistic: the sense of valuing the gift of this resource; the obligation to preserve it; equitable sharing among human and nonhuman users; religious importance of god's creation and/or gift; water in most religions as the original source of life

Scientific: lessons of aquatic chemistry, ecology, and biology; hydraulics of stream flow; hydrology of groundwater transport

Symbolic: a brook communicating to us through the gurgling of its tumbling waters; the strength and power of the flow of a mighty river

Utilitarian: transportation; recreation; food production; waste disposal; hydropower

What an extraordinary catalog for an element that essentially has no persona of its own. Water as a liquid has no shape, yet it is readily defined by its surroundings. Water has no hardness; it is completely yielding to the touch, yet is hard as concrete when impacted at high speed. Water has no color when viewed in a transparent container, yet becomes vividly green or blue as an ocean, and readily reflects at its surface everything around it. Pure water has no taste, yet it readily absorbs and transmits the taste of any suspended or dissolved substances. It has no smell, yet, as atmospheric humidity, readily distributes the aromas of its surroundings. This ubiquitous part of our environment truly has a protean personality, readily changing to assimilate its surroundings.

On the other hand, water surprisingly has strong animistic traits that give it lifelike qualities, which strongly reinforce our humanistic bond. These traits come primarily from motion, power, change, and sound. The turbulence of a surging river in flood, the vortices (whirlpools and eddies) creating whitewater in a boulder-strewn river, the rhythm of a pounding ocean surf, endows water with a great sense of power. A sculptor's description of a bronze sculpture reads: "Water streams and patters playfully over the curved collar from one column to the next, then falls into the pool. When the wind makes the curtains of water into thin veils, the static columns seem to become figures dancing with each other—an image of community" (Dreiseitl 2005). (See Figure 4-1.) A flowing river is constantly changing, a quality of living things: "You could not step twice into the same river; for other waters are ever flowing on to you" (Greek philosopher Heraclitus, 540 BC–480 BC). There is a fascinating body of writing on the inspirational power of water's flowforms, including notable volumes by Leonardo da Vinci. As summarized by West Marrin,

Figure 4-1: Water animates a bronze sculpture in Ittigen residential park, Berne, Switzerland.

"da Vinci was convinced that the vortical motion of fluids was a key to understanding and utilizing the power of the universe" (Marrin 2002).

Water in motion in natural settings makes a range of musical sounds, being capable of varying in volume, pitch, timbre, texture, and rhythm. Music is fundamental to the human soul—it is found in all cultures and all epochs. Anthropologist Steve Mithen writes that music may in fact be a genetically encoded element of the human psyche embedded many hundreds of thousands of years ago, just as was biophilia (Mithen 2006). If so, water's ability to create musical sounds comprehensible to humans is itself a highly biophilic humanistic trait.

In addition to the fundamental utilitarian services, water provides an extensive catalog of services that enhance the quality of human life. Water provides a substantial part of our recreational life. Swimming, canoeing and kayaking, rafting, motorboating, sailing, scuba diving, downhill and cross-country skiing, ice skating, fishing, and water-themed entertainment parks are enjoyed by almost all. Even a state as small as Connecticut has 112,319 registered power boats (Connecticut State Department of Environmental Protection, personal communication, 2007), in addition to countless sailboats, canoes, kayaks, tubes, rowboats, racing shells, and rafts. A boat is often described as a hole in the water into which money is poured. What inspires us to endure this financial burden for an opportunity that may be enjoyed only a few times a year?

Water provides many venues for contemplation and spiritual restoration, from the paths circumscribing the lake in New York's Central Park to the lounge chairs at the beach to a graceful public fountain providing an oasis of comfort in a sea of urban concrete.

Water has provided the basis for countless works of art, literature, and music. Rivers form the backbone of many paintings of the Hudson River School and English landscape painters, and both rivers and ponds are fundamental to the impressionistic works of Claude Monet. Frank Herbert in *Dune*, one of the most popular works of science fiction, writes of a world where water is so scarce that it must be preserved in insulated body suits. *A Perfect Storm*, a popular movie, takes man's relationship with weather and the sea as the setting for human struggle, survival, and tragedy. Claude Debussy's *La Mer (The Sea)* animates the ocean with movement titles "The Dialogue of the Wind and the Sea" and "Play of the Waves." *Jean de Florette* and its sequel *Manon de la Source*, two extraordinary French films of the 1980s, use the importance of spring water for survival in rural France to both an individual farmer and the town as a whole as the framework for a drama of extraordinary power and tragedy. *Chinatown* does likewise with the public water supply system in Southern California.

The restorative powers of water have been long recognized. At least since ancient Greek and Roman times, healing baths have been popular. Roman scholar Pliny the Elder (AD 23–79) wrote about healing spas. Geothermal springs, such as those in Japan, Iceland, Lake Tiberias (Israel), the banks of the Danube in Hungary, and natural springs rich in minerals have been valued for their medicinal and healing powers. Saratoga Springs, New York, was founded as a resort town for visitors coming to the mineral spring baths. It became so popular that by 1900 the town had 10,000 hotel beds. Hydrotherapy has long been a part of both mainstream and alternative medicine.

Water has held a central place in virtually every religion. Christianity adopted water when Jesus was baptized by John the Baptist in the River Jordan. Baptism now marks the entry to a Christian life. In some churches, it has evolved from actual immersion to sprinkling with a few drops of holy water from a font, but the principle endures. The progression of life, death, and Christian rebirth is accomplished by immersion. Hindus believe that immersion in the Ganges River at least once is a sacred obligation, and they also make pilgrimages to the holy river Narmada. Judaism has celebrated the purifying waters of the ritual bath *(mikvah)* for more than 4,000 years. Visitors to Delphi in ancient Greece had to perform ablutions at the Castalian fountain prior to admission to the oracle. The River Styx held a special place in Greek mythology as the boundary of life and death. Muslims perform a ritual foot-cleansing in water from a clean source before entering a mosque. The Japanese Shinto have a water bowl in every shrine for purification. The writings of Muhammad discuss the inalienable water rights of Muslims. The well of Zemzem at Mecca is considered sacred water.

In addition to its biophilic contribution, a short list of water's services to civilization includes agriculture, through irrigation and rainfall; transportation; recreation; food; waste conveyance and disposal; industrial cooling; hydropower; and moderation of climate.

Venice, Italy, stands as a compelling example of the power of water to define a city. While the architecture is extraordinary, the water canals pervading the city transform it into a biophilic site of tremendous appeal. The appeal is strong enough for many other cities to claim they are the "Venice of the [North, South, East, West]." The list of claimants includes Stockholm; Saint Petersburg; Bruges; Amsterdam; Birmingham, UK; Hamburg; Ottawa; Bangkok; Lijiang, China; Nantes; Galway, Ireland; San Antonio, Texas; Recife, Brazil; Aveiro, Portugal; Bydgoszcz, Poland; Zakynthos, Greece; Udaipur, India; and, among many others, the eponymous Venice, California.[1]

WATER IN CONTEMPORARY WESTERN CIVILIZATION

The twentieth century is destined to become known as the age of engineering. We combined our evolving knowledge of materials and construction techniques with our knowledge of science and computation to produce extraordinary new creations, from towering skyscrapers to genetic engineering to aircraft and spacecraft. We celebrated these technological accomplishments, and our collective ego basked in the glow that humans were capable of such invention. We began to see engineering as the only body of knowledge necessary for problem solving. In the built environment at both large and intimate scales, roadblocks presented by the natural world to human ambition could simply be engineered over, around, or through.

Most of nature, and especially water, fell victim to this "advance" of civilization. Water in nature is never linear. Water bodies always have curvilinear boundaries. Rivers, left to themselves, will develop a meandering, hairpin-dominated course. The engineered world, however, thinks in straight lines. Water, whether as fresh supply or wastewater, is universally contained in straight pipes. Many, many river courses have been straightened for the convenience of abutting landowners. The Los Angeles River, among many others, was not only confined to a linear channel but also a concrete-lined one, so that it became an inanimate object in the landscape and suffered the loss of both life-sustaining and biophilic potential. (See Figure 4-2.) Rivers, which in the springtime freshet naturally overflow across irregular floodplains, were confined to channels with dikes and levees. Indeed, we are even trying to channelize a river as mighty as the lower Mississippi with enormous engineered control structures (McPhee 1989).

For the most part, we spent the twentieth century sequestering our natural water resources, depriving them of biophilic potential for human celebration. In Connecticut, we buried downtown rivers in Hartford (Park River) and Meriden (Harbor Brook) so that they forfeit their opportunity to contribute to the quality of life. We constructed our interstate highways with no regard to the absolute barrier to access they create. New Haven was cut off from its ocean (Long Island Sound) and Hartford from its major river (the Connecticut). The major river traversing Hamden, a New Haven suburb, is almost completely obscured for most of its passage. Not even a sign alerts a passerby that he is in the company of a lovely suburban water place.

Figure 4-2: The Los Angeles River, channelized and lifeless in a concrete entombment.

Our engineering arsenal ultimately enabled us to pollute our environment on a global scale. Water has an extensive resumé of use as waste conveyance. In the dense urban areas of Europe, human waste was simply dumped in the street to be carried away by whatever rain fell. In the 1800s, underground sanitary sewers were installed, again using water as the vehicle, but at least out of human sight and contact. As urban areas grew in the eastern United States with increasing immigration during the first half of the twentieth century, vast quantities of urban detritus was loaded on barges to be dumped several miles offshore. Industrial waste, much of it toxic, was deposited in open sludge pits, or, worse, dumped directly into the nearest watercourse, which was hydraulically connected to ground and surface water drinking sources. Many rivers became unsafe for fishing or swimming; our ocean beaches were closed due to heavy concentrations of pathogens from human waste. Our attraction to water as an extraordinary biophilic element turned to revulsion. In the words of psychologist/philosopher Ivan Illich, "H_2O is a social creation of modern times, a resource that is scarce and that calls for technical management. It is an observed fluid that has lost the ability to mirror the water of dreams" (Illich 1985).

Fortunately, there are individual instances where water has been recently clearly recognized for its biophilic value. San Antonio, Texas, has turned its Paseo del Rio riverfront into an extraordinary destination for both tourists and residents. Shops and restaurants line several miles of the San Antonio River, and sightseeing boats cruise the narrow waterway. The city's website calls it "the pride of the city." New York City is now making an extensive, coordinated effort to rehabilitate the entire area along the Hudson and East rivers. Providence, Rhode Island, has restored the downtown area along the three rivers and celebrates the area with Waterfire, an extraordinary water event held throughout the summer, which serves as a "moving symbol of Providence's renaissance." The Detroit airport installed a magnificent indoor fountain in one terminal, which has itself become a major tourist attraction. The airport built a very large constructed wetland 12 miles away to compensate for marshland taken for airport expansion. Open to the public, it has hiking trails, fishing, and canoe rentals. West Edmonton, Alberta, built one of the largest shopping malls in the world. As part of the entertainment section, wave action generators in one of the pools allow visitors to actually experience surfing. Whether this is seen as a significant or trivial attempt to connect with water, installation of the wave pool is certainly evidence of its importance. Barcelona, Spain, has begun a project to reconnect with the sea with promenades and parks. Many cities in Europe, such as London, Hamburg, Berlin, and Lisbon have plans to connect with the rivers that defined these cities centuries ago.

There are very few places left on earth that have enough fresh water, all the time, to meet all demands. We are coming very slowly to appreciate the economic value of water. In almost no market, with the exception of bottled water sold in extremely poor urban areas, does the cost of water reflect its economic value. Large multinational corporations now bid for the rights to privatize water delivery in urban areas, raising fundamental issues of water as a human right versus water as an economic commodity. Water diversion rights are mostly based on historical entitlements rather than on balanced needs analysis. In some political jurisdictions, water users have, on paper, rights to divert more water than actually exists in the river. Water is becoming recognized by many as the looming acute environmental crisis, far more immediate than global warming.

In sum, over the past century, we have exercised our engineering prowess to defeat and devalue nature, and we have degraded our water resources rather than celebrating them. With notable exceptions, biophilic opportunities for water in the built environment have been generally ignored. We have largely forfeited the gifts water offers to improve our quality of life, and ignored the opportunities to make the built environment a far more satisfying realm for us. In doing so, we have alienated ourselves from an emotional attachment to water, and have brought many to regard water as a contaminated and unhealthy element. Pollution and strictly utilitarian values have crippled our emotional attachment to water.

But, slowly, within the past 40 years, the tide has begun to turn. As we embrace sustainable and environmentally cooperative strategies, we find an opportunity to renew our biophilic connection with this extraordinary resource. The challenge before us now is to move from our strongly utilitarian connection with water in the built environment to one that celebrates our aesthetic, symbolic, naturalistic, and humanistic attachments. Water is a part of the built environment, but, throughout the age of engineering, has been mostly hidden. Every site must provide for potable water supply, wastewater conveyance, humidity regulation, landscaping, storm-water management, and site hydrology. We must now turn to using the biophilic qualities water offers to maximize the opportunities to create pleasant, satisfying, stimulating, and profitable environments for human occupancy.

TWENTY-FIRST-CENTURY OPPORTUNITIES

Sadly, water has been regarded in the past century as at worst an alien visitor in the built environment, at best as a token accoutrement. Water offers a host of opportunities for enhancement of the built environment. However, most recent efforts utilize water mainly for its superficial entertainment value. Many urban plazas include a fountain, perhaps with animated sprays and color. The water here softens the harshness of the surrounding stone and concrete and lowers the perceived temperature of a hot summer day, but there is little attempt at integration with the local landscape of open space and buildings. The fountain endures the same lonely existence as the captive tree, rising as a solitary protest to the sea of surrounding hardspace. The casinos of Las Vegas serve as extreme examples of water as entertainment. The animated, colorful movement appeals to observers, but there is no attachment formed to the water, which is simply the vehicle for the animation. Worse, these fountains thrive in a region with hopelessly inadequate local water sources, evaporating huge quantities of water, all of it imported. This profligate use of water "borrowed" from other regions actually sends a disturbing message to any observer at all aware of water shortage crises across the globe.

Water parks are becoming popular across the country. These are recreation venues that celebrate the opportunities water offers for the sensual pleasures of immersion and contact, adrenaline highs, and socialization. Unfortunately, they offer visitors little insight into more fundamental biophilic values of water.

As explained above, a small fraction of the total water volume on earth is readily apparent to us. This is the fresh water that participates in the hydrologic cycle, and is the water with the greatest potential for biophilic attachment. With few exceptions, there is no attempt to link the fountain to the hydrologic cycle water all around it, reinforcing the idea that the water is there not as an opportunity for onlookers to revel in this natural, connected resource, but as an introduced artifact imposed on the landscape.

Water can be incorporated at different scales. In the following discussion, the word *landscape* is used to refer to the totality of natural open space, engineered open space (gardens, hardscape plazas), buildings, and the boundaries marking the intersection of buildings and their surrounding open space. It is critically important that biophilic design recognize three opportunities for design: the urban and natural surroundings, the structure(s), and the connection between the two. The scale of water can be as a strictly interior element, with no tie to the architecture of the building, with its removal of no more consequence than rearranging the furniture. Water can form a structural component of the fundamental design of the building, with its absence requiring a significant compensatory effort. Water can serve to connect the building to the immediately surrounding environment. Water can exist as an element of the exterior immediate environment, without connection either to the building itself or to the wider landscape. Finally, water can serve to integrate the large-scale landscape—to connect the building to its environment, whether the local city block, the neighborhood, or the immediately surrounding countryside and abutting natural features. As water integrates the natural environment in many ways, it also integrates the urban environment in ways we are just coming to understand and appreciate.

It is critically important to recognize the core of water's own existence, and how that is supplemented by interactions with its natural surroundings. Water, by itself, has a discrete set of physical and chemical properties, with scant biophilic content. It is through its interactions that water expresses its biophilic identity, and enables us to experience that identity. The strongly biophilic character of water is expressed in many ways:

The *surface* of water, when calm, has extraordinary reflective capacity. We can see in water all that surrounds it, whether the natural world, an adjacent building, or our own reflection. A mild disturbance of the surface turns water into an editor, a critic, a filter, a commentator on the surroundings, and gives water one of its animistic qualities.

Water interacts with *sunlight*, varying its hue and mood with the amount and incidence of solar rays. Deprived of a brightly lit source, water becomes somber and recessive, refusing to communicate. Water in bright sunlight tells many stories of its surroundings and becomes animated as the reflected sparkles dance across its surface.

Water interacts with many *natural materials* to enhance the experience of contact. The grain and texture of many rocks, such as granite, are vividly enhanced by water. Water moving across the surface of rock creates an animated dance. The grain of wood is similarly enhanced. As most wood floats on water, the sense of water as a natural resource giving sustenance and life to wood as another natural resource can engender powerful emotions in us. One of the most powerful posters this author has experienced is a photo of several wood canoes from the Seattle Center for Wooden Boats, titled "Wood on Water."

Moving water has very strong biophilic attraction. Motion adds an element of animism and life to the water, whether it is a waterfall, a cascade, a steeply descending stream, or pond with circulation.

As a primary life-sustaining force, water's significance is dramatized by the addition of *flora*. Facultative wetland vegetation, riparian plantings, and immersed water-based plants such as water lilies combine with water to express strongly biophilic elements of life. Similarly, the addition of *fauna* enhances water's biophilic attraction. Of course, in the context of the built environment, this is mainly limited to fish.

There follows a modest catalog of specific strategies for bringing water to the built environment in ways that enhance our biophilic experience. The list is by no means complete, and is drawn from the author's subjective experience, but is offered to highlight examples for opportunities for incorporation in architectural design at different scales.

While this chapter is intended to be more inspirational than instructive, a brief list of the challenges in introducing water in the built environment should be noted. These include unwanted growth, such as algae; increased interior humidity (which also, depending on conditions, can be a plus); moisture damage from leakage; the need for filtration equipment; ice in exterior locations; energy requirements for pumps; mineral deposits; insect-borne disease; necessary periodic maintenance; and, of course, the ubiquitous environmental hazard in our culture, legal liability.

1. *Roof gardens and green roofs* provide an opportunity to integrate biophilic vegetation with the building, and offer three opportunities for addition of water. The vegetation requires water to live. The roof absorbs and processes storm water. Fountains, ponds, and stormwater cisterns can all be incorporated in the roof design.
2. *Indoor plumbing* has represented the most utilitarian aspect of water. Little attention has been given to how we deliver water in indoor applications. For the most part, our association is limited to the expectation of water emerging every time the faucet handle is turned. Yet one of the easiest ways to raise appreciation of our biophilic attachment with water is to ask: why is it that we stay in the shower long after the soap has done its job? Many appreciate the restorative benefits of jacuzzis or hot tubs. Yet the basic delivery of water has opportunities of its own. Kohler, for example, has taken notable steps to draw attention to water delivery by its design of faucets, water flows, and baths. Not only does the shape of the fixtures inspire biophilic attachment, but water as laminar sheet flows through the air and across horizontal surfaces, and delivery in streams suggestive of waterfalls adds an aesthetic dimension.
3. *Water as interior pools or basins* can provide a design element that is integrated with the structure of the room. Water becomes an equal component of the

room's elements along with the floor, walls, and furniture, whether or not it is supplemented by vegetation. (See Figure 4-3 in color insert.)

4. *Water as recreation* provides unlimited possibilities. It is perhaps this strategy that is best understood and commonly found as swimming pools, spas, jacuzzis, and hot tubs.
5. *Providing natural function in concert with aesthetics* creates many opportunities to appreciate water as a life force. *Wetlands* incorporated as an integral element of the outdoor landscape provide biological function as well as placing water in the setting. (See Figure 4-4.) Open or closed *cisterns or rain barrels* for rainwater capture and reuse not only provide an opportunity for water placement, but vividly illustrate the connection of hydrologic water and human consumptive needs.

Figure 4-4: Pool and wetland, providing both biological function and biophilic aesthetics. East Entrance to the National Museum of the American Indian, Washington, DC.

6. *Interior ecosystems* provide excellent opportunities for introducing both water and vegetation. The scale could extend from a small area of plantings sustained by a modest pond to an entire atrium filled with substantial trees, extensive vegetation, watercourses with circulation, walkways integrated into the scene, and fish living in the water. (See Figure 4-5 in color insert.)
7. *Exterior water gardens* offer opportunities limited only by imagination to add water to the building setting. Perhaps the most effective model is the oriental garden, composed of water, stone, vegetation, and graceful structures. (See Figure 4-6 in color insert.)
8. A freestanding or embedded *aquarium* provides a living aquatic ecosystem with many attractive components: water, fauna, vegetation, and possibly motion.
9. *A waterfall* has universal appeal as flowing water in motion. The sound of water, especially when enhanced by large objects breaking the fall in the plunge pool, adds a dimension of biophilic appeal. A setting that mimics a natural waterfall, complete with surrounding landscape and vegetation, would add even more. (See Figure 4-7.)
10. *A cascade* of water descending a series of steps would be an effective emulation of a rocky mountain stream, with elements of motion, sound, and natural mimicry. (See Figure 4-9.)
11. *Blurring the distinction between inside and outside*, although difficult to achieve, can provide a powerful biophilic effect. A reflecting pool, a transparent exterior wall adjacent to the water feature, or an actual flow of water between the interior and exterior can do this effectively. (See Figure 4-10.)
12. *Using water to produce sounds*, as mentioned above, helps to create biophilic attachments. The sounds could be the trickling of an emulated descending mountainside brook, the soft patter or voluptuous explosions of a fountain, or water-activated musical chimes.
13. *Fountains* provide the most popular presence of water in the built environment. The scale extends from many acres of multiple fountains, such as Peterhof in Saint Petersburg, Russia, and the Villa d'Este in Tivoli, Italy, to individual fountains

Figure 4-7: Waterfall incorporated as a striking design element in the building exterior. National Museum of the American Indian.

Figure 4-8: Water cascade softens an urban hardscape plaza, providing motion and animation. Sun Life Plaza, Vancouver, British Columbia, Canada.

relieving the expanse of an urban plaza, to a modest recirculating font on a residential lawn.

14. *Accessible fountains* multiply the biophilic appeal by allowing visitors to interact physically with the water. Pictures of children playing among the sprays of low-pressure fountains in public parks are compelling scenes, teaching us that mere contact with water as a play venue provides biophilic satisfaction for both participant and observer.
15. Water can serve as *an element of a work of art and sculpture*. While having little or no connection to the natural world, water as a component of human-created art can greatly enhance the appeal and

Figure 4-9: Cascade adds surprising motion and life to an interior setting. Entelechy II, Sea Island, GA.

Figure 4-10: Standing water adjacent to an office complex blurs the boundaries between inside and outside. McLaren Technology Centre Research Centre, London.

effectiveness of the piece. The opportunities here to add water to a project are extensive, and can range from an isolated piece to an entire site.

16. Water in motion creates powerful human attachments. *Kinetic sculpture* driven by a water source, rather than wind, becomes a visual draw to an observer. Water becomes the driver, the agent of motion, and thus acquires animistic qualities.

Figure 4-11: Columns of water enclosed in transparent casings provide an unexpected biophilic addition to the room. TropWorld Casino. Atlantic City, NJ.

17. *Extended waterscapes*, as part of the immediate landscape of a structure, may require significant planning and financing, but add enormously to the appeal of the siting. An extended waterscape may encompass a multiacre park, or provide water-oriented connected space between a building and a nearby feature such as a watercourse.
18. *Integration with the earth* projects may take significant commitment and imagination, but can provide an intense experience with water in the context of the built environment. Stepwells in India that provide river access for ritual and cleansing activities are excellent examples. (See Figure 4-12.) The Water Crater at the regional garden show in Westphalen, Germany, features a staircase descending 18 meters into an excavated crater with a groundwater pool and geyser at the bottom. The designers describe this site: "One can immerse oneself in this closed space, this *hortus conclusus*, to experience that astonishing and vital element of water with the whole of one's body and all of one's senses."
19. A water element can be added through a *connection with adjacent existing natural features*. A nearby watercourse can supply opportunities for water. The building can be placed adjacent to the river, or the layout and design carefully tailored to emphasize the connection. (See Figure 4-13 and Figure 4-14 in color insert.)

Figure 4-12: Stepwell at Chand Baori, Abhaneri, India, showing an extensive architectural structure specifically designed for access to water.

Figure 4-13: Connection with a nature center and adjacent water body, Peggy Notebaert Center, Chicago, IL.

Figure 4-15: Rainfall is processed through small wetlands on the building shell, to an interior pool, then to a cistern beneath the building and an exterior pond. The water then supplies on-site greenhouses. Interior waterfalls create sound and cool the air in summer. Prisma, Nuremberg, Germany.

20. *Engineered emulations of natural settings* can have strong biophilic appeal. Re-creations of river systems with moving water, boulders, and cascades, or ponds with vegetation and fish place the natural world in juxtaposition to the built environment.
21. *Hydromimicry* takes its name from biomimicry, modeling engineering solutions after natural processes, essentially using solutions found in nature to construct processes that work cooperatively with nature. For example, blowing air over exposed water and water held by vegetation will achieve evaporative cooling.
22. *Interior water handling* provides opportunities to add an aesthetic element to a utilitarian function. Currently handled by solid pipes buried in walls, supply water and gray water from sinks and showers could be routed in clear conveyances, treating building occupants to both the motion and sounds of flowing water, without compromising water-handling standards.
23. *On-site stormwater routing* has traditionally involved gutters and downspouts, which as closed systems deprive us of any biophilic benefit. There are now alternatives that make the storm water visible and audible. Rain chains are an effective example; water is channeled through the gutter to the vertical metal chain, which then guides the water to the ground. (See Figure 4-17.) Storm water at the Sid-

well Friends Middle School is led to overhanging fixtures at the roof edge, which guide the water to open downspouts. The water is collected in an open channel paralleling a walkway, then channeled to a constructed wetland near the building. (See Figure 4-16 in color insert.)

24. *Water handling on an urban neighborhood scale* provides opportunities to integrate storm-water disposal, water treatment, storm-water use for landscaping and for interior gray-water purposes, and visible water features. It requires considerable commitment, planning, and financing but offers extensive biophilic, engineering and environmental benefits. Herbert Dreiseitl has established considerable expertise here, as evinced by his design for the Potsdamer Platz in Berlin. (See Figure 4-18.)

Figure 4-17: Residential gutters lead, not to enclosed downspouts, but to rain chains that conduct the water to ground level, creating animation and music on the way.

25. *Stormwater handling on a small neighborhood scale* may involve a small creek traversing the neighborhood that allows storm water to flow off-site, and to permeate to groundwater along the way. The creek may be a perennial stream, or might be wet only after storm events. Swales, linear vegetated depressions that hold storm water, are also effective means of promoting infiltration, and are now used as environmentally preferable roadside alternatives to curbs and storm drains. (See Figure 4-19 in color insert.)
26. *Stormwater handling on a site scale* may involve rain gardens and retention ponds to hold storm water, rather than transporting it for disposal to the nearest watercourse. The water is absorbed by vegetation and infiltrates to the underlying groundwater. Standing water is visible for some period after the rain event. This method not only places water in the immediate environment, but helps to emphasize the natural hydrologic cycle.
27. *Biological wastewater treatment* consists of a series of biologically active tanks, cells, or other structures,

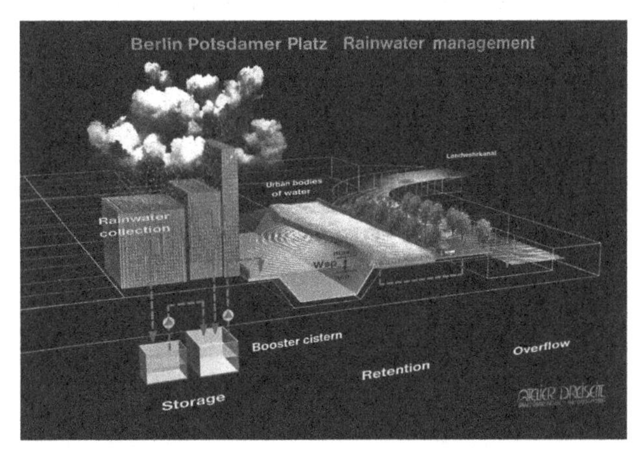

Figure 4-18: Schematic for water handling on an extensive site scale, involving roof collection, building downspouts, underground storage, urban lakes, and hydrologic connection with a nearby canal. Potsdamer Platz, Berlin, Germany.

which process and purify wastewater on-site. The tanks, containing plants, algae, microorganisms, bacteria, and fish, use bioremediation techniques to break down organic contaminants in the black-water or gray-water waste stream. The site could be housed in a greenhouse-like structure, or could extend to several acres in an outdoor setting. The highly purified water could be discharged to an adjacent wetland, or recycled to the building for non-potable uses.

28. *Traversing a watercourse* with a building, bridge, or cantilevered structure provides one of the most dramatic ways of integrating water and the built environment. (See Figure 4-20.)
29. Of course, the ultimate in bringing water to the built environment is simply to make the whole thing from water. Constructing life-size replicas of buildings from ice is exactly what ice carnivals in Sapporo, Japan; Montreal, Canada; and other cities have done since Russia started the craze in the 1700s. An ice hotel is built every winter in Jukkasjarvi, Sweden, on the River Torne, hosting thousands of overnight guests in rooms where even the furniture is carved from ice. Of course, there is the Inuit (Eskimo) igloo, built not out of biophilic exuberance but out of stark necessity.

Clearly, the presence of water in the built environment can substantially enhance the biophilic qualities of the site. While a few designers now regularly incorporate water in their designs, the vast majority do not. Unfortunately, with the exception of books on fountains and pre-nineteenth-century Europe, there have been only a handful of publications dedicated to water and the built environment. However, there are sufficient examples of modern effective practice to demonstrate the extraordinary opportunities water offers. As our knowledge of biophilic theory and practice grows, and our efforts to value water as a biophilic, natural, and life-sustaining resource continue to increase, we come to see water as an integral component of the built environment.

NOTES

1. http://en.wikipedia.org/wiki/List_of_places_known_as_%27the_Venice_of_something%27.

Figure 4-20: Pods holding the theater, restaurants, and other visitor attractions, cantilevered over the water at Ontario Place, a cultural, leisure and entertainment parkland in Toronto, Ontario, Canada.

REFERENCES

Anderes, Fred, and Ann Agranof. 1983. *Ice Palaces.* New York: Abbeville Press.

Bachelard, Gaston. 1983. *Water and Dreams: An Essay on the Imagination of Matter.* Dallas: Pegasus Foundation.

Black, Maggie. 2004. *Water, Life Force.* Toronto, ON: Between the Lines.

Campbell, Craig S. 1978. *Water in Landscape Architecture.* New York: Van Nostrand Reinhold.

Dreiseitl, Herbert. 2005. *New Waterscapes.* Basel: Birkhauser.

Farber. Thomas. 1994. *On Water.* Hopewell, NJ: Ecco.

France, Robert Lawrence. 2003. *Deep Immersion: The Experience of Water.* Sheffield, VT: Green Frigate.

French, Herbert E. 1970. *Of Rivers and the Sea.* New York: Putnam.

Great Rivers of the World. 1984. Washington, DC: National Geographic Society.

Hegewald, Julia A.B. 2002. *Water Architecture in South Asia: A Study of Types, Development, and Meanings.* Boston: Brill.

Illich, Ivan. 1985. *H_2O and the Waters of Forgetfulness: Reflections on the Historicity of Stuff.* London: Boyars.

Kellert, Stephen. 1996. *The Value of Life: Biological Diversity and Human Society.* Washington, DC: Island Press.

Kellert, Stephen. 1997. *Kinship to Mastery: Biophilia in Human Evolution and Development.* Washington, DC: Island Press.

Kellert, Stephen. 2005. *Building for Life.* Washington, DC: Island Press.

Kellert, Stephen, and Timothy J. Farnham, eds. 2002. *The Good in Nature and Humanity: Connecting Science, Religion, and Spirituality with the Natural World.* Washington, DC: Island Press.

Kellert, Stephen, and Edward O. Wilson, eds. 1993. *The Biophilia Hypothesis.* Washington, DC: Island Press.

Lanz, Klaus. 1995. *Greenpeace Book of Water.* New York: Sterling Publishing Co.

Larkin, David. 2000. *Mill: The History and Future of Naturally Powered Buildings.* New York: Universe/Rizzoli.

Livingston, Morna. 2002. *Steps to Water: The Ancient Stepwells of India.* New York: Princeton Architectural.

MacBroom, James Grant. 1998. *The River Book.* Hartford: Connecticut Department of Environmental Protection.

Marks, William E. 2001. *Holy Order of Water: Healing the Earth's Waters and Ourselves.* Great Barrington, MA: Bell Pond.

Marrin, West. 2002. *Universal Water: The Ancient Wisdom and Scientific Theory of Water.* Maui: Inner Ocean Publishing.

McPhee, John. 1989. *The Control of Nature.* New York: Farrar, Straus & Giroux.

Mithen, Steven. 2006. *The Singing Neanderthals: The Origins of Music, Language, Mind, and Body.* Cambridge, MA: Harvard University Press.

Moore, Charles W. 1994. *Water and Architecture.* New York: Abrams.

Rothenberg, David, and Marta Ulvaeus. 2001. *Writing on Water.* Cambridge, MA: Terra Nova/MIT.

Schwenk, Theodor, and Wolfram Schwenk. 1989. *Water, the Element of Life.* Herndon, VA: Anthroscopic Press.

Simonds, John Ormsbee. 1961. *Landscape Architecture, A Manual of Site Planning and Design.* 2nd ed. NY: McGraw Hill.

Spring, Bob, Ira Spring, and Harvey Manning. 1970. *Cool, Clear Water: The Key to Our Environment.* Seattle: Superior-Salisbury.

Swanson, Peter. 2001. *Water: The Drop of Life.* Minnetonka, MN: NorthWord.

Thoreau, Henry David. 2001. *Reflecting Heaven: Thoreau on Water.* Edited by Robert France. Boston: Houghton Mifflin.

Thoreau, Henry David. 2003. *Profitably Soaked: Thoreau's Engagement with Water.* Edited by Robert France. Sheffield, VT: Green Frigate.

"Wasser Water. Designing with Water: Promenades and Water Features." 2003. Topos magazine.

Water Spaces, vol. 1. 1997. Mulgrave, Victoria, Australia: Images Publishing Group.

Water Spaces, vol. 2. 1999. Mulgrave, Victoria, Australia: Images Publishing Group.

Water Spaces, vol. 3. 2002. Mulgrave, Victoria, Australia: Images Publishing Group.

Water Spaces, vol. 4. 2006. Edited by Joseph Boschetti. Mulgrave, Victoria, Australia: Images Publishing Group.

Wilson, E. O. 1984. *Biophilia.* Cambridge, MA: Harvard University Press.

Wylson, Anthony. 1986. *Aquatecture: Architecture and Water.* London: Architectural Press.

chapter

5

Neuroscience, the Natural Environment, and Building Design

Nikos A. Salingaros and Kenneth G. Masden II

INTRODUCTION

Our mental processes enable us to interact with and adapt to our environment. We instinctively crave physical and biological connection to the world. The human perceptual mechanisms through which these processes work establish our relationship and response to both architecture and the built environment. The basis for this interaction is human nature itself: the end result of the evolution of our neural system in response to external stimuli such as the informational fields present in the natural environment.

Humans, seeking shelter from the elements, are compelled to construct buildings and cities. Historically, the form of those structures arose from within the material logic of their immediate surroundings and from the spatial ordering processes of their minds (through biological necessity). Utilizing what was at hand to give structure to existence, people instinctively constructed places that provided the constituent information, form, and meaning that their sense of well-being required. Design decisions occurred as a natural extension of the neurological processes that make us alive and human.

Not consciously aware of the nature of these processes, humankind simply built its buildings and cities in this manner without question for millennia. Over the course of time, however, the relationship to the physical world began to take on a greater complexity through applied meaning, that is, local mythology, symbolisms, and social structures. As the *process of building* was usurped by the *process of design*, architecture as a tectonic expression of innate human ideas about form, space, and surface became more difficult to grasp. People's relationship to the physical world was further complicated with twentieth-century advances in technology and industrialization.

This is clearly evident in the practice of architecture today. Following several centuries of refinement and addition to the traditional vocabulary of architecture, the design process, once the exclusive domain of the master builder, has taken root in a different soil altogether. As architecture shifted from the domain of craft into the intellectual property of the university, the study of architecture began to align itself with other academic disciplines, although incompletely. While architecture mimicked the academic realm of philosophy, it reinvented itself as a new discipline detached from its own evolution. Over time, architects effectively disconnected themselves from their history, which was thenceforth treated more like archaeology: interesting, but irrelevant to present-day design concerns. In the years that followed, architectural design and the study of design methodologies were all but severed from those processes that had served for millennia to render the built environment as something intrinsically human.

We contend here that processes underlying human engagement with the physical world support biophilic design as a reconnective methodology. Furthermore, we believe that this knowledge can guide current and future architecture toward a more intrinsically human expression. The following is an overview of our exposition.

Related scientific research establishes the positive physiological effects of particular types of environments, such as those constructed within the concept of biophilia ("Biologically Based Design"). These respond in their form to the human need for intimate contact with living forms. Explaining biophilia ("Biophilic Architecture and Neurological Nourishment"), we outline two distinct, convergent approaches to its interpretation and architectural implementation. This body of knowledge is then contextualized within a broad, unifying movement. We review techniques ("An Architecture That Arises from Human Nature") that seek to establish a method or process for architectural design relating directly to human sensibilities. A body of compelling research supports this way of thinking about the built environment. Practical information given here and in the closing sections is meant to help and inspire architects wishing to implement these ideas. As this dialogue on reconnecting the built environment to humans and their everyday lives continues to grow, we are confident that the discipline of biophilic design will find its way into the mainstream education and practice of architecture.

In the remaining sections of this chapter, we will discuss how human nature directly affects architecture. The key here is informational connectivity, which our research establishes as the mechanism by which humans relate to biological forms and connect with the physical world. We define three different conceptions of human beings: mechanical, biological, and transcendental ("Three Different Conceptions of Being Human"). The abstract human being of the twentieth century ("Level One: The Abstract Human Being") is an ideal inhabitant of places that are designed according to strictly formal criteria. In contrast, the biological human being of both the preindustrial era and the new millennium requires a particular type of sensory feedback from the environment ("Level Two: The Biological Human Being"). This type of feedback/information is becoming harder and harder to find in contemporary cities. We will show how the precise nature of structures that provide the appropriate feedback can be discovered in the unselfconscious traditional and vernacular built environments ("Extending Level Two: Expert Knowledge and Patterns"). We identify a part of this stored information with "expert knowledge" that supports *pattern language* as an essential design tool. Furthermore, we argue that when human beings experience emulated biological qualities such as in human-computer interfaces, they engage in a natural way ("Further Extending Level Two: Human-Machine and Human-Animal Interactions"). This is the same type of connection observed with animals, such as pets, and suggests the possibility of an intimate neurological connection with architecture.

Toward the end of the chapter, we delve into the highest conception of human nature; the transcendent human being possesses qualities that seem to transcend our biological nature ("Level Three: The Transcendent Human Being"). We contend that transcendence is generated via connection through higher-level neurological processes. Those qualities make possible our greatest intellectual and creative achievements. Philosophy and religion enter into this discussion unavoidably. Accepting this ultimate capacity of human beings leads us to questions about re-creating architecture that transcends its materiality ("Architecture That Transcends Materiality"). Certain buildings—some of them religious, others quite modest—achieve such an intense degree of connection that they can induce a state of healing in us. The informational content in this type of structure is simply so successful in its conception that it connects more directly to neurological processes. It requires far less translation and interpretation by the mind and thus presents itself as inspired or divine. Our aim is to understand how that mechanism arises, as it relates to the concept of biophilic design.

Finally, a closing section ("Fourteen Steps Toward a More Responsive Design") gives a list of practical design techniques. These are meant to help practitioners who might wish to engage architectural design in a more human manner. With the addition of some forward-thinking speculations, we consider how computers and robots might create those humanlike qualities in architecture that once breathed life into the built environment. The final section ("Some Patterns from *A Pattern Language*") summarizes several patterns from Alexander's *Pattern Language* (Alexander et al. 1977) that are relevant to biophilic design. These practical design patterns anticipate and support the message of this chapter.

BIOLOGICALLY BASED DESIGN

The positive effects of biophilic design must be understood in architectural terms: as form and form-making principles, and structural systems. Biologically based design utilizes observed effects and tries to document them into an empirical and tested body of knowledge. At the same time, an extensive research program is beginning to uncover the deeper causes for these effects: that is, a possible innate reaction to the specific geometry of natural forms, detail, hierarchical subdivisions, color, et cetera. Since this project is far broader than the traditional study of architecture, designers must actively solicit help from other disciplines whose knowledge can help to explain human response to design. It is essential not to be partial in any way, since, in addition to known factors, there are clearly unknown factors playing a role yet to be discovered.

Recent investigations lead us inescapably to the fact that we engage emotionally with the built environment through architectural forms and surfaces. We experience our surroundings no differently than we experience natural environments, other living creatures, and other human beings. We relate to details, surfaces, and architectural spaces in much the same way as we relate to domestic animals such as our pets. The mechanism through which we engage with subjects outside ourselves relies on a connection established via information exchange. Our neurological mechanism reacts to the information field (the transmission component), while inducing a reaction in the state of our body (the physiological component). Some of the highest levels of sensory connection to the built environment have been evidenced in the great buildings and urban spaces of the past (Alexander 2002–2005; Salingaros 2005, 2006). Both natural and built environments possess intrinsic qualities that enable such a strong connection, and that in turn can be healing. This works through the sense of well-being established and maintained in the life of those who engage with such a structure. Great architects in the past were better able to discern those qualities, and to reproduce them in their buildings, because they were more engaged with their immediate surroundings.

What we are depends on the natural environment that shaped our bodies and senses (Kellert 2005; Kellert and Wilson 1993; Orians and Heerwagen 1992). Far from being able to liberate our modern selves from our historical development, we inherit our biological origin in the structure of our mind and body. Nature has built on top of this over successive millennia, in increasing layers of sophistication. Evolution works by using what is already there, extending and recombining existing pieces to make something new. We thus depend on the presence of certain determinant qualities in the environment not only for our existence but also equally for our sense of belonging and well-being. Denying this genetic dependence is akin to denying our necessity for food and air. The typologies of traditional and vernacular architectures are predicated on biological necessity. They are not romantic expressions (as some would have us believe), but in fact a primal source of neurological nourishment.

A new chapter in scientific investigation is beginning to document environmental factors that affect our physiological well-being. Going beyond the century-old debates on aesthetics, a neurological basis for aesthetic response is now being established (Ramachandran and Rogers-Ramachandran 2006). The mechanism for neurological nourishment was recently discovered in studies using Functional Magnetic Resonance Imaging. Humans have an innate hunger for certain types of information: the circuits for this have been associated with the brain's pleasure centers, which also control the reduction of pain (Biederman and Vessel 2006). It is easy to hypothesize that this neurophysiologic mechanism is the result of an advantageous evolutionary adaptation.

A growing amount of research finds that fractal qualities in our environment (i.e., ordered details arranged in a nested scaling hierarchy) contribute positively to human well-being (Hagerhall, Purcell, and Taylor 2004; Taylor 2006; Taylor et al. 2005). Gothic architecture is intrinsically fractal, and has been conjectured to be an externalization of the fractal patterns of our brain's neural organization (Goldberger 1996). The parallel between built fractal patterns and possible cerebral organization is too strong to be a coincidence (Salingaros 2006). This idea is supported independently by the way we perceive and find meaning in patterns in our environment (Kellert 2005; Salingaros 2006). It is no surprise then that humans build those patterns into their creations. Investigations of all traditional architectural and urban forms and ornamentation confirm their essentially fractal qualities (Crompton 2002; Salingaros 2005, 2006).

Another direction of research has uncovered undisputed clinical advantages (faster hospital healing) of natural environments, including artificial environments mimicking geometrical qualities of natural environments (Frumkin 2001; Ulrich 1984, 2000). Pain relief in hospital settings is significantly improved by viewing natural (or videos of natural) environments (Tse et al. 2002), thus confirming the link between specific types of informational input and pain reduction. These developments have sparked the interest of organizations concerned with improving the positive human qualities of their spaces. Much of this research has started to be applied in the field of interior design rather than architecture (Augustin and Wise 2000; Wise and Leigh-Hazzard 2002). There are two principal reasons for this: first, interiors are much easier to manipulate than entire buildings; and second, environments for work, leisure, or health care can make a more immediate and substantive difference in human well-being and performance.

Reviewing the positive effect that fractals and natural complexity have on humans, Yannick Joye (2006, 2007a, 2007b) reinforces our own conclusions on the essential hardwired nature of the process. This is not the result of a conscious response to recognizing fractal or complex patterns in the environment: it is built into our neural system. Reaction to a neurologically nourishing environment is physiological (i.e., emotional) rather than intellectual. There is mounting evidence of an innate information-processing system that has evolved along with the rest of our physiology (Joye 2006, 2007a, 2007b). This system is acutely tuned to the visual complexity of the natural environment, specifically to respond positively to the highest levels of organized complexity (Salingaros 2006).

Some researchers concentrate on human response to fractal qualities, whereas others measure the benefits of the complex geometry found in natural forms. Fractals are an important component of this effect, but by no means represent the full gamut of connective qualities. Additional geometrical properties of natural/biological forms clearly contribute to a positive physiological response in humans (Alexander 2002–2005; Enquist and Arak 1994; Kellert 2005; Klinger and Salingaros 2000; Salingaros 2005, 2006). Symmetry—more precisely, a hierarchy of subsymmetries on many distinct scales—plays a crucial role. The overall perceived complexity is better understood using a multidimensional model rather than the simplistic one-dimensional model of plainness versus complication. Not only the presence of information, but especially how that information is organized, produces a positive or negative effect on our perceptive system (Klinger and Salingaros 2000; Salingaros 2006).

We assume an underlying genetic factor as the basis for why the ordered geometry of biological forms connects with and leads to healing effects on human beings. Many scientists now believe that evolution has a direction: the increasing complexity from emergent life forms in a primordial soup to human beings is not random (Conway-Morris 2003). While not speaking of "purpose," we may discern a flow of organization toward a very specific type of organized complexity (Carroll 2001; Valentine, Collins, and Meyer 1994). As such, evolution becomes understandable in informational terms, where adaptive forces act in a fairly restricted direction (though without an end result in sight). Some species do reach a complexity plateau, and individual organismic components may simplify as a result of adaptation, yet the strand of human evolution has moved toward increasing complexity. A corollary to this conclusion is that all life-forms share an informational kinship based on very special geometrical complexity, which builds up in a cumulative process. The built environment, considered as an externalization of intrinsic human complexity, fits better in the larger scheme of things whenever it follows the same informational template. The design of our buildings and cities should therefore try to adapt to the evolutionary direction of biological life in the universe.

BIOPHILIC ARCHITECTURE AND NEUROLOGICAL NOURISHMENT

Human beings connect physiologically and psychologically to structures embodying organized complexity more strongly than to environments that are either too plain, or present disorganized complexity (Salingaros 2006). It follows that the built environment performs a crucial function—in some instances to the same degree—as does the natural environment. The connection process (outlined in the following sections) plays a key role in our lives, because it influences our health and mental well-being. Studying the geometrical characteristics of the type of visual complexity responsible for positive effects reveals its commonality with biological structures. Applying such concepts to architecture leads to two distinct conclusions: first, that we should bring as much of nature as we can into our everyday environments so as to experience it first-hand; and second, that we need to shape our built environment to incorporate those same geometrical qualities found in nature.

Human beings are biologically predisposed to require contact with natural forms. Following the arguments of Edward Wilson (1984), people are not capable of living a complete and healthy life detached from nature. By this Wilson means that we benefit from direct contact with living biological forms, and not the poor substitute we see in so many urban and architectural settings today. Wilson's *biophilia hypothesis* asserts that we need contact with nature and with the complex geometry of natural forms, just as much as we require nutrients and air for our metabolism (Kellert 2005; Kellert and Wilson 1993).

One aspect of biophilic architecture, therefore, is the intimate merging of artificial structures with natural structures. This could involve bringing nature into a building, using natural materials and surfaces, allowing natural light, and incorporating plants into the structure. It also means setting a building within a natural environment instead of simply erasing nature to erect the building (Kellert 2005). While many architects may indeed claim to practice in this way, they more frequently replace nature by a very poor image of nature: an artificial representation or substitute that lacks the requisite complexity. That is in keeping with the abstract conception of architecture that was applied throughout the twentieth century and continues today. Strips of lawn and a few interior potted plants do not represent anything but an abstraction of nature, not the real thing. This is a minimalist image lacking complexity and hierarchy. Biophilia demands a vastly more intense connection with plant and animal life, leading to the support of ecosystems and native plant species whenever possible.

Some good solutions incorporate small ecosystems consisting of a rich combination of plants within a building or in a building's garden or courtyard. A flat lawn, by contrast, while better than a rectangular concrete slab, represents the same visual purity (emptiness) as the plain slab. Our senses perceive it as a single scale and are unable to connect to it fractally. Moreover, lawn is an ecological monoculture irrelevant to local ecology, because it exists on a single ecological scale. Nature exhibits ecological complexity: interacting plants that in turn provide visual complexity, which is a source of neurological nourishment. Not surprisingly, this way of thinking leads to buildings that are more sustainable and incorporate natural processes that help in energy efficiency. Sustainability goes hand-in-hand with a new respect for nature coming from biophilia (Kellert 2005).

For all its benefits of helping users to connect with nature in their everyday interior work environment, this first approach is only a partial solution. The biophilic element here is plant life brought next to and into a building, but the building itself could still be made in an alien or artificial form and built using artificial materials. Human connection is then possible only with the plant forms, but never with the building itself. This problem is particularly acute in an age where the majority of architects use industrial materials and modernist typologies without question. This practice only serves to undermine the requisite natural connections that humans need. The natural aspect of an industrial building-plus-garden is simply a biological component grafted onto an armature that is fundamentally hostile to human sensibilities. There is always a sharp contrast between the building and the natural elements that it encloses. It still triggers an underlying neurological disconnection on a basic level.

A second, and much deeper, aspect of biophilic architecture requires us to incorporate the essential geometrical qualities of nature into the building and urban structure. This implies a more complex built geometry, following the same complexity as natural forms themselves. Once again, there is a danger of misunderstanding this geometry and superficially copying shapes that are irrelevant to a particular building or city. Architectural magazines are full of images of organic-looking (and unrealizable) buildings, whereas we actually mean ordinary-looking buildings that are more adapted to human sensibilities. For example, making a giant copy of an organism out of industrial materials becomes an iconic statement that fails to provide any level of connectivity. The shape of a giant mollusk, crab, amoeba, or centipede is still an abstract concept imposed on a building, little better in quality of abstraction from a giant box or rectangular slab. That belies a fundamental misconception about living structure, which connects on the human levels of scale through organized details and hierarchical connections (Alexander 2002–2005; Salingaros 2005, 2006).

Neurological nourishment depends upon an engagement with information and its organization. This connective mechanism acts on all geometrical levels, from the microscopic through increasing physical scales up to the size of the city. The correct connective rules were rediscovered repeatedly by traditional societies and are applied throughout historic and vernacular architectures. Traditional ornamentation, color, articulated surfaces, and the shape of interior space helped to achieve informational connectivity. Long misinterpreted as a copy of natural forms, ornamentation in its deepest expressions is far more than that: it is a distillation of geometrical connective rules that trigger our neurophysiology directly. These qualities are emphatically not present in the dominant architectural ideology of the twentieth century.

Some biophilic architects consider that neurological nourishment comes strictly from living biological forms. In their view, ornamented forms and surfaces are derivative of natural forms, and thus provide only a secondhand (i.e., vicarious) experience. We, on the other hand, believe that the underlying geometrical complexity of living structures is what nourishes humans. This geometry could be equally expressed in biological organisms as in artifacts and buildings; the difference is merely one of degree (Alexander 2002–2005). If implemented correctly, it is not neurologically discernable, only more or less intense. Every living being incorporates this essential geometry to an astonishing degree in its physical form, whereas only the greatest of human creations even come close. In this view, the distinction between the living and the artificial is left intentionally vague, and life itself is drawn closer to geometry. At the same time, this approach helps to explain the intense connection people feel with certain inanimate objects, e.g., the artifacts and creations of our human past.

Traditional techniques for creating neurologically nourishing structures are wedded to spiritual explanations, which are often unacceptable to contemporary architects (and to business clients). Not surprisingly, the most intense connection is achieved in historic sacred sites, buildings, and artifacts. It is only in recent times that a scientific explanation has been given for what were originally religious/mystical practices of architecture and design (Alexander 2002–2005; Salingaros 2006). Today, it is finally possible to build an intensely connective building and justify it scientifically, by extending the geometrical logic of the natural world into the built world.

To summarize, two branches of contemporary biophilic architecture are beginning to be practiced today (Kellert 2005). One basically continues to use industrial typologies but incorporates plants and natural features in a nontrivial manner; while the other alters the building materials, surfaces, and geometry themselves so that they connect neurologically to the user. This second type ties in more deeply to older traditional, sacred, and vernacular architectures. So far, the first (high-tech) method has an advantage over the second (mathematical/sacred) method, because it is already in line with the industrial building/economic engine of our global society. Visually and philosophically very distinct, nevertheless, these two movements are contributing to a rediscovery of our immediate connection to the environment.

Perhaps the greatest impact of the biophilic movement is to establish a value system for a particular group of essential geometric qualities. Living forms and the geometrical characteristics they embody must be protected from destruction, because they provide us with neurological nourishment (Wilson 1984). This is the seed for conservation, both of biological species and of historic and traditional architectures.

AN ARCHITECTURE THAT ARISES FROM HUMAN NATURE

The desire to overcome nature, to separate man from the universe by placing him above natural constraints, led to the ultimate architectural assertion of the twentieth century, one expressing total autonomy. Adaptive processes were replaced by a formalized, self-referential, autonomous architectural order. The degree of separation that architecture placed between itself and nature was celebrated as a great accomplishment. This architectural movement culminated in the 1970s with a declaration made about an exhibition of current design work: "This spectacularly beautiful work, elegant, formal, and totally detached from the world around it, represents a kind of counterrevolution in today's educational thought and practice" (Huxtable 1999). Indeed, the value of twentieth-century architecture was now solely predicated on its degree of separation from the world around it, the world in which humans seek comfort and shelter (Masden 2006).

To consider the service of architecture as something other than human seems contradictory to its very inception, for it was human nature that first gave it form by compelling humans to build. If we are to consider whom architecture should serve and reestablish the relationship between architecture and humanity, then we must consider the essence of human nature and grasp how human beings came to create particular kinds of structures. We must account for the neurological processes that operate as our interface with the physical world, and ask why, if these processes are intrinsically human, were we ever able to stray so far away from this human dimension.

Edward Wilson's seminal book *On Human Nature* (Wilson 1978) laid the groundwork for understanding our biological nature, explaining how our actions are determined to a large part by genetic structure and evolution. Wilson thus places human actions on a sound biological foundation. Even so, people often contradict their biological nature by acting against it without any apparent logic, as when they join a mass movement (Hoffer 1951). People are sometimes manipulated into adopting an ideology that then controls their actions in violation of their biological nature (Salingaros 2004).

These ideas are relevant to architecture in a positive sense. The early stages of the artistic process are a result of a vast number of unconscious forces and impulses. To initiate this process toward a healthier architecture, we need to ask: what are the tactile, perceptual, and mental processes necessary for a human sense of well-being? We are not going to describe how to incorporate biological elements into the built environment—the principal component of biophilic design—since that is dealt with by other authors (Kellert 2005). Rather, we have developed techniques for design and construction that use materials to create a source of neurological nourishment. We draw from comprehensive architectural design methods developed only recently (Alexander 2002–2005; Salingaros 2005, 2006; Salingaros and Masden 2007).

Several suggestions can help to implement this program. A closing section of this chapter ("Fourteen Steps Toward a More Responsive Design") summarizes some of the underlying principles that we and others are utilizing to design and build new enriching and engaging environments. Although built today with the latest technological materials, these environments reproduce with great effect the best that older built environments were able to offer. We, working today, strive for the same neurological nourishment from what we build as did historical architects working in centuries past. In the past, techniques for achieving this goal were learned intuitively. Modern science is revealing the mechanisms whereby neurological nourishment acts, so that we can learn to use it in a more controlled manner. Today, we are once again aware of the physical properties and natural geometries that architects in centuries past called upon to create the great human places we now wish to emulate.

Biophilic design's principal contribution is to make use of plants and complex natural settings as much—and as intimately—as possible in the built environment (Kellert 2005). While our design approach does not focus specifically on the biophilic component, it supports it in a fundamental manner. By reorienting design away from formal or ideological statements and toward a process of optimizing neurological engagement, we are setting up the conditions for accepting biophilia. Otherwise, the conceptual distance between nonresponsive architecture and the natural environment is so vast that most people simply cannot bridge the gap. We are presently living in an alternative mental universe where human creations are forever distanced from natural forms. This gap is spreading daily, as the progressive development of new technologies rewards us with useful gadgets that are increasingly unnatural.

To implement biophilic design, we need to create a conceptual framework based upon informational connection. This program goes against the current trends of academic specialization, since it requires the cooperation of many different disciplines. Present ways of thinking about architecture are inadequate: the representation of architectural problems has to change from an abstract domain to the natural domain dominated by human physiology and positive emotions. The forces pushing for a reorientation necessarily come from outside architecture, and may even be resisted by architectural academia. If we are successful in this, then future architects will conceive architecture in a fundamentally different manner.

THREE DIFFERENT CONCEPTIONS OF BEING HUMAN

Biophilic design techniques depend upon the mental processes and physical mechanisms that people have evolved in response to the natural environment (Kellert 2005). It is now necessary to consider the nature of human beings, which underpins biophilic design as a necessity and not an option. Many readers could misinterpret the biophilic focus on nature as diverting attention away from human beings themselves, even though its goal is to enhance human life on earth. This discussion is needed to prevent our work (and our colleagues' work) from being branded as just another architectural "style" that can be applied or ignored depending on the prevailing fashion.

We identify three fundamentally different conceptions of human nature, summarizing each of these levels in turn. In the first level, a human being is regarded as a component placed into an abstract, mechanical world. Here, human beings interact only minimally (superficially) with the natural world, a condition of being disconnected. This is an abstract conception of humanity, yet it is representative of much of contemporary thinking. It is the world of the contemporary architect, in which humans participate only as sketches, intentionally blurry photos, or indistinct shadows on a computer screen. The imageability of the design is primary, with the occupant either absent or represented only symbolically. A human here is not even biological: he or she exists as an inert passenger in a fundamentally sterile and noninteractive world.

In the second level, a human being is an organism made of sensors that interact with its environment. Here, humans are biological entities, animals that possess a sensory apparatus enabling them to receive and use measurable input. This is a condition of biological connectedness to the world, that is, *situatedness* (Salingaros and Masden 2006a). In this richly biological view, a human being represents a biological system that has evolved to perceive and react with inanimate matter and especially with other organisms. Humans are considered as animals (not meant in any negative way), sharing all the evolved neural apparatus necessary to make sense of the natural world. Human modes of interaction are those we understand through nerves and sensors.

In the third level, a human being is something much more than a biological neural system. The third conception corresponds to the much older metaphysical picture of humans as spiritual beings, connected to the universe in ways that other animals are not. This is a condition of transcendental engagement with the world. The definition of human essence extends into realms more properly covered by humanistic philosophy and religion. Much of what it means to "be human" lies in this domain, and these additional qualities distinguish us from other animals. To dismiss all of this as "unscientific" would be to miss the point of humanity. In the prescientific ages—as for example, the Middle Ages in Europe—our conception of what we were as human beings was almost exclusively based upon insight that came from internal development. Transcendental engagement anchored our sense of *self*, and continues to do so for the majority of people in the developing world today. Mystical and religious, this intuitive understanding serves to tie human beings to their world in a manner independent of science. The connection, moreover, is believed to have been much stronger than the later development of a strictly scientific framework linking human beings to the rational dimension of the physical universe.

Curiously, the three levels of being human, going from detachment (disconnection), to a biological connection (situatedness), and finally to a more profound transcendental engagement, correspond to going backward in historical time as it pertains to human existence. This seems counterintuitive at best. If one were to reword this observation, it could be said that humankind has regressed in the depth of its connection to its surroundings (i.e., the universe) over the past decades and centuries. Just because we have increased our scientific knowledge of the world, this does not guarantee that we maintain our connection to it in the human dimension. Indeed, the Cartesian method required us to detach ourselves from our world in the name of scientific enquiry, in order to be able to perform unbiased experiments. This may be fine for scientific experimentation, but it is certainly no way to maintain our human nature and to effectively operate within the world as human beings.

LEVEL ONE: THE ABSTRACT HUMAN BEING

The "modern" human being inhabits an industrialized, technological world. Since this world has become an ever vaster and encompassing machine, so too its human inhabitants have become but an ever smaller (and, by implication, less significant) component of that machine. The biological constitution of these contemporary human beings has little relevance to their situatedness in the universe: such a person could just as well be made out of metal, wires, and a minimal number of electronic sensors—a robot. The biological (not to mention the transcendental) nature of humanity is herein denied. A human being is simply a neutral cog in the machinery of the universe. It doesn't help that contemporary physics paints precisely such a hopeless picture of cosmic irrelevance for human nature and the human spirit.

In contemporary architecture, reluctant acknowledgment is sometimes made of the genetic structure of a human being, but it is far less than would at first appear. Too often, even the most rudimentary neural capacity of humans does not enter into play when designing buildings and urban environments. Human physiological and psychological response seldom figures in design discussions today. Architects pretend to have surpassed human nature. Instead, certain formal and abstract notions about space, materials, and form are of primary concern. Those do not arise, however, from a full understanding of the processes at work that give human beings their existential foothold on earth.

A movement to mold human beings into manipulable consumers of industrial products has been taking place for many decades. Much broader in scope than architecture and urbanism, these two disciplines have nevertheless played a significant role in an era of massive social engineering. In the drive to transform human beings into controllable objects, people's connection to nature is suppressed. Modern individuals—at least in the more developed countries—live in a physical world defined by machines and industrial materials, and their information fields come from media images and messages. Nature is either eliminated from the human environment, or has been relegated to a purely decorative role. Evolutionary developed sensibilities have been numbed. The world's remaining population is no better off, because it aspires to emulate this unnatural state as a sign of progress. An automated, disconnected population is insensitive to the healing effects of natural environments.

A more benign, but nevertheless equally effective, transformation led to the abstraction/mechanization of the human environment. Early-twentieth-century advances in microbiology and sanitary practices coincided with the introduction of industrial materials. A "healthy" environment became associated with the visually sterile, industrial look of polished metal or porcelain surfaces. For example, kitchens changed from being geometrically messy to looking like sterile factory environments; and from being made from soft and natural materials to being built using hard industrial materials (Salingaros 2006). Plants (not to mention domestic animals) had no place there. People's preoccupation with improved health made them suspicious of all life, not just the harmful microbes and fungi that cause disease. This was a great misunderstanding, since microbes can thrive on any surface, even ones that look sterile to the naked eye. But the clean, industrial look became part of our worldview, and we are still threatened by signs of life that violate it.

This contemporary condition demonstrates that human beings can be psychologically conditioned to act against their biological nature (Hoffer 1951, Salingaros 2004). We are now facing a population whose sensibilities have been detached from most other life forms, and oriented principally toward an artificial world of images and machines. Explaining the benefits of biophilic design to such individuals—who no longer see relevance in real trees, animals, and ecosystems—presents a serious challenge.

LEVEL TWO: THE BIOLOGICAL HUMAN BEING

We are biological creatures made of sensors that enable us to interact with our surroundings. Intelligence and consciousness are evolutionary products of our sensory systems. Up to a certain point (more than we care to admit), we share this neurological basis with other creatures of the earth (Wilson 1978, 1984). In the past, an innate understanding of how forms, spaces, and surfaces affect us was used to design the built environment, aimed at maximizing its positive effect on us. That changed when formal criteria and abstractions were introduced, replacing those of an older, humanistic architecture. By coincidence, societal discontinuities leading into the twentieth century made this replacement possible, a change that could not have taken place before then (Salingaros 2006).

However, this does not mean that our sensory apparatus has changed in any way. We still have the same genetic structure, and our physical and psychological needs have remained the same over many millennia (Wilson 1978, 1984). Our neurophysiologic requirements have been tempered to some extent by fashionable ideas, images, and ideologies, yet our response mechanisms still operate automatically. Therefore, we will instinctively react in a negative manner to a built environment that is neurologically non-nourishing or actually causes physical anxiety and distress. It is very easy to understand the type of environment that is healthy for us—or, conversely, is unhealthy—based upon our sensory apparatus. We need only to pay heed to the signals from our own body, unencumbered by psychological conditioning.

Empirical evidence continues to accumulate toward a greater understanding of how humans operate physiologically in the built environment (Frumkin 2001). In hospital design, the geometry of the environment plays a significant role in how long it takes for a patient to be cured. Roger Ulrich has done pioneering work in this topic (Ulrich 1984, 2000). Surprisingly, schools do not show a strong enough interest in human physiological and psychological response to the built environment, despite decades of experimental findings on this topic. Architects instead seek greater distance and obscurity in the ethereal terrain of contemporary philosophies (Salingaros 2004). Departments of Architecture around the world still train students in hospital design based on formal stylistic ideas of spaces and materials, not paying attention to Ulrich's work.

Our eye/brain system has evolved to perceive fine detail, contrast, symmetries, color, and connections. Symmetry, visual connections, ornament, and fine detail are necessary on buildings—not for any stylistic reason, but because our perception is built to engage with those features (Enquist and Arak 1994; Salingaros 2003, 2006). The physiological basis for sensory experience is the ultimate source of our being, which thus relies strongly on certain geometric elements to which we connect. Creating an environment that deliberately eschews these elements (visual elements that are found in nature and in all traditional architectures) has negative consequences for our physiology, and thus for our mental health and sense of well-being (Joye 2006, 2007a, 2007b; Kellert 2005).

Environments devoid of neurologically nourishing information mimic signs of human pathology. For example, colorless, drab, minimalist surfaces and spaces reproduce clinical symptoms of macular degeneration, stroke, cerebral achromatopsia, and visual agnosia (Salingaros 2003, 2006). We feel anxious in such environments, because they provoke in us a similar sensation as sensory deprivation and neurophysiologic breakdown. It is curious that architectural design in the past several decades has incorporated more and more such alarming elements and devices as part of its stylistic vocabulary. Some architectural critics attempt to portray these in a positive light using seductive images, and defend them by employing specious references to technological progress (Salingaros 2004).

The discipline of environmental psychology actually began in faculties of architecture, as a natural investigation of how built environments were affecting people. As soon as the first results (several decades ago) indicated that some of the most fashionable contemporary architectural and urban typologies, spaces, and surfaces might in fact be generating physiological and psychological anxiety in their users, fellow architects lost interest. Environmental psychologists moved (or were systematically relocated) outside architectural academia, into Departments of Psychology, which is where they can be found today.

Ironically, to understand the environmental aspect better, we turn to studies on higher mammals. Judith Heerwagen has studied zoo animal behavior in naturalistic versus more artificial environments (Heerwagen 2005). Starting from substantial observations of zoo animals, she reports the results of implementing a transformation toward more naturalistic habitats. As a consequence, the animals' psychological and social well-being has been drastically improved. Zoo animals kept in drab, monotonous, and minimalist environments (i.e., those that we humans also perceive as boring and depressing) exhibited neurotic, aberrant, and antisocial behavior never observed in the wild. Moved to more naturalistic and stimulating habitats, the animals returned to more normal, healthier behavior patterns.

This body of results has dramatic implication for our children. Evidence has been accumulating since the 1960s that complexity and stimulation in the environment can lead to increased intelligence of a developing animal. Incontrovertible results are obtained with young rats raised in information-rich environments, whose brains increase in size, and whose neural connectivity can improve by up to 20 percent (Squire and Kandel 1999, 200). This represents much more than just an anatomical change in the brain, because it optimizes the cortical physiology responsible for intelligence. Rats raised in enriched environments are then observed to do much better on intelligence tests (such as solving complex maze problems) and in training. We interpret this result as the fulfillment of a necessary external component in the brain's development. It also raises questions of collective culpability for neglecting or minimizing neurological connective structure.

We need to point out the importance of relying on clinical studies rather than on surveys. Many studies recording user preferences have been done over decades, some of them uncovering the advantages of natural environments and of environments mimicking those geometrical qualities (Joye 2006, 2007a, 2007b; Kellert 2005; Kellert and Wilson 1993). Nevertheless, a large number of these studies showed only moderate preferences or were inconclusive. A recent experiment raises the possibility that the earlier results may in fact reflect conditioned response. In a clinical comparison of two distinct environments, one a plain room, and the other with wooden beams added to create hierarchical scaling, the subjects did not express any preference. Yet the physiological monitors recorded a marked response in favor of the room with hierarchical subdivisions and natural detail (Tsunetsugu, Miyazaki, and Sato 2005). We (and the study's authors) conclude that the physiological effects of the environment cannot always be consciously recognized.

EXTENDING LEVEL TWO: EXPERT KNOWLEDGE AND PATTERNS

A major question in cognitive neuroscience is: which components of the brain's wiring are innate (genetic) and which components are acquired through interaction with the environment (learned)? There is a dimension of being human that goes further than direct sensory perception, yet remains within biology. It is simply sensory experience on a higher hierarchical level. That mechanism is a product of learning, and is vital in distinguishing human beings from machines. It is also of crucial importance to the arguments raised here about architectural connection to the *self*. Human existence, and the projection of the *self* into the world, is formulated from within the individual through perception of the outside world, thus generating an interpretative framework.

This is the domain of "expert knowledge," where complex data about the environment have become so internalized that perception seems almost extrasensory (but is not). Experience represents a sensory response that has become too complex for us to easily describe, categorize, or understand in an analytic manner. Experience provides us with a repertoire of patterns that we then use to unconsciously match unfamiliar situations (Klein 1998). Many qualities often attributed to intelligence are in fact the result of well-developed perceptual skills at the level of expert knowledge.

Our basic neurophysiologic makeup is genetically determined. After birth, however, our neural network is shaped by the environment and learning, thus acquiring additional, nongenetic properties. These properties include the recognition of structural and functional patterns. The genetic basis makes learning structures possible, but privileges a certain type of learning structure that is based upon the genetic template. Learning, in turn, helps propagate our genes; thus these two informational components are interdependent. Altogether, the genetic and learned components of our memory and sensory systems work as one seamless whole, acting as a set of innate responses.

Emotional learning is the result of sensory input, but remains subconscious (i.e., stored in nondeclarative memory). It works independently of conscious (declarative) memory, since much of the information that we process is not accessible to conscious awareness (Squire and Kandel 1999). Patterns learned emotionally through perception act in the same way as inherited (genetically based) responses. The reason they evoke a positive emotion to begin with, is because they satisfy an internal template. As a result of our evolution, our internal template is very specific. Many aspects of our behavior and personality are either acquired in this manner or are innate, and both are stored as unconscious knowledge (Squire and Kandel 1999, 173).

Andrius Kulikauskas (2006) makes the following perceptive statement about behavioral patterns that have a biological origin:

> Patterns also can help us make sense of the social importance of our body language, for (from videos taken by sociologists) it seems that we have a "sixth sense" by which we literally dance in relationship to each other (shifting our body at speeds faster than we are aware) and which I imagine we cue against our environment (which is why ornamentation may be very important). This is a faculty that I believe autistic people do not have (as if they were blind or deaf in this regard) and so must focus their conscious mind on cues that most of us find simple

to read (such as when are others interested or not in what we have to say).

Patterns recognized by our neurophysiologic apparatus are a key to understanding humanity and its connection to the universe. Patterns organize individual actions into more complex wholes. While this is a process well understood in a language, where words are combined to achieve a meaningful message, it remains outside most people's analytical understanding of the world. Cognitive psychologists recognize patterns as schemata that identify certain preferred sensory inputs. Patterns also control coordinated body movements. Almost every human activity will be found to contain patterns, and those patterns generate the forms and connective complexity of traditional architecture and urbanism (Alexander et al. 1977). We will discuss later how humans connect to particular robots and computers that mimic human patterns of speech or behavior.

Expert knowledge in architecture and urbanism is embedded in traditional environments. Whereas some design components are contextual (i.e., cultural, temporal, or location-specific), many are indeed universal. All we need to do is to "read" them from the unselfconscious built environment. Christopher Alexander's *Pattern Language* codified evolved patterns of how humans interact with their environment and with each other (Alexander et al. 1977). This prescient book established a practical combinatoric framework for design, based on evolved solutions. Incidentally, it already contains many of the key concepts that later came together to define biophilic design. Although these concepts were originally expected to generate a more human architecture, academic architects showed little interest in this information (Salingaros 2005, 2006). Instead, the patterns framework was picked up by the computer software community, which now uses it routinely to handle the complexity found in large software programs.

In the last section of this chapter, we have summarized several Alexandrine patterns. The reader can readily see how these design patterns anticipate and support biophilic design. Architects can draw upon the pattern language, combining that helpful knowledge with the latest notions of human adaptivity into an innovative design method. In turn, the value of the pattern language can be truly appreciated only now, in the context of biophilic design.

When *A Pattern Language* (Alexander et al. 1977) was first published, the most important supporting results from evolutionary biology were not yet widely available. Today, we understand evolutionary convergence as a fundamental indicator of the parallel, independent evolution of specific patterns (Conway-Morris 2003). Faced with a vast solution space, evolution has repeatedly found a relatively small number of working prototypes. Those are characterized by morphological similarity. They have been rediscovered by distinct genetic strains converging toward the same solution by exploring adaptive possibilities. In the same way, a small number of architectural and urban patterns combining social and geometrical elements have arisen spontaneously in different cultures and at different times. Their appearance is evolutionary, since they are the end result of typological exploration via trial and error over generations. Out of an uncountably infinite number of possible typologies, the adaptive ones are relatively few, and can be classified. Obviously, there are rules (whose precise nature we ignore at present) operating at a high level of selection, so that design of the human habitat is far from random (Alexander 2002–2005).

FURTHER EXTENDING LEVEL TWO: HUMAN-MACHINE AND HUMAN-ANIMAL INTERACTIONS

In the effort to reconnect architecture to human sensibilities, it seems appropriate to learn from other fields where such connection is achieved. Any explanation of how natural environments influence human beings must uncover what exactly is being transmitted, and what effect that information has upon our physiology. It thus makes sense to study human-machine interactions, which rely on analogous mechanisms. Biophilia works through information fields, but how do human beings really connect to nonhuman systems? Can we tell whether a system we actually connect to is human or nonhuman? From within contemporary technologies of computers and the science of robotics, we can pick up clues about our own interactions on the level of being human.

Alan Turing (1950) devised the first test meant to distinguish a human being from inanimate information processors (i.e., computers). Its basic premise was that one should be able to determine if a respondent is a computer or a real person from the responses to questions in a conversation. The annual Loebner Prize awards the robot (or rather, its builder) that comes closest to acting human. Just in case, there is a large amount of cash on reserve for when the Turing Test is eventually passed. Even so, we have the example of the notorious ELIZA program written by Joseph Weizenbaum in 1963, where a piece of software emulated a psychiatrist so accurately that many of its respondents were convinced there was a real person at the other end of the computer terminal (Weizenbaum 1976). And things have progressed remarkably since those early times in computing.

In a separate development, Rodney Brooks builds mobile robots that can mimic many nonverbal human qualities (Brooks 2002). Even though they make no attempt to physically resemble human beings in form, they are programmed to "engage" humans by means of behaviors such as eye and head movement (moving what we might identify as their "eyes" and "head"). Those robots are able to express emotions through movement in ways that mimic human behavior, and are capable of doing so because Brooks has programmed varying facial expressions. They have an "aliveness" to them that is most unusual in inanimate objects. People respond involuntarily in a way that engages the robots, and seem disappointed or shocked when these robots occasionally act in a nonhuman manner (Brooks 2002, 149).

As the above examples make clear, it is possible to emulate human qualities and emotions, at least partially, by means of programs that mimic patterns of speech, movement, and behavior. The observer interacts through patterns of a very specific type of complexity that is characteristic of living beings, and specifically of human beings. We are describing complex connections established on an altogether higher level, beyond simple sensory input such as visual stimuli. Such patterns identify human qualities, even though it may only be a machine mimicking a human being.

Increasing the complexity of interaction in a definite direction (defined precisely by what our neurophysiology and sensory apparatus are built to detect) eventually leads to higher degrees of human connectivity. We may connect only partially to a robot exhibiting certain human responses, but we fully connect to a real human being. Human patterns come together and cooperate much better than computers or robots have been able to do so far. We interact with a close friend or family member on yet a different level, since we share extra layers of commonality. This goes even further than genetics. Acquaintance has given us knowledge and experience of that person's behavior that has become intuitive: expert knowledge of their thinking and behavioral patterns enables us to "read that person's mind."

A separate topic of relevance concerns human-animal interactions. Human beings co-evolved with the other animals, and domesticated some of them. Since the beginnings of history, people have benefited from (and documented) the positive emotional and health effects of contact with domestic animals. This is one of the dimensions of biophilia. In recent years, more rigorous evidence has been accumulating on the therapeutic effects of animal contact (Barker 1999). There is a growing industry in animal-assisted therapy (Roth 2000). While we don't wish to enter into this topic's scientific foundations here, we single out the connective channel as a key aspect to our own discussion. Whatever positive physiological and psychological effects are observed to result from human-animal interactions, they must certainly occur via information exchange. And such information is richly complex and pattern-based.

The reason we are talking about animals, robots, and artificial intelligence is to establish the human need and capacity for information exchange. What makes us recognizably human is a set of complex, organized informational patterns that evolved along with our body. Our sensory apparatus instantly detects the degree of kinship of any perceived patterns to our own selves. The processes at work in our neurological hardware require a far greater degree of information than the abstract forms of architecture currently provide. Informational coding is missing from today's architecture. Within the intentional condition of contemporary architectural environments we are detached from the perceivable world.

LEVEL THREE: THE TRANSCENDENT HUMAN BEING

Exploring human nature more deeply leads us to understand humanity as something more than a mass of intelligent animals that reproduce licentiously, and thus destroy the natural world by exhausting all of its resources. In times past, humanity had a more noble conception of itself. An anthropocentric view, yes, but also one endowed with responsibility toward a natural world that was itself alive. This was a more authentically sustainable form of being. To advance our idea, we resurrect the old romantic worldview of a past in which people felt connected to the universe in terms of religion, mythology, societal kinship, traditional values, et cetera. We are not trying to discount how far anthropology is based on genetics, only trying to recapture something lost.

Whatever one may say, there once was a more profound conception of a human being's connection with the universe, and it was not based on theoretical presupposition (or science in the strictest sense). Curiously, the early development of modern science tended to question and therefore weaken our valuation of this connection. Our place in the universe was nevertheless based on empirical observation, which is coincidentally the basis of all science. People experienced a deep connection with each other, with living beings, and with nature. They experienced a sense of wonder at the Creation (Wilson 2006). This was as evident as data collected from an experiment, although the connection is not measurable on a quantitative scale. Traditional explanations for the connective process did not come from science, but from inner beliefs. Expressed in terms of emotions, those truths could not survive the rise of science.

Our present understanding of biological and ecological interdependence is only very recent. Wilson has made remarkable progress in providing a biological basis for what was previously attributed to the supernatural aspects of human nature (Wilson 1978, 1984). A real phenomenon such as our connection to the physical world, experienced beyond any reasonable doubt, is nevertheless vulnerable if its explanation is not grounded in science. This is one reason that the mechanism of neurological nourishment and engagement was dismissed at the beginning of modern (industrial) times. We are finally accumulating scientific evidence to support conclusions reached much earlier by traditional societies.

Christopher Alexander (2002–2005) raises the same issue about our loss of fundamental connectivity, in the context of architecture and urbanism. Alexander argues for an underlying and far-reaching interconnectedness based upon fundamental geometrical properties. He also shows how that has been severely, sometimes catastrophically, damaged (Salingaros and Masden 2006b). This work is starting to become better known with profound effect, as people realize that the twentieth century lost a major component of human connection to the universe. Steps toward disconnection were taken voluntarily, sometimes even eagerly, in the name of technological progress. Unfortunately, such traditional knowledge and beliefs as had sustained the built environment for millennia were readily discarded.

Much of what we take to be uniquely human, such as our emotions and higher aspirations, is a manifestation of our transcendental engagement. The emotion of love has throughout the ages generated a strong attachment to other individuals. The love of one's creator anchors our religions, and created the greatest buildings the world has known. Even though romantic love can be partially explained in biological terms (the search for compatible genes), that surely misses the essence of the experience. All the world's poetry, songs, music, and literature that have been generated by love cannot be explained away as the primal sexual urge of biological reproduction. And even those aspects that have a biological explanation are better described in their own domain: the connective dimension of human nature.

That also holds true for our place in the universe. We connect with our universe through the animating aspect of *self*, filtered through culture and religion. People's behavior, values, and concept of self are thought to be learned from their relationship to the world, through existence. It is existence that gives form to reality through human perception, whereas the body and mind simultaneously manufacture that which we know as reality. For many human beings today, and for the vast majority in earlier times, this connection was deep and profound. In intensity and meaning, it goes far beyond (in terms of complexity) what our direct physiological sensors are programmed to reveal (Masden 2005, Salingaros and Masden 2006a 2007).

The theological concept of "mystery" is relevant here, as something that is sensed to be both true and internally rational, except that its totality cannot be fully grasped (McGrath 2005). In this sense, mystery is not irrational, but inevitable. Both biophilia and architecture have components that belong to this category, and to dismiss them would be to impoverish our conception of those disciplines. The scientist's interpretation is to be cautiously optimistic that with improved experimental techniques, effects that the natural environment is observed to have on human beings will be more completely explained in due time. The nonscientist may be content to consider the possibility that not all of the universe's mysteries are open to human comprehension. Either way, we should not ignore observed effects just because we don't understand the mechanism by which they act. Arrogance (or fear) ignores observations when they threaten an established but narrow conception of the world.

People's attachment to their universe, and to their beliefs, is as deep as their attachment to life itself. In traditional preindustrialized cultures, the awe and fascination with natural forms and with deities is indivisible. Nevertheless, history is a sequence of human mass activity, sometimes violent as in uprisings and wars, driven by beliefs in how the world should be structured and connected. People are willing to sacrifice their lives in order to achieve a certain type of connection to their mental world, to impose a particular structuring for those left behind, or to prevent what they perceive to be a disconnection (a detachment from their picture of the world). Ironically, they will readily detach themselves from the real world in order to follow an abstract ideology (Hoffer 1951, Salingaros 2004). Rational thinking in a technological age did not save humanity from such aberrations, and it certainly has not preserved our deeper connection to nature.

ARCHITECTURE THAT TRANSCENDS MATERIALITY

On many levels, what it essentially means to be human is lost in the practice of architecture today. The denial of human nature acquired greater authority at the turn of the twentieth century, coinciding with the rise in scientific and technological applications. A likely explanation is that people became infatuated with early scientific advances, which confused technology with science itself. They misinterpreted crude technological applications as a substitute for a more complex reality. The promise of science—but a promise based on false premises, eagerly followed by people who did not understand science—has over time stripped humanity of some of its most important nonmeasurable qualities. What could not be quantitatively measured was presumed not to exist, and was relegated to superstition; a vestige of the past that merited only contempt.

Architects throughout the world—those teaching in universities or working in professional offices to produce commercial buildings, modest apartments, and private houses—thirst for some signs of truth in their profession. They dream of a new architecture they can use to overcome the limitations of what they are doing, and to broaden their horizons with infinite newfound possibilities. New forms, new ideas, and new concepts—that's what keeps architecture perpetually moving forward and keeps architects emotionally alive. Newness, moreover, is most invigorating when it can be applied to one's everyday practice. Accepting every architect's thirst for truth, biophilic design offers a more genuine and healthier alternative to what architects currently embrace.

The manifestation of life transcends both material and spiritual realms of thought. Living structure is animated in a mysterious manner. Traditional categories, such as physical versus spiritual and inert versus alive, become somewhat blurred as inexplicable phenomena occur to make things alive. We now understand life as a state of matter possessing certain very special chemical/organizational properties, and are discovering more and more of those properties in the laboratory. We may know many of their details, but we cannot be entirely sure we comprehend how they all work together. It helps to discuss these matters in a culture where science and religion are not kept strictly separate, because religion serves most effectively to keep alive a sense of wonder at living forms.

Christopher Alexander has investigated these fundamental questions (Alexander 2002–2005). Alexander's results reveal how living structure may be conceived as crossing over between animate and inanimate forms. Physical matter does transcend its inert materiality through very special informational configurations. This process can endow physical forms with the characteristics of life—certainly not all of them (i.e., not including metabolism and replication), but moving in a direction toward structures that we identify with living forms. Parallel with this solidly geometrical process, the closer we approach our goal of creating "life," the closer we seem to be moving toward traditional extra-scientific ways of interpreting the world.

Human beings feel most alive in their spiritual moments. In such instances, we feel connected to our environment, in a deep sense belonging to it and to the universe. This stage of inseparable reality has been described in spiritual terms. The experience is unmistakable. It enables us to inhabit the material and spiritual worlds at the same time. The impression of material transcendence is connected with the sacred. Religious architecture of the past helps us to achieve this type of connective experience; indeed, that was its original purpose. The only problem is that traditional explanations of what is going on tend to be nonscientific. Alexander's life work provides a scientific foundation for this observed phenomenon. His results raise many questions about the nature of reality (Alexander 2002–2005, Salingaros and Masden 2006b).

As far as architecture is concerned, we accept the highest level of connectivity of human beings to the material world as real. When this occurs, the built environment may be said to transcend its materiality. All traditional cultures have built sacred spaces in which one experiences an unusually high degree of connection. Sacred spaces are nourishing to whoever occupies them. How is this achieved? We believe that it's the same process that underlies the biophilic phenomenon. Rather than any mysterious force field unknown to physics, informational fields act to establish a manifestation of the requisite connections. Those who love nature can experience a transcendent communion with it. Ancient religions explain this mystery as sacred communion with nature. Consciously working with the mechanism of informational exchange, we can re-create buildings having the same intense degree of connection. Such buildings will provide the highest level of neurological nourishment.

Hassan Fathy grew increasingly to interpret architectural and urban form in sacred terms. He was not referring to religious buildings, but to everyday dwellings for the poor, a project that occupied him throughout his entire life (Fathy 1973). Fathy saw in simple built spaces, surfaces, textures, and configurations an expression of the sacred. This unfortunately brought him into conflict with postwar industrialization, which his colleagues adopted as the only rational solution to the world's housing problems (Pyla 2007). Many other architects, including Louis Sullivan and Frank Lloyd Wright, were likely to talk in mystical terms about their architecture, trying to express something they felt instinctively—and could build—but could not formulate very clearly. Our explanation of how architecture connects to human beings therefore rests on considerable precedent and can now be more clearly understood in neurological terms.

We are aware, however, of a tremendous existing confusion on how to actually achieve architectural transcendence. This is most evident in contemporary religious architecture. According to their architects, some new churches built in a stripped, minimalist style are supposed to represent transcendence. They do nothing of the sort. Without natural elements, figurative art, or ornament, they fail to engage the user in any positive way. Their empty informational field only communicates sensory deprivation, provoking physiological unease. Far from working on the transcendent level of human existence, this design style is a throwback to the mechanical conception of humans. We see a form imposed on top of this presumption ignoring human connective needs. Despite any probable good intentions, the result amounts to a triumph of the architect's will over human nature.

Coming around in a reinforcing circle of reasoning, the effort to create "life" in architecture teaches us a new and welcome humility. Once we focus our efforts on technically establishing neurological nourishment, we cannot fail to notice that nature achieves this effortlessly. Nature also does it so much better than we could ever hope to do. A single live flower can humble most structures built by humans. Interpreted correctly, this calls for a drastic reorientation, not only in how we build, but also in our basic value system. We should simply put to use what nature already provides for our neurological connectivity and nourishment. Plants, animals, and ecosystems thus assume a priority over our own constructions. This is the essence of biophilia.

Possibly our fellow biophilic designers might feel that we have crossed too far into philosophy/religion in trying to support an innovative design method. We insist, however, that we are merely following the thread of thought to its inevitable and logical conclusion. Both Christopher Alexander (2002–2005) and Edward Wilson (2006) have been led independently, at the summit of their careers, to reconsider the meaning of life and human existence. We (and they) see the future as viable only if humankind reattaches to biological life and to the life-generating geometry of the universe. For this reason, Alexander and Wilson have called for a rapprochement between science and religion (proposing two very different types of alliance) in order to save the Creation.

Architecture, as an activity to house human beings, acquires deeper meaning in the world depending upon the human vision of the nature of God. Does God exist in an abstract, minimalist geometry? Or is God instead to be found where there is also evidence of life? In the latter view, which is supported by the world's main religions, God is manifest in a natural geometry—in living structure (Alexander 2002–2005). God is more likely to reside in the highly organized complex geometry of the fundamental structure of matter. But these two types of architectural geometry (minimalist and biophilic) are opposite in their mathematical qualities. There exists a basic incompatibility between two opposite geometries preferred by human beings. A deep theological question we must nevertheless raise it here—because it leads to a separate philosophical validation of biophilia.

CONCLUSION

Human beings have evolved the ability and the need to process information embedded in their environment. Architects, on the other hand, in the process of distancing their work from what is natural, have come to rely increasingly on artificial criteria and the superficial manipulation of images. When images and surface effects supplant everyday human desires and sensibilities in the name of artistic endeavor, humans are left to live out their lives in a series of ill-fitting, overexaggerated, and often idiosyncratic formal architectural schemes. Ordinary people see this trend—architecture turning away from human qualities—as the imposition of building design against their most basic instincts. But they have been able to do little about it, given the nature of the business of architecture and the seduction of technological progress.

There is a neurological and physiological necessity to engage the environment. Architects today can accomplish this by recognizing the operations that connect humans with their environment, and by distinguishing among distinct levels of being human. Biophilic design reorients architecture toward a world governed by coherent information; it also leads people to think on many levels of complexity (which is the way nature works). Reinforcing this tendency, architects can now adopt a higher standard: one that asserts that buildings are by their very nature human. Students, academicians, and practitioners of architecture wishing to contribute to environmental regeneration must therefore ascribe to the essential qualities of human nature, that is, to the physical and mental processes that allow us to occupy our world.

The information necessary for humans to connect to the world around them can take many forms, including calligraphy, representational ornament, and abstract geometrical ornament, with the physical object varying in size from an architectural detail up to architectural structures and urban spaces. In a fundamental sense, therefore, the natural and traditional built environments rich in informational content make a place more intrinsically human. The natural world interfaces smoothly with human creations, but only when those are built in the same coherent manner. By emphasizing informational content, we can shape the built environment according to the constituent logic and order necessary to provide neurological connection at a human scale and thus emotional nourishment.

FOURTEEN STEPS TOWARD A MORE RESPONSIVE DESIGN

The following are some practical techniques that can be used to implement a more responsive and natural approach to design. They form part of a recently developed comprehensive method for architectural design (Alexander 2002–2005; Salingaros 2005, 2006; Salingaros and Masden 2006a, 2007). We emphasize that these points do not simply represent our personal preferences; nor do they include all the supporting rationale that leads up to them. They are the outcome of a theoretical and scientific methodology that is too voluminous to be reviewed here. These design steps support biophilic design. The logic is clear in that they do not *arise* from the biophilia hypothesis, but instead *support* it from independent directions.

We could publish these points separately as a design method meant for practicing architects. They might be accepted, or not, based upon the novelty of the "look and feel" of the resulting buildings. This is the manner in which today's architects adopt new styles and initiate new movements in design. Certainly, the sensory quality of the type of buildings we propose is strikingly different from the crystalline, bloblike, jagged, or minimalist environment produced by some contemporary designers. Nevertheless, some architects may resist the implementation of a so-called biophilic style, precisely because it serves to displace their preferred style. This may spark a heated polemic driven by ideology, politics, and idiosyncratic preferences. To prevent the debate from getting stuck in such an unproductive direction, the body of this chapter is necessary, because it presents an architecture devoid of stylistic predilections.

1. The smallest perceivable scale is established with either the microstructure of natural materials, or by using very fine-grained texture/ornament. The ordered complexity of natural structure cannot easily be duplicated on this scale. The region containing fine detail has to be immediately accessible to human contact (and not lost at a distance). A universal rule for the distribution of sizes in a complex system suggests that there should be very many identifiable components on the smaller scales, several on the intermediate scales, and only a few on the largest scales. The smaller the scale, the larger is the number of elements contributing to that scale. Fractals obey such an "inverse-power-law" distribution. This rule implies an enormous amount of necessary ordered detail on the smallest scales, just as seen in nature. It also implies the necessity for articulated texture and ornament—not on every surface, but prominent and accessible nevertheless.

2. Design that adapts to human sensibilities should have a very definite scaling hierarchy. Obvious differentiated subdivisions or components need to obey a scaling rule, where elements on the next larger scale are roughly 3 times (more accurately, 2.7 times) larger than those on the immediately smaller scale. Although the dimension of each scale can be very approximate, so that the ratio between successive scales could be anywhere from 2 to 5, no scale should be missing. This is essential. A very different concern is to avoid scales intermediate to the scaling hierarchy, since those would distort the ratios. All fractals have a precisely defined scaling hierarchy (each with its own scaling ratio). Despite the widespread use of natural materials such as wood and grained stone, however, the intermediate and larger scales have not been designed coherently in recent decades, so the fractal connective effect (which emerges only with the proper scaling hierarchy) is lost.

3. Symmetry is essential in design, not as expressed by an overall scale, but rather by the richness of subsymmetries on smaller and intermediate scales. Connective symmetry is an extensive quality, ideally acting throughout all levels of a scaling hierarchy. The density of subsymmetries, and their intensity of interaction inside a particular scale and across distinct scales, is what leads to visual coherence. Many of those symmetries are

going to be approximate, and a nonmodular use of materials accommodates such broken symmetries. Here we apply results on symmetry breaking, where small individual variations in a module contribute to create an informational higher scale when modules are combined. This emergent phenomenon is impossible to achieve with repeated identical components or modules.

4. Small-scale construction systems should be inverted—conceptually turned inside out—to optimize informational load. Nowadays, wooden studs and beams are built inside walls and covered with industrial sheetrock (plasterboard). This type of construction hides the materials with the greatest informational content, presenting instead an abstracted geometry to which we cannot connect. Innovative architects can and should develop new structural systems that preserve natural materials for the visible surfaces, to be used in regions that human beings can directly access through sense or touch. This being said, however, care should also be taken in how those materials are placed within the structure. Despite Frank Lloyd Wright's habit of using rough stone and brick for interiors, which at some level do provide a more intense informational experience, their surface is hostile to the touch; thus they should be located out of immediate reach. We also don't mean to imply that all wood should be cut into sheets of veneer. The standard 2-by-6-inch boards could still be used to bear loads, but in such a manner or configuration so as not to hide their natural grain and soft acoustical properties. Load-bearing wall interiors such as concrete or steel (with nonconnective surfaces and textures) can replace the current misuse of more natural materials such as wood.

5. Large-scale construction requires different techniques altogether. But we have to be smarter about how we use industrial methods for larger buildings, since there is no connective value in an "industrial look". We can learn a lot from nineteenth-century modular production of ornamental panels and building components. Going back to the precedent of Islamic tiling patterns, industrial modules such as those used by Hector Guimard, Louis Sullivan, and Frank Lloyd Wright represent an effective extension of the requisite types of neurologically engaging patterns. Though it hasn't been used for many decades, this form of architectural expression contains a high degree of encoded information and is thus very useful for establishing human well-being. Some of the older buildings that we most admire as being "handmade" are in fact the products of a modular construction process. With today's advanced technology, this method in the hands of an architect who understands organized complexity can provide endless architectural possibilities.

6. Natural materials from older building should be reused. Architects must train their eye to look for those materials that help to establish the scaling hierarchy and deliver a high informational content. Every consideration should be given to incorporating materials found in architectural salvage yards into new buildings. Their informational load cannot be duplicated in new materials without incurring a considerably higher cost. This requires adjusting the design to accommodate locally found components. Another technique is to use natural unfinished materials where appropriate. We should not try to overcontrol construction by practices such as cutting everything to a uniform modular size. To save a natural material (which is usually both expensive and "unsized"), we adjust our design to use the available sizes (or variety of sizes) with minimal waste. This implies developing a respect for the material over and above the authority of the design, in much the same way as found in the early practices of Shinto carpenters. The material logic of these natural materials provides yet another level of neurological connectivity.

7. When the use of concrete is necessary, and when it will be present at a human level, the most should be made out of its intrinsic plasticity. Concrete is not a natural material, and therefore has to be manipulated before it can give positive sensory feedback. Instead of letting its surface present either a random texture or an unfriendly flat panel, we can mold it with patterns (as did Frank Lloyd Wright). Unfinished concrete has an informationally frustrating surface because it lacks ordered microstructure. We could moderate the unfriendly surface of raw concrete with a permanent surface or aggregate added while the concrete is still wet (a pioneer in this technique is Antoni Gaudí). On the other hand, using wooden planks to form the concrete

does leave a wood grain impression, but this does not produce a visually coherent surface because it does not have the correct fractal scaling into microstructure, and at best seems unnatural. We have in mind more ordered patterns formed into the concrete, as well as a more "natural" surface, both visual and tactile.

8. The kind of architect who builds biophilic buildings must have a full understanding of how human qualities reveal themselves through the construction process. Whenever possible, we should give a free hand to the workmen to find their own expression, such as in laying tile and adjusting dimensions of a window. The craft of building should once again be recognized, and craftsmen should be given the authority to mold the smaller scales of the building as it develops, so as to foster the most pleasurable feedback as they see it. Natural structure shows an infinite and subtle variation, and this potential should be extended into the use of construction materials and methods. Expression through materials requires an intimate working knowledge of the nature of these materials. This freedom extends the design process out of the hands of the architect, and was practiced in recent times by Friedrich Hundertwasser. By allowing individual input in this manner, we imbue the architecture with a life outside the frozen expression typically conceived on the drawing board. This was the way of the Master Builder, as seen in most preindustrial buildings.

9. The same idea can now be implemented via high technology, using computers and robotic manufacturing to generate individual components of a building. It's a similar freedom as that given to craftsmen, except that it is now made industrial through a technological basis. We can program robotic fabricators to emulate the variations in the physical mechanics of material placement, so as to endow components and materials with a similar degree of life to that found in preindustrial buildings. The technology exists to create an enormous variation, which can serve in the same way as individual hands-on design created by human craftsmen. We are taking advantage of a newfound capacity for mimicking the necessary variability inherent in nonindustrial processes. The same degree of life would be impossible to achieve with repeated identical components such as the standardized modules now available for construction.

10. Unlike previous concerns about the cost of custom work and custom-made components, we have found that the technology needed to administer this work is becoming more available. We have also come to realize that this type of work need only constitute a small part of the construction to still provide the desired effect. With just a little imagination, as much as 95 percent of the materials used could be off-the-shelf materials and standard components. Of course, such standard materials must be reconsidered in innovative ways to provide the greatest degree of neurological connectivity. Standard forms and building components whose dimensions and surfaces fit into the biologically structured fractal scaling scheme can be used without alterations. The mathematical coherence established through the effective application of fractal geometry elicits the neurological engagement required for a sense of well-being.

11. Several possibilities help to achieve biophilic design on the scale of a room. One is geometrical interweaving of plant life and natural features with the fabric of the building itself. We abandon the rectangular or convex footprint of a building, for a more meandering or crenellated boundary that partially surrounds gardens, verandas, and patios. If the building is large enough, then an indoor garden is possible. We focus here on a key concept. Intensify a fairly intimate scale: a complex piece of nature existing on the scale of a human being (1 to 2 meters in size) can make a substantial difference to our biophilic connection. We ask clients to resist their conditioned impulse toward the "purity" that is so often associated with the abstract aesthetization of many modernist designs, and to allow a high level of natural engagement to be present throughout the inhabited space.

12. The issue is simply that of an ingrained idea about what the geometry of the environment should be. Native plants growing wild define a particular complex geometry, and this is the geometry that can best serve to keep us healthy and make us well. An unfortunate practice throughout much of the twentieth century was to identify natural connective geometry as something that

should be discarded. Instead, we must focus our intelligence and technological power toward establishing and creating natural geometry where it is absent, and reinforcing it where it already exists. The development of a natural geometry and life on buildings, via weathering and via the invasion of plants, is nowadays seen as an unwelcome intrusion—as a sign of decay. We, on the contrary, see these as symptoms of increasing life. We can protect the built environment from physical decay while letting it evolve in a more viscerally responsive direction.

13. Human beings can interact with nature only if the urban geometry permits such interactions. In addition to visual line-of-sight, we pay attention to pedestrian access and the formation of urban space. Having some plant life available is only a first step: we need to make it accessible to pedestrians and design an environment in which such an interaction can be maintained and enjoyed. Frequently, ornamental plants may be seen but not approached. We must create gardens that are physically accessible, designed coherently so that it is pleasant to enter them. The worst disaster is suburban space, in which vast expanses of flat lawn and asphalt define a psychologically hostile environment for the pedestrian. Sidewalks are exposed in the middle of this space, between the asphalt road and the forbidden lawn. Private lawns are out-of-bounds, while any bushes and trees form a protective wall around a house, instead of belonging to the public realm. We have to question this habit, breaking up such no-man's-land into well-defined urban space crisscrossed by paths.

14. On the broader urban scale, we should again focus any distribution of units or uses away from a uniformity that privileges the largest scale. Moving away from large, purely decorative lawns, we try to design many complex natural areas, resisting the amalgamation of every plant into one "park," resisting the alignment of everything according to a simplistic geometry, and avoiding the homogenization of green spaces to a single plant species. We should seek instead to preserve and reproduce visual and biological complexity such as is found in natural environments. A natural (fractal) distribution of sizes applied to green spaces in the city implies the existence of very many small ones, several ones of intermediate size, and only a few large ones. Each of these green spaces should in turn have its own internal distribution, which can be achieved only by having internal complexity and variety.

To apply these and similar guidelines, many building systems and practices will have to be reconsidered. The construction industry will have to overcome its built-in modularity in systems, accepted methods, and practices. For example, the building industry often keeps the architect legally removed from the building process, sometimes not even allowing the architect access to the site. But the architect must be fully engaged from start to finish and allowed a more active role in the processes of construction and assembly. This moves closer to the historical model of the Master Builder, which also predicates a more responsible role from the architect in achieving well-being for the building's users. We want to pull out of existing systemic connections in construction practice, and reorient architecture toward the highest degree of ordered information.

SOME PATTERNS FROM *A PATTERN LANGUAGE*

We have selected 15 patterns from Alexander's *Pattern Language* (Alexander et al. 1977) to summarize here. A common thread running through Alexander's work is the need to connect human beings with nature, looking to nature as a source of mental and physical nourishment. That work anticipates and supports biophilic design. Like the concept of biophilia, patterns have meaning for human life, and are not simply someone's individual preference. Thirty years after their publication, architects know about patterns without really understanding what they mean. Many patterns seem irrelevant when interpreted, as they often are, in the framework of a formal architecture; they make sense only within the context of biophilic design. This chapter gave scientific validation for these and other patterns, which should prompt their reconsideration by the architectural community. That prescient design framework contains 253 patterns in all, which can be used for generating environments adapted to human sensibilities. The following brief pattern descriptions are our own: we urge the reader to read the original several-page description of each pattern.

Pattern 3: City Country Fingers. Build a city radially instead of concentrically, with fingers of green space and farmland coming to its center.

Pattern 7: The Countryside. Reconceive unbuilt land as one whole, encompassing farms, parks, and wilderness, and provide access to all of it.

Pattern 24: Sacred Sites. Identify and protect sites having extraordinary importance to the community, whether they are located in a built or green area.

Pattern 51: Green Streets. Don't automatically build low-density/low-speed local roads out of asphalt, but instead use paving stones and gravel set into grass.

Pattern 60: Accessible Green. People will only use green spaces when those are very close to where they live and work, accessible by a pedestrian path.

Pattern 64: Pools and Streams. People need contact with natural streams, ponds, and reservoirs, so these must not all be covered up.

Pattern 74: Animals. People need contact with animals, both domestic and wild, so the city must accommodate instead of discourage them.

Pattern 104: Site Repair. When siting a building, put it on the least attractive part of the lot, preserving the best of the natural environment.

Pattern 111: Half-Hidden Garden. For a garden to be used, it must not be too exposed by being out front, nor completely hidden by being in the back.

Pattern 171: Tree Places. Trees shape social places, so shape buildings around existing trees, and plant new trees to generate a usable, inviting urban space.

Pattern 172: Garden Growing Wild. To be useful, a garden must be closer to growing wild, according to nature's rules, than conforming to an artificial image.

Pattern 176: Garden Seat. One cannot enjoy a garden if it does not have a semisecluded place to sit and contemplate the plant growth.

Pattern 245: Raised Flowers. Flowers provide maximum benefit when they grow along frequently used paths; they must be protected and near eye level.

Pattern 246: Climbing Plants. A building connects to its surroundings when plant life grows into it, with the plants climbing up walls and trellises.

Pattern 247: Paving with Cracks Between the Stones. Paving stones laid directly onto earth, with gaps between them, allow growing plants to create a half-natural environment.

We will not undertake here the task of combining the pattern language framework (these and other patterns) with our 14 steps from the previous section into a humanly adaptive design tool; yet it should be obvious that this can and should be done. Whoever is interested in this project should further refer to results on the combinatorial nature of patterns (Salingaros 2005, 2006). It is necessary to understand those properties—their linguistic component—before patterns can be most effectively used in design applications. Patterns combine to form more complex coherent wholes, precisely the way matter organizes to form higher-level complex entities. We can apply patterns to generate an adaptive, living environment, while the patterns themselves (their evolution in solution space, and combinatorial properties) mimic the geometry of life.

REFERENCES

Alexander, C. 2002–2005. *The Nature of Order,* books 1 to 4. Berkeley, CA: Center for Environmental Structure.

Alexander, C., S. Ishikawa, M. Silverstein, M. Jacobson, I. Fiksdahl-King, and S. Angel. 1977. *A Pattern Language.* New York: Oxford University Press.

Augustin, S., and J. A. Wise. 2000. "From Savannah to Silicon Valley." *IIDA Perspective,* Winter/Spring, 67–72. http://www.haworth.com/haworth/assets/From%20Savannah%20to%20Silicon%20Valley.pdf.

Barker, S. B. 1999. "Therapeutic Aspects of the Human-Companion Animal Interaction." *Psychiatric Times* 16(2). http://www.psychiatrictimes.com/p990243.html.

Biederman, I., and E. A. Vessel. 2006. "Perceptual Pleasure

and the Brain." *American Scientist* 94 (May–June): 247–253.

Brooks, R. A. 2002. *Flesh and Machines*. New York: Pantheon Books.

Carroll, S. B. 2001. "Chance and Necessity: The Evolution of Morphological Complexity and Diversity." *Nature* 409: 1102–1109.

Conway-Morris, S. 2003. *Life's Solution: Inevitable Humans in a Lonely Universe*. Cambridge: Cambridge University Press.

Crompton, A. 2002. "Fractals and Picturesque Composition." *Environment and Planning B* 29:451–459.

Enquist, M., and A. Arak. 1994. "Symmetry, Beauty, and Evolution." *Nature* 372:169–172.

Fathy, H. 1973. *Architecture for the Poor*. Chicago: University of Chicago Press.

Frumkin, H. 2001. "Beyond Toxicity: Human Health and the Natural Environment." *American Journal of Preventive Medicine* 20:234–240.

Goldberger, A. L. 1996. "Fractals and the Birth of Gothic: Reflections on the Biologic Basis of Creativity." *Molecular Psychiatry* 1:99–104.

Hagerhall, C. M., T. Purcell, and R. Taylor. 2004. "Fractal Dimension of Landscape Silhouette Outlines as a Predictor of Landscape Preference." *Journal of Environmental Psychology* 24:247–255.

Heerwagen, J. H. 2005. "Psychosocial Value of Space." *Whole Building Design Guide*. www.wbdg.org/design/psychspace_value.php.

Hoffer, E. 1951. *The True Believer: Thoughts on the Nature of Mass Movements*. New York: HarperCollins.

Huxtable, A. L. 1999. "New York Times Review of the Museum of Modern Art Exhibition and Foreword." In *Education of an Architect: A Point of View. The Cooper Union School of Art and Architecture 1964–1971*. New York: Monacelli Press.

Joye, Y. 2006. "An Interdisciplinary Argument for Natural Morphologies in Architectural Design." *Environment and Planning B*, 33:239–252.

Joye, Y. 2007a. "Toward Biophilic Architecture: Drawing Lessons from Psychology." *Review of General Psychology* 11: in press.

Joye, Y. 2007b. "Fractal Architecture Could Be Good for You." *Nexus Network Journal* 9(2): 311–320.

Kellert, S. R. 2005. *Building for Life: Designing and Understanding the Human-Nature Connection*. Washington, DC: Island Press.

Kellert, S. R. and E.O. Wilson, eds. 1993. *The Biophilia Hypothesis*. Washington, DC: Island Press.

Klein, G. 1998. *Sources of Power: How People Make Decisions*. Cambridge, MA: MIT Press.

Klinger, A., and N. A. Salingaros. 2000. "A Pattern Measure." *Environment and Planning B: Planning and Design* 27: 537–547.

Kulikauskas, A. 2006. "How Might We Create a Really Human Environment?" *Global Villages*, March 10, 2006, http://groups.yahoo.com/group/globalvillages/message/1015.

Masden, K. G., II. 2005. "Being There." *Catholic University of America Summer Institute for Architecture Journal* 2: 51–56.

Masden, K. G., II. 2006. "The Education of an Urbanist: A Real Point of View." In *The Teaching of Architecture and Urbanism in the Age of Globalization*, edited by José Baganha. Casal de Cambra, Portugal: Caleidoscopio Ediçao e Artes Graficas, 173–179.

McGrath, A. 2005. *Dawkins' God: Genes, Memes, and the Meaning of Life*. Oxford: Blackwell Publishing.

Orians, G. H., and J. H. Heerwagen. 1992. "Evolved Responses to Landscapes." In *The Adapted Mind*, edited by J. H. Barkow, L. Cosmides, and J. Tooby. New York: Oxford University Press, 555–579.

Pyla, P. I. 2007. "Hassan Fathy Revisited." *Journal of Architectural Education* 60:28–39.

Ramachandran, V. S., and D. Rogers-Ramachandran. 2006. "The Neurology of Aesthetics." *Scientific American Mind* 17(5): 16–18.

Roth, J. 2000. "Pet Therapy Uses with Geriatric Adults." *International Journal of Psychosocial Rehabilitation* 4:27–39.

Salingaros, N. A. 2003. "The Sensory Value of Ornament." *Communication and Cognition* 36(3–4): 331–351. Revised version: Salingaros, N. A. (2006), chapter 4.

Salingaros, N. A. 2004. *Anti-Architecture and Deconstruction*. Solingen, Germany: Umbau-Verlag. Second enlarged edition, 2007.

Salingaros, N. A. 2005. *Principles of Urban Structure*. Amsterdam: Techne Press.

Salingaros, N. A. 2006. *A Theory of Architecture*. Solingen, Germany: Umbau-Verlag.

Salingaros, N. A., and K. G. Masden, II. 2006a. "Architecture: Biological Form and Artificial Intelligence." *Structurist* 45/46: 54–61.

Salingaros, N. A., and K. G. Masden, II. 2006b. "Review of Christopher Alexander's *The Nature of Order, Book Four: The Luminous Ground*." *Structurist* 45/46: 39–42.

Salingaros, N. A., and K. G. Masden, II. 2007. "Restructuring 21st-Century Architecture Through Human Intelligence", *ArchNet International Journal of Architectural Research* 1(1): 36–52. http://archnet.org/library/documents/one-document.tcl?document_id=10066.

Squire, L. R., and E. R. Kandel. 1999. *Memory: From Mind to Molecules*. New York: Scientific American Library.

Taylor, R. P. 2006. "Reduction of Physiological Stress

Using Fractal Art and Architecture." *Leonardo* 39(3): 245–251.

Taylor, R. P., B. Newell, B. Spehar, and C. Clifford. 2005. "Fractals: A Resonance Between Art and Nature?" In *Mathematics and Culture II: Visual Perfection*, edited by Michele Emmer. Berlin: Springer-Verlag, 53–63.

Tse, M. M. Y., J. K. F. Ng, J. W. Y. Chung, and T. K. S. Wong. 2002. "The effect of Visual Stimuli on Pain Threshold and Tolerance." *Journal of Clinical Nursing* 11:462–469.

Tsunetsugu, Y., Y. Miyazaki, and H. Sato. 2005. "Visual Effects of Interior Design in Actual-Size Living Rooms on Physiological Responses." *Building and Environment* 40: 1341–1346.

Turing, A. A. 1950. "Computing Machinery and Intelligence." *Mind* 59:433–460.

Ulrich, R. S. 1984. "View Through Window May Influence Recovery from Surgery." *Science* 224:420–421.

Ulrich, R. S. 2000. "Evidence Based Environmental Design for Improving Medical Outcomes." In *Healing By Design: Building for Health Care in the 21st Century*. Montreal: McGill University Health Center.

Valentine, J. W., A. G. Collins, and C. P. Meyer. 1994. "Morphological Complexity Increase in Metazoans." *Paleobiology* 20:131–142.

Weizenbaum, H. 1976. *Computer Power and Human Reason*. San Francisco: W. H. Freeman.

Wilson, E. O. 1978. *On Human Nature*. Cambridge, MA: Harvard University Press.

Wilson, E. O. 1984. *Biophilia*. Cambridge, MA: Harvard University Press.

Wilson, E. O. 2006. *The Creation*. New York: W. W. Norton.

Wise, J. A. and T. Leigh-Hazzard. 2002. "Fractals: What Nature Can Teach Design." *American Society of Interior Designers ICON*, March, 14–21.

PART II

The Science and Benefits of Biophilic Design

chapter 6

Biophilic Theory and Research for Healthcare Design

Roger S. Ulrich

This is an era of vast investment internationally in new healthcare buildings. More than $40 billion was spent in the United States in 2006 on new healthcare construction, and annual U.S. spending is projected to reach $61 billion by 2010 (Jones 2007). The United Kingdom has begun a program to create upwards of 100 hospitals and thousands of clinics. This surge of construction provides a major opportunity to use biophilia research to inform the design of better, more healing healthcare environments. There are more than 50 rigorous studies relevant to understanding the influences of such biophilic elements as nature views and daylight in healthcare settings on patients, family, and staff. This growing literature indicates that evidence-based biophilic design can have a positive impact by reducing stress, improving emotional well-being, alleviating pain, and fostering improvements in other outcomes.

This chapter describes biophilia theory and selectively reviews scientific research pertinent to designing healthcare settings that reduce stress and promote better health outcomes. Patients and other users of healthcare facilities can potentially derive benefits from widely different types of encounters with biophilic elements or nature including physically active experiences such as horticultural therapy (Wichrowski et al. 2005); less physically active contacts, for instance, sitting and talking in a garden; and passive interactions such as looking at nature through a window (Ulrich 1999). The discussion concentrates mainly on the effects of passive visual experiences with nature and exposure to sunlight on patient outcomes. Although the amount of research relevant to healthcare design is limited, there are a growing number of scientifically rigorous studies, making this one of the most rapidly developing and coalescing areas of biophilia/design research.

The next section defines two key terms that are important throughout the chapter, *health outcome* and *stress*. The following discussion covers biophilia theory relating to the proposition that exposure to nature and sunlight helps to mitigate stress, reduce pain, and improve other outcomes. Later sections survey research findings and list evidence-based biophilic design recommendations for hospitals and clinics.

WHAT ARE HEALTH OUTCOMES?

Health outcome broadly refers to an indicator or measure of healthcare quality. There are many different types of health or medical outcomes, including

- Observable signs and symptoms relating to patients' conditions (examples: intake of pain medication, blood pressure, length of hospital stay)
- Satisfaction and other reported outcomes (examples: patient satisfaction, health-related quality of life, staff satisfaction)
- Safety outcomes (examples: infection rate, medical errors, falls)
- Economic outcomes (examples: cost of patient care, recruitment or hiring costs due to staff turnover, revenue from patients choosing a facility)

Outcome studies have major importance in healthcare because they provide the most sound and credible basis for evaluating whether a particular medical intervention, treatment, or service is medically effective and cost-efficient. An important related point is that outcome research methods can be adapted to evaluate to what extent biophilic design measures in healthcare facilities are beneficial and cost-effective compared to creating conventional hospitals and clinics designed without biophilic features. Healthcare providers everywhere are under strong pressures to control costs yet increase care quality, and they face mounting financial demands such as paying for costly imaging technology and recruiting employees with key skills. Intuitive or qualitative arguments in favor of biophilic design carry little weight with administrators forced to pay close attention to the bottom line. There can be no question that the resources accorded to biophilic design in the healthcare sector will be heavily affected by the extent to which sound studies demonstrate that biophilic measures provide actual gains through improving outcomes and reducing costs compared to alternatives such as not providing nature (Ulrich 2002).

The priority of specific health outcomes used for measuring the effects of biophilic design can vary in different categories of patients. Suppose, for example, that a hospital is considering whether to install a garden designed to benefit patients recovering from cardiac surgery. Here, administrators would be more likely to allocate space and funding for the project if credible research showed that a well-designed garden would improve outcomes relevant for the surgery patients, such as reducing intake of pain drugs, lowering anxiety or stress, increasing satisfaction with the care experience, improving the capacity to move or walk independently at time of discharge, and reducing the length of hospital stay. By contrast, the selection of outcomes would be different for gauging the effects of a garden, for instance, on terminally ill persons in a hospice, and could focus on an evaluation of whether garden exposure improved reported quality of life and reduced depression, pain, and family stress. As another example, assessing the effects of a hospital garden designed for use by nurses and other staff would measure such outcomes as absenteeism, turnover, work stress, and satisfaction (Ulrich 1999, 2002).

STRESS: A MAJOR PROBLEM IN HEALTHCARE

Stress is defined here as a process of responding to events, environmental features, or situations that are challenging, exceed coping resources, or threaten well-being. Stress is central to understanding how biophilic design, and healthcare physical environments more generally, can influence outcomes (Ulrich 1991, 1999, 2006). A vast body of research has documented that patients experience stress, and that a large proportion suffer acute stress. Examples of the many stressful aspects of hospitalization include fear about impending surgery, lack of information, painful medical procedures, reduced physical capabilities, depersonalization, loss of control, and disruption of social relationships. Much added stress, unfortunately, stems from poorly designed healthcare environments that are noisy, hinder the presence of family and friends, deny privacy, prevent patients from seeing out windows, or force bedridden persons to stare directly at glaring ceiling lights (Ulrich 1991; Ulrich et al. 2006).

In addition to afflicting patients, stress is also a burden for families of patients and visitors, and a pervasive problem among healthcare staff. Occupations such as nursing are stressful because they often involve high work demands, low sense of control, stressful events such as the death of a patient, noise, fatigue, inadequately designed work and care settings that force nurses to spend much of their time walking up and down halls engaged in wasteful fetching, and a lack of break rooms or respite spaces (Ulrich et al. 2006).

The stress experienced by a patient is an important negative outcome in itself, and directly and adversely affects many other outcomes (Ulrich 1991). These unhealthy effects stem from detrimental psychological, physiological, neuroendocrine, and behavioral changes linked to stress. Examples of psychological manifestations of patient stress responses include feelings of fear or anxiety, sadness, and a sense of helplessness. Physiological accompaniments often include, for instance, elevated blood pressure and heart rate. The neuroendocrine component produces increased levels of a steroid (cortisol) and stress hormones (such as epinephrine) that tax the heart and other major organs. Much research has shown that stress-related neuroendocrine and physiological mobilization suppresses immune system functioning, thereby decreasing resistance to infection and worsening recovery indicators such as wound healing (Kiecolt-Glaser et al. 1998; Kiecolt-Glaser et al. 1995). Behavioral effects of stress range from social withdrawal, verbal outbursts, and sleeplessness to a failure to take medications. Given these findings, the contention is justified that biophilic design should foster improved outcomes to the extent that the presence of nature and other biophilic elements in healthcare environments is effective in reducing stress.

BIOPHILIA THEORY: WHY NATURE SHOULD FOSTER RESTORATION FROM STRESS

The intuitive belief that contact with nature promotes psychological well-being and physical health dates back at least two thousand years and has appeared widely in Western and Asian cultures (Ulrich et al. 1991). Until recent years, many writers ascribed this belief to culture and individual learning, often asserting that societies inculcate their inhabitants to like nature but associate cities with stress (e.g., Tuan 1974). It can also be argued that people learn positive and restorative associations with nature through personal experiences such as vacations in rural environments, but acquire negative associations with cities because of work pressures, noise, crime, and other urban stressors.

These interpretations based on learning or culture fail to explain adequately, however, evidence from a large scientific literature showing that diverse cultures and socioeconomic groups show high similarity in responding positively to nature views (Ulrich 1993). Evolutionary theory more easily accounts for this widespread agreement by proposing that millions of years of evolution have left modern humans with a partly genetic predisposition for responding positively to nature (Appleton 1975; Orians 1986; Kaplan and Kaplan 1989; Ulrich 1983). In this vein Wilson's biophilia hypothesis (1984) holds that humans have a partly genetic tendency to pay attention to, affiliate with, and otherwise respond positively to nature. The notion that biophilia is represented in the human gene pool carries with it the proposition that certain types of positive responses were adaptive for early humans and increased fitness or chances for survival (Ulrich 1993).

Considerable evolutionary writing has discussed survival-related benefits of preferences or aesthetic liking for nature (Appleton 1975; Coss 2003). Orians and Heerwagen have convincingly argued that savanna-like views should be preferred by modern humans because savannas were superior during evolution to other habitats for providing primary necessities such as food, water, and security (Orians 1980, 1986; Heerwagen and Orians 1993). Apart from preferences, evolutionary theory is also important for explaining why certain types of nature views and content (vegetation, water, sunlight) should have stress-reducing and healthful influences.

Detailed conceptual arguments have been developed elsewhere as to why a capability for fast recovery from stress following a demanding episode was so vital for the survival of early humans as to favor the selection of individuals with a biologically prepared capacity for responding restoratively to many nature settings (Ulrich et al. 1991; Ulrich 1993). Daily living for premodern people was demanding and stressful, and involved encounters with threats or risks. Acquiring a partly genetic capability for faster recuperation from stress would have had several key advantages, including faster replenishment of energy expended in the physiological arousal and behavior involved in stress responding to threats ("flight-or-fight") and other challenging situations. Other health-related benefits should include rapid declines in stress-related negative emotions such as fear and anger, increases in positive feelings, and salutary changes in bodily systems indicative of lessened physiological and neuroendocrine mobilization (lower blood pressure, reduced levels of stress hormones, enhanced immune function). Physiological restoration should be expected to include prominent reduction of fatigue and deleterious autonomic/sympathetic nervous system activity, because sympathetic mobilization is centrally involved in stress responses for dealing with taxing situations and threats (Ulrich et al. 1991; Ulrich 1993, 1999). A testable hypothesis that follows from this evolutionary-functional reasoning is that restorative responses to nature should occur rapidly, usually within minutes—or even as fast as several seconds in certain bodily systems—rather than in several hours or days (Ulrich et al. 1991; Ulrich 1993). The theory also expressly contends that reductions in stress should directly and indirectly promote improved health outcomes, such as lessened pain and faster wound healing in connection with enhanced immune function (Parsons 1991; Ulrich 1991, 1999).

This restoration theory further holds that modern humans, as a genetic remnant of evolution, have a capacity for acquiring stress-reducing responses to certain nature settings and content (vegetation, water), but have no such disposition for most built or artifact-dominated environments and materials (concrete, glass, metal, plastic) (Ulrich 1993, 1999). Properties of nature settings that should be effective in producing restoration include security linked to spatial openness that fosters visual surveillance; sunshine or good light in contrast to poor light or threatening weather; and qualities linked with high habitability and food availability, including calm or slowly moving water, verdant vegetation, flowers, savanna-like or parklike properties (scattered trees, grassy understory), and unthreatening wildlife such as birds (Ulrich 1993, 1999; Ulrich and Gilpin 2003). These conceptual arguments have a practical implication, which is that designing healthcare buildings to incorporate nature features such as those listed can harness therapeutic responses and influences that are carryovers from evolution, resulting in more restorative and healing patient care settings.

Evolutionary theory is also useful for identifying specific types of nature features and configurations that should be *avoided* when designing healthcare environments to reduce stress and foster better outcomes (Ulrich and Gilpin 2003). It has been proposed elsewhere that people have a partly genetic disposition not only for biophilic or positive responses to nature features that were advantageous during evolution, but also for negative/avoidance responses to certain nature stimuli that signaled *threats or dangers* for early humans (Öhman 1986; Ulrich 1993; Coss 2003). These stressful and potentially dangerous stimuli included shadowy enclosed spaces, snakes and spiders, reptilian-like tessellated scale patterns, pointed or piercing forms, and angry and fearful human faces (Öhman 1986; Coss 2003; Ulrich 1993). Findings from scores of conditioning experiments, and behavior-genetic investigations and other research on human twins, leave no doubt that genetic factors play a major role in fear and stress responses to certain visual features such as snakes and angry faces (Ulrich 1993). The partly genetic underpinning of negative responses emphasizes the importance of excluding views or images of such phenomena from healthcare settings where stress is a problem (Ulrich and Gilpin 2003).

Theory: Daylight, Restoration, and Health

Another carryover of evolution is that modern humans are psychologically and biologically attuned to light and changing cycles of light and darkness. Daylight and sun exposure were critically important for the day-to-day well-being and survival of premodern people. In line with earlier conceptual arguments regarding nature scenes, it is proposed that positive responses to nature settings should be enhanced during good lighting or sunny conditions. Daylight and sunshine enabled visual surveillance of surroundings, finding food and water, locating and pursuing game, and avoiding threats such as predators that would be concealed in darkness. Clear or sunny conditions, compared to overcast conditions or dark clouds with thunder, signaled less short-term risk from adverse weather.

Further, human physiology evolved to require sun exposure for metabolism of vitamin D, which is vital for overall health. Vitamin D is important for the development of a healthy musculoskeletal system, preventing rickets and osteoporosis, maintaining muscle strength, and preventing chronic diseases such as type 1 diabetes and rheumatoid arthritis (Holick 2005). Also, daylight is the main environmental stimulus for regulating circadian or body clock rhythms that cycle approximately every 24 hours, and synchronize the sleep and awake cycle with night and day. Daylight exposure affects levels of the hormone melatonin, which influences levels of energy, alertness, and activity. When exposure to daylight or artificial light is inadequate, melatonin levels increase and cause drowsiness and depression. In sum, an evolutionary conceptual perspective predicts that well-lighted or sunny nature settings should be more effective than dark or overcast nature scenes in eliciting positive responses, improving emotional well-being, fostering restoration, and promoting health. The least effective scenes should be built or artifact-dominated spaces in shadow or overcast conditions.

RESEARCH: NATURE VIEWS AND RESTORATION

Consistent with the evolutionary restoration theory outlined above, several studies of nonpatient groups using prospective experimental designs indicate that even briefly viewing nature settings can produce substantial and rapid psychological and physiological restoration from stress (Ulrich 1979; Ulrich et al. 1991; Hartig et al. 1995; Parsons et al. 1998; Parsons and Hartig 2000; Hartig et al. 2003; Van den Berg et al. 2003). Restorative or stress-reducing effects of looking at nature are manifested as a constellation of beneficial changes that include reduced levels of negatively toned emotions (fear, anger), elevated positive emotions (pleasantness), and changes in physiological systems indicative of diminished stress mobilization or arousal (autonomic/sympathetic, electrocortical, neuroendocrine, musculoskeletal) (Ulrich et al. 1991). Studies in both laboratories and real environments have consistently found that viewing nature produces significant physiological restoration within three to five minutes at most, as evidenced, for example, in brain electrical activity, blood pressure, heart activity, and muscle tension (Ulrich 1983; Ulrich et al. 1991; Parsons et al. 1998; Hartig et al. 2003; Laumann et al. 2003). (See Figures 6-1 to 6-3.) Fredrickson and Levenson (1998) exposed participants to a fear-eliciting film, and reported that those persons assigned randomly to view a nature film (water) exhibited significant recovery from cardiovascular stress in only 20 seconds. These rapid and beneficial psychological and physiological changes can be accompanied by sustained yet nontaxing or nonvigilant attention and perceptual intake with respect to the nature setting, as indicated by heart rate deceleration and reduced autonomic/sympathetic nervous system activity (Ulrich et al. 1991; Laumann et al. 2003). Although most nature views are stress reducing, most built or urban settings lacking nature (streets, parking lots, building exteriors without nature, parking lots, windowless rooms) are unsuccessful in producing restoration, and in some instances worsen stress (Ulrich 1979; Van den Berg et al. 2003).

Research: Effects of Viewing Nature on Patient Stress

Survey research on bedridden hospital patients suggests that they assign high preference and importance to having a bedside window view of nature (Verderber 1986). A study of elderly in urban long-term care facilities has similarly found that residents attach considerable importance to having access to window views of outdoor

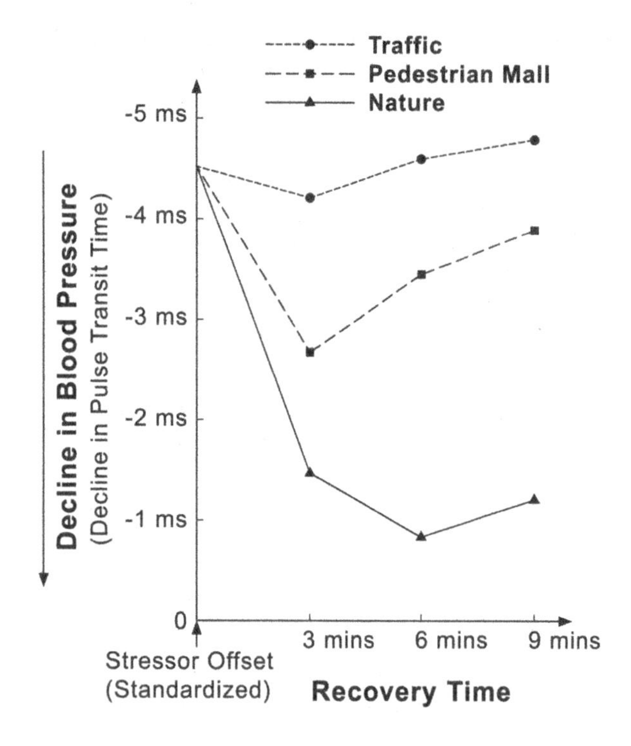

Figure 6-1: Systolic blood pressure (via pulse transit time) during recovery from stress in persons exposed to nature settings or urban settings lacking nature.

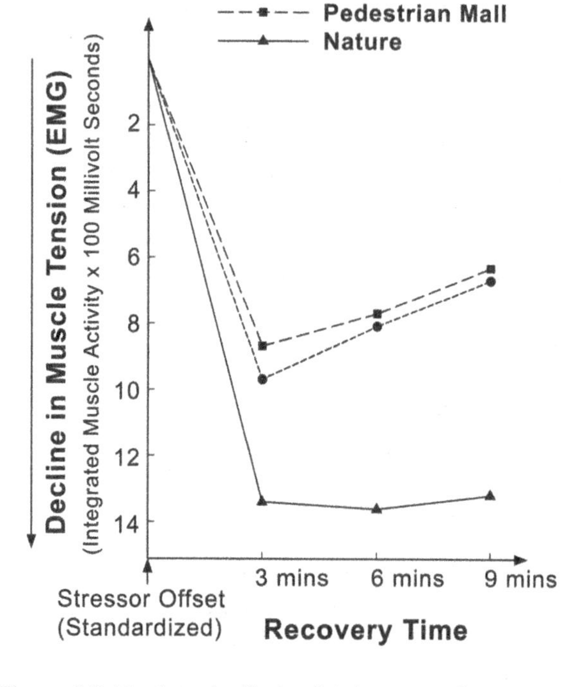

Figure 6-2: Muscle tension (forehead) during recovery from stress in persons exposed to nature settings or urban settings lacking nature.

spaces with prominent nature such as plants, gardens, and birds (Kearney and Winterbottom 2005). In the same study, long-term care elderly expressed dislike for window views of built content lacking nature, such as rooftops and building walls.

In view of the earlier discussion of restoration, it is worth emphasizing that a few studies in healthcare settings have found that visual exposure to nature can effectively lower stress. An early investigation by Katcher and his associates measured restoration from anxiety in patients waiting to undergo dental surgery in a room with or without an aquarium with fish (Katcher et al. 1984). They found that anxiety was lower on days when the aquarium was present, and scores for patient compliance during surgery were higher. A study by Heerwagen (1990) suggested that patients in a dental clinic were less stressed on days when a large nature mural was hung in the waiting room, in contrast to days when there was no nature scene. A prospective randomized experiment focusing on stressed blood donors found that participants had lower blood pressure and pulse rates when a wall-mounted television displayed a nature videotape, compared to when the television showed either an urban videotape or daytime television programs such as game or talk shows (Ulrich et al. 2003). A quasi-experimental study of patients with dementia, including Alzheimer's disease, suggested that adding large color nature pictures and recorded nature sounds (birds, brook) to a shower room lessened stress and cut incidents of agitated aggressive behavior such as hitting, kicking, and biting (Whall et al. 1997). (See Figure 6-4.)

A growing body of research suggests that visual exposure to nature not only reduces patient stress but also improves other important outcomes such as pain. Before reviewing these studies, it is useful to digress briefly and examine theory relevant to understanding why nature exposure could be expected to decrease pain.

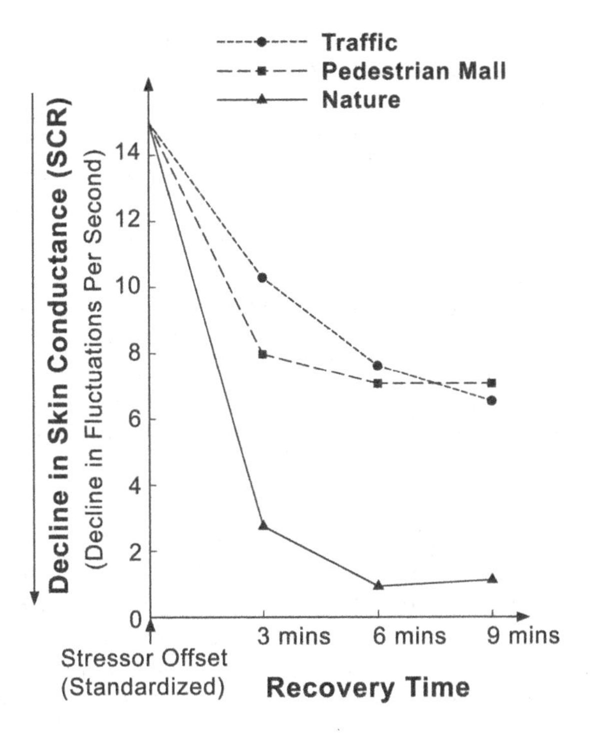

Figure 6-3: Changes in skin conductance (SCR) during recovery from stress in persons exposed to nature settings or urban settings lacking nature. Skin conductance activity is controlled by the autonomic nervous system. Greater decline of autonomic activity indicates larger reduction in physiological stress mobilization.

Figure 6-4: Waiting area to calm stressed patients and family. Doernbecher Children's Hospital, Portland, Oregon.

BIOPHILIA THEORY: WHY NATURE EXPOSURE SHOULD REDUCE PAIN

It is reasonable to propose that nature exposure should mitigate pain in patients through different mechanisms, including stress reduction and distraction. The most influential model in medicine and health psychology for explaining pain is *gate control* theory (Melzack and Wall 1965, 1982). According to this theory, neural structures or mechanisms in the spinal cord act as a gate in the transmission of sensory input or pain impulses through the spinal cord to the brain. When the gate is open, impulses flow to the brain and pain is experienced. When the gate is closed, pain impulses are inhibited from reaching the brain and pain is diminished or not felt. A key premise of gate control theory is that the gate can be closed by messages that descend from the brain and are influenced by psychological or emotional factors (Melzack and Wall 1965, 1982). Negative emotions such as anxiety and depression, and focusing on an injury, can open the gate and thereby increase pain. Positive feelings such as relaxation, or distracting the patient's focus away from an injury, close the gate and thus reduce pain.

Studies were reviewed earlier indicating that viewing nature reliably produces restoration from stress, as manifested by declines in negative emotions such as anxiety, enhanced positive feelings, and physiological changes indicative of diminished stress mobilization. These positive changes, according to gate control theory, should close the gate and inhibit pain impulses from reaching the brain, thereby alleviating pain. Furthermore, restorative psychological and physiological effects of viewing nature can be accompanied by sustained yet nontaxing attention and perceptual intake that should reduce pain via distraction. To the extent a nature view holds the patient's focus and attention, and diverts their focus away from pain, gate control theory predicts that nature distraction will tend to close the gate and reduce pain. Accordingly, gate control theory suggests that a nature view can close the gate and alleviate pain by distracting the patient, reducing stress, and increasing positive emotions.

In addition to gate control theory, another prominent pain perspective, *distraction* theory, offers a rather different explanation for pain-reducing effects of viewing nature (Ulrich et al. 2006). Distraction is defined as concentrating on aspects of the environment that are outside oneself (Brewer and Karoly 1989). According to distraction theory, persons have a limited amount of available conscious attention (McCaul and Malott 1984). Pain requires considerable conscious attention and draws upon much of the limited amount available. Distraction theory proposes that as the amount of conscious attention directed to pain increases, the intensity of experienced pain will correspondingly rise (Brewer and Karoly 1989). However, if patients become engrossed in an external pleasant distraction such as a nature scene, they will have less conscious attention to direct to their bodily sensations of pain, and experienced pain therefore will diminish. The theory explicitly predicts that the more engrossing and diverting a distraction, the greater the pain reduction (McCaul and Malott 1984). Traditional distraction theory implies that both emotionally positive and negative distractions, if highly engrossing, should be effective in alleviating pain. However, findings from one strong study suggest that emotionally pleasant picture distractions are more effective than unpleasant visual stimuli in increasing tolerance for pain (Wied and Verbaten 2001). This implies that many nature views may be effective in reducing pain because they are emotionally pleasant distractions as well as capable of eliciting sustained attention and perceptual intake.

Research Findings: Effects of Nature Exposure on Pain

Several studies using experimental or quasi-experimental designs have shown convincingly that nature distraction can produce substantial and clinically important alleviation of pain. A study of patients recovering from abdominal surgery found that those assigned to rooms with a bedside view of nature (trees) had better postoperative recovery courses than matched patients assigned to identical rooms with windows overlooking a brick building wall (Ulrich 1984). Patients with the nature window view, compared to those with the wall view, suffered significantly less pain, as indicated by needing far fewer doses of strong narcotic pain medications than their matched counterparts with the wall view. Furthermore, the nature view patients had shorter postsurgery stays, better emotional well-being, and fewer minor complications such as persistent nausea or headache (Ulrich 1984). (See Figures 6-5 to 6-7.)

In another study (Ulrich et al. 1993) patients who had undergone heart surgery were assigned to view color photos mounted directly in their line of vision. Patients shown a picture of a spatially open, well-lighted view of trees and water required fewer doses of strong pain drugs and suffered less anxiety than groups exposed to abstract images or patients assigned to a control group with no picture (Ulrich et al. 1993). A picture of a spatially enclosed and shadowy forest setting, however, did not significantly lessen pain or anxiety. In the same study, patients assigned an abstract picture with straight-edged or rectilinear forms had worse outcomes than the control group with no pictures. A well-controlled experiment in a hospital by Tse and colleagues found that volunteers had much greater pain tolerance and a higher threshold for detecting pain

Figure 6-5: Hospital window view of trees.

Figure 6-6: Hospital window view of brick wall.

when assigned to view a nature videotape on an eyeglass display (waterfall, mountains, landscapes) in contrast to looking at a blank display (Tse et al. 2002).

As noted, distraction and gate control theory predict that the more engrossing a nature distraction, the greater the pain alleviation. This implies that nature exposures may tend to be more diverting and hence pain reducing if they involve sound as well as visual stimulation, and are high in realism and "immersion" (Wismeijer and Vingerhoets 2005). A study of burn patients suffering acute pain showed that distracting patients during burn dressings with a bedside television screen displaying nature scenes (forest, flowers, waterfalls, ocean, wildlife) accompanied by music lessened both pain and anxiety/stress (Miller et al. 1992). A randomized prospective trial of patients undergoing painful bronchoscopy found that compared to a control group who looked at a blank ceiling during the procedure, pain was lower when patients were assigned to look at a ceiling-mounted nature scene and listen to sounds recorded at the nature setting (moving water, birds, breezes through wildflowers and grass) (Diette et al. 2003).

Lee and colleagues carried out a randomized controlled trial of the effects of nature distraction on pain and patient-controlled sedation during colonoscopy (Lee et al. 2004). They reported that visual distraction alone lowered pain but did not reduce intake of sedative medication during colonoscopy. However, a more engrossing combination of visual and audio distraction (nature scenery with classical music) significantly lowered both pain and self-administered sedation during colonoscopy, a finding consistent with the predictions of distraction and gate control theory (Lee et al. 2004). Kozarek and colleagues (1997) investigated the effects of seeing and listening to a travelogue with nature scenes on patients undergoing unpleasant gastric laboratory procedures. Patient reports and nurse observations were in accord in suggesting that the audiovisual distraction improved comfort and tolerance for the procedures, compared to when the patients previously had the procedures without distraction (Kozarek et al. 1997).

It is evident from this review that the majority of pain studies to date have simulated nature using equipment such as television screens, eyeglass displays, or virtual reality, rather than exposing patients to real nature, for instance, a garden or window view of trees. One reason is that simulations make it easier to carry out

Analgesic Strength	Number of Pain Drug Doses (days 2-5 after surgery)	
	Nature View Patients	**Wall** View Patients
Strong	0.96	2.48
Moderate	1.74	3.65
Weak	5.39	2.57

after Ulrich, 1984

Figure 6-7: Pain drug intake of patients with window view of nature versus brick wall. Examples in the strong analgesic category were potent narcotics administered by injection. The weak category consisted of oral doses of drugs such as acetaminophen.

prospective randomized clinical trials that achieve the rigor required in medical research. Also, there are certain highly stressful medical settings where it is not feasible to provide visual contact with real nature even through windows, such as an underground shielded room for radiation therapy or the tight confines of an MRI scanner, and simulations may be the only viable option for exposing patients to nature. These exceptions notwithstanding, creative evidence-based designers have demonstrated in many healthcare projects that window views of gardens and other real nature can be successfully provided for challenging and restrictive medical environments, for instance, treatment spaces where measures such as pressurized air, HEPA filtration, and sealed windows are needed to protect immunocompromised and other acutely vulnerable patients from infection.

There is a clear need for controlled experimental studies to evaluate the extent to which real nature environments may outperform simulations in reducing pain and stress. Do simulations, compared to real nature settings, lose much of their effectiveness in distracting and calming patients during longer hospital stays? It seems likely that over long-term healthcare exposures, real nature should be more effective than simulations in sustaining distraction and positive responding owing to greater authenticity, immersiveness, and the multisensory stimulation and ongoing visual change inherent in real nature settings (Ulrich 1993). It is important that additional rigorous research be conducted to clarify how integrating nature into healthcare buildings is medically beneficial and cost-effective compared to conventional design approaches that tend to omit nature, so that administrators are equipped to make well-informed decisions benefiting their patients, staff, and budgets.

NATURE ART IN HEALTHCARE SETTINGS

There has been considerable research on people's responses to art, but most has examined the art preference of nonpatient adults, rather than art's effects on patient stress, recovery, pain, or other outcomes. Although the relationship between preference responses and restoration effects is not well understood, preference studies are nonetheless useful for identifying types of art that are most liked by patients, family members, and staff in healthcare facilities (Ulrich and Gilpin 2003). Limited research using path analysis suggests that emotional components of restoration drive or strongly influence preferences, raising the possibility that art preferences are linked to and reflect restorative responses (Van den Berg et al. 2003).

Studies have shown that the vast majority of adults across different cultures prefer nature over other art subject matter (Wypijewski 1997; Kettlewell 1988; Winston and Cupchik 1992). Adults internationally also reflect strong similarity in *disliking* abstract art (Wypijewski 1997). A few studies of patient art preferences have produced results that closely parallel those for the nonpatient public (Ulrich and Gilpin 2003). Carpman and Grant (1993) studied 300 randomly selected hospital inpatients and found they overwhelmingly preferred realistic nature art and disliked abstract images. Hathorn and Ulrich surveyed small samples of black and white Americans in a large urban hospital to determine their preferences for a highly varied collection of several hundred art images. Consistent with the prediction from evolutionary theory that nature scenes should be liked across different groups, both blacks and whites accorded high preference to paintings of nature landscapes, and judged nature as the most appropriate subject matter for art in patient rooms (Ulrich and Gilpin 2003). Blacks and whites were also similar in especially liking nature paintings depicting spatially open settings in sunny or well-lighted conditions, with green vegetation and water features. Art depicting gardens with flowers also received consistently high preference scores (Ulrich and Gilpin 2003). A study of adults in a Scandinavian mental health facility found that patients reported positive emotional responses to nature paintings and prints, but consistently evidenced negative, stressful reactions to abstract artworks in which the content was ambiguous and disordered (Ulrich 1991).

A recent study by Eisen (2006) is among the first to investigate scientifically the art preferences of schoolchildren and hospitalized pediatric patients. The research compared the art preferences of children across four different age groups: 5–7, 8–10, 11–13, and 14–17. Findings suggested that across all age groups and both genders, the great majority of schoolchildren (n=129) and hospitalized pediatric patients (n=79) were similar in preferring nature art over abstract arrays that varied in complexity, color brightness, and presence versus absence of a cartoonlike image. In the case of the schoolchildren, for example, a total of nearly 75 percent accorded highest preference either to a representational nature artwork (forest with lake and deer) or an impressionistic nature scene (beach with waves) (Eisen 2006). These findings are broadly consistent with biophilia or evolutionary theory, but run counter to traditional intuition-based design guidelines that often recommend abstract or cartoonlike images for healthcare spaces for children.

GARDENS IN HEALTHCARE FACILITIES

Evidence from a few studies suggests that well-designed gardens can be efficacious settings in healthcare facilities for fostering restoration among stressed patients, family, and staff (Ulrich 1999; Marcus and Barnes 1999). Gardens not only provide restorative and pleasant nature views, but can also reduce stress and improve outcomes through other established mechanisms, such as fostering access to privacy and social support, creating opportunities for restorative escape (and control) with respect to stressful clinical environments, and providing settings that enable physically active pursuits in pleasant nature surroundings ranging from active play to physical rehabilitation (Ulrich 1999; Marcus and Barnes 1995; Whitehouse et al. 2001; Hartig and Marcus 2006). If viewing a garden produces restoration and improved mood in patients, they may be more likely to engage in other healthy and stress-reducing activities such as walking in the garden or talking with a friend (Ulrich 1999). The assumption that pleasant gardens help to motivate patients to engage in physical activity, as well as alleviate their emotional stress, has led some hospitals to design rehabilitation gardens that enable physiotherapists to treat specific categories of patients, such as those recovering from stroke, fractures, and burns. Notable examples of such gardens include those at Legacy Health in Portland, Oregon, and the Rusk Institute of Rehabilitation Medicine in New York City. (See Figure 6-8.)

Marcus and Barnes (1995) used a combination of behavioral observation and interview methods in postoccupancy studies of four hospital gardens in California. Their findings suggested that recovery from stress was the most important category of benefits realized by nearly all users of the gardens. Similarly, a postoccupancy evaluation of a garden in a children's hospital identified restoration from stress and improved emotional well-being as the primary benefits for users (Whitehouse et al. 2001). This finding was supported by convergent evidence from observation of users, interviews with on-site users, and a hospital-wide questionnaire survey of staff and patients' parents. Another investigation of three gardens in a pediatric cancer center found that emotional stress was lower for all types of users when they were in the gardens rather than inside

Figure 6-8: Children's Garden at Legacy Good Samaritan Hospital, Portland, Oregon. The garden was designed to be restorative and promote improvements in specific health outcomes for pediatric patients and their families. The participatory design process involved nurses and physicians, a landscape architect, horticultural therapists, and artists. The garden provides a variety of spaces and features to support active play by children in addition to restorative sedentary activities (viewing nature, socializing) by adult family and staff. Other spaces support rehabilitation activities such as horticultural therapy. The design creates easy access to shade, as the medical conditions and treatments of some patients make them negatively sensitive to direct sun exposure.

the hospital (Sherman et al. 2005). A recurring finding across these studies is that adult garden users, including family and staff, engage mostly in sedentary activities such as passive relaxation, socializing, and eating. By contrast, children interact actively with garden features much more than adults (Sherman et al. 2005). The research implies that designers can create healthcare gardens that benefit both adults and children by providing a variety of spaces, ranging from active play features and spaces for children to calm refuges for adults. Concerning the latter, restorative refuges, the limited research suggests that gardens will tend to mitigate stress effectively for adult users when spaces contain verdant foliage, flowers, water, grassy spaces with trees or large shrubs and a modicum of spatial openness, compatible pleasant nature sounds (birds, breezes, water), and comfortable movable seating (Ulrich et al. 2006; Ulrich 1999; Marcus and Barnes 1995, 1999; Rodiek 2005).

Broadly similar findings have emerged from research on gardens and other outdoor spaces in assisted living facilities for elderly residents. Based on studies of fourteen assisted living facilities, Rodiek (2005) reported that elderly residents preferred outdoor spaces with abundant greenery, flowers, birds, water features, and fresh air. In contrast to these positive characteristics, research on hospital gardens suggests that the following environmental qualities can elicit negative responses in patients and other users, and may hinder restoration or even worsen stress: predominance of hardscape rather than nature (concrete, for example); intrusive urban or mechanical sounds (such as traffic or an air-conditioning compressor); crowding; and ambiguous or abstract art and design features that are readily interpreted in multiple ways and may elicit negative reactions in some stressed patients (Ulrich 1991, 1999; Ulrich and Gilpin 2003).

Other research suggests that gardens can also be important for reducing stress in healthcare workers and increasing job satisfaction (Marcus and Barnes 1995; Whitehouse et al. 2001; Sherman et al. 2005). Although staff research is limited, the findings are noteworthy in light of the serious and widespread problems in healthcare of mounting work demands and pressures, high staff stress, low job satisfaction, and high turnover. Findings from other research on stressed employees in non-healthcare workplaces appear relevant to healthcare workers such as nurses and physicians. A study of European white-collar and blue-collar employees in different non-health work settings found that window views of nature buffered job stress and enhanced health-related well-being (Leather et al. 1998). Research by Kaplan (1993) found that office workers with a window view of nature reported lower frustration and higher life satisfaction and overall health.

EFFECTS OF DAYLIGHT EXPOSURE IN HEALTHCARE FACILITIES

As mentioned in an earlier theory section, daylight and sun exposure were vital for the well-being, health and survival of early humans. As a legacy of evolution, modern humans are psychologically and physiologically attuned to full-spectrum light and changing cycles of light and darkness. Consistent with evolutionary arguments, studies across a variety of environments (hospitals and workplaces) suggest patients and other groups accord even higher preference and importance to a nature window view when the outdoor setting is illuminated by clear light conditions or sunlight rather than shade (Verderber 1986; Leather et al. 1998; Ulrich et al. 2006). In healthcare facilities, evolutionary carryover arguably is also evident in beneficial effects of daylight and sun exposure on patient and staff outcomes, including facilitation of critical chemical reactions within the body, influences on circadian or body clock rhythms, and positive effects on emotional well-being and pain (Boyce et al. 2003; Joseph 2006; Ulrich et al. 2006).

A well-documented effect of daylight in healthcare facilities is preventing jaundice in newborns by fostering excretion of bilirubin. The potentially harmful impact of depriving infants in hospitals of light is evident in research showing that incidence of jaundice increases when windows in maternity units are covered or shaded and no full-spectrum artificial light exposure is provided (Barss and Comfort 1985; Giunta and Rath 1969). Also, light radiation absorbed through the skin stimulates other beneficial chemical reactions such as metabolism of vitamin D, which is important for preventing osteoporosis and certain chronic diseases in groups such as elderly in long-term care facilities (Holick 2005). Additionally, research on residents in Alzheimer's disease care units has linked facilities designed for higher light exposure with lower patient agitation levels (Sloane et al. 1998). This finding is reinforced by results from a prospective study of dementia patients showing that two 10-day periods of exposure to bright morning light reduced agitation (Lovell et al. 1995). The same patients became significantly more agitated on nontreatment days.

As previously noted, exposure to daylight is important for regulating circadian or body clock rhythms, and synchronizing the sleep and awake cycle. Limited research suggests that exposure to higher levels of daylight or white artificial light may improve sleep in community-dwelling older adults and persons in dementia facilities (Van Someren et al. 1997). At least three studies of preterm infants have found that exposure to daylight and night, or cycled artificial light, improves sleep and weight gain (Mann et al. 1986; Blackburn and Patteson 1991; Miller et al. 1995). For day-shift healthcare staff, morning daylight exposure is the primary environmental stimulus for entraining or regulating circadian rhythms, thereby fostering daytime alertness, cognitive performance, and nighttime sleep quality (Rea 2004). The key role of daylight for regulating body clock rhythms implies the importance of providing windows in healthcare workspaces and break areas for nurses and other staff. Tinted windows that attenuate daylight exposure for staff, however, may hinder circadian entrainment and erode alertness and sleep quality (Rea 2004). A study of staff in a Turkish hospital suggested that nurses who were exposed to daylight for three or more hours a day experienced less work stress and had higher job satisfaction (Alimoglu and Donmez 2005).

Research: Effects of Daylight on Depression and Pain

Findings from several rigorous studies indicate that exposure to light—daylight or bright artificial light—is effective in reducing depression and improving mood, even for patients hospitalized with severe depression. The mechanisms of action of light treatment on depression are partly but not fully understood. Light falling on the retina influences activity of the pineal gland and by this pathway suppresses or delays secretion of melatonin, thereby reducing depression, increasing daytime alertness, and fostering sleep quality (Martiny 2004).

A meta-analysis of randomized controlled studies published in the *American Journal of Psychiatry* concluded that light treatment for nonseasonal and seasonal depression is "efficacious, with effect sizes equivalent to those in most antidepressant pharmacotherapy trials" (Golden et al. 2005, 656). Compared to antidepressant drugs, light exposure offers the advantage of being faster acting. In this regard, several studies have found that light can significantly alleviate depression after approximately two weeks of treatment, while antidepressant drugs have a delayed onset of at least four to six weeks. Some studies suggest that exposure to morning light is more effective than afternoon or evening light, although light exposure occurring in the middle of the day or afternoon also significantly reduces depression (Martiny, 2004).

Although artificial light is often used to treat depression, a few studies suggest that architectural design and siting decisions for healthcare facilities can affect depression levels and other outcomes by influencing the amount of daylight exposure patients receive. Beauchemin and Hays (1996) reported that adult patients hospitalized for severe depression in a Canadian hospital had substantially shorter stays if they were assigned to sunny rooms rather than rooms that were always in shade. Similarly, a study in an Italian facility found that adult patients hospitalized for bipolar depression stayed an average of 3.7 days less if they were assigned east-facing rooms exposed to bright morning light, compared to similar patients in west-facing rooms with less sunlight (Benedetti et al. 2001). Apart from mental health patients, depression is a serious problem across several other categories of patients, such as those with cardiovascular disease and cancer. A retrospective study of myocardial infarction patients in an intensive care unit in a Canadian hospital suggested that female patients had shorter stays if their rooms had sunny versus shaded or dim window exposures (Beauchemin and Hays 1998). In the same study, mortality in both sexes was lower in sunny rooms than in north-facing shaded rooms.

In addition to reducing depression and shortening length of stay, there is some evidence that higher daylight exposure alleviates pain. The presumed mechanism for pain reduction is that higher sunlight exposure influences levels of serotonin, a neurotransmitter known to inhibit pain pathways. Walch and colleagues (2005) carried out a strong prospective study focusing on patients undergoing lumbar spinal or cervical surgeries who were admitted postoperatively to rooms either on the bright or shaded side of an inpatient surgical ward. Patients in the bright rooms, compared to those in the more shaded rooms, were exposed on average to 46 percent higher sunlight intensity. Findings indicated that patients in rooms with more sunlight reported less pain and stress, took 22 percent less analgesic medication and had 21 percent lower medication costs. It should be mentioned that the shaded patient rooms—and heightened pain—resulted from construction of a new building 25 meters away that blocked sunlight on one side of the older building with surgical wards. This episode underscores for architects and healthcare administrators the importance of paying close attention to building orientation in new projects, and avoiding site plans where some buildings block light from others (Ulrich et al. 2006). (See Figure 6-9.)

Figure 6-9: Hospital rooftop garden at Legacy Health, Salmon Creek, Washington. The garden enables users to be exposed to sun and engage in sedentary restorative activities such as viewing distant hills and forests. A meditation room juts into the garden (right center). The garden provides restorative garden window views for bedridden patients, and the berm prevents persons in the garden from looking into patient rooms and violating privacy (patient rooms are out of picture to right).

SUMMARY AND DESIGN IMPLICATIONS

The chapter discusses evolutionary or biophilia theory proposing that exposure to nature and sunlight in healthcare settings should reduce stress, lessen pain, and foster improvements in other health outcomes. The conceptual arguments have a practical implication, which is that designing healthcare environments to incorporate nature and daylight can harness therapeutic influences that are carryovers from evolution, resulting in more restorative and healing settings for patients, family, and staff.

The theory contends that a capability for fast recovery from stress following demanding episodes was so critical for enhancing survival chances of early humans as to favor individuals with a partly genetic predisposition for restorative responding to many nature settings. Such stress reduction should directly and indirectly promote improvements in other health outcomes, such as enhanced immune function and reduced pain. Regarding pain, evolutionary theory is integrated with gate control and distraction theory from medicine to explain why nature exposure can be expected to alleviate pain in patients. The evolutionary framework holds that modern humans, as a genetic remnant of evolution, have a capacity for readily acquiring restorative and other healthful responses to certain nature scenes and content (vegetation, water), but have no such predisposition for most built or artifact-dominated environments and materials (concrete, glass, metal, for example).

In addition to nature, daylight and sun exposure also were critical for the well-being, health, and survival of early humans. As a legacy of this importance, modern humans are psychologically and biologically attuned to sunlight and changing cycles of light and darkness. The evolutionary framework predicts that exposure to sunny or well-lighted nature in healthcare buildings should be more effective than dark or overcast nature scenes in fostering restoration, improving emotional well-being, and promoting health. The least effective physical settings should be built or artifact-dominated spaces that lack nature and have overcast or dim light conditions.

In accord with the conceptual position regarding nature, empirical studies of nonpatient groups using prospective experimental designs have shown that even briefly viewing nature can produce rapid and substantial psychological and physiological recovery from stress. Rigorous studies in laboratories and real environments have reliably found that viewing nature produces significant physiological restoration within a few minutes. Limited research in healthcare settings has similarly found that viewing nature fosters restoration in stressed patients. Importantly, several well-controlled prospective investigations of patients have shown convincingly that nature distraction can produce substantial and clinically important pain reduction.

Findings from a few studies suggest that well-designed gardens in healthcare facilities can be efficacious restorative settings for stressed patients, family and staff. Limited research suggests the possibility that well-designed gardens are important for reducing stress and increasing job satisfaction in nurses and other healthcare workers. Regarding art, empirical studies have shown that adults across cultures prefer nature over other art subject matter, a finding broadly consistent with biophilia or evolutionary theory. Recent research on art preferences of children across different age groups suggests the great majority of hospitalized pediatric patients and nonpatient schoolchildren likewise prefer nature art.

A growing amount of research has shown that daylight or sun exposure in healthcare settings has beneficial influences on patient and staff. Daylight stimulates metabolism of vitamin D, and plays a central role in regulating body clock rhythms and synchronizing sleep and awake cycles. Exposure to daylight and night, or cycled artificial light, improves sleep and weight gain in preterm infants and appears to reduce agitation and improve sleep in persons with Alzheimer's disease. For day-shift healthcare staff, morning daylight helps regulate circadian rhythms, and thereby may foster daytime alertness, better cognitive performance, and improved nighttime sleep. Higher daylight exposure levels in healthcare buildings may lessen work stress and increase job satisfaction among nurses.

Several well-controlled studies of patients have produced strong evidence that exposure to light—bright artificial light or daylight—is effective in reducing depression and improving mood, even in persons suffering severe depression. A few investigations suggest that architectural design and siting decisions for healthcare buildings, by influencing the amount of daylight exposure patients receive, can impact depression levels and outcomes such as length of stay and pain. Concerning the last, pain, a strong study found that surgical patients assigned to bright rooms, compared to those with rooms in shade, reported less pain and required fewer analgesic medications.

The priority and resources accorded to biophilic healthcare design will be heavily influenced by the extent to which rigorous research demonstrates that biophilic measures improve outcomes and are cost-effective. Although the amount of biophilia/health research is steadily increasing, and several sound studies already are available on issues such as reduction of stress, pain and depression, there is a clear need for additional research to address gaps in knowledge. Rigorous prospective investigations are needed to deepen understanding of such topics as the effectiveness of daylight exposure and real nature views in alleviating pain across diverse categories of patients, the impacts of physically active and passive garden experiences on outcomes, and the extent to which real nature environments may outperform simulations in fostering gains in clinical outcomes. A priority need is for research to develop the business or financial case for biophilic healthcare design. Optimism seems warranted for pursuing this key direction, as some credible research already implies that by designing hospitals to provide nature views and daylight exposure, substantial cost savings can be achieved because, for instance, intake of costly pain drugs is reduced, and stays are shortened for some categories of patients (Berry et al. 2004).

Implications for Evidence-Based Biophilic Design of Healthcare Buildings

Despite the research needs just noted, there is now enough sound evidence available to support the following biophilic design recommendations for healthcare environments:

- Architectural siting and design should provide restorative window views of nature and gardens from patient rooms, waiting areas, staff work spaces, and other interior areas where stress is a problem. Patient rooms and windows should be designed to make it possible for bedridden persons to view outdoor nature. Affording nature window views in treatment and procedure spaces where stress and pain are problems warrants high priority.
- Provide nature views with characteristics identified by research as effective in alleviating stress and improving outcomes, including green foliage, flowers, water, savanna-like or parklike characteristics (trees with grassy understory, some visual depth), unthreatening wildlife such as birds, and sunshine or good light in contrast to dim light or shadow. Avoid window views of outdoor spaces with the following properties, which can hinder restoration or even aggravate stress in some patients: spaces dominated by hardscape or starkly built content (such as concrete); roof tops and parking lots lacking vegetation; walls of other buildings; and abstract or ambiguous sculpture that can be interpreted in multiple ways by stressed patients (Ulrich 1999).
- The evidence linking higher daylight or sun exposure to reduced patient depression, pain, and other improved outcomes underscores the importance of giving careful consideration to healthcare building orientation and site planning in new healthcare projects (Ulrich et al. 2006). Avoid site plans where some buildings block light from others. Hospitals and mental health facilities should be sited and designed to ensure that depressed patients have abundant natural light.
- Avoid deep plan building layouts and floor plans—with a large proportion of windowless rooms—as these may tend to worsen patient and staff outcomes. Also, hospitals should not be designed with patient windows looking out into an enclosed and roofed atrium with few skylights and little natural light, as this architectural approach virtually eliminates natural light exposure in patient rooms. Atrium-facing patient rooms can have the additional drawback of requiring that patient windows be heavily tinted to prevent persons in the atrium from looking into rooms and violating privacy—a possible infringement of federal patient privacy regulations.
- Larger windows should be provided to permit more exposure to daylight and restorative nature views in patient rooms and other spaces where depression, pain, and stress are problems. Avoid designs, however, that create sun glare patches. Biophilic considerations favor patient rooms designed with the bathroom located on the hallway or headwall sides, rather than the window or outboard wall, to facilitate larger exterior windows, greater daylight exposure, and better visual access to nature or gardens for bedridden patients.
- Provide well-designed outdoor gardens for patients, family, and staff. Evidence-based design characteristics for successful healthcare gardens include prominent real nature content (such as verdant vegetation, water); convenient way-finding to the garden; accessibility; movable seating that facilitates social interaction; access to privacy; congruent nature sounds (birds, water, breezes) rather than intrusive urban or machine sounds (traffic, air-conditioning compressors); and opportunities for physical activity, movement, or exercise (Marcus and Barnes 1995; Ulrich 1999). Gardens should provide users with easy access to shade, as some patients' medical conditions or treatments make them negatively sensitive to direct sun exposure. Garden spaces intended for adult family members and staff should support restorative sedentary activities such as viewing nature and socializing. In the case of gardens for children, it is important to include active play features and spaces in addition to calm refuges for adults. In large healthcare facilities, provide a number of decentralized gardens located conve-

niently close to patient care units, waiting areas, and staff work spaces, to increase garden usage and benefits.

- It is recommended that visual art (paintings, prints, and photographs) displayed in patient rooms and other healthcare spaces where stress and pain are problems give priority to representational nature subject matter with unambiguously positive content. Designers should consult the evidence-based guidelines for selecting healthcare art used by several hospitals and university medical centers (Ulrich 1991; Ulrich and Gilpin 2003). The following are examples of subject matter categories recommended by these guidelines: waterscapes with calm or nonturbulent water; landscapes with visual depth or openness in the immediate foreground; nature settings depicted during warmer seasons when vegetation is verdant and flowers are visible; scenes with positive cultural artifacts such as barns and older houses in nature surroundings; garden scenes; people at leisure in places with prominent nature; and outdoor scenes in sunny conditions, not overcast or foreboding weather (Ulrich and Gilpin 2003, 134–136). Designers and healthcare administrators should avoid abstract, emotionally negative, or surreal artwork, as it can aggravate stress in some patients.
- Consider providing technology to enable patients to experience simulated nature (television screens, eyeglass displays, virtual reality) in highly stressful medical settings where it is not feasible to provide visual contact with real nature, including shielded rooms for radiation therapy, imaging, or procedures such as cardiac catheterization. Nature simulations that involve both visual stimulation and sound may tend to be more engrossing and hence more effective for alleviating severe pain.

REFERENCES

Alimoglu, M. K., and L. Donmez. 2005. "Daylight Exposure and Other Predictors of Burnout Among Nurses in a University Hospital." *International Journal of Nursing Studies* 42(6): 549–555.

Appleton, J. 1975. *The Experience of Landscape*. London: Wiley.

Barss, P., and K. Comfort. 1985. "Ward Design and Neonatal Jaundice in the Tropics: Report of an Epidemic." *British Medical Journal* 291:400–401.

Beauchemin, K. M., and P. Hays. 1996. "Sunny Hospital Rooms Expedite Recovery from Severe and Refractory Depressions." *Journal of Affective Disorders* 40(1–2): 49–51.

Beauchemin, K. M., and P. Hays. 1998. "Dying in the Dark: Sunshine, Gender and Outcomes in Myocardial Infarction." *Journal of the Royal Society of Medicine* 91:352–354.

Benedetti, F., C. Colombo, B. Barbini, E. Campori, and E. Smeraldi. 2001. "Morning Sunlight Reduces Length of Hospitalization in Bipolar Depression." *Journal of Affective Disorders* 62(3): 221–223.

Berry, L. L., D. Parker, R. C. Coile, D. K. Hamilton, D. D. O'Neill, and B. L. Sadler. 2004. "The Business Case for Better Buildings." *Frontiers of Health Services Management* 21(1): 3–24.

Blackburn, S., and D. Patteson. 1991. "Effects of Cycled Light on Activity State and Cardiorespiratory Function in Preterm Infants." *Journal of Perinatology and Neonatal Nursing* 4(4): 47–54.

Boyce, P., C. Hunter, and O. Howlett. 2003. *The Benefits of Daylight through Windows*. Troy, NY: Rensselaer Polytechnic Institute.

Brewer, B. W., and P. Karoly. 1989. "Effects of Attentional Focusing on Pain Perception." *Motivation and Emotion* 13(3): 193–203.

Carpman, J. R., and M. A. Grant. 1993. *Design That Cares: Planning Health Facilities for Patients and Visitors*. 2nd ed. Chicago: American Hospital Publishing.

Coss, R. G. 2003. "The Role of Evolved Perceptual Biases in Art and Design." In *Evolutionary Aesthetics*, edited by E. Voland and K. Grammer, 69–130. New York: Springer.

Diette, G. B., N. Lechtzin, E. Haponik, A. Devrotes, and H. R. Rubin. 2003. "Distraction Therapy with Nature Sights and Sounds Reduces Pain During Flexible Bronchoscopy: A Complementary Approach to Routine Analgesia." *Chest* 123(3): 941–948.

Eisen, S. 2006. "Effects of Art in Pediatric Healthcare." Ph.D. diss., Texas A&M University.

Fredrickson, B. L., and R. W. Levenson. 1998. "Positive Emotions Speed Recovery from the Cardiovascular Sequelae of Negative Emotions." *Cognition and Emotion* 12(2): 191–220.

Giunta, F., and J. Rath. 1969. "Effect of Environmental Illumination in Prevention of Hyperbilirubiemia of Prematurity." *Pediatrics* 44(2): 162–167.

Golden, R. N., B. N. Gaynes, R. D. Ekstrom, R. M. Hamer, F. M. Jacobsen, T. Suppes, K. L. Wisner, and C. B. Nemeroff. 2005. "The Efficacy of Light Therapy in the

Treatment of Mood Disorders: A Review and Meta-Analysis of the Evidence." *American Journal of Psychiatry* 162(4): 656–662.

Hartig, T., A. Book, J. Garvill, T. Olsson, and T. Gärling. 1995. "Environmental Influences on Psychological Restoration." *Scandinavian Journal of Psychology* 37:378–393.

Hartig, T., G. W. Evans, L. D. Jamner, D. S. Davis, and T. Gärling. 2003. "Tracking Restoration in Natural and Urban Field Settings." *Journal of Environmental Psychology* 23:109–123.

Hartig, T., and C. C. Marcus. 2006. "Healing Gardens: Places for Nature in Health Care." *Lancet* 368:S36–S37.

Heerwagen, J. H. 1990. "The Psychological Aspects of Windows and Window Design." In *Proceedings of 21st Annual Conference of the Environmental Design Research Association,* edited by K. H. Anthony, J. Choi, and B. Orland, 269–280. Oklahoma City: Environmental Design Research Association.

Heerwagen, J., and G. H. Orians. 1993. "Humans, Habitats, and Aesthetics." In *The Biophilia Hypothesis,* edited by S. Kellert and E. O. Wilson, 138–172. Washington, DC: Shearwater/Island Press.

Holick, M. F. 2005. "The Vitamin D Deficiency Epidemic and Its Health Consequences." *Journal of Nutrition* 135(11): 2739–2748.

Jones, H. 2007. *FMI's Construction Outlook: First Quarter 2007.* Raleigh, NC: FMI Corporation.

Joseph, A. 2006. *The Impact of Light on Outcomes in Healthcare Settings.* Issue Paper No. 2. Concord, CA: The Center for Health Design.

Kaplan, R. 1993. "The Role of Nature in the Context of the Workplace." *Landscape and Urban Planning* 26:193–201.

Kaplan, R., and S. Kaplan. 1989. *The Experience of Nature.* New York: Cambridge University Press.

Katcher, A., H. Segal, and A. Beck. 1984. "Comparison of Contemplation and Hypnosis for the Reduction of Anxiety and Discomfort During Dental Surgery." *American Journal of Clinical Hypnosis* 27:14–21.

Kearney, A. R., and D. Winterbottom. 2005. "Nearby Nature and Long-Term Care Facility Residents: Benefits and Design Recommendations." *Journal of Housing for the Elderly* 19(3/4): 7–28.

Kettlewell, N. 1988. "An Examination of Preferences for Subject Matter in Art." *Empirical Studies of the Arts* 6:59–65.

Kiecolt-Glaser, J. K., G. C. Page, P. T. Marucha, R. C. MacCallum, and R. Glaser. 1998. "Psychological Perspectives on Surgical Recovery: Perspectives from Psychoneuroimmunology." *American Psychologist* 53:1209–1218.

Kiecolt-Glaser, J. K., P. T. Marucha, W. B. Malarkey, A. M. Mercado, and R. Glaser. 1995. "Slowing of Wound Healing by Psychological Stress." *Lancet* 346:1194–1196.

Kozarek, R. A., S. L. Raltz, L. Neal, P. Wilber, S. Stewart, and J. Ragsdale. 1997. "Prospective Trial Using Virtual Vision(r) as Distraction Technique in Patients Undergoing Gastric Laboratory Procedures." *Gastroenterology Nursing* 20(1): 12–18.

Laumann, K., T. Gärling, and K. M. Stormark. 2003. "Selective Attention and Heart Rate Responses to Natural and Urban Environments." *Journal of Environmental Psychology* 23:125–134.

Leather, P., M. Pyrgas, D. Beale, and C. Lawrence. 1998. "Windows in the Workplace: Sunlight, View, and Occupational Stress." *Environment and Behavior* 30(6): 739–762.

Lee, D. W. H., A. C. W. Chan, S. K. H. Wong, T. M. K. Fung, A. C. N. Li, S. K. C. Chan, L. M. Mui, E. K. W. Ng, and S. C. S. Chung. 2004. "Can Visual Distraction Decrease the Dose of Patient-Controlled Sedation Required During Colonoscopy? A Prospective Randomized Controlled Trial." *Endoscopy* 36(3): 197–201.

Lovell, B. B., S. Ancoli-Isreal, and R. Gervirtz. 1995. "Effect of Bright Light Treatment on Agitated Behavior in Institutionalized Elderly Subjects." *Psychiatry Research* 57(1): 7–12.

Mann, N., R. Haddow, L. Stokes, S. Goodley, and N. Rutter. 1986. "Effect of Night and Day on Preterm Infants in a Newborn Nursery." *British Medical Journal* 293(6557): 1265–1267.

Marcus, C. C., and M. Barnes. 1995. *Gardens in Healthcare Facilities: Uses, Therapeutic Benefits, and Design Recommendations.* Concord, CA: The Center for Health Design.

Marcus, C. C., and M. Barnes M., eds. 1999. *Healing Gardens: Therapeutic Benefits and Design Recommendations.* New York: Wiley.

Martiny, K. 2004. "Adjunctive Bright Light in Non-Seasonal Major Depression." *Acta Psychiatry Scandinavia* 110 (Supplement 425): 7–28.

McCaul, K. D., and J. M. Malott. 1984. Distraction and Coping with Pain. *Psychological Bulletin* 95(3): 516–533.

Melzack, R., and P. D. Wall. 1965. "Pain Mechanisms: A New Theory." *Science* 150:971–979.

Melzack, R., and P. D. Wall. 1982. *The Challenge of Pain.* New York: Basic Books.

Miller, A. C., L. C. Hickman, and G. K. Lemasters. 1992. "A Distraction Technique for Control of Burn Pain." *Journal of Burn Care and Rehabilitation* 13(5): 576–580.

Miller, C. L., R. White, T. L. Whitman, M. F. O'Callaghan, and S. E. Maxwell. 1995. "The Effects of Cycled Versus Non-Cycled Lighting on Growth and Development in Preterm Infants." *Infant Behavior and Development* 18(1): 87–95.

Öhman, A. 1986. "Face the Beast and Fear the Face: Animal and Social Fears as Prototypes for Evolutionary Analyses of Emotion." *Psychophysiology* 23:123–145.

Orians, G. H. 1980. "Habitat Selection: General Theory and Applications to Human Behavior." In *The Evolution of Human Social Behavior*, edited by J. S. Lockard, 49–66. New York: Elsevier.

Orians, G. H. 1986. "An Ecological and Evolutionary Approach to Landscape Aesthetics." In *Meanings and Values in Landscape*, edited by E. C. Penning-Rowsell and D. Lowenthal, 3–25. London: Allen and Unwin.

Parsons, R. 1991. "The Potential Influences of Environmental Perception on Human Health." *Journal of Environmental Psychology* 11:1–23.

Parsons, R., and T. Hartig. 2000. "Environmental Psychophysiology." In *Handbook of Psychophysiology*, 2nd ed., edited by J. T. Cacioppo, L. G. Tassinary, and G. Berntson, 815–846. New York: Cambridge University Press.

Parsons, R., L. G. Tassinary, R. S. Ulrich, M. R. Hebl, and M. Grossman-Alexander. 1998. "The View from the Road: Implications for Stress Recovery and Immunization." *Journal of Environmental Psychology* 18:113–140.

Rea, M. 2004. "Lighting for Caregivers in the Neonatal Intensive Care Unit." *Clinical Perinatology* 31:229–242.

Rodiek, S. 2005. "Resident Perceptions of Physical Environment Features That Influence Outdoor Usage at Assisted Living Facilities." *Journal of Housing for the Elderly* 19(3/4): 95–107.

Schneider, S. M., M. Prince-Paul, M. J. Allen, P. Silverman, and D. Talaba. 2004. "Virtual Reality as a Distraction Intervention for Women Receiving Chemotherapy." *Oncology Nursing Forum* 31(1): 81–88.

Sherman, S. A., J. W. Varni, R. S. Ulrich, and V. L. Malcarne. 2005. "Post-Occupancy Evaluation of Healing Gardens in a Pediatric Cancer Center." *Landscape and Urban Planning* 73:167–183.

Sloane, P. D., C. M. Mitchell, J. Preisser, C. Phillips, C. Commander, and E. Burker. 1998. "Environmental Correlates of Resident Agitation in Alzheimer's Disease Special Care Units." *Journal of the American Geriatrics Society* 46:862–869.

Tse, M. M. Y., J. K. F. Ng, J. W. Y. Chung, and T. K. S. Wong. 2002. "The Effect of Visual Stimuli on Pain Threshold and Tolerance." *Journal of Clinical Nursing* 11(4): 462–469.

Tuan, Y. R. 1974. *Topophilia: A Study of Environmental Perception, Attitudes, and Values.* Englewood Cliffs, NJ: Prentice Hall.

Ulrich, R. S. 1979. "Visual Landscapes and Psychological Well-Being." *Landscape Research* 4(1): 17–23.

Ulrich, R. S. 1983. "Aesthetic and Affective Response to Natural Environment." In *Human Behavior and Environment, Vol. 6: Behavior and the Natural Environment*, edited by I. Altman and J. F. Wohlwill, 85–125. New York: Plenum.

Ulrich, R. S. 1984. "View Through a Window May Influence Recovery from Surgery." *Science* 224:420–421.

Ulrich, R. S. 1991. "Effects of Health Facility Interior Design on Wellness: Theory and Recent Scientific Research." *Journal of Health Care Design* 3:97–109.

Ulrich, R. S. 1993. "Biophilia, Biophobia, and Natural Landscapes." In *The Biophilia Hypothesis*, edited by S. Kellert and E. O. Wilson, 74–137. Washington, DC: Shearwater/Island Press.

Ulrich, R. S. 1999. "Effects of Gardens on Health Outcomes: Theory and Research." In *Healing Gardens*, edited by C.C. Marcus and M. Barnes, 27–86. New York: John Wiley.

Ulrich, R. S. 2002. "Communicating with the Healthcare Community About Plant and Garden Benefits." In *Interaction by Design: Bringing People and Plants Together for Health and Well-Being*, edited by C. Shoemaker. Ames, Iowa: Iowa State University Press.

Ulrich, R. S. 2006. "Evidence-Based Healthcare Design." In *The Architecture of Hospitals*, edited by C. Wagenaar, 281–289, 345–346. Belgium: NAI Publishers.

Ulrich, R. S., and L. Gilpin. 2003. "Healing Arts." In *Putting Patients First: Designing and Practicing Patient-Centered Care*, edited by S. B. Frampton, L. Gilpin, and P. Charmel, 117–146. San Francisco: Jossey-Bass.

Ulrich, R. S., O. Lundén, and J. L. Eltinge. 1993. "Effects of Exposure to Nature and Abstract Pictures on Patients Recovering from Heart Surgery." Paper presented at the Thirty-Third Meeting of the Society for Psychophysiological Research. *Psychophysiology* 30 (Supplement 1): 7.

Ulrich, R. S., R. F. Simons, B. D. Losito, E. Fiorito, M. A. Miles, and M. Zelson. 1991. "Stress Recovery During Exposure to Natural and Urban Environments." *Journal of Environmental Psychology* 11:201–230.

Ulrich, R. S., R. F. Simons, and M. A. Miles. 2003. "Effects of Environmental Simulations and Television on Blood Donor Stress." *Journal of Architectural & Planning Research* 20(1): 38–47.

Ulrich, R. S., C. Zimring, X. Quan, and A. Joseph. 2006. "The Environment's Impact on Stress." In *Improving Healthcare with Better Building Design*, edited by S. Marberry, 37–61. Chicago: Health Administration Press.

Van den Berg, A., S. L. Koole, and N. Y. Van der Wulp. 2003. "Environmental Preference and Restoration: How Are They Related?" *Journal of Environmental Psychology* 23:135–146.

Van Someren, E. J. W., A. Kessler, M. Mirmiran, and D. F. Swaab. 1997. "Indirect Bright Light Improves Circadian Rest-Activity Rhythm Disturbances in Demented Patients." *Biological Psychiatry* 4(19): 955–963.

Verderber, S. 1986. "Dimensions of Person-Window Transactions in the Hospital Environment." *Environment & Behavior* 18(4): 450–466.

Walch, J. M., B. S. Rabin, R. Day, J. N. Williams, K. Choi, and J. D. Kang. 2005. "The Effect of Sunlight on Post-Operative Analgesic Medication Usage: A Prospective Study of Patients Undergoing Spinal Surgery." *Psychosomatic Medicine* 67:156–163.

Whall, A. L., M. E. Black, C. J. Groh, D. J. Yankou, B. J. Kupferschmid, and N. L. Foster. 1997. "The Effect of Natural Environments upon Agitation and Aggression in Late Stage Dementia Patients." *American Journal of Alzheimer's Disease and Other Dementias*, September–October, 216–220.

Whitehouse, S., J. W. Varni, M. Seid, C. Cooper-Marcus, M. J. Ensberg, J. R. Jacobs, et al. 2001. "Evaluating a Children's Hospital Garden Environment: Utilization and Consumer Satisfaction." *Journal of Environmental Psychology* 21(3): 301–314.

Wichrowski, M., J. Whiteson, F. Haas, A. Mola, and M. J. Rey. 2005. "Effects of Horticultural Therapy on Mood and Heart Rate in Patients Participating in an Inpatient Cardiopulmonary Rehabilitation Program." *Journal of Cardiopulmonary Rehabilitation* 25:270–274.

Wied, M. D., and M. N. Verbaten. 2001. "Affective Picture Processing, Attention, and Pain Tolerance." *Pain* 90:163–172.

Wilson, E. O. 1984. *Biophilia*. Cambridge, MA: Harvard University Press.

Winston, A. S., and G. C. Cupchik. 1992. "The Evaluation of High Art and Popular Art by Naive and Experienced Viewers." *Visual Arts Research* 18:1–14.

Wismeijer, A. J., and J. J. M. Vingerhoets. 2005. "The Use of Virtual Reality and Audiovisual Eyeglass Systems as Adjunct Analgesic Techniques: A Review of the Literature." *Annals of Behavioral Medicine* 30(3): 268–278.

Wypijewski, J., ed. 1997. *Painting by the Numbers: Komar and Melamid's Scientific Guide to Art*. New York: Farrar, Straus, & Giroux.

chapter 7

Nature Contact and Human Health: Building the Evidence Base

Howard Frumkin

A wise man proportions his belief to the evidence.

—David Hume, *An Enquiry Concerning Human Understanding*

People like contact with nature. And because people like contact with nature, they may tend to believe that contact with nature is salubrious.

This ought to concern us. Demosthenes, in the fourth century BC, said, "Nothing is so easy as to deceive one's self; for what we wish, that we readily believe." We like nature, we want to believe that nature is good for our health, and presto! We believe it. But if asked to prove the link, most people would have difficulty doing so.

Meanwhile, back at the regulatory agencies, pharmaceutical companies, hospitals, and clinics, the importance of evidence-based thinking has grown steadily during the last few decades—recalling David Hume's adage "A wise man proportions his belief to the evidence." Most of us would not want to take a medication, or submit to a surgical procedure, if it didn't have strong evidence of efficacy and safety. We have come to expect that the doctors who recommend such treatments, and the regulatory agencies that permit their use, base their decisions on strong evidence. And those who pay for health care—employers and insurance companies in the United States, national health care systems in most other countries, and patients themselves in some places—don't want to pay without evidence of value. (An important and fascinating exception is alternative medicines, which many otherwise judicious people are willing to purchase without any evidence that they work or even that they are safe.)

In this chapter I hope to bring the perspectives of evidence-based medicine and public health to the theme of biophilic design. At the center of this discussion is a simple pair of questions: How do we know what we know about health benefits of nature? And how can we improve our knowledge? I begin by introducing the paradigm of evidence-based medicine, focusing on a discipline called clinical epidemiology. Next, I assess the available evidence for the health benefits of nature contact in terms of prevailing standards of clinical epidemiology. Finally, and without further courtship, I propose marriage—making the case that clinical epidemiology can and should be applied to nature contact, to investigate systematically the human health benefits of this contact—and I suggest some of the ways this might occur.

THE PRACTICE OF CLINICAL EPIDEMIOLOGY

> If the thing believed is incredible, it is also incredible that the incredible should have been so believed.
>
> —St. Augustine, *City of God*

In July 2002, millions of women were faced with ("clobbered with" might be more accurate) troubling evidence about a medication they were taking. Thirty-eight percent of U.S. postmenopausal women were taking estrogen replacement (Keating et al. 1999), believing that it would relieve their menopausal symptoms, prevent osteoporosis, prevent heart disease, and/or benefit them in other ways. Several decades of clinical experience and observational evidence had suggested these benefits. But the results of a careful study told a different story (Writing Group 2002). Over 16,000 women entered the study between 1993 and 1998, and were randomly assigned to receive either an estrogen-progestin combination or a placebo. Neither the women nor their physicians knew which pill they were taking, a procedure called double-blinding that reduces the probability of biased observations. After an average of 5.2 years of follow-up, a surprising result emerged. The women on estrogen replacement therapy had an increased risk of coronary heart disease, an increased risk of breast cancer, an increased risk of stroke, and an increased risk of pulmonary embolism. On a more cheerful note, they had a decreased risk of colorectal cancer, endometrial cancer, and hip fracture. The data suggested that for every 10,000 women on estrogen replacement therapy, each year there would be seven more heart attacks, eight more strokes, eight more pulmonary embolisms, and eight more cases of breast cancer, along with six fewer colorectal cancers and five fewer hip fractures, than if they were not on this treatment.

Newspaper headlines across the world announced the finding, physicians braced themselves for questions from concerned patients, and women themselves struggled with decisions about whether to discontinue their medication. Many did. But an interesting feature of the discussion was what went unsaid: nobody seriously challenged the results. The rigorous design of the study (called a double-blind randomized controlled trial), its large size, the careful definition of the health outcomes being studied, the control of potential confounders such as smoking, body mass index, and diabetes, and the inclusion of various alternative data analysis strategies, made the results incontrovertible. Women taking replacement estrogens were not left with an easy decision, but at least they could start with sound evidence.

This phenomenon, the reversal of commonly accepted medical beliefs by evidence from a randomized controlled trial, had plenty of precedent (Sackett et al. 1991, 193). In the early 1960s, a former president of the American College of Surgeons reasoned that cooling the stomach lining could decrease acid secretion and therefore cure ulcers. He developed a new technique, placing balloons in the stomachs of ulcer patients and filling them with liquid cooled to –10°C. Patients seemed to improve (Wagensteen et al. 1962), and "gastric freezing" took off. Within a few years over 2,500 gastric freezing machines were sold and an estimated 15,000 patients were treated. Then came a randomized controlled trial (Ruffin et al. 1969). Sixty-nine patients underwent gastric freeze and 68 had a sham procedure. The proportions that went on to subsequent surgery, hospitalization for intractable pain, or bleeding from their ulcers, were 51 percent in the treated patients, and 44 percent in the placebo patients. Gastric freezing did not work.

In the early 1950s, an innovative operation was introduced for the treatment of angina, the pain caused by ischemic heart disease. The idea was that blood would be diverted to the heart if a nearby artery, the internal mammary artery, were closed. Internal mammary artery ligation caught on rapidly. In 1957, *Reader's Di-*

gest published an enthusiastic accolade, "New Surgery for Ailing Hearts." Thousands of patients had their internal mammary arteries ligated, and in some reports up to 90 percent of patients reported symptom relief. But by the end of the decade, two randomized controlled trials gave a clearer picture. In one (Dimond et al. 1960), 18 patients were randomized: 13 to the surgery and 5 to a sham operation. There was improvement in 10 of the 13 who underwent internal mammary ligation and in all 5 of those who underwent sham surgery! In another, 8 patients were randomized to surgery, and there was a 34 percent improvement in their symptoms. But 9 patients were randomized to a sham procedure, and they had a 42 percent improvement in their symptoms (Cobb et al. 1959). Internal mammary artery ligation did not work.

Randomized controlled trials do not just debunk mistaken practices. They can also establish the value of an intervention, sometimes in equally surprising ways. Two examples are physical activity and aspirin. In the 1980s, investigators at Northwestern University randomized 102 people to a physical activity program and 99 to a control group. The people in the physical activity group had an 8.8 percent incidence of high blood pressure, while those in the control group had a 19.2 percent incidence (Stamler et al. 1989). This important result helped establish the value of physical activity for hypertension prevention. During that same decade, the Physicians' Health Study gave either daily aspirin or placebo to over 20,000 physicians, followed them for an average of just over five years, and found a 44 percent reduction in the risk of myocardial infarction in those who took aspirin (Steering Committee, 1989). There was also a small increase in the risk of stroke and of ulcer disease. This result, and others like it, helped establish the value of aspirin for heart disease prevention.

A leading group of clinical epidemiologists at McMaster University has written that we can support our health beliefs with one of three kinds of evidence (Sackett et al. 1991, 191). One is *induction*, concluding from unsystematic observations, or from general principles, that something *ought to work*. Another is *deduction*, concluding that something works when it successfully withstands formal attempts to demonstrate its worthlessness. The third might be called *seduction*—concluding based on faith, or the assurances of other people, that something works.

For too long, medical practice has relied on some combination of inductive reasoning and seductive beliefs. But in recent years, more and more treatments have been subjected to rigorous deductive study. The goal here is evidence-based medicine, defined as "the conscientious, explicit and judicious use of current best evidence in making decisions about the care of individual patients or the delivery of health services" (Sackett et al. 1996). The discipline that has propelled this trend is called clinical epidemiology, and the method that epitomizes it is the randomized controlled trial, the method described in the foregoing examples.

Clinical epidemiology is a subfield of the larger field of epidemiology. Epidemiology is the study of the distribution and determinants of disease in human populations. It was epidemiologic research that showed us that smoking causes cancer, that lead causes neurological impairment, that high cholesterol is a risk factor for heart disease, that cholesterol-lowering agents reduce the risk of heart disease, that regular colonoscopy reduces mortality from colon cancer. Epidemiology is a standard, albeit small, part of the medical and nursing curriculum, but it is taught principally in schools of public health, generally at the graduate level. Epidemiology is well established both as a profession and as a research discipline. It has its own doctoral degrees, professional societies, meetings, and journals, with names like *Epidemiology*, *Annals of Epidemiology*, *American Journal of Epidemiology*. But it is not a ghettoized specialty; epidemiologic articles appear regularly in major medical journals such as *Lancet*, the *Journal of the American Medical Association*, and the *New England Journal of Medicine*.

Clinical epidemiology focuses on clinical issues including diagnostic tests, medical treatments, and outcomes. It has its own journals, such as the *Journal of Clinical Epidemiology*, and clinical epidemiology articles appear regularly in other leading medical and public health journals. It has a selection of excellent textbooks (Sackett et al. 1991; Fletcher and Fletcher 2005; Hulley et al. 2001). And numerous academic research units in leading medical centers around the world now specialize in clinical epidemiology.

The randomized controlled trial, a principal tool of

clinical epidemiologists, is conceptually nothing more than a true experiment, familiar to anyone trained in any of the sciences. Start with a carefully defined group of patients. Randomly allocate them into two or more different groups, so that potentially important factors—behavioral, genetic, psychological, and others—are likely to be evenly distributed. Give one group treatment A and the other treatment B (or a placebo). Do not reveal to the patients, or to their caregivers, which treatment they are receiving—a procedure known as blinding—so they cannot differentially influence the outcomes by their knowledge or preconceptions. Systematically observe the outcomes of the treatments. Draw conclusions, and generalize those conclusions to patients similar to those who were studied.

In assessing the results of a randomized controlled trial, several important questions should be asked, questions in which every epidemiology student is drilled. These are shown in Table 7-1.

Indeed this form of critical thinking, and the loyalty to deductive evidence that it reflects, are increasingly becoming the dominant paradigm of health research. Some might view it as reductionist and rigid. Some point out, correctly in my view, that systematic empirical evidence is not the only way of knowing the world; qualitative research, common sense, and inspired epiphanies all have their place. But based on the track record of clinical epidemiology in correcting errors and establishing safe and effective treatments, it represents an enormous and welcome advance. How might we apply this approach to our understanding of nature contact?

TABLE 7-1 Evaluating a randomized clinical trial

- What was the patient population?
- What was the treatment and how was it defined and measured?
- What was the health outcome and how was it defined and measured?
- What was the statistical power of the study to detect an effect?
- What were the potential sources of selection bias and how were they managed?
- What were the potential sources of information bias and how were they managed?
- What were the potential confounders and how were they managed?
- What was the result? Was it valid? Statistically significant? Clinically significant?
- To whom may the result be generalized?

HEALTH BENEFITS OF NATURE CONTACT: THE EVIDENCE

Should we think of nature contact as a health intervention? If so, has it been studied adequately? Do we have enough evidence to offer recommendations to patients, to architects and designers, and/or to the public at large?

We do have a few randomized controlled trials. Not surprisingly, these tend to be small, and to look at short-term outcomes. For example, investigators at Johns Hopkins University were interested in controlling the pain associated with bronchoscopy, a diagnostic procedure in which a fiber-optic tube is inserted down into the lungs (Diette et al., 2003). They randomly assigned 80 bronchoscopy patients to one of two groups. Forty of the patients viewed a pristine nature scene before the procedure and listened to the sounds of a bubbling brook during the procedure. The other 40 patients had no such intervention. There was a 50 percent increase in the level of self-reported "very good" or "excellent" pain control among the intervention patients compared to the control patients.

But it is difficult to assign people randomly to a natural setting or one devoid of nature, especially with respect to long-duration exposures. Instead, some astute investigators have recognized situations in which something close to random assignment was performed for them, and have taken advantage of these situations—an approach that, if it weren't for the double entendre, we would call a natural experiment. Frances Kuo, William Sullivan, and their colleagues at the University of Illinois have brilliantly demonstrated this approach in their studies at Chicago's Robert Taylor Homes. This public housing complex consisted of 28 identical high-rise buildings arrayed along a three-mile stretch of land, bounded by busy roadways and railway lines. Some of the buildings were surrounded by pleasant stands of trees, whereas others opened onto barren stretches of ground. Residents were assigned essentially at random to a building with one landscape type or the other, because assignment depended on where a vacancy existed when their names came up on the Housing Authority list. The research compared residents of the buildings with and without trees, and was limited to those who lived on the lower floors (to ensure that participants in buildings surrounded by trees did have tree views from their windows). The results were striking; nearby trees are associated with higher levels of attention and self-discipline, less violence and other aggressive behavior, lower crime rates, and better interpersonal relations (Kuo 2001; Kuo and Sullivan 2001a, 2001b; Taylor et al. 2002).

In 1981, Ernest Moore, a University of Michigan architect, took advantage of a natural experiment at the State Prison of Southern Michigan, a massive Depression-era structure. Half the prisoners occupied cells along the outside wall, with a window view of rolling farmland and trees, while the other half occupied cells that faced the prison courtyard. The prisoners in the interior cells had a 24 percent higher frequency of sick call visits, compared to those in exterior cells. Moore could not identify any design feature to explain this difference, and concluded that the outside view "may provide some stress reduction" (Moore 1981).

Another classic example came from a healthcare setting. On the surgical floors of a 200-bed suburban Pennsylvania hospital, some rooms faced a stand of deciduous trees, while others faced a brown brick wall. Postoperative patients were assigned to one or the other kind of room, apparently based on nothing other than room availability. The investigator, Roger Ulrich, reviewed records of all cholecystectomy patients over a 10-year interval, restricted to the summer months when the trees were in foliage. Endpoints were the length of hospitalization, the need for pain and anxiety medications, the occurrence of minor medical complications, and nurses' notes. Patients with tree views had statistically significantly shorter hospitalizations (7.96 days compared to 8.70 days), less need for pain medications, and fewer negative comments in the nurses' notes, compared to patients with brick views (Ulrich 1984).

A third study design also retains some of the advantages of a randomized trial, but instead of comparing two groups of people, exposed and unexposed, it compares people to themselves, before and after an intervention. Good examples come from the study of "green exercise"—exercise that takes place in natural settings, combining (in theory) the benefits of physical activity with the benefits of nature contact. Terry Hartig has given us a nice example of this approach (Bodin and Hartig 2003). In this study, 12 runners went on two runs each, one through a nature reserve featuring pine-birch forest and open fields, the other along sidewalks and streets in an area of mid-rise apartment houses and commercial development. The runners rated their emotions, in the categories of revitalization, tranquility, anxiety/depression, and anger, after each of the runs, permitting a comparison of the two environments. There were small advantages in favor of the natural setting.

Observational cohort studies that follow a population over time, either prospectively or retrospectively, are a fourth study design, and one that is commonly used in environmental epidemiology. Suppose we hypothesize that asbestos exposure increases the risk of lung cancer. We identify a cohort of asbestos-exposed workers and follow them forward in time. We might characterize each worker's exposure, and compare the highly exposed workers with the less exposed workers, to assess whether more exposure is associated with more risk. We might also compare the worker cohort to an unexposed comparison group, such as the general population, to assess whether the exposed are at higher risk than the unexposed. One major challenge in such studies is potential confounding. If the exposed and unexposed populations differ with respect to some extraneous factor, such as cigarette smoking, then it might be that factor, and not the exposure, that accounts for observed differences in the health outcome. Epidemiologists typically try to manage confounding by collecting information on potential confounders, if available, and carrying out multivariate statistical analyses that examine each potential predictor independent of the others.

A final study design to mention, also a form of observational study, is the case-control design. Case-control studies are the logical reverse of cohort studies. Instead of comparing exposed and unexposed people with respect to a health outcome of interest, we compare ill and well people with respect to antecedent exposures. For example, suppose we hypothesize that drinking coffee increases the risk of pancreatic cancer. We identify a group of patients with this disease and a control group free of this disease, and query both groups about their coffee-drinking histories. If the pancreatic cancer patients report more coffee drinking than the controls, then other things being equal, this would support the hypothesis of an association between coffee drinking and pancreatic cancer.

None of these designs is perfect, and studying nature contact poses some challenges not present in studies of medications. For example, in the experimental or quasi-experimental designs, it is impossible to blind the subjects to their exposures, so there is potential for information bias—differential reporting of outcomes based on the subject's preferences or preconceptions. It is very difficult to blind the investigators to the exposure status of subjects, increasing the potential for information bias. Subjects may selectively enroll in such studies, and if they are not typical of the larger population—say, if they are unusually responsive to nature contact—then conclusions may not be generalizable. But these sorts of designs are probably the best we have. And they certainly provide better information than the anecdotal account that seems to typify much of the literature in this field.

A detailed analysis of one example is illustrative. This was a study of the effects of participating in a Master Gardener Program on female inmates in a federal prison in Texas (Migura et al. 1996), presented at a research meeting on people-plant interactions and published in 1996. The study hypothesis was that participating in the Master Gardener Program would enhance psychological well-being, specifically locus of control, self-esteem, and life satisfaction. Thirty-six inmates who volunteered for the Master Gardener program were compared to an unspecified number of inmates who did not participate. All subjects were studied with a before and after survey, including selected items from four published psychological scales. The published report provided no information on the content, frequency, or duration of the Master Gardener Program, nor did it indicate how many of the participants dropped out prior to completion. The reported results indicated small, statistically nonsignificant improvements in locus of control, self-esteem, and life satisfaction among participants in the Master Gardener Program. No results were provided for the nonparticipants. Based on these results, the investigators recommended continued implementation of the Master Gardener Program, together with continued study.

This was not a randomized study; inmates sorted themselves, based on interest and other factors, into participants and nonparticipants. As a result, if the participants showed any advantage over the nonparticipants in psychological well-being, that advantage may have predated the intervention. However, the investigators did not report a comparison of participants and nonparticipants, so this criticism is moot. Instead, the study was analyzed and reported as an uncontrolled study—essentially a case series. It is impossible to know if the Master Gardener participants fared better than, worse than, or the same as nonparticipants. In fact, without a comparison group, the simple fact of participating in a program, whether Master Gardener, carpentry, or high school equivalency, might have accounted for any improvement seen. We are not told what the Master Gardener program consisted of—how many encounters, what kinds of plants, in what kind of facility, with what kinds of instructors, over how much time—so even if the study showed a benefit, it would be impossible to replicate it or to implement the program in other settings. The measurement of outcome used existing questionnaire instruments, a methodologic strength, but since many of the inmates were uneducated, and some may not have spoken English well, we need to worry about their ability to complete the questionnaires reliably. The study was small, with only 36 program participants. The statistical power of a study of 36 people to show clinically important changes in the outcome measures used, over the time frame of the study, was not reported, but it is possible that 36 was too small a number to reveal important changes. Information bias could have occurred, for example, if subjects wanted to please the investigators by reporting improvements on the posttest compared to the pretest. Selection bias could have occurred if participants who disliked the program selectively dropped out, leaving a sample enriched with those who benefited. Confounding could have occurred if any other changes during the time of the study—a change in seasons, changes in prison food, changes in the break schedule—improved the prisoners' well-being. Overall, then, it is impossible to draw any conclusions from this study.

NATURE AND HUMAN HEALTH: BUILDING THE EVIDENCE BASE

> Believe nothing, O monks, merely because you have been told it . . . or because it is traditional, or because you yourselves have imagined it. Do not believe what your teacher tells you merely out of respect for the teacher. But whatsoever, after due examination an analysis, you find to be conducive to the good . . . that doctrine believe and cling to, and take it as your guide.
>
> —Buddha

We need to establish a tradition of health research within the community of scientists interested in nature contact (Frumkin 2003, 2004). This tradition is strikingly absent now. In fact, most of the leading researchers who have contributed to this body of knowledge are represented in this book—and it's not a very thick book! What might this research look like?

It would study well-defined populations with specific, well-defined health conditions, recruited in large numbers to achieve a high level of statistical power. It would use randomized controlled trials whenever possible, and similar study designs, such as natural experiments, otherwise. Both the "exposure"—nature contact—and the outcome would be carefully defined and operationalized.

These two elements of the research—the exposure and the outcome—deserve attention. In a clinical trial of, say, a medication, the researchers need to obtain standardized, high-quality preparations of the medication, and administer them to subjects at known doses. If there is something that impairs the absorption of the medication, such as a full stomach, then subjects need to avoid that something, to achieve standard dosing. Scrupulous attention to these factors is necessary in drawing firm conclusions. Similarly, in an observational study of a potentially harmful environmental exposure, careful exposure assessment is critical—so critical that it forms its own subfield of environmental epidemiology. If we want to study the health effects of exposure to a pesticide, we quantify each subject's exposure profile, through a careful history, through environmental measurements of the pesticide, or ideally through biomarkers—blood or urine assays that specify each individual's exposure.

How do we measure the exposure when the exposure is nature? How do we quantify the dose of what has been called "vitamin G" (for greenspace) (Groenewegen et al. 2006)? To do so we need to deconstruct nature contact. In a walk through the park, are the key parameters the density of trees? The species of trees? The greenness of the trees? The presence or absence of birds? Of flowers? Of certain smells? The design of the path? Does it matter if it is a sunny day or a cloudy day? Is the length of time spent in the park a critical factor? If we study short-term exposures such as a walk in the park, are the results generalizable to long-term exposures such as residential or workplace character? Some answers can be ventured based on insights from environmental psychology research; we know some of the features of natural places that make them attractive to people (Kaplan and Kaplan 1989; Kaplan, Kaplan, and Ryan 1998), and a priori these features might be studied as independent variables. But we have a long way to go in identifying what dimensions of nature contact are most likely to promote health and well-being.

We also have challenges in measuring the health outcomes of interest. Investigators have studied a wide range of health outcomes, from violent behavior to postoperative recovery. We probably need to learn to measure positive outcomes and not just negative outcomes—health and well-being, and not just pathology—a challenge for both psychology and medicine. What are the health outcomes most likely to improve following nature contact? In part, this may be postulated based on theoretical approaches such as biophilia and attention restoration. People can be asked to report how they feel using survey techniques, but even with standardized, validated surveys, it is difficult to eliminate bias. Tests of psychological function such as attention are an important option, in view of the theoretical model of attention restoration. Some investigators, in an effort to understand the effects of stress, aging, and poverty, have been looking at biomarkers that vary with these conditions. Psychological stress has been associated with measurable decrements in immune function, such as reduced immune response to influenza (Rosenkranz et al. 2003) and pneumonia (Glaser et al. 2000) vaccinations, slower wound healing (Kiecolt-Glaser et al. 1995), and increased proinflammatory cytokines (Kiecolt-Glaser et al. 2003). Stress has also been associated with shortening of telomeres (Epel et al. 2004), the DNA-protein complexes that cap chromosomal ends, and telomere shortening in turn seems to reflect cell aging (von Zglinicki and Martin-Ruiz 2005). And stress has been associated with other physiological changes, such as increases in steroid and insulin levels (McEwen 1998). Could these approaches to measuring the effects of stress suggest a strategy for studying the benefits of nature contact, by measuring changes in the opposite direction? And could such research reveal insights into the mechanisms of these benefits?

Imagine a clinical trial such as the following. There is an inpatient mental health facility with a forested area on its grounds. Patients with a particular diagnosis—schizophrenia, bipolar disorder, dementia—are randomly assigned to one of two groups, which are equivalent in terms of age, gender, medication use, diagnosis, and other factors. But they differ in an important way: one group spends an hour each day in the forested area, while the other spends the same hour in an interior room. The activity levels in each setting are similar; they differ only by the presence of trees. The patients in each group are studied over time for clinical improvement, using both biomarkers and standard clinical measures of disease activity.

Even when we can't randomly assign people to different conditions, we can take advantage of natural experiments, such as Kuo's Chicago housing study or Ulrich's study of postsurgical patients. Suppose we identify two urban neighborhoods, similar in socioeconomic status and other demographic factors, topography, and building type, but differing in the presence of trees. One neighborhood has plenty of trees on its streets and in its parks, while the other has few trees. We might observe the level of walking in each neighborhood to test the hypothesis that trees promote walking. We might observe the quality of social interaction in each neighborhood to test the hypothesis that trees promote social capital. We might observe the level of crime, or the driving habits of drivers passing through, or the level of cleanliness of the streets, to test the hypothesis that trees promote better civic behavior. Residents of the two neighborhoods would not have been randomly assigned, so we would need to worry that those who chose to live in the neighborhood with trees somehow differed from their counterparts in the barren neighborhood ("selection bias"), but with careful study design this bias can be limited.

Research such as this requires collaboration between investigators who know trees and investigators who know health. How often has a forester, or a botanist, or an architect, or a landscape architect, worked with an epidemiologist? We need a clinical epidemiology of nature contact.

This has implications for training. Students of environmental studies, architecture, design, and allied fields need to be taught about the possible health implications of the work they do, and those with a research bent need to be taught research methodology. Similarly, clinical epidemiologists and health scientists need to be taught about the potential health benefits of nature contact, including contact with trees, and encouraged to partner with colleagues from other fields. Promising venues for these training initiatives are institutions that house more than one kind of training program.

Is there funding to support such research? I believe so. The major federal agencies that support environmental health research, such as the National Institute of Environmental Health Sciences, have begun to take an interest in environmental factors other than toxic chemicals, such as the built environment. Other institutes within the NIH, which are organized around specific illnesses or body systems, have a stronger and stronger track record of funding sound research focusing on health outcomes of interest, including research that centers on innovative variables. Other federal agencies, such as the Forest Service, have funded some of this research in the past, and should take an interest in continuing and expanding this support, since the resulting insights could powerfully advance their mission. The private sector offers other funding possibilities. While the pharmaceutical industry is unlikely to support research on the health benefits of nature contact, a potentially low-cost and nontoxic alternative to some medications, other business sectors such as nurseries and gardening supply firms can and should be approached.

Finally, as rigorous research results emerge, they need to be published in high-quality health science journals. Again, I do not for a moment mean to disparage the professional publications of forestry, horticulture, landscape architecture, and allied fields. But if we generate important health information, it needs to breach the disciplinary walls and penetrate the world of those who make health decisions, set health policy, and treat patients.

LIMITS TO THESE CLAIMS

There are several limits to my claims. First, there are important health benefits of nature contact that operate on a larger-than-clinical scale and can be characterized through means other than clinical studies. Consider, for example, the benefits of nearby trees. Trees offer indirect health benefits by

- Reducing air pollutants, especially ozone, nitrogen dioxide, and to a lesser extent particulate matter (McPherson et al. 1997) (although trees are sources of certain harmful pollutants, primarily hydrocarbons such as terpenes and pinenes, but also pollen and other allergens).
- Reducing greenhouse warming by fixing carbon dioxide during photosynthesis. One hectare of forest can remove 10 to 15 tons of carbon dioxide from the air each year, and approximately half the dry weight of wood is carbon (Brown et al. 1996; Nowak and Crane 2002).
- Reducing the demand for air conditioning during warm weather by shading buildings (although they slightly increase the demand for heating during cold weather) (Heisler 1986; Simpson and McPherson 1998), thereby reducing energy demands on power plants.
- Reducing heat concentrations over parking lots (McPherson et al. 2001), sidewalks, and streets, helping to mitigate the urban heat island effect (Weng and Yang 2004) and to avoid the direct hazards of heat, a well-recognized danger (Vandentorren et al. 2004; CDC 2004; Johnson et al. 2005) and one that will tend to become worse with global warming (Keatinge and Donaldson 2004).
- Providing medications such as quinine from Cinchona bark, linden from the lime tree, cold remedies from the eucalyptus, and Paclitaxel from the Pacific yew (Grifo et al. 1997).
- Serving as a noise barrier in urban areas, or along roadways.
- Producing shade and protecting people from sunlight.

A second limit to my claims builds on the first. Clinical research, I have suggested, is a very good thing—but health benefits do not tell the whole story. Protecting natural assets offers other benefits, such as environmental protection, sustainability, and economic payoffs. This does not challenge the value of clinical research, but it reminds us that clinical research does not answer every important question.

Third, my analogy to clinical research may be overwrought. Perhaps we don't need such rigorous evidence when it comes to nature contact. After all, if women take postmenopausal estrogen replacement therapy without a sound evidence base, the results of being wrong may be dire—excess cases of heart attacks, strokes, and breast cancer. In contrast, if children are sent outside to play in nature in the belief that they will benefit, the risk of being wrong is minimal (other than the occasional case of poison ivy).

This objection is well-founded. Nevertheless, better evidence will give us a deeper understanding of the benefits of nature contact, a firmer basis for making recommendations, and a better claim on scarce resources. These are worthwhile goals.

A fourth limit to my claims may seem inconsistent with my prolonged paean to empirical research. But it must be said that there are other ways of apprehending reality than empirical research—as anyone who has fallen in love, or thrilled to a beautiful painting or sonata, or achieved a spiritual insight, knows. Nevertheless, given the track record of empirical research in establishing new insights, correcting time-honored errors, and supporting policy changes, this research needs to be at the center of our efforts to move forward.

A final limit to my claims goes as follows: Maybe we don't know everything there is to know about human benefits of nature contact, but we have a pretty fair idea, and we know a lot about designing nature into the built environment. And given the pace at which decisions are being made and places built, there is a pressing need to implement what we know. We can't wait for the research.

Fair enough. But this is not an either-or dilemma. We can move ahead based on what we know, acknowledging as we do the limits of our knowledge—a common practice in clinical medicine, public health, and many other arenas. But at the same time we can and should press for better research.

CONCLUSION

Anecdotal experience, common sense, evolutionary theory, and even some empirical evidence suggest that contact with nature confers health benefits. But there is very little rigorous evidence of this association. At this time we can only offer limited data-based recommendations about what kinds of nature contact will be beneficial, among which patients, with which medical conditions, and under what circumstances. Such recommendations are now the norm with respect to medications, surgical procedures, and other health interventions. A collaboration involving environmental scientists, biologists, and their colleagues; architects, planners, and their colleagues; and health researchers such as epidemiologists, physicians, and psychologists, offers great promise for filling these data gaps, through well-designed, carefully executed studies. Such research will deepen our understanding of the human relationship with nature, increase the reverence we feel for nature, and help us improve human health.

REFERENCES

Bodin, M., and T. Hartig. 2003. "Does the Outdoor Environment Matter for Psychological Restoration Gained Through Running?" *Psychology of Sport and Exercise* 4: 141–153.

Brown, S., Sathaye, J., Cannell, M., and Kauppi, P. 1996. Management of Forests for Mitigations of Greenhouse Gas Emissions. In: Watson, R.T., Zinyowere, M. C. & Moss, R. H. (Eds.) *Climate Change 1995. Impacts, Adaptations and Mitigation of Climate Change: Scientific-Technical Analyses.* IPCC Second Assessment Report. Cambridge University Press. pp. 773–797.

CDC (Centers for Disease Control and Prevention). 2004. "Impact of Heat Waves on Mortality: Rome, Italy, June–August 2003." *Morbidity and Mortality Weekly Report* 53(17): 369–371.

Center for Watershed Protection. 2003. *Impacts of Impervious Cover on Aquatic Systems.* Washington, DC: Center for Watershed Protection.

Chung, S. H. 1996. "Therapeutic Effect and Evaluation of a Horticultural Therapy Program in a Korean Psychiatric Ward." In *People-Plant Interactions in Urban Areas: Proceedings of a Research and Education Symposium, May 23–26, 1996, San Antonio, Texas,* edited by P. Williams and J. Zajicek, 92–97. College Station: Texas A&M University.

Cobb, L. A., G. I. Thomas, G. H. Dillard, R. A. Merendino, and R. A. Bruce. 1959. "An Evaluation of Internal Mammary Artery Ligation by a Double Blind Technic." *New England Journal of Medicine* 260:1115–1118.

Curriero, F. C., J. A. Patz, J. B. Rose, and S. Lele. 2001. "The Association Between Extreme Precipitation and Waterborne Disease Outbreaks in the United States, 1948–1994." *American Journal of Public Health* 91:1194–1199.

Diette, G. B., N. Lechtzin, E. Haponik, A. Devrotes, and H. R. Rubin. 2003. "Distraction Therapy with Nature Sights and Sounds Reduces Pain During Flexible Bronchoscopy: A Complementary Approach to Routine Analgesia." *Chest* 123(3): 941–948.

Dimond, E. G., C. F. Kittle, and J. E. Crockett. 1960. "Comparison of Internal Mammary Artery Ligation and Sham Operation for Angina Pectoris." *American Journal of Cardiology* 5:483–486.

Epel, E. S., E. H. Blackburn, J. Lin, F. S. Dhabhar, N. E. Adler, J. D. Morrow, and R. M. Cawthon. 2004. "Accelerated Telomere Shortening in Response to Life Stress." *Proceedings of the National Academy of Sciences* 101:17312–17315.

Epstein, P. R. 2001. "Climate Change and Emerging Infectious Diseases." *Microbes and Infection* 3(9): 747–754.

Feinstein, A. R. 1985. *Clinical Epidemiology: The Architecture of Clinical Research.* Philadelphia: Saunders.

Flagler, J., and R. P. Poincelot, eds. 1994. *People-Plant Relationships: Setting Research Priorities.* Binghamton, NY: Food Products Press.

Fletcher, R. H., and S. W. Fletcher. 2005. *Clinical Epidemiology: The Essentials,* 4th ed. Philadelphia: Lippincott, Williams & Wilkins.

Francis, M., P. Lindsey, and J. S. Rice, eds. 1994. *The Healing Dimensions of People-Plant Relations: Proceedings of a Research Symposium. March 24–27, 1994, University of California–*

Davis. Davis: University of California–Davis, Center for Design Research.

Frumkin, H. 2001. "Beyond Toxicity: Human Health and the Natural Environment." *American Journal of Preventive Medicine* 20:234–240.

Frumkin, H. 2003. "Healthy Places: Exploring the Evidence." *American Journal of Public Health* 93(9): 1451–1455.

Frumkin, H. 2004. "White Coats, Green Plants: Clinical Epidemiology Meets Horticulture." *Acta Horticulturae* 639: 15–26.

Glaser, R., J. F. Sheridan, W. B. Malarkey, R. C. MacCallum, and J. K. Kiecolt-Glaser. 2000. "Chronic Stress Modulates the Immune Response to a Pneumococcal Pneumonia Vaccine." *Psychosomatic Medicine* 62:804–807.

Grifo, F., D. Newman, A. S. Fairfield, B. Bhattacharya, and J. T. Grupenhoff. 1997. "The Origins of Prescription Drugs." In *Biodiversity and Human Health*, edited by F. Grifo and J. Rosenthal, 131–163. Washington, DC: Island Press.

Groenewegen, P. P., A. E. van den Berg, S. de Vries, and R. A. Verheij. 2006. "Vitamin G: Effects of Green Space on Health, Well-Being, and Social Safety." *BMC Public Health* 6:149. http://www.biomedcentral.com/1471–2458/6/149.

Haines, A., and J. A. Patz. 2004. "Health Effects of Climate Change." *Journal of the American Medical Association* 291(1): 99–103.

Heerwagen, J. H., and G. H. Orians. "Humans, Habitats, and Aesthetics." 1993. In *The Biophilia Hypothesis*, edited by S. R. Kellert and E. O. Wilson, 138–172. Washington, DC: Island Press.

Heisler, G. M. 1986. "Energy Savings with Trees." *Journal of Arboriculture* 12:113–125.

Hulley, S. B., S. R. Cummings, W. S. Browner, D. Grady, N. Hearst, and T. B. Newman. 2001. *Designing Clinical Research*, 2nd ed. Philadelphia: Lippincott, Williams & Wilkins.

Johnson, H., R. S. Kovats, G. McGregor, J. Stedman, M. Gibbs, H. Walton, L. Cook, and E. Black. 2005. "The Impact of the 2003 Heat Wave on Mortality and Hospital Admissions in England." *Health Statistics Quarterly* 25: 6–11.

Kaplan, R. 1983. "The Role of Nature in the Urban Context." In *Behavior and the Natural Environment*, edited by I. Altham and J. Wohlwill. New York: Plenum.

Kaplan, R., and S. Kaplan. 1989. *The Experience of Nature: A Psychological Perspective*. Cambridge: Cambridge University Press.

Kaplan, R., S. Kaplan, and R. L. Ryan. 1998. *With People in Mind: Design and Management of Everyday Nature*. Washington, DC: Island Press.

Keating, N., P. Cleary, A. Aossi, A. Zaslavsky, and J. Ayanian. 1999. "Use of Hormone Replacement Therapy by Postmenopausal Women in the United States." *Annals of Internal Medicine* 130:545–553.

Keatinge, W. R., and G. C. Donaldson. 2004. "The Impact of Global Warming on Health and Mortality." *Southern Medical Journal* 97(11): 1093–1099.

Kiecolt-Glaser, J. K., P. T. Marucha, W. B. Malarkey, A. M. Mercado, and R. Glaser. 1995. "Slowing of Wound Healing by Psychological Stress." *Lancet* 346:1194–1196.

Kiecolt-Glaser, J. K., K. J. Preacher, R. C. MacCallum, C. Atkinson, W. B. Malarkey, and R. Glaser. 2003. "Chronic Stress and Age-Related Increases in the Proinflammatory Cytokine Interleukin-6." *Proceedings of the National Academy of Sciences* 100:9090–9095.

Kuo, F. E. 2001. "Coping with Poverty: Impacts of Environment and Attention in the Inner City." *Environment and Behavior* 33(1): 5–34.

Kuo, F. E., and W. C. Sullivan. 2001a. "Aggression and Violence in the Inner City: Effects of Environment via Mental Fatigue." *Environment and Behavior* 33(4): 543–571.

Kuo, F. E., and W. C. Sullivan. 2001b. "Environment and Crime in the Inner City: Does Vegetation Reduce Crime?" *Environment and Behavior* 33(3): 343–367.

McEwen, B. S. 1998. "Protective and Damaging Effects of Stress Mediators." *New England Journal of Medicine* 338:171–179.

McPherson, E. G., D. Nowak, G. Heisler, et al. 1997. "Quantifying Urban Forest Structure, Function, and Value: The Chicago Urban Forest Climate Project." *Urban Ecosystems* 1:49–61.

McPherson, E. G., J. R. Simpson, and K. I. Scott. 2001. "Actualizing Microclimate and Air-Quality Benefits with Parking Lot Tree Shade Ordinances." *Wetter und Leben* 4(98): 353–369. http://cufr.ucdavis.edu/products/11/cufr_69.pdf.

Migura, M. M., L. A. Whittlesey, and J. M. Zajicek. 1996. "Effects of the Master Gardener Program on the Self-Development of Female Inmates of a Federal Prison Camp." In *People-Plant Interactions in Urban Areas: Proceedings of a Research and Education Symposium, May 23–26, 1996, San Antonio, Texas*, edited by P. Williams and J. Zajicek, 72–76. College Station: Texas A&M University.

Mooney, P. F., and S. L. Milstein. 1994. "Assessing the Benefits of a Therapeutic Horticulture Program for Seniors in Intermediate Care." In *The Healing Dimensions of People-Plant Relations: Proceedings of a Research Symposium. March 24–27, 1994, University of California–Davis*, edited by M. Francis, P. Lindsey, and J. S. Rice, 173–194. Davis: University of California–Davis, Center for Design Research.

Moore, E. O. 1981–1982. "A Prison Environment's Effect on Health Care Service Demands." *Journal of Environmental Systems* 11:17–34.

Nowak, D. J., and D. E. Crane. 2002. "Carbon Storage and Sequestration by Urban Trees in the USA." *Environmental Pollution* 116:381–389.

Perlman, M. 1994. *The Power of Trees: The Reforesting of the Soul.* Dallas: Spring Publications.

Pretty, J., M. Griffin, M. Sellens, and C. Pretty. 2003. "Green Exercise: Complementary Roles of Nature, Exercise and Diet in Physical and Emotional Well-Being and Implications for Public Health Policy." CES Occasional Paper 2003-1, University of Essex, March 2003. http://www2.essex.ac.uk/ces/ResearchProgrammes/CESOccasionalPapers/GreenExercise.pdf.

Relf, D., ed. 1992. *The Role of Horticulture in Human Well-Being and Social Development: A National Symposium, 19–21 April 1990, Arlington, Virginia.* Portland, OR: Timber Press.

Rosenkranz, M. A., D. C. Jackson, K. M. Dalton, I. Dolski, C. D. Ryff, B. H. Singer, D. Muller, N. H. Kalin, and R. J. Davidson. 2003. "Affective Style and in vivo Immune Response: Neurobehavioral Mechanisms." *Proceedings of the National Academy of Sciences* 100:11148–11152.

Rossouw, J. E., G. L. Anderson, R. L. Prentice, A. Z. LaCroix, C. Kooperberg, M. L. Stefanick, R. D. Jackson, S. A. Beresford, B. V. Howard, K. C. Johnson, J. M. Kotchen, and J. Ockene. Writing Group for the Women's Health Initiative Investigators. 2002. "Risks and Benefits of Estrogen Plus Progestin in Healthy Postmenopausal Women: Principal Results from the Women's Health Initiative Randomized Controlled Trial." *Journal of the American Medical Association* 288:321–333.

Ruffin, J. M., J. E. Grizzle, N. C. Hightower, et al. 1969. "A Cooperative Double-Blind Evaluation of Gastric 'Freezing' in the Treatment of Duodenal Ulcer." *New England Journal of Medicine* 281:16–19.

Sackett, D. L., R. B. Haynes, G. H. Guyatt, and P. Tugwell. 1991. *Clinical Epidemiology: A Basic Science for Clinical Medicine*, 2nd ed. Philadelphia: Lippincott, Williams & Wilkins.

Sackett, D. L., W. M. Rosenberg, J. A. Gray, R. B. Haynes, and W. S. Richardson. 1996. "Evidence-Based Medicine: What It Is and What It Isn't." *British Medical Journal* 312: 71–72.

Simpson, J. R., and E. G. McPherson. 1998. "Simulation of Tree Shade Impacts on Residential Energy Use for Space Conditioning in Sacramento." *Atmospheric Environment: Urban Atmospheres* 32(1): 69–74.

Stamler, R., J. Stamler, F. C. Gosch, J. Civinelli, J. Fishman, P. McKeever, A. McDonald, and A. R. Dyer. 1989. "Primary Prevention of Hypertension by Nutritional-Hygienic Means. Final Report of a Randomized, Controlled Trial." *Journal of the American Medical Association* 262:1801–1807.

Steering Committee of the Physicians' Health Study Research Group. 1989. "Final Report on the Aspirin Component of the Ongoing Physicians' Health Study." *New England Journal of Medicine* 321:129–135.

Taylor, A. F., F. E. Kuo, and W. C. Sullivan. 2002. "Views of Nature and Self-Discipline: Evidence from Inner City Children." *Journal of Environmental Psychology* 22(1–2): 49–63.

Ulrich, R. S. 1984. "View Through a Window May Influence Recovery from Surgery." *Science* 224:420–421.

Vandentorren, S., F. Suzan, S. Medina, M. Pascal, A. Maulpoix, J. C. Cohen, and M. Ledrans. 2004. "Mortality in 13 French Cities During the August 2003 Heat Wave." *American Journal of Public Health* 94(9): 1518–1520.

von Zglinicki, T., and C. M. Martin-Ruiz. 2005. "Telomeres as Biomarkers for Aging and Age-Related Diseases." *Current Molecular Medicine* 5:197–203.

Wagensteen, O. H., E. T. Peter, D. M. Nicoloff, et al. 1962. "Achieving 'Physiological Gastrectomy' by Gastric Freezing: A Preliminary Report of an Experimental and Clinical Study." *Journal of the American Medical Association* 180:439–445.

Weng, Q., and S. Yang. 2004. "Managing the Adverse Thermal Effects of Urban Development in a Densely Populated Chinese City." *Journal of Environmental Management* 70(2): 145–156.

Williams, P., and J. Zajicek. 1996. *People-Plant Interactions in Urban Areas: Proceedings of a Research and Education Symposium, May 23–26, 1996, San Antonio, Texas.* College Station: Texas A&M University.

Wilson, E. O. 1984. *Biophilia: The Human Bond with Other Species.* Cambridge, MA: Harvard University Press.

Wilson, E. O. 1993. "Biophilia and the Conservation Ethic." In *The Biophilia Hypothesis*, edited by S. R. Kellert and E. O. Wilson, 31–41. Washington, DC: Island Press.

chapter 8

Where Windows Become Doors

Vivian Loftness with Megan Snyder

Biophilic design recognizes that
the line between indoors and outdoors must be rethought;
that indoor rooms must communicate with outdoor rooms;
that windows must become doors.

Biophilic design must be achieved in
a regionally rich design paradigm,
with an understanding of the physiological and the psychological,
the mechanisms and the boundaries,
through a transdisciplinary design process.

As architects, we are often shocked by windowless buildings; by brutalist design that eliminates access to the street or garden; by fashion design that introduces punched openings, slits, and slashes to create compositional interest but little connection to the nature and life that surrounds each building. If architecture is to have greater demands for the design of openings, proponents of biophilic design must define the levels of connection to be pursued, and to what extent these goals are regional or cultural. Biophilic design proponents need to uncover the physiological and psychological mechanisms that mandate these connections, and the regional boundaries that bracket success.

This chapter will address the importance of a rich and informed design process for the introduction of windows in buildings, to guarantee that "windows become doors" that ensure access to views, daylight, sunlight, fresh air, breezes, natural comfort, passive survivability, outdoor spaces and activities, extended space, circadian regulation, seasons, climate, and nature's sounds, smells, and life. Windows connect building occupants with a richness that may be critical to the individual; at the same time, they also provide those outside the building with a level of transparency, oversight, and contact with life's activities that is critical to community. Together "windows that become doors" define the spirit of place, central to timeless architecture.

VIEWS

While there is significant debate about the importance of *indoor* daylight or sunshine for human health and performance, there seems to be a growing consensus that access to a view of nature is significant. Beginning with the seminal work of Roger Ulrich (1984), then Mendell (1991), Heschong Mahone Group (2003), and now Kellert (2005), seated views of nature and proximity to windows are being linked to reduced length of stay after surgery, reduced sick building syndrome (SBS), increased performance at task, and overall improved emotional health.

In a 1990 survey of over 2,000 employees in two buildings at the U.S. Department of Energy, Carnegie Mellon University's Center for Building Performance and Diagnostics identified 10–20 percent lower sick building symptoms among employees with seated views of windows, controlling for rank (see Figure 8-1) (CBPD/DOE 1994). Whether user perception of personal health was improved due to the light, view, perimeter conditioning systems, or increased level of environmental control at the window (blinds, HVAC controls) is unclear. Regardless, there is significant benefit in a workforce that has fewer health symptoms across the board.

Two field studies frame the conclusion that views are a significant factor in health and productivity. In a 1984 observational field study of 23 matched pairs of patients at a Pennsylvania hospital, Ulrich identified an 8.5 percent reduction in postoperative hospital stay (7.96 days versus 8.7 days) for gall bladder surgery patients who had a view of a natural scene from their hospital room, as compared to those with a view of a brick wall. Patients with a view of nature also received fewer negative evaluations from nurses and took fewer strong analgesics. Recovery data were extracted from hospital records by a nurse with extensive surgical floor experience who had no knowledge of which scene was visible from each patient's window (Ulrich 1984).

In 2003, Heschong Mahone Group conducted a field study of 100 full-time customer service representatives at the Sacramento Municipal Utility District (SMUD) Call Center to investigate the influence of windows and daylight on worker productivity. The researchers identified a 6–7 percent faster average call

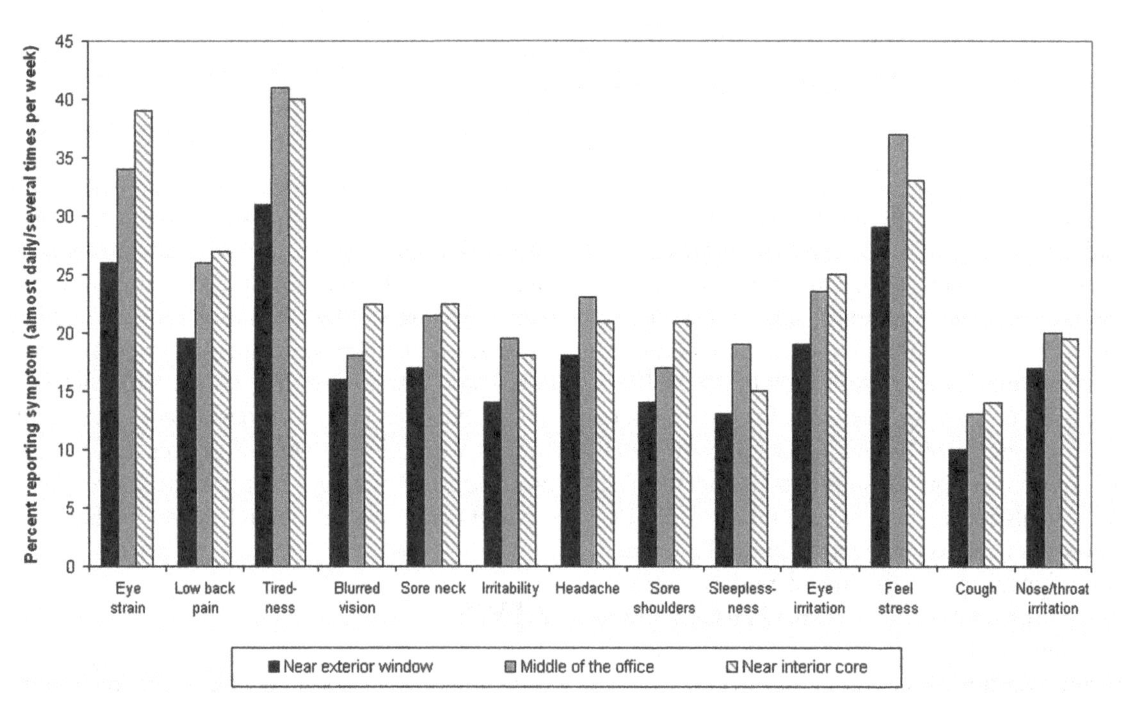

Figure 8-1: Comparison between window proximity and health complaints at two USDOE offices (Center for Building Performance and Diagnostics/Department of Energy [CBPD/DOE] 1994)

Figure 8-2: Three workstations without and with outdoor views. Left: A typical windowless cubicle. Middle/right: Workstations with views of buildings and landscape.

handling time—a standard measure of call center productivity—for employees with seated access to views with vegetation content through large windows from their cubicles, as compared to employees with no view of the outdoors (Heschong Mahone Group 2003).

Whether these health, performance, and satisfaction improvements are the result of views, daylight, infiltration of higher quantities of air, or heightened levels of control is unclear. In addition to identifying which quality of the windowed environment is responsible for these gains, research is critically needed to understand the physiological and psychological mechanisms that might explain these benefits, the importance of the content of a window view (e.g., landscape versus sky versus building walls), and the importance of controlling conditions such overheating and glare that might accompany proximity to a window.

For the design community, we argue that the existing literature is robust enough to mandate the inclusion of windows with views for all occupied spaces, alongside effective design of window size and location in the wall including sill height, sight lines, and view content. At the same time, it will be imperative for designers to address regional and cultural boundaries on window design for views, to ensure that requirements for glare, heat gain, noise control, privacy, and security are equally met—the definition of quality design.

DAYLIGHT/SUNLIGHT/CIRCADIAN RHYTHMS/SEASONS/CLIMATE

In addition to views, the introduction of well-designed windows can provide the benefits of daylight, daylight variability over time and season, and sunlight in occupied spaces. While daylight can provide higher light levels that support improved performance at visual tasks, it also can create glare that compromises performance. As a result, while the debate continues as to the mechanisms whereby daylight improves performance or health outcomes, research is revealing that the natural variability of daylight and sunlight, especially morning sunlight, reduces length of stay for patients recovering from surgery, bipolar treatment, and seasonal affective disorder (SAD) treatment (Beauchemin and Hays 1996; Benedetti et al. 2001; Choi 2005; Walch et al. 2005).

Research at the Lighting Research Center at Rensselaer Polytechnic Institute has begun to reveal the relationship between exposure to ultraviolet light and melatonin production, which controls circadian rhythms, sleep cycles, performance at task, and even cancer cell development (Bullough et al. 2006). In a 1997 controlled experiment of 20 night-shift workers, Boyce et al. revealed the importance of time-of-day light intensities to performance at task. Participants demonstrated statistically significant improvement on short-term memory and grammatical reasoning tasks under specific lighting conditions from large skylight-simulating fixtures with hidden fluorescent lamps. Steadily decreasing illuminance that simulated natural daylight variation from midday to dusk (2800 to 200 lux) and fixed high illuminance of 2800 lux improved the performance of night workers relative to fixed low illuminance of 250 lux or steadily increasing illuminance (Boyce et al. 1997).

In a 2002 presentation at the EPRI symposium on Lighting and Health, Joan Roberts explained the mechanisms: the internal circadian clock is set externally by visible light, most critically ultraviolet light, and circadian rhythms control the ebb and flow of most hormones in the pituitary, pineal, adrenal, and thyroid glands. With circadian imbalance due to lack of UV-triggered melatonin production, humans experience loss of sleep, carbohydrate cravings, poor coordination, depression, and susceptibility to disease. For good health, Roberts argues that it is equally important to have bright visible light during the day and to have darkness at night (Roberts 2003).

While spending more time outdoors in direct sun, especially early in the day, might be a preferred vehicle for ensuring UV-induced melatonin production, it is not always possible for patients in hospitals or for schoolchildren who begin and end school in the dark. Four field studies frame the conclusion that time-of-day light intensities—most directly through sunlight penetration—are a significant factor in hospital recovery rates.

In a 1996 observational field study of 174 patients at a hospital in Edmonton, Alberta, Canada, Beauchemin and Hays identified a 2.6-day reduction in length of stay among seasonal affective disorder (SAD) patients located in sunny rooms, as compared to those in sunless rooms. Patients were randomly assigned to rooms, and the difference in length of stay was consistent across seasons (Beauchemin and Hays 1996). Benedetti et al. found similar benefits with respect to hospitalization for bipolar disorder in a field study of 187 inpatients at San Raffaele Hospital in Milan, Italy, in 2001. The researchers identified a 30 percent reduction in length of stay in summer and a 26 percent reduction in length of stay in autumn among patients in eastern rooms (exposed to direct sunlight in the morning) as compared to patients in western rooms (exposed to direct sunlight in the evening) (Benedetti et al. 2001).

Additional research has shown that the benefits of sunlight extend beyond treatment of patients with psychological disorders . In a 2005 field study of 141 patients at Inha University Hospital in Korea, Choi identified a 41 percent reduction (3.2 days) in average length of stay among gynecology patients in brightly daylit rooms, as compared to those in dull rooms, in the spring, and an average 26 percent reduction (1.9 days) in average length of stay among surgery ward patients in bright rooms, as compared to those in dull rooms, in the fall. Across all seasons, the average daylight illuminance in bright rooms was 317 lux, compared to only 190 lux in dull rooms (Choi 2005). Walch et al. identified a 22 percent reduction in analgesic medication use among patients in "bright" rooms who were exposed to more natural sunlight after surgery (average 73,537 lux-hrs), as compared to patients located in "dim" rooms after surgery (average 50,410 lux-hrs of sunlight) in a 2005 prospective study of 89 elective cervical and lumbar spinal surgery patients at Montefiore Hospital in Pittsburgh (Walch et al. 2005).

Given the effective design of windows for views, the additional task of designing to bring time-of-day lighting variability and sunshine into occupied spaces is not difficult. In addition to the regionally appropriate design of size, location in the wall, sill height, sight lines, and view content, the architect must address orientation of the windows and skylights, UV transparency of the glass, and the geometry of the room to ensure sunlight penetration without overheating and glare. As with design for views, it is imperative for designers to address regional and cultural boundaries on window design for sunlight, and to ensure that ultraviolet fading is not an issue and that brightness contrast and adaptation are not disabling, in addition to the previously addressed issues of glare, solar overheating, noise, privacy, and security.

Alternatively, the design of windows as doors and the creation of outdoor rooms commensurate with indoor rooms might ensure that building occupants spend critical time outside in early morning sun each day. Finnish architect Alvar Aalto achieved international acclaim for his Paimio Sanatorium, which was designed to support outdoor sunshine therapy by permitting each tuberculosis patient's bed to be wheeled directly out onto adjacent southeastern terraces—even in the harsh Scandinavian climate.

FRESH AIR AND NATURAL VENTILATION

The value of high outside air delivery rates to improved health and performance is becoming increasingly evident in research. The depth of the evidence has resulted in updated ASHRAE standards, from 10 cfm to 20 cfm per person minimum in offices, and a growing interest in CO_2 sensors and demand-controlled actuators to ensure higher ventilation rates wherever occupants congregate in buildings or where activities raise indoor pollution levels.

In a 1996 field study of 690 residents at a four-building nursing home facility in Wisconsin, Drinka et al. identified an 87.3 percent reduction in the incidence of influenza in a building with 100 percent outside air ventilation and local filtration for each room, as compared to three buildings with 30–70 percent recirculated air and central filtration only. A total of 65 positive influenza cultures were taken in the buildings with recirculated air and central filtration (12 percent attack rate), while only 3 positive cultures were taken in the building with 100 percent fresh air and local filtration (1.6 percent attack rate) (Drinka et al. 1996). In 2004, Shendell et al. of Lawrence Berkeley National Laboratory compared school attendance data and measured CO_2 concentrations from 436 classrooms in Washington and Idaho, and determined that a 1,000 ppm increase in net (indoor minus outdoor) classroom CO_2 concentration is associated with an average 0.7 percent decrease in annual average daily student attendance, most likely due to health consequences of poor ventilation, indicating that attendance may be improved by an increased ventilation rate and lower CO_2 concentrations (Shendell et al. 2004).

For many practitioners, however, it is not clear whether increased levels of outside air are more effectively delivered through operable windows or through mechanical systems that incorporate filtration, dehumidification, and thermal conditioning of outside air. To this end, one must look to the increasing number of studies, principally from Europe and Scandinavia where natural ventilation still dominates, that compare occupant health in naturally ventilated buildings versus mechanically ventilated and air-conditioned buildings. As shown in Table 8-1, Seppanen and Fisk have identified over a dozen studies revealing the benefits of natural ventilation in reducing headaches, mucosal symptoms, colds, coughs, circulatory problems, and sick building syndrome (SBS) (Seppanen and Fisk 2002).

In a 1990 multiple-building field study of 86 workers in 43 office buildings in the UK, Robertson et al. identified a 9 percent reduction in sickness absence and a 59 percent reduction in self-reported SBS symptoms among workers in naturally ventilated buildings, as compared to workers in air-conditioned buildings (Robertson et al. 1990). In 1992, Kelland surveyed 110 employees at each of two London hospitals—one naturally ventilated and one mechanically ventilated—and identified a 40 percent lower rate of self-reported SBS symptoms in the naturally ventilated building (Kelland 1992). In a 2004 analysis of monthly surveys of 920 professional middle-aged women in France, Preziosi et al. identified a 57.1 percent reduction in sickness absence, a 16.7 percent reduction in doctor visits, and a 34.8 percent reduction in hospital stays among participants with natural ventilation in their workplace, as compared to those with an air-conditioned workplace, after adjustment for demographic variables (Preziosi et al. 2004).

While many of these studies use SBS symptoms as an index of health, newer studies related to measured respiratory illnesses and asthma are raising questions as to whether increased outside air rates are critical for dilution of indoor pollutants; and/or higher levels of oxygen and other outdoor gases are important to human health and performance; and/or naturally delivered outside air might indeed be "fresher" than mechanically delivered outside air, given mixing with return air and the HVAC pathways.

TABLE 8-1 Comparison of SBS symptom prevalence in buildings with natural ventilation, mechanical ventilation, and air conditioning

Reference		Study and Building Characteristics					Ventilation System Type								Results	
								Mechanical Without AC			Air Conditioning					
First Author	Year	Controlled confounders#	No of respondents in comparison	Sealed or openable windows*	Smoking	Recirculation*	Natural Ventilation	Mechanical, exhaust	Simple mechanical, no humidification	Simple mechanical with humidification	No Humidification	Steam Hum.**	Evaporative Hum.	Spray Hum.	Number of Symptoms with significantly higher prevalences in assessment^^	Range of risk ratio or (odds ratio) for outcomes
Jaakkola	95	P,W,B	868	O		Y/N	7 o	—	—	—	*9				2 of 14 S	1.5-2.6
Mendell	96	P,W	710	S	N	Y	3 o	—	—	—	*6				6 of 7 S	1.6-5.40
Burge^	87	none	1459	S/O			11 o	—	—	—	*10				10 of 10 S	(1.3-2.1)
Harrison^	87	none	1044	S		Y/N	8 o	—	—	—	*6				6 of 6 S	(1.7-2.9)
Zweers	92	P,W,B	2806	S/O	Y		21 o	—	—	—	*				5 gr. of S	1.5-1.7
Jaakkola	95	P,W,B	335	O		Y	7 o	—	—	—	—	*2			3 of 14 S	(1.9-2.5)
Burge^	87	none	863	S/O			11 o	—	—	—	—	*4			8 of 10 S	(1.3-2.1)
Zweers	92	P,W,B	3573	S/O	Y		21 o	—	—	—	—	*			5 of 5 gr. of S	1.3-1.9
Jaakkola	95	P,W,B	559	O		T/N	7 o	—	—	—	—	—	*3		3 of 14 S	(2.0-2.7)
Teeuw	94	none	927	S/O		T/N	7 o	—	—	—	—	—	*7		6 of 8 S	1.4-2
Burge^	87	none	1991	S/O			11 o	—	—	—	—	—	■	■ 15	10 of 10 S	(1.4-2.2)
Finnegan^	87	none	787	S	Y	T/N	3 o	—	—	—	—	—	■	■ 3	6 of 11 S	(2.5-4.8)
Harrison^	87	none	2080	S		T/N	8 o	—	—	—	—	—	■	■ 13	5 of 6 S	(2.1-3.2)
Hedge^	84	none	1214				2 o	—	—	—	—	—	■	■ 2	2 of 2 S	(2.7-3.0)
Zweers	92	P,W,B	3846	S/O	Y		21 o	—	—	—	—	—	—	*	5 of 5 gr. of S	1.5-2.1
Brasche	99	P,W					o	—	—	—	■	■	■	■	3 of 7 S	(1.4-1.4)
Hawkins	91	P	255		N	Y	6 o	—	—	—	—	□	□	□ 6	S score	
Jaakkola	95	P,W,B	1828	O		Y/N			18 o	—	*9				2 of 14 S	(1.3-1.7)
Jaakkola	95	P,W,B	1295	O		Y/N			18 o	—	—	*2			1 of 14 S	(1.8-1.8)
Jaakkola	95	P,W,B	1519	O		Y/N			18 o	—	—	—	o 3			

^ as reanalyzed by Mendel (1990) #P = personal factors, W = work factors, B = building factors
*In mechanically-ventilated buildings **Hum = Humidification ^^gr = groups

Key: o—o, o—□ } No statistically significant difference in symptoms; o—*, o—■ } Statistically significant difference in symptoms

Source: Seppanen and Fisk 2002

As proponents of biophilia, we would argue that fresh air is more effectively delivered through operable windows than through mechanical systems except in periods with very high temperature, high humidity, outdoor pollution, or noise. We make this case based on the quantities of "free" breathing air available outside our windows, and the access to nature that an open window provides. Moreover, windows are very effective rapid cooling devices for classrooms and meeting rooms during the dominant cool and cold periods of the year. Operable windows also reduce our vulnerability to mechanically transmitted toxins and to loss of breathing air during power outages.

The pros and cons of increasing outside air rates

TABLE 8-2 Should windows open?

No	Yes
Avoid outdoor pollution	Dilute indoor pollution-HVAC Dilute indoor pollution-materials/activities
Avoid outdoor humidity	Diffuse indoor humidity build-up
Avoid outdoor noise (traffic, mowers)	Connect to nature-air, sounds
High-quality HVAC provides control	Increase local thermal control in cool periods
Avoid rain penetration	Design windows to shed rain Increase local ventilation rates without heat recovery

through natural ventilation versus mechanical means are outlined in Table 8-2. There is a definite emphasis on the value of natural ventilation, especially given weak investments in the long-term field maintenance of our HVAC systems and controls.

Natural ventilation has its limits in very high temperature, high humidity or high pollution periods, which has led to the development of mixed-mode cooling and ventilation HVAC systems. Mixed-mode HVAC supports the use of both natural and mechanical ventilation. Natural ventilation, as the dominant ventilation and cooling source through operable windows and vents, is complemented by mechanical ventilation and cooling and a closed building when natural conditioning is not possible.

Designing for effective natural ventilation and natural cooling through operable windows and vents is challenging for the architectural community. It is dependent on both a rich understanding of cross-ventilation, stack ventilation, and thermally induced ventilation, and on the ability to define solution sets that are regional and even site specific.

At Carnegie Mellon University, the BIDS™ team (see sidebar) has identified a number of field studies that demonstrate HVAC energy savings, health improvements, and individual productivity gains due to mixed-mode or natural ventilation. By combining the findings from these research studies, Carnegie Mellon's BIDS team has determined that natural ventilation and mixed-mode conditioning systems can provide 47–79 percent HVAC energy savings, 0.3–3.6 percent health cost savings, and 0.2–18 percent productivity gains, for an average return on investment of 120 percent.

Indeed, Carnegie Mellon's BIDS team continues to pursue studies that link access to the natural environment—including daylight, window views, natural ventilation, and indoor nature—to energy, health, and productivity benefits. Figures 8-3 and 8-4 illustrate the strength of the evidence demonstrated in published research to support design that provides access to nature for building occupants.

Carnegie Mellon's BIDS Tool

A new building investment decision support tool—BIDS™—has been developed by the Center for Building Performance and Diagnostics (CBPD) at Carnegie Mellon University, with the support of the Advanced Building Systems Integration Consortium (ABSIC).

Drawing on the results of over 270 field, laboratory, and simulation studies, this cost-benefit decision support tool quantifies the financial benefits of selecting advanced building systems to deliver privacy and interaction, ergonomics, lighting control, thermal control, network flexibility, and access to the natural environment.

The BIDS tool allows ABSIC members to build life-cycle justifications for high-performance design innovations, and illustrates the return on investment possible through a range of cost savings from the "immediate dollars" of energy efficiency, waste management, and churn, to the "long-term dollars" of indoor environmental quality, productivity, and health.

With DOE funding, a summary of these studies related to several major energy design decisions can be found at http://cbpd.arc.cmu.edu/ebids.

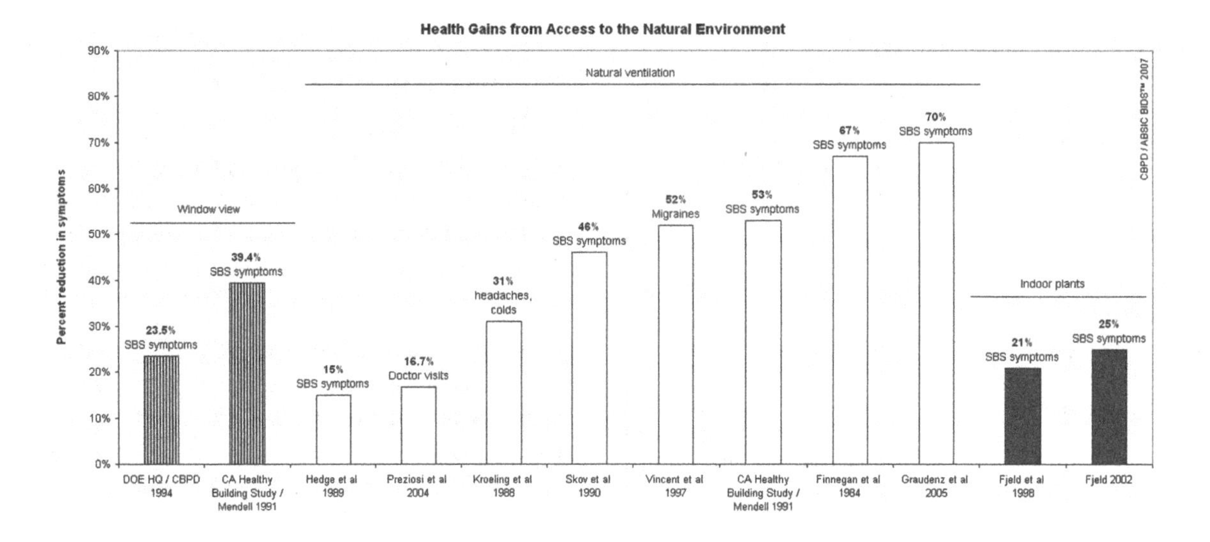

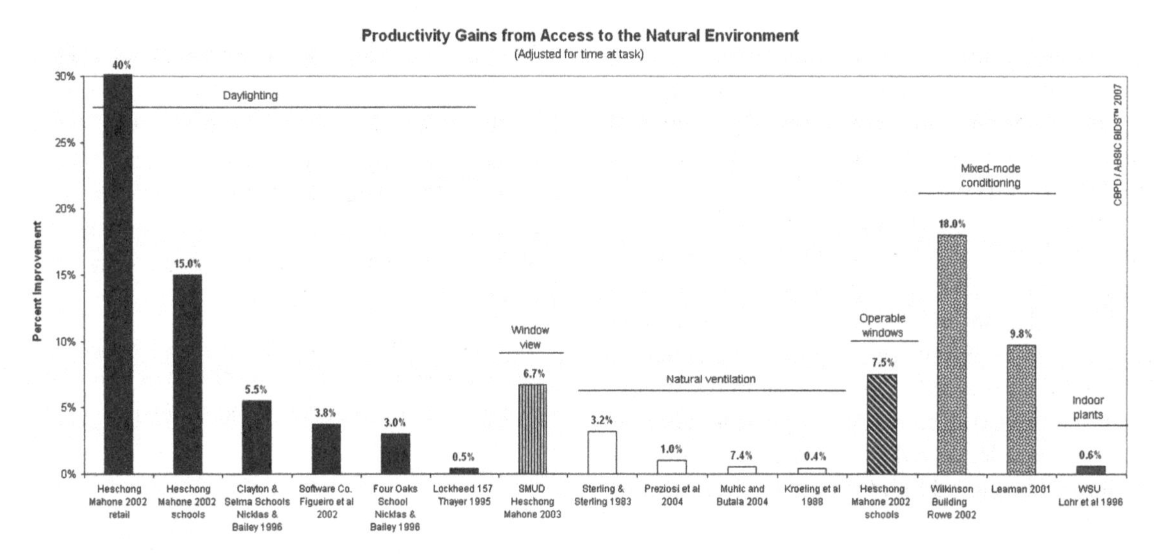

Figures 8-3 and 8-4: Health and productivity gains from access to the natural environment

NATURAL CONDITIONING AND PASSIVE SURVIVABILITY

There are significant additional arguments for a commitment to enhanced connection to the outdoors, including the benefits of passive solar heating, passive cooling, and the assurance that you can survive, or even thrive, in a power outage. However, most of the justifications for these biophilic actions are based on energy and resource savings, since the scientific and health literature seems to be silent on the benefits of passive heating and cooling.

Natural Heating

One of the most natural forms of heat is the use of solar heat gain directly in buildings. Through direct gain, indirect gain, or isolated gain passive solar systems, solar energy is collected, absorbed, stored, and distributed in buildings, typically without mechanical energy, to provide free heat even when the sun goes down. In colder climates, the advantages of passive solar heating include the ability to be comfortable in lower air temperatures, as long as you are sitting in the sun or adjacent to warm surfaces. As a consequence, the air will be less dry, reducing static electricity and excessively dry skin and mucous membranes. Research could confirm whether warmer (solar-heated) surfaces and sunshine also help to reduce mold growth and offer a modest sterilization effect, and whether the warmer surfaces reduce stiff and arthritic conditions, since both radiant and conductive heat loss from the body is curtailed. One only has to see a cat curled in the sun on a window seat to surmise that there are both physiological and psychological benefits of embracing sunshine in building design.

Natural Cooling

In addition to natural ventilation, the use of operable windows can provide natural cooling as an alternative to the prolonged periods of air-conditioning that are becoming more prevalent in buildings today—often right through the winter. Natural cooling is achieved by the simple indoor-outdoor air exchange of room air when temperatures outdoors are cold (often through high transoms or planned infiltration); by the convective benefits of natural breezes when temperatures outdoors are moderate; and in innovative projects, by the use of earth tubes or evaporatively cooled chimneys to precool and humidify outside air before delivery in summer. In addition, natural cooling can be achieved through the use of thermal mass (heavy construction) that diminishes the impact of day-night temperature swings, takes advantage of earth-sheltered temperatures, or captures night-sky radiant cooling—especially in desert climates.

The advantages of natural cooling for human comfort and long-term health should be compared to the impacts of variable and constant cold air blown through diffusers. The quantity and quality of outdoor air that can be delivered through natural cooling should be compared to that delivered by a range of mechanical systems, over time.

However, the limits of natural cooling in hot humid and hot dry climates must be accommodated, especially in the summer months. Mixed-mode conditioning has emerged to extend natural cooling for as long as possible, in as many spaces as possible, while ensuring that mechanical cooling and filtration is provided only when and where it is needed.

Passive Survivability

A term introduced by Alex Wilson of *Environmental Building News*, *passive survivability* is "a building's ability to maintain critical life-support conditions if services such as power, heating fuel, or water are lost" (Wilson 2006). The combination of daylighting, natural ventilation, passive solar heating, and passive cooling, as well as rainwater collection and gravity-fed water utilities, supports human activities independent of the grid. This independence is becoming more critical as we face world energy and water shortages, global warming, and increasingly extreme weather. Moreover, the independence that self-sufficiency offers to children and adults is a value that may well have been exemplified by wilderness survival camps and Amish homesteading experiences for the uninitiated. For architects and engineers, the challenge is to first design "architecture unplugged" to realize the maximum time that a high quality of life can be free of utility grids, and then carefully introduce the most resource-effective solutions to ensure this quality of life year round.

The strongest justifications that we may have for the design of "diaphanous" buildings—which are characterized by an extreme delicacy of form, so as to be almost transparent to light and air—is the realization that we are running out of fossil fuels. Over 10 percent of all U.S. energy use is in lighting buildings, much of this during the daytime when daylight is abundant. Add to this the 6 percent of all U.S energy use spent cooling buildings in both summer and winter, and you have a significant argument for the environmental benefits of windows for daylighting and natural ventilation. Given the dominant number of existing buildings—schools, hospitals, offices, manufacturing facilities—that were originally designed for effective daylighting and natural ventilation, the erosion of natural conditioning is a serious energy cost to the nation. Effective daylighting can yield 10–60 percent reductions in annual lighting energy consumption, with average energy savings for introducing daylight dimming technologies in existing building at over 30 percent. Emerging mixed-mode HVAC systems that interactively support natural ventilation and air conditioning are demonstrating 40–75 percent reductions in annual HVAC energy consumption for cooling (see CBPD eBIDS). Moreover, design for access to the natural environment, including daylighting and natural ventilation strategies, has shown measurable gains for productivity and health in the workplace. The United States needs to meet European and Scandinavian standards that ensure that every worker is within seven meters of a window wall, for views, light, and air. The effective use of natural conditioning with well-designed windows, window controls, and mechanical and lighting system interfaces promises to yield major energy efficiency gains of up to 5 percent of *all* U.S. energy use, to reduce risk in power outages, as well as to provide measurable health and quality-of-life gains.

In many respects, sustainable, healthy buildings have many of the characteristics of sustainable, healthy humans—they are physically fit rather than obese (thin floor plans, finger plans, and courtyard buildings); they have circulatory systems that take the heat from the core out to the surface (e.g., air flow windows and water flow mullions); and they absorb sunlight and breathe fresh air. At the same time, sustainable buildings are designed to reduce climate stresses—rain, cold and hot temperatures, diurnal temperature swings, excessive sun, freeze-thaw—with completely regional design solutions.

ACCESS TO THE OUTDOORS: EXTENDED SPACE, NATURE'S SEASONS / TEXTURES / SOUNDS / SMELLS / FLORA / FAUNA, AND CELEBRATION OF PLACE

Other chapters in this book introduce the importance of access to nature for human health, education, and inspiration. However, the physiological and psychological benefits of an ongoing connection to the locally unique and seasonally dynamic natural environment still need to be quantified. To what extent is science learning enhanced by direct contact with nature's textures, sounds, smells, flora, and fauna? To what extent does the visual and physical extension of indoor space into natural settings reduce claustrophobia or promote the health of seniors?

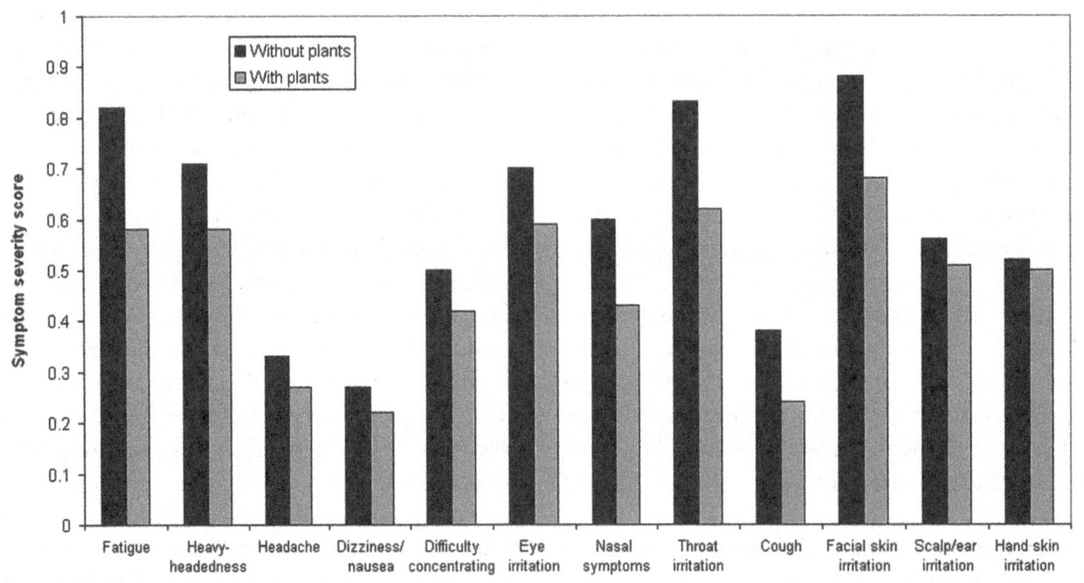

Figure 8-5: Severity of symptoms in the presence and absence of indoor plants

In addition to the studies cited by Robin Moore on early childhood development in natural playgrounds instead of hard-top play yards, a 1998 field experiment by Fjeld et al. revealed that even potted "nature" benefits building occupants (Fjeld et al. 1998). Using a randomized crossover design, the researchers investigated the impact of potted plants on self-assessed health symptoms of 51 employees at a Norwegian oil company, and found an average 21 percent reduction in reported SBS symptoms among occupants when a substantial number of plants were present in their offices (Figure 8-5).

The challenge to designers is not only to connect indoor spaces with the outdoors, but to carefully integrate the natural diversity of the region—its unique climate and seasons, textures, sounds, smells, and diversity of landscape and species. The gardens of Frank Gehry's Disney Hall in Los Angeles capture all of these attributes, while providing workstation views of nature, outdoor meeting areas, and eating spaces for the office occupants (see Figure 8-6 in color insert).

TRANSPARENCY: ACCESS TO LIFE'S ACTIVITIES

Access to human diversity and activity must also be a central tenet of the biophilia community. Our final argument for the introduction of windows in buildings reverses the direction of the access needed. While a direct connection from the indoors to the natural diversity of outdoor places may be critical for human health and inspiration, the direct connection from outdoors to inside is equally critical.

Picture three different coffeehouses: one has no windows or very dark windows that are effectively opaque from the street; one has clear but sealed windows; and one has a glass wall that rolls up into the ceiling so the separation between indoors and outdoors is eliminated (Figure 8-7). Which of these coffeehouses supports the ongoing growth of community spirit? Which ensures safety for those on the street? Which provides a sense of belonging and inclusion that may be critical to avoiding isolation and depression? At the same time, which of these coffeehouses enables the occupants to access the richness of their urban or natural setting—the sights, sounds, smells, and activities?

Since the early work of Oscar Newman on *Creating Defensible Space* (1972), little research has explored the importance of building transparency and street accessibility on human and community health. Since every opening in a building has the potential to connect humans as well as nature, the importance of those connections should be equally addressed.

Figure 8-7: Three coffeehouses with varying degrees of transparency

Figure 8-9: Hospital waiting rooms. Left: A typical ER waiting room where stress and aggressive behavior are often evident. Middle/Right: An ER waiting room with garden views, Providence St. Vincent Hospital, Portland, Oregon (ZGF/Robert Murase)

WINDOWS REVEAL THE SPIRIT OF PLACE

Research is critically needed to support an informed and regional design process to guarantee that "windows become doors" to views, daylight, sunlight, fresh air, breezes, natural comfort, passive survivability, outdoor spaces and activities, extended space, circadian regulation, seasons, climate, and nature's sounds, smells, and indigenous fauna and flora. Collectively, the studies presented in this chapter suggest the importance of access to the natural environment to health or performance outcomes; however, the credibility of the individual studies varies. The difficulty lies in the need to control physical and organization variables for a robust field experiment, and the need to prolong the duration of study for a robust lab experiment. To address these issues, built environment researchers must begin to take cues from the medical community, which consistently pursues longitudinal studies toward the development of health standards. Funding is critically needed for controlled and longitudinal field studies that reveal the mechanisms and the impacts of the quality of the built environment on health and productivity. The import of this long-term research commitment to the study of biophilia and the built environment would ensure that we design the most effective learning environments, healing environments, and working environments.

In spite of all that is unknown in this arena, we feel that sufficient evidence exists to support the following: south-facing hospital rooms with direct sunlight penetration; abundant, natural ventilation in workplaces; daylighting with glare control in schools and workplaces; and views of nature for building occupants, as well as critical access to and engagement of nature in design. The medical community is quickly taking the lead to include natural elements and other "evidence-based" design features into new hospitals and medical centers. Exemplary among these are the new facilities being developed in partnership with the Pebble Project of the Center for Health Design, which incorporate design features known to influence quality of care and patient outcomes (Center for Health Design 2006). (See Figure 8-8 in color insert.) The difference between conventional hospital design and evidence-based design with nature is illustrated in Figure 8-9.

The importance of biophilic design should be quantified, potentially codified, and certainly celebrated. Windows connect building occupants with a richness that is critical to human health and inspiration. Windows provide those outside the building with the level of transparency, oversight, and contact with life's activities that is critical to community. Together, "windows that become doors" define the spirit of place, central to a timeless built environment.

REFERENCES

Beauchemin, K. M., and P. Hays. 1996. "Sunny Hospital Rooms Expedite Recovery from Severe and Refractory Depression." *Journal of Affective Disorders* 40:49–51.

Benedetti, F., et al. 2001. "Morning Sunlight Reduces Length of Hospitalization in Bipolar Depression." *Journal of Affective Disorders* 62:221–223.

Boyce, P. R., J. W. Beckstead, N. H. Eklund, R. W. Strobel, and M. S. Rea. 1997. "Lighting the Graveyard Shift: The Influence of a Daylight-Simulating Skylight on the Task Performance and Mood of Night-Shift Workers." *Lighting Research and Technology* 29(3): 105–134.

Bullough, J. D., M. S. Rea, M. G. Figueriro. 2006. "Of Mice and Women: Light as a Circadian Stimulus in Breast Cancer Research." *Cancer Causes and Control* 17(4): 375–383.

Center for Building Performance and Diagnostics (CBPD). Carnegie Mellon University. BIDS: Building Investment Decision Support. http://cbpd.arc.cmu.edu/bids.

Center for Building Performance and Diagnostics (CBPD). Carnegie Mellon University. eBIDS: Energy Building Investment Decision Support. http://cbpd.arc.cmu.edu/ebids.

Center for Building Performance and Diagnostics/Department of Energy (CBPD/DOE). 1994. "Field Studies of the Major Issues Facing Existing Building Owners, Managers and Users." Pittsburgh: Carnegie Mellon University, Center for Building Performance and Diagnostics, Department of Energy Building Studies.

Center for Health Design. 2006. "The Pebble Project(r) Overview." http://www.healthdesign.org/research/pebble/overview.php.

Choi, Joonho. 2005. Study of the relationship between indoor daylight environments and patient average length of stay (ALOS) in healthcare facilities. Master's thesis, Texas A&M University.

Drinka, P., P. Krause, M. Schilling, B. Miller, P. Shult, and S. Gravenstein. 1996. "Report of an Outbreak: Nursing Home Architecture and Influenza-A Attack Rates." *Journal of the American Geriatrics Society* 44:910–913.

Fjeld, T., B. Veiersted, L. Sandvik, G. Riise, and F. Levy. 1998. "The Effect of Indoor Foliage Plants on Health and Discomfort Symptoms Among Office Workers." *Indoor & Built Environment* 7:204–209.

Heschong Mahone Group. 2003. *Windows and Offices: A Study of Office Worker Performance and the Indoor Environment.* California Energy Commission Technical Report. http://www.h-m-g.com/projects/daylighting/projects-PIER.htm.

Kelland, P. 1992. "Sick Building Syndrome, Working Environments and Hospital Staff." *Indoor Environment* 1:335–340.

Kellert, Stephen. 2005. *Building for Life: Designing the Human-Nature Connection.* Washington, DC: Island Press.

Mendell, Mark J. 1991. Risk factors for work-related symptoms in Northern California office workers. Ph.D. diss., University of California.

Newman, Oscar. 1972. *Defensible Space: Crime Prevention Through Urban Design.* New York: Macmillan.

Preziosi, P., S. Czerniichow, P. Gehanno, and S. Hercberg. 2004. "Workplace Air-Conditioning and Health Services Attendance Among French Middle-Aged Women: A Prospective Cohort Study." *International Journal of Epidemiology* 33(5): 1120–1123.

Roberts, J. E. 2003.[AU: Date per EPRI web site.] "Light Interactions with the Human Eye as a Function of Age." In *The Fifth International LRO Lighting Research Symposium, Light and Human Health, November 3–5, 2002. Report No. 1009370.* Palo Alto, CA: Electric Power Research Institute.

Robertson, A. S., K. T. Roberts, P. S. Burge, and G. Raw. 1990. "The Effect of Change in Building Ventilation Category on Sickness Absence Rates and the Prevalence of Sick Building Syndrome." In *Proceedings of Indoor Air '90* 1:237–242.

Seppanen, O., and W. J. Fisk. 2002. "Association of Ventilation System Type with SBS Symptoms in Office Workers." *Indoor Air* 12(2): 98–112.

Shendell, D., R. Prill, W. Fisk, M. Apte, D. Blake, and D. Faulkner. 2004. "Associations Between Classroom CO_2 Concentrations and Student Attendance in Washington and Idaho." *Indoor Air* 14:333–341.

Ulrich, R. 1984. "View Through a Window May Influence Recovery from Surgery." *Science* 224(4647): 420–421.

Walch, Jeffrey, et al. 2005. "The Effect of Sunlight on Postoperative Analgesic Medication Use: A Prospective Study of Patients Undergoing Spinal Surgery." *Journal of Psychosomatic Medicine* 67:156–163.

Wilson, A. 2006. "Passive Survivability: A New Design Criterion for Buildings." *Environmental Building News* 15(5).

chapter

9

Restorative Environmental Design: What, When, Where, and for Whom?

Terry Hartig, Tina Bringslimark, and Grete Grindal Patil

An extensive built environment stands as a signal human achievement. Roofs, rooms, roadways, windows, walls—myriad features of the built environment attest to the ability of people to shelter themselves from the elements and, in addition, to create enjoyable and inspiring settings for ever more diverse human activities. Yet in many societies, the ongoing extension of the built environment has also seemed like a sustained and coordinated attack on the natural environment. Around the globe today, the built environment overtaxes ecological systems and ignores geophysical realities, thereby increasing the risk of catastrophic events and in other ways diminishing the advantages sought through its construction. Also, by consuming habitat, the development of the built environment continues to undermine the conditions for survival of many nonhuman species.

To help resolve these problems, a growing number of professionals have developed design and technological innovations that reduce harm to the natural environment caused by the construction and operation of buildings. Yet, while commending these innovations, Kellert (2005) has also warned that low-impact or "green" buildings will not suffice if their users do not enjoy them, see fit to maintain them, and so keep them in use over the long run. In identifying ways to make the built environment more pleasing and enjoyable, he invokes an innate human affinity for the natural world, *biophilia*, which he sees rooted in our evolutionary origins in the natural environment (after Wilson 1984; see also Kellert and Wilson 1993; Heerwagen and Orians 1993; Joye 2006). Kellert argues that visual representations of nature, symbols of nature, views to nature, indoor plants, and other natural objects and design elements appeal to this innate affinity and so can evoke positive experiences in the built environment. Indeed, he points out, they have done so for centuries. Kellert (2005) advocates an approach to building that combines biophilic features with low-impact, environmentally sensitive technologies. He has named this approach *restorative environmental design*. The name emphasizes the aim of the approach to "reestablish positive connections between nature and humanity in the built environment" (Kellert 2005, 4; see Chapter 1).

In this chapter, we consider restorative environmental design from a general perspective on relations between people and the environment. The *restoration perspective* has roots in an area of environmental psychology from which restorative environmental design draws theoretical and empirical support. By considering restorative environmental design in light of knowledge organized under the restoration perspective, our intention here is to support the theoretical as well as the practical development of the design approach.

In the following, we first elaborate on the restoration perspective and some theorizing, empirical research, and conceptual issues that fall within its scope. We then consider what restorative environmental design involves, with particular regard to its psychological aspects. We also take up some issues regarding when, where, and for whom it may work. We conclude with comments on prospects and challenges for restorative environmental design.

THE RESTORATION PERSPECTIVE

Much of the research and theory in environmental psychology has concerned problems in the adaptation of people to the environment. Three of the field's core concerns complement each other; they have to do with closely related aspects of adaptation but they differ in their emphases. Work in one of the areas, *environmental stress*, has emphasized the demands from the environment that challenge adaptation, as well as the changes that take place in people as they face those demands with the resources that they have available. Work in another area, *coping*, has emphasized the psychological, social, material, and other resources that people use in their efforts to meet environmental demands, as well as the various strategies that they adopt to deploy their resources. Work in the third area, *restorative environments*, has emphasized the processes through which people restore the resources that they have depleted in meeting environmental demands, and in particular the characteristics of environments that promote restoration of the depleted resources (cf. Saegert and Winkel 1990).

Each of the three research areas builds on distinctive theoretical and practical premises. Those premises constitute particular perspectives on adaptation as a fundamental issue in relations between people and the environment. The theoretical premise of the stress perspective is that when a person continuously faces heavy demands, then adaptation can fail, as reflected for example in poor physical or mental health. To prevent that failure, interventions can seek to reduce the burden of demands that people face. In contrast, the theoretical premise of the coping perspective is that a person can meet even very heavy demands over long periods if he or she has sufficient physiological, psychological, material, and social resources. Measures that make resources more readily available to people, or that help people to make better use of the resources already available to them, can help them to better maintain adaptation. For its part, the theoretical premise of the restoration perspective acknowledges that a person can have ample protection from environmental demands and abundant resources at hand and yet still need periodic restoration. In pursuing goals, in sustaining social relations, in playing and creating, in doing many of the activities that add meaning to life above and beyond merely surviving, a person inevitably depletes some of his or her resources. To continue with his or her activities while also maintaining adaptation to the environment, he or she must restore resources that have become depleted. Interventions that enhance people's opportunities for restoration can help them to more readily, quickly, and completely restore those resources. We summarize the premises of the three perspectives in Table 9-1 (cf. Hartig 2001).

Each of the practical premises lends itself to a general approach to environmental design, and has done so since people first began to build. Examples of what we might call *protective environmental design* include walls and roofs that shelter people from the elements, and so

TABLE 9-1 Perspectives on human adaptation to the environment

	Stress perspective	Coping perspective	Restoration perspective
Theoretical Premise	Heavy demands can undermine adaptation.	Readily available resources support adaptation.	Adaptation requires periodic restoration.
Practical Premise	Interventions can eliminate or mitigate demands.	Interventions can enhance the availability of resources.	Interventions can enhance opportunities for restoration.

eliminate or mitigate some demands, such as exposure to rain and snow, the sun and the cold. Examples of *instorative environmental design* involve some means to deepen or strengthen the ability of people to meet subsequent demands. They include arrangements for heat and water indoors that make those resources available to people where and when they need them most; paths and roads that extend people's reach to other places and so open up access to social contacts, material resources, and innovations; and playgrounds and gardens that stimulate learning and challenge people to develop new capabilities.[1]

For its part, *restorative environmental design*—as it follows from the premises stated for the restoration perspective—may in some respects seem indistinguishable from protective environmental design. For example, the roofs and walls that shelter people from the elements also help them to sleep more soundly. However, restorative environmental design has an important distinguishing characteristic: it goes beyond the elimination or mitigation of demands to the provision of features that promote restoration (cf. Hartig 2004). An urban park ordinarily does more than provide residents with a place to escape from their everyday demands, one that is quieter and less hectic than other outdoor urban spaces; it also provides pleasing distractions that pull visitors' thoughts away from the demands they face, helping them to renew a depleted capacity to direct attention (Kaplan and Kaplan 1989), overcome negative emotions, and wind down physiologically (Ulrich 1983).

Other features of the built environment may have restorative as well as protective and instorative value. Windows provide protection from the elements while also letting in light that people can use to more effectively carry out activities indoors. If a window opens onto pleasing natural scenery during respites from work, then it may also promote restoration during breaks from demanding tasks (e.g., Kaplan 1993, 2001). Still other features may not serve any protective function, but they may promote restoration for people in need of restoration and they may also give an instorative boost of positive feelings to those not in need of restoration at the moment. Indoor plants and pictures of nature in windowless interior spaces might provide such dual benefits (e.g., Bringslimark, Hartig, and Patil 2007a, 2007c; cf. Heerwagen and Orians 1986).

That the different general design approaches overlap should not come as a surprise; the premises on which they rest concern inseparable and to an extent mutually defining aspects of adaptation. Recognizing this, one can see that environmental design can simultaneously serve protective, instorative, and restorative functions in support of adaptation; one can look on the built environment as a complementary set of adaptive capabilities, continuously employed or available on demand or with need. At the same time, one can see that design measures guided by one perspective on adaptation may work against the requirements for adaptation indicated by another perspective. A new building may, for example, protect its occupants from some environmental demands and it may make some resources available to them for their activities indoors, but at the same time it may disallow them access to other important resources and reduce the restorative quality of the surrounding environment (cf. Hartig 2007a).

This brings us back to Kellert's (2005) characterization of restorative environmental design as a combination of biophilic features and environmentally sensitive technologies. The design approach aims not only to reduce the harm that stems from the built environment, but also to make the built environment more pleasing and enjoyable. It seeks both to avoid and minimize harmful impacts on the natural environment, as well as to provide and restore beneficial contacts between people and nature in the built environment (see also Chapter 1). This characterization might appear at first glance to differ from the one indicated by our presentation of the restoration perspective, which emphasizes environmental features that promote the restoration of personal and social resources depleted by individuals, dyads, and perhaps small groups. Yet there are three important commonalities. First, plants, views of natural scenery, and other natural environmental features that Kellert identifies as pleasing, that he sees as evoking biophilic responses, are also thought to promote restoration. Second, what Kellert sees as in need of restoration—positive connections between nature and humanity in the built environment—can be construed as a form of adaptive resource. Third, like other design measures indicated by the restoration perspective, restorative environmental design as Kellert has presented it may also serve protective and instorative functions.

We expand on these three commonalities and related points below; they concern the "what" of restorative environmental design. Before doing so, however, we will elaborate on some theoretical and empirical work in environmental psychology from which restorative environmental design draws support. That will also prepare the way for our discussion of the "what" of restorative environmental design, as well as the "where," "when," and "for whom."

THEORY AND EMPIRICAL RESEARCH ON RESTORATIVE ENVIRONMENTS

The knowledge organized under the restoration perspective primarily concerns restoration and the environmental requirements for restoration. It stems from the unavoidable fact that people must contend with numerous demands in everyday life, either self-generated or imposed upon them by the environment. The demands take countless forms, but all involve a call to mobilize resources. A woman late for a train to the airport calls up the physical energy and muscular strength she needs to run. An office worker faced with a boring task musters the ability to filter out coworkers' voices in order to get it done by the end of the day. A man changing residences asks some friends to help him carry boxes out to the moving van. In such cases, when people use their physiological, cognitive, or social resources to a lesser or greater degree, they potentiate some corresponding degree of restoration. The different forms of resource depletion in turn imply different restoration processes. These may proceed alone or in tandem, some requiring more time, some less. They may occur during waking hours as well as with sleep. And they may each have distinctive environmental requirements.

Psychological theories about the environmental requirements for restoration are of particular interest here because some beneficial effects of biophilic features in the built environment, including restorative effects, are assumed to be mediated by psychological processes (Kellert 2005). Two theories currently dominate the psychological literature on restorative environments. They provide different perspectives on what happens during restoration, in that they deal with different forms of resource depletion and they emphasize different outcomes. However, both of the theories assign particular significance to trees, water, and other natural elements as environmental features that promote restoration, and they build on evolutionary assumptions in doing so. We will briefly overview the two theories before going on to discuss empirical research that they have guided.

Attention restoration theory (Kaplan and Kaplan 1989; Kaplan 1995) deals with attentional fatigue, or a depleted capacity for directing one's attention. It sees restoration from attentional fatigue occurring when a person can gain psychological distance from tasks, the pursuit of goals, and the like, in which he or she routinely must direct attention (being away). When away in this sense, restoration is promoted if the person can rely on effortless, interest-driven attention (fascination) in the encounter with the environment. When the person can let his or her attention go to that which is interesting, he or she can rest the cognitive mechanism that would otherwise work to filter out things that are more interesting than the task at hand. If at the same time the person experiences the environment as coherently ordered and of substantial scope (extent), then fascination with the environment can be sustained. The theory also acknowledges the importance of the match between the person's inclinations at the time, the demands imposed by the environment, and the environmental supports for intended activities (compatibility). The Kaplans argue that these four restorative factors commonly hold at high levels in natural environments. That people are so readily fascinated by natural features in particular is thought to have an evolutionary basis; it would have been adaptive in a biological sense for protohumans to have had their attention rapidly and effortlessly captured by environmental features relevant for survival. However, the Kaplans do not claim that only natural environments are restorative (e.g., Ouellette, Kaplan, and Kaplan 2005). Whether restoration takes place in a natural environment or some other environment, it becomes manifest in a renewed ability to focus and so, for example, in an improved ability to complete tasks that require concentration.

The other of the two theories, psychoevolutionary theory (Ulrich, Simons, Losito, Fiorito, Miles, and Zelson 1991; see also Ulrich 1983), concerns stress reduction rather than attention restoration. It emphasizes the beneficial changes in physiological activity and emotions that occur as a person views a scene. For someone experiencing stress after a situation that involved challenge or threat, viewing a scene might open into restoration. This initially depends on visual characteristics of the scene that can very rapidly evoke an emotional response of a general character, such as interest or fear. This response is thought to take place without a conscious judgment about the scene, and indeed it can occur before a person can formulate such a judgment. The characteristics of the scene that elicit the response include gross structure, gross depth properties, and some general classes of environmental content. In the case of restoration, the process would go something like this: a scene with moderate and ordered complexity, moderate depth, a focal point, and natural contents such as vegetation and water would rapidly evoke positive emotions and hold attention, displacing or restricting negative thoughts and allowing a reduction in arousal that had been heightened by stress. The roles of natural contents and visual characteristics in this process have evolutionary underpinnings, according to Ulrich; humans are biologically prepared to respond rapidly and positively to environmental features that signal possibilities for survival. Restoration becomes manifest in emotions and in physiological parameters such as blood pressure, heart rate, and muscle tension.

One might emphasize particular differences between the theories, but here we think it suffices to say that they provide complementary views on what can happen in a restorative experience, in that psychophysiological stress and attentional fatigue will sometimes arise alone while in other circumstances they may coincide (for more details, see Hartig 2007b). Departing from one or both of these theories, empirical studies have typically built on their common view of environmental features that promote restoration, and so have estimated the effects of different amounts of natural features in actual or photographically simulated environments.

The empirical studies have concerned either the effects of discrete restorative experiences, isolated in time, or the cumulative effects of multiple restorative experiences (Hartig 2007b). The studies of discrete restorative experiences, usually true experiments, have aimed at understanding just what happens between a person and an environment that helps restoration proceed in a given instance. They have tested hypotheses about the emergence of particular kinds of restoration outcomes within particular amounts of time in particular environments, in the interest of assessing the validity of theoretical claims and establishing an empirical basis for practical measures. Experiments guided by psychoevolutionary theory have tested predictions about immediate physiological and emotional effects of viewing natural versus other kinds of environments following exposure to a stressor. They have shown that, within a matter of minutes, looking at scenes of nature can more completely bring physiological arousal back toward prestressor levels than can looking at ordinary urban outdoor scenes (Ulrich et al. 1991) or sitting in a room without a view (Hartig, Evans, Jamner, Davis, and Gärling 2003a). Scenes of nature can also quickly evoke more positive emotions and reduce negative emotions compared to scenes of urban outdoor spaces (e.g., Ulrich 1979; Van den Berg, Koole, and Van der Wulp 2003). In contrast, experiments guided by attention restoration theory have tested predictions about environmental effects on the performance of tasks that require directed attention, though with less certainty about how long it should take for those effects to emerge. Enhanced restoration with photographic simulations of natural versus urban environments has not consistently emerged after 7–20 minutes (cf. Hartig, et al. 1996; Van den Berg, et al. 2003; Berto 2005), but in field experiments differential effects have appeared after longer periods, from 20–50 minutes, spent walking in either a natural or urban environment (Hartig, Mang, and Evans 1991; Hartig et al. 2003a).

Knowing what happens in a discrete restorative experience is important, but one such experience will ordinarily do little to support adaptation in the long run. For this reason, researchers have also tried to measure cumulative effects of environments that varied in restorative quality. The operating assumption behind such work is that people who access environments of high restorative quality during those periods when restoration can occur will realize greater restorative benefits over the long run than they would by spending the time in environments of lesser restorative quality. Working from this assumption, researchers have focused their attention on people in their everyday contexts, where they would ordinarily and regularly find possibilities for restoration over an extended span of time.

The residential context has come into focus in several studies of cumulative effects, since people ordinarily spend a large proportion of their waking as well as sleeping hours within their dwelling or the area around it. For example, Kuo and Sullivan (2001; Kuo 2001) studied low-income urban residents of multifamily buildings that had varying amounts of trees and other vegetation nearby. They uncovered plausible evidence that repeated instances of attention restoration, supported by access to nearby greenery, can have important cumulative effects, including lower domestic violence and better management of major life issues. Other studies have also produced evidence that speaks to possible cumulative effects of restorative experiences supported by natural features in the residential context, as reflected in outcomes such as higher levels of residential satisfaction (Kaplan 2001) and lower psychological distress in children (Wells and Evans 2003).

The workplace is another context in which many people regularly and over an extended span of time come to need and find opportunities for restoration. Here, too, researchers have taken interest in possible cumulative effects of repeated restorative experiences and the influence of environmental variations on those effects. For example, Kaplan (1993) discussed the potential cumulative value of "micro-restorative experiences" in workplaces; a worker might more effectively restore cognitive resources needed for work by periodically looking out a window onto natural features such as trees and vegetation versus onto other view contents. Results from Kaplan's workplace studies suggest that workers with window views onto natural features were more satisfied with their jobs. More recently, Bringslimark and colleagues (2007b) found that self-reported sick leave was negatively associated with the presence of indoor plants within view from the workstation.

Healthcare settings have also drawn the attention of researchers interested in cumulative effects of restorative experiences. Hospitals, clinics, and doctors' offices do not belong to the everyday life of most people, but many people can count on spending some time in such settings at some point. Ulrich's (1984) seminal study of environmental effects on recovery from surgery started from awareness of the stress and anxiety that people often face when receiving treatment. He studied the records kept for patients who, after surgery, were placed in a room that had a window view of either trees or a brick wall. Those with the tree views used fewer potent painkillers than similar patients who had a view of a brick wall. They also had shorter postoperative stays and fewer negative evaluations from nurses. This study, although modest in size, has proved influential in discussions of hospital design, perhaps because the outcomes are important to patients, staff, administrators, and insurers alike.

How do the theories and empirical studies that we have just overviewed aid understanding of restorative environmental design as presented by Kellert (2005; see also Chapter 1)? Most apparently, they speak to an important basis for positive relations between people and nature in the built environment: some natural features can promote psychological restoration. Less apparently, they direct attention to a number of theoretical, conceptual, and practical issues, including the need to specify the resources restored with the design approach, the possibilities for multiple restorative processes to run in tandem, the distinction between effects of discrete restorative experiences and their cumulative benefits, and the different ways in which biophilic design measures can support human adaptation over time. These issues figure in the "what" of restorative environmental design, to which we turn in the next section. Other issues that we have identified in our overview, such as the time required for restoration and the fact that people vary in their restoration needs over time, will be taken up in the subsequent sections on the "where," "when," and "for whom" of restorative environmental design.

RESTORATIVE ENVIRONMENTAL DESIGN: WHAT

What, then, does restorative environmental design involve? The practical rationale, theorizing about biophilia, and design recommendations have been presented in detail by Kellert (2005), Heerwagen and Hase (2001), and elsewhere in this volume; we need not repeat the details here. Instead, to answer the question posed, we comment on defining characteristics of the design approach in light of knowledge organized under the restoration perspective. Informed by theory, empirical research, and conceptual analyses concerning restoration and restorative environments, our discussion focuses on three characteristics: the particular view of which natural and built features to combine in human environments; the different processes of restoration engaged; and the interest in promoting benefits of varying kinds, at different levels, over different spans of time. The restoration perspective provides insights on each of these characteristics of restorative environmental design, particularly with regard to their psychological aspects.

Inclusion of the Benign, Protection from the Dangerous

To begin with, restorative environmental design involves a particular view on what natural and built features to integrate in human environments. It combines low-impact technologies with diverse natural features, from indoor plants to design forms that mimic forms found in nature (see Chapter 1). Importantly, not only the built features of interest to the design approach have a benign character. That the natural features built into or brought into human environments should be called "biophilic" is of course telling. They are features of the natural environment that many people like. Implicitly, features of the natural environment that people dislike, that may awaken disgust or fear (see Ulrich 1993) or otherwise challenge individual adaptation, are to be kept outside and at a safe distance. As already indicated, people may like some natural features in otherwise built environments because they promote psychological restoration (see also Purcell, Peron, and Berto 2001). Conversely, disliked features may cause a need for restoration or disallow it, and excluding them from the setting might at least permit restoration, if not promote it. In this respect, restorative environmental design as presented in this volume is like other design approaches that follow from a restoration perspective on people's adaptation to the environment; that is, it serves protective as well as restorative functions. The positive connections between nature and humanity that it seeks to reinforce in the built environment appear to involve each enriching the other, and neither causing the other harm. Rather than aggressive separation from the natural environment and the threats it presents, the stance seems to be one of making a secure place within it, with notions of what is secure informed by the knowledge that harm to the natural environment can open people to new and greater dangers. Thus, in addition to serving restoration, the design approach is, in its regard for the natural environment, protective in both direct and indirect ways; direct, in that it provides protection from what is potentially dangerous in the natural environment, and indirect, in that it reduces impacts on the natural environment that would otherwise subsequently increase the risk of harm to people.

Multiple Restoration Processes Running in Tandem

Although it also serves protective functions, a defining characteristic of restorative environmental design is of course the emphasis that it places on processes of restoration. Each of its main components serves some form of restoration. Low-impact environmental design serves ecological restoration in a broad sense. By reducing consumption of energy and materials, and by reducing waste, air pollution, and so forth, it helps to restore the integrity of compromised ecological systems. Biophilic design serves psychological restoration. By providing opportunities for contact with nature in the built environment, it enhances opportunities for different kinds of restoration that individuals will regularly need, such as psychophysiological stress recovery and directed attention restoration. Biophilic design also serves what we might call biocultural restoration. It seeks to reestablish the close and rich connection between nature and humanity thought to have existed previously. This connection is expected to arise anew from the positive experiences that people will have in buildings with biophilic design features, including but not limited to restorative experiences.

Although they clearly differ in kind, these three restoration processes have some important conceptual similarities and some interrelations. Considered from the restoration perspective, at least four aspects of the processes have relevance for the characterization of restorative environmental design, particularly in psychological terms. They are the adaptive resources involved, the manner in which psychological processes are engaged, the time span over which restoration occurs, and the determination of restoration success (cf. Hartig 2004). We take each of these in turn.

1. *The adaptive resources involved.* Recall that the theoretical premise of the restoration perspective emphasizes the depletion and restoration of adaptive resources by individuals. Psychological restoration apparently fits with that emphasis, and the knowledge organized under the restoration perspective mainly concerns the renewal of psychological resources. Ecological and biocultural restoration, on the other hand, do not appear to fit with that emphasis, as the adaptive resources in question are not held by individuals and they are not depleted and restored in the same ways or with the same frequency and regularity as psychological resources.

Yet, both ecological and biocultural restoration do affect resources of use to individuals, and those resources do have psychological aspects. All individuals rely on ecological services as fundamental, if poorly acknowledged, adaptive resources; without those services, individuals would likely find life much more difficult and uncertain. To the extent that people recognize that ecological services aid their adaptation, they evaluate them in relation to the demands they face in everyday life. Positive connections between nature and humanity in the built environment can also be construed as a form of adaptive resource; conceivably, they can help individuals to manage the demands of everyday life by boosting positive emotions, providing a sense of perspective on their life circumstances, encouraging learning, helping them feel more secure in wild surroundings, and so on.

2. *The manner in which psychological processes are engaged.* Psychological processes are of course integral to psychological restoration. They also figure prominently in biocultural restoration, which is carried forward by restorative and other positive experiences promoted by biophilic features in the built environment. Less apparently, psychological processes also play a role in the pursuit of ecological restoration through restorative environmental design. Even though low-impact environmental design technologies such as passive solar heating may not engage psychological processes in their ongoing operation, psychological processes still come into the picture. The risk perceptions, environmental attitudes, beliefs about being able to change undesirable circumstances, and ultimately the behaviors of the people who invent, market, and buy those technologies are also important. Moreover, some research suggests that people's motivation to behave in such environmentally friendly ways stems in part from their use of natural environments for psychological restoration (Hartig, Kaiser, and Bowler 2001; Hartig, Kaiser, and Strumse 2007). Thus, psychological restoration may reinforce ecological as well as biocultural restoration, just as ecological and biocultural restoration may enhance people's opportunities for psychological restoration.

3. *The time span over which restoration occurs.* Psychological, biocultural, and ecological restoration run in tandem, but they concern different entities—individuals, populations, and ecological systems—and they run over different spans of time. This fact is important in light of the different premises of the restoration perspective. Although its theoretical premise emphasizes the depletion and restoration of adaptive resources by individuals, its practical premise concerns interventions that typically affect an environment shared by multiple

individuals. The interventions may have relatively immediate effects in terms of enhanced opportunities for psychological restoration, but as some changes remain in place for many generations, so also may their effects persist and accumulate within and across generations. For example, once created, urban parks tend to remain where people put them. Trees may grow old, die, and be replaced, the flower beds may be changed each spring, rain may drive away snow, but the park as a place for restoration remains available to urban residents and visitors. Generation after generation, users of the park may realize the same kind of restorative benefits envisioned by those who originally created it (cf. Olmsted 1997/1870), both in discrete restorative experiences and more cumulatively. An intriguing possibility is that cumulative effects realized by individuals work across generations to make for a more stress-tolerant, better adapted population.

The same type of restorative intervention may be adopted by successive generations and implemented in new locations, as with the common provision of rooms dedicated to sleeping with the design and construction of new housing. At any given time, individuals positively value these environmental design features for their service to restoration, among other possible reasons. Over time, possibly generations, individuals continue to value the design features and so conserve them, in place, as with parks, or through practice, as with biophilic features introduced in new buildings. By promoting psychological restoration in everyday life, biophilic design measures may thus not only reinforce ecological and biocultural restoration at a given time, but also help to sustain those processes over generations, a span of time that they may require.

4. *The determination of restoration success.* Logically, restoration concludes with the return to some initial condition or earlier state of affairs. For some forms of restoration it is difficult to specify that initial condition. The problem lies in the fact that the entities involved—for example, persons, the natural environment—are complex and change in numerous ways over even brief spans of time within the normal course of development. For example, a woman may come face-to-face with death, and while her blood pressure and heart rate decline to some initial level after the terrible moment has passed, she is no longer the same in some fundamental way, nor can she be, having gained an insight on her own mortality. It follows that a determination of when restoration has been achieved can refer only to some subset of attributes for some initial condition at some prioritized historical moment. This problem is well recognized as a challenge to assessments of psychological restoration (e.g., Linden, Earle, Gerin, and Christenfeld 1997; Hartig et al. 1996) and of ecological restoration (e.g., Hobbs and Harris 2001; Hobbs and Norton 1996).

Consider biocultural restoration in light of this problem. Here, too, the initial condition is difficult to define, so it will be difficult to know when restoration is achieved. What seems apparent to us now, though, is that in many societies, few people will want a close and rich connection between nature and humanity to arise anew under exactly the same circumstances in which such a connection may have existed at some much earlier time. People in postprimitive societies have over millenia erected structures and in other ways changed the environment to reduce stressful demands and improve their access to adaptive resources. We doubt that many of the people living in those societies today want to reverse all of those changes and give up the protective and instorative functions of their built environment.

Proponents of restorative environmental design also share this view; the emphasis is not on leaving the built environment, but on reconciling it with the natural environment in a way that is psychologically, culturally, and ecologically sound. It thus seems that the design approach seeks to achieve a state of affairs that never existed. How then should one conceive of the biocultural restoration goal it is meant to serve? The question comes down to the attributes used to describe the initial condition of positive relations between nature and humanity. It would seem that a commonly shared view of humans as part of rather than dominant over nature would be one such attribute (cf. Kluckhohn 1953). Other attributes would concern the ways in which people experience nature. In any case, it seems necessary to distinguish what has existed from what has never existed in describing the success of restoration.

All of this said, it is worth remembering that the success of restoration has prospective as well as retrospective referents. Whether ecological, psychological, or biocultural, people value restoration efforts because they value what has been lost or depleted in light of what is yet to come. Completion of restoration may be defined with respect to the past, but its success is important only with respect to the future. In this future orientation, the intention to serve restorative functions through environmental design has much in common with the intention to serve protective and instorative functions.

Multiplicity of Benefits

A third defining characteristic of restorative environmental design is the set of benefits that it seeks to provide. Those benefits vary in kind, and they are to be realized at different levels over different spans of time. As already indicated in the foregoing, some of those benefits are ecological, some are cultural, and some are psychological. Some are realized by individual persons, others by populations of humans and other species. Some emerge within moments, as within the restorative experiences of individuals, while others may emerge as cumulative effects only after years of repeated restorative experiences. Still others may be realized only through persistent efforts over generations to reduce human impacts on ecological systems and otherwise encourage different patterns of exchange between humanity and the rest of nature. Some benefits are provided continuously, as with relatively low consumption of energy with the ongoing operation of "green" buildings. Others are provided on demand, in a sense, as when a person in need of restoration more quickly recovers depleted psychological resources because views onto natural features promote restoration.

The different benefits speak to the different ways in which restorative environmental design can support the adaptive processes we described at the outset. Through its low-impact and biophilic design measures, the approach can enhance protective, instorative, and restorative functions of the built environment while also working to address the challenges to human adaptation that follow from damage to ecological systems. These different benefits relate to one another, with the restorative benefits that individuals realize from biophilic design features possibly supporting the biocultural transformation and helping to motivate people to behave more ecologically. In fact, it may be difficult to distinguish some benefits as restorative versus protective or instorative. Conceivably, some newly added biophilic design features may mitigate the harmful effects of existing features. For example, trees planted along urban streets may reduce traffic noise entering adjacent residences and workplaces while also enhancing the restorative quality of the views available to residents and workers. Cumulative effects of the intervention could be due to improved protection from noise as well as enhanced restorative quality.

We have more to say about benefits in the following sections on the "when," "where," and "for whom" of restorative environmental design. As for this section, we have sufficiently demonstrated how the restoration perspective can support restorative environmental design, beyond guiding research on how biophilic features might promote positive experiences in built environments. We have also shown how the restoration perspective directs attention to a number of issues relevant to the theoretical development and application of the design approach, such as the specification of the resources restored; how restorative processes of different kinds can run in tandem; the role of psychological processes in ecological and biocultural restoration; the specification of criteria for success; and the interrelations among restorative, instorative, and protective functions. In the coming sections, we discuss a number of additional issues to which the restoration perspective directs attention.

RESTORATIVE ENVIRONMENTAL DESIGN: WHEN AND WHERE

A discussion of the "when" and "where" of restorative environmental design could approach those intertwined topics on vastly different scales. Approaching them on a grand scale, it could look to the historical moment in a group of societies where ecological and demographic trends, scientific and technological advances, and emerging ethical and aesthetic sensibilities converged to provide the impulse for change in some long-standing environmental practices. It could take into account how inexorable population growth, urbanization, global climate change, habitat destruction, and other trends impelled calls for new ways of building and another way of placing much of humanity in relation to nature. It could acknowledge how the major architectural movement of the last century stranded large numbers of people in buildings that they found uninspiring at best and dehumanizing at worst, despite their functionality. Such a discussion could even consider how environmental psychology emerged in response both to architecture that neglected user needs and to destruction of the natural environment.

However, our discussion here only touches on the broad contextual aspects of restorative environmental design; it does not focus on them. Rather than the grand scale, our discussion of the "when" and "where" of restorative environmental design concentrates on matters of time and place that, although important, are modest and mundane. We make the general point that, to promote psychological restoration with biophilic features in the built environment, restorative environmental design would do well to consider when people need restoration, when they can take the time for restoration, where they are at those times, how much time they can dedicate to restoration while there, and other such matters. In the following, we first discuss the distribution of people's needs and opportunities for restoration over places and time. We then consider implications of their distribution for the practice of biophilic design.

The Social Ecology of Stress and Restoration

The first and foremost point for us to make is that restoration needs and opportunities are distributed systematically over time and places. Chance does play a role, in that restoration needs and opportunities do sometimes arise at random, as when an unanticipated argument leaves one needing time to shed the anger, or when rushing out to the next meeting, one is captured momentarily by the beauty of a flowering plant newly placed in the entrance of one's workplace. To a substantial degree, however, restoration needs and opportunities come along as regularly and predictably as the sun goes up and then down again.

The sources of this regularity are encompassed by a social ecological model of stress and restoration (Hartig, Johansson, and Kylin 2003b). This model aids recognition of how discrete restorative experiences come to be regularly repeated and so come to have cumulative effects. It also aids recognition of how aspects of a broader context influence possibilities for realizing restorative benefits from different environments. In presenting the model here, we use "stress" as shorthand for the mobilization and depletion of resources by a person facing demands of one kind or another. As before, "restoration" covers processes like psychophysiological stress recovery and attention restoration through which a person renews depleted resources.

The model consists of three assumptions. The first is that people continuously cycle through stress and restoration processes; that is, they deplete some set of resources as they try to meet demands, then they renew those resources to some greater or lesser degree, then they deplete them again, and so on. The second assumption is that these cycles of stress and restoration are regulated by activity cycles, or the patterns of activities that people ordinarily perform within allocated periods of time. Daily and weekly cycles of activity in particular are routinized and planful; people commonly proceed with some understanding that about how they will participate in different activities in different places at different times. Some activities in some places at some times require that people mobilize resources to meet demands. Other activities, in other places and/or times, allow for restoration of depleted resources. The model's third assumption is that economic, technological, and other processes that work above the individual level influence people's activity cycles. Such processes affect, among other things, the times of the day, days of the week, and weeks of the year that people have available for work and rest; the places they move among in their cycles of activity; and the degree to which particular kinds of places engender stress or promote restoration. It follows that, through their influence on activity cycles, processes operating above the individual level influence stress-restoration cycles.

The model's third assumption deals with events that unfold on a large scale. It encompasses the development of restorative environmental design as well as the processes now at work in its emergence. That is, facing the facts of ecological destruction and seeking to establish more positive relations between nature and humanity in the built environment, design professionals and behavioral researchers are working with new technology, new theory and research, changed ethical sensibilities, and so on to bring low-impact and biophilic features together in a new design approach. Conceivably, some of the design features can function in the intended way independently of how people behave. Yet, in important respects, the success of the design approach is predicated on people using, responding to and coming to value particular design features. Such is the case for the biophilic features, which should promote positive experiences, including restorative experiences. For this reason, restorative environmental design, and more specifically its biophilic component, can be more effective if it is grounded in the matters encompassed by the second assumption in the social ecological model, that is, in the activity cycles that regulate people's cycles of stress and restoration.

Implications of Activity Cycles for Biophilic Design

With a grounding in the mundane matters of people's activity cycles, biophilic design can build on the way in which people's restoration needs and opportunities are systematically distributed over places and time. To illustrate this point, we comment on the practical implications of two general aspects of activity cycles.

1. *Multiple settings.* One important aspect of activity cycles is that they incorporate multiple settings, or places "characterized by recurring patterns of behavior and by widely recognized place meanings (e.g., functional orientation)" (Stokols and Shumaker 1981, 483–484; see also Schoggen 1989). Settings are arranged and furnished to support particular activities by particular people who interact in fulfilling particular roles. Those people occupy them at particular times for particular durations. Among the meanings or values that they attach to a setting, some may reflect on the demands faced when there or on its capacity for supporting restoration. Thus, settings may come to have distinct stress or restoration valences.

One practical implication of this aspect of activity cycles concerns whether or not to include biophilic features in a given setting. Including elements of nature in a setting dedicated to restorative activities may increase the degree to which people actually restore. On the other hand, biophilic features in a setting that is dedicated to intensive work activities may do little to promote positive experiences. Some research suggests that they may even have negative effects. In a study by Larsen and colleagues (1998), productivity on a simple task declined in a room furnished with a large number of potted plants. By interfering with the performance of work tasks, the attention-capturing qualities of some biophilic features might ultimately increase the degree of stress experienced by a worker, who must exert more effort to concentrate on the task at hand. Mitigation of psychophysiological stress or attentional fatigue due to the work might be more effectively accomplished by enhancing the restorative quality of a break room or some other nearby setting available for restorative activities (cf. Shibata and Suzuki 2001). Thus, a decision about whether or how to include biophilic features in any given setting should take into consideration how its occupants organize demanding and restorative activities within a constellation of settings.

2. *Pathways between settings.* As just indicated, another important aspect of activity cycles is that people move among settings. Movements relate to the experience of stress and restoration in a number of ways. Here we focus on the fact that some recurrent patterns of movement can themselves define a type of setting, situated along some form of pathway. The commute from home to work and back is a good example. Some people may regard their commute as a restful interlude between household and work demands, while others may only regard it as travel fraught with major and minor annoyances. The stress or restoration valences that people assign to their commute vary with factors such as distance, time required, and predictability of commute conditions (e.g., Kluger 1998; Stokols and Novaco 1981). Presumably, what people see along a pathway between settings also plays a role; if they like what they see, they may restore more effectively (cf. Parsons, Tassinary, Ulrich, Hebl, and Grossman-Alexander 1998).

This aspect of activity cycles has a number of practical implications for biophilic design. In general, it encourages a view not only to the interiors and immediate exteriors of buildings as potential locations for biophilic features, but also to the pathways that join buildings. More specifically, it can influence a number of design choices that take into consideration how people use particular pathways. Decisions about the routing of some pathways can ensure that those who travel along them have views to existing natural features that will promote positive, perhaps restorative, experiences. The selection of biophilic features to put in along a pathway can be guided by expectations as to whether travel along it will be by foot, bicycle, private car, or some collective transportation mode. For example, along roads trafficked by private cars moving at high speed, biophilic features should perhaps be of a simple and uniform sort, so that they do not exert too strong a pull on the attention of drivers. Thus, design decisions can acknowledge the fact that people are often not stationary in the built environment, but rather moving between settings within their activity cycles. In contrast to the biophilic features in places where people remain relatively stationary, those placed along pathways may be noticed only in passing if at all. Yet, those brief encounters may positively color both a person's experience while traveling and in turn how he or she feels at the destination (cf. Novaco, Kliewer, and Broquet 1991).

We have more to say about activity cycles in the next section, where we discuss "for whom" restorative environmental design can provide benefits. To conclude here, we acknowledge that some if not all of the practical points just made have already been translated into features of the built environment. The translations have been made by people with widely varying degrees of professional training and over many years. Instead of working from an explicit model of the ecology of stress and restoration, they may simply have applied good "common sense." In any case, they presumably acted on the practical premise of the restoration perspective by seeking to enhance opportunities for restoration where and when they thought people could take time for it. Nonetheless, we anticipate that the practice of biophilic design will benefit from deliberate consideration of the systematic way in which restoration needs and opportunities get distributed over places and time within activity cycles. Reference to activity cycles can conceivably lead to novel interventions. Discovering them will however require knowledge not only of places and times, but also of the people involved. In the next section, we turn to consider individual differences in restoration needs and possibilities.

RESTORATIVE ENVIRONMENTAL DESIGN: FOR WHOM

Restorative environmental design places a high priority on the experiences that people have in the built environment. People should enjoy the buildings they live in and move among, and not simply exist or function in them with little impact on the natural environment. A similar prioritization is one of the cornerstones of environmental psychology, a discipline dedicated to improving the fit between people and the sociophysical environment. An ongoing task for many environmental psychologists is to bridge the gap between those who design and those they design for. Over decades of research and consultancy, environmental psychologists have frequently found that architects, urban planners, civil engineers, and other professionals have shaped the built environment in ways that annoy, thwart, threaten, and otherwise dissatisfy many of those who were to benefit from the change (e.g., Sommer 1983). Such unanticipated and unwanted results have occurred despite the good intentions of those responsible for the design and construction. Some of these problems are due to the assumptions they have made about the people that they would come to affect through their design efforts. Of course, their assumptions have often been accurate, and designers have helped and pleased many people through their work. Yet they have not always gotten things right; assumptions about what people like or should like, their adaptability, and so on, have sometimes been off base. For this reason, environmental psychologists have long advocated bringing the users and knowledge of them into the design process (Sommer 1983; Cherulnik 1993). This holds for restorative environmental design as for other design approaches; promoting positive experiences in the built environment will require input from and knowledge about the users.

Proponents of restorative environmental design assume that biophilic features will promote positive experiences in the built environment. They have grounded this assumption in a large body of diverse kinds of evidence. We previously discussed some of that evidence, namely, the empirical research that indicates that natural environments better promote psychological restoration than predominantly built environments lacking natural features. In the following, we add some nuance to our earlier discussion by pointing out some differences among people that may influence the extent to which biophilic design features engender restorative benefits. Our intent here is to illustrate, not to provide an exhaustive treatment of the topic; we only look at four broad sociodemographic characteristics: gender, occupation, place of residence, and socioeconomic position. For each of these characteristics, we make some observations about needs and opportunities for restoration within activity cycles, and we note some implications for restorative environmental design.

1. *Gender*. Men and women commonly have different activity cycles, with different kinds and amounts of demands distributed across settings and times in different ways. The differences reflect differences in their social roles, and they appear particularly pronounced among adults with children. For example, over the past several decades, in Sweden, Norway, the United States and elsewhere, the proportion of women who work outside the home while their children are still young has increased tremendously (e.g., Barnett and Hyde 2001). Yet neither the mother's role as primary caregiver nor the division of domestic labor in two-parent households has changed as quickly. Women still assume more responsibility for domestic work, including child care. Among the full-time employed, as the number of children in a household increases, the combined burden of paid and unpaid work also increases, and more so for mothers than for their partners (e.g., Lundberg, Mårdberg, and Frankenhaeuser 1994). This translates into different patterns of activity within and across settings. For example, in their analyses of Canadian time-use data, Ahrentzen, Levine, and Michelson (1989) found that full-time employed married women spent twice as much time alone in the kitchen and more time with children in bedrooms and bathrooms than did full-time employed married men. In his analyses of Swedish time-use data, Rydenstam (1992) found that women's leisure time was divided into shorter periods than men's, and their leisure activity episodes were more frequently broken off to do work in the home. Both of these studies suggest that women in dual-income families with children have their attempts at restoration in the home frustrated more frequently than do their partners, just as they take on a greater burden of domestic work. Such findings raise questions about whether biophilic design features in the home will engender greater restorative benefits for men or women. They also prompt questions about the amount of effort required for maintenance of some biophilic features; those that require high maintenance may provide little benefit in some settings if those responsible for their maintenance already feel overworked in those settings (cf. Hartig and Fransson, in press).

2. *Place of residence*. Within their activity cycles, people commonly spend many of their waking hours and most of their sleeping hours in their residence. The location of the residence therefore has important implications for everyday demands and opportunities for restoration (Hartig et al. 2003b). For example, compared to urban residents, people living in small towns and rural areas have closer access to natural areas and may otherwise find the built and natural environment more evenly interwoven. They may, of course, differ from urban residents in many other ways relevant to restoration needs and opportunities, but one can reasonably ask whether biophilic design features in their built environment would provide less cumulative benefits for them than for urban residents, who have less contact with the natural world in their everyday lives. With much of the world's population settled or settling in urban areas and likely to remain in or around them (United Nations 2002), urban living conditions are recognized as having great importance for the pursuit of sustainability, not only in ecological terms, but also in social and psychological terms (Van den Berg, Hartig, and Staats 2007). On the one hand, high-density urban living appears to offer a variety of ecological advantages, such as less automobile-based travel and less consumption of fuel for heating apartments. On the other hand, urban life often involves heavy psychological and social

demands. In the United States and other urbanized societies, urban stressors such as noise from traffic, fear of crime, and crowding in public and private spaces, continue to motivate movement toward the urban periphery, closer to nature but still not too far from the amenities and opportunities that the city offers. This pattern of residential development and mobility engenders planning and transportation practices that thwart sustainability and degrade access to or the quality of those experiences originally sought at the outskirts of town. Biophilic design may help to realize the ecological advantages of urbanicity while mitigating the psychological and social costs.

3. *Occupation.* During a normal week, many adults spend a large number of their hours, and sometimes the largest number of their waking hours, engaged in some form of paid work. The occupations they pursue differ in many ways, including the demands that they impose and the settings in which they are carried out. In modern societies, many people work in urban office spaces. Office work typically involves mental rather than physical demands, so the restoration needs that arise differ qualitatively from those experienced with other kinds of work. At the same time, the cost of office space and the manner in which office buildings have been constructed have led to the placement of office workers in indoor spaces that exacerbate mental demands and offer little support for psychological restoration. Office workers must often share their workspaces with others, and they must struggle against the sound of telephones ringing, others talking, and so forth as they try to concentrate on their work (Evans and Johnson 2000). At the same time, their work spaces commonly lack views to the outdoors, and particularly views of natural features, that might promote psychological restoration during brief respites from work (Kaplan 1993). To the extent that they promote restoration, biophilic design features suitable for windowless indoor spaces might prove particularly beneficial for such workers (cf. Heerwagen and Orians 1986; Bringslimark et al. 2007c). However, as noted previously, the attention-capturing qualities of some biophilic features might ultimately make it more difficult to concentrate on the task at hand. Thus, efforts might instead be concentrated on enhancing the restorative quality of break rooms or other nearby settings to which office workers can go for a respite away from their desks.

4. *Socioeconomic position.* People with few economic resources generally have poorer health outcomes than people who have more economic means (Adler et al. 1994). This disparity in health may be due in part to a greater burden of demands faced in everyday life, from problems meeting basic expenses to more stressful conditions in the residential context, such as noise, crowding, and a lack of security (Saegert and Evans 2003). At the same time, people living in poverty, particularly in urban areas, may have relatively little access to places that promote restoration. For example, low-income multifamily housing may lack surrounding green spaces and other leisure amenities (cf. Kuo 2001). Also, poverty may translate into limited means for travel to more suitable places during leisure time. Given the implications of poverty for demands and restoration opportunities in everyday life, biophilic design features can work to narrow socioeconomic disparities in health. On the other hand, biophilic design might work to preserve or even expand those disparities. Promoted as a fashionable "new" approach, depicted in the glossiest magazines and available at substantial cost only to the well-to-do, biophilic features may enhance the already good restorative quality of the settings occupied by people who have relatively weak needs for restoration and who otherwise can access a wide range of positive experiences. One broader issue is whether biophilic design will extend the upper bound of restorative quality in those settings available to a few, or shift the bottom level upward. Pursuit of the latter objective would likely provide a greater net benefit.

Certainly, one could discuss other differences among people with regard to restoration needs and opportunities. Our discussion here nonetheless suffices to illustrate that knowledge of such differences can be important to the implementation of restorative environmental design. Before leaving this discussion, however, we want to again note the importance of differences *within* persons. People change over time, and sometimes quite rapidly, as with the depletion and restoration of adaptive resources. A person feeling exhausted and having difficulty focusing may behave quite differently from the way he or she would when not in need of restoration. Furthermore, that person's liking for different environments may differ substantially (Staats, Kieviet, and Hartig 2003; Staats and Hartig 2004), and he or she may be more attuned to those features of the environment that are relevant for restoration (Hartig and Staats, 2006). Thus, although biophilic design features may be constantly present in some physical sense, their prominence in a person's experience of the environment may depend on that person's need for restoration. Analogous points might be made with respect to other kinds of change within individuals over time, as with development over the lifespan. For example, Moore and Marcus (see Chapter 10), Louv (see Chapter 11), and Pyle and Orr (see Chapter 12) discuss how positive experiences in natural environments may play an important function in early development. Others have commented on the particular value of contact with nature for people at a late stage in life, particularly if they live in some kind of residential facility (e.g., Grahn and Bengtsson 2005). From a restoration perspective, the basic issues remain the same: people differ in their restoration needs and opportunities, and biophilic design interventions will do well to attend to those differences.

CONCLUDING COMMENTS: SOME PROSPECTS AND CHALLENGES FOR RESTORATIVE ENVIRONMENTAL DESIGN

We have in this chapter considered restorative environmental design in light of knowledge organized under the restoration perspective. Building on theory, empirical research, and conceptual analyses concerned with psychological restoration and restorative environments, we have offered some thoughts on what restorative environmental design involves, and when, where, and for whom it might best work. In closing, we wish to offer some views on its prospects and on challenges to its widespread implementation, looking beyond those points already raised in the foregoing.

In one sense, the outlook for restorative environmental design appears excellent, simply because the need is so great. More and more people recognize that modern ways of living involve practices that increasingly threaten the ability of ecological systems to sustain large portions of humanity and a host of other species. Concerned about the trends, many people, companies, and institutions are now acting to change lifestyles, technologies, and relations between humanity and nature more generally. A fundamental shift is taking place in the way that many basic human activities are evaluated, from goals of subjugating and improving on nature to goals of cooperation and reunification with nature.

Yet the possibility exists that the low-impact component of restorative environmental design will crowd out the biophilic component. Fascination with technological capabilities has, in the past, blinded designers to important needs of eventual users, as with maximal exploitation of the ability to build high despite the potential psychological costs to residents living on upper floors (cf. Evans, Wells, and Moch 2003). Some people will surely take delight in the low-impact technologies represented in a building that they use; for some of them, a positive experience of the building may have little or nothing to do with the inclusion of biophilic features. However, not all of those people who come to live or work in low-impact buildings will be similarly fascinated by the technology, and to the extent that the technology is "laid bare" as in a showcase, users may find it too technical and off-putting. A few potted plants thrown in as token biophilic elements may not suffice to counteract the technical display and engender the positive experiences envisioned by Kellert (2005). It seems to us that a challenge faced by restorative environmental design will be to find ways to integrate biophilic and low-impact features, so that what helps to reduce human impacts on ecological systems also contributes to positive experiences aside from those grounded in fascination with the technology per se.

The prospect for the success of the design approach also carries a challenge in one other major respect. The biophilia hypothesis asserts that people have an evolved affinity for other life, and the implementation of biophilic design may also serve to further acceptance by laypeople of that essentially biological hypothesis. Yet, the validity of the hypothesis does not rest on the beliefs of the general public, but on a body of scientific evidence. At present, the scientific evidence cannot say to what extent any common human affinity for other life is due to biology. In his seminal work on the topic, Wilson (1984) acknowledged the role of culture in shaping human responses to other life forms and the natural environment more generally. Similarly, Kellert has treated biophilia as "a 'weak' genetic tendency whose full and functional development depends on sufficient experience, learning, and cultural support" (Kellert 2005, 4). Proponents of restorative environmental design will do well in reminding clients and other members of the public that expectations of positive responses to biophilic features have much to do with cultural forces. By doing so, they will help to ward off unreasonable expectations regarding one-size-fits-all solutions grounded in expectations of biological determinism. Indeed, proponents of the design approach might point out that, just as natural selection has conserved adaptive genetic mutations over the long course of human evolution, so has a process of cultural selection conserved those products of the human intellect, like architectural forms, that have helped people to better live in an often demanding natural world. Through its biophilic component, restorative environmental design can encourage a future cultural fitness that builds on some echo of a biological past.

NOTES

1. The word *instorative* has been used to distinguish a particular family of benefits that people may realize through their experience in a particular environment (Hartig, et al.1996). In contrast to restorative benefits, which involve the renewal of depleted resources, some benefits involve the acquisition of new resources; a person may, for example, become more self-reliant or self-confident, acquire new skills, or gain in physical fitness. Such benefits are appropriately described as "instorative." Despite its recent origin, the word *instorative* does have etymological grounds. Just as *restorative* means "of or relating to restoration," the word *instorative* means "of or relating to instoration." For its part, *instoration* is a simple modification of the word *instauration*, one meaning of which is "an act of instituting or establishing something." The same modification brought *restoration* from *restauration*. The word *instore* exists in English, but it is described as obsolete. It means "furnish" or "provide" (compare with "store"). (For further details, see *Webster's Third New International Dictionary of the English Language, unabridged*). In Sweden, Grahn has built on this concept in describing how gardens might provide instorative benefits (see Grahn and Bengtsson, 2005).

REFERENCES

Adler, N. E., T. Boyce, M. Chesney, S. Cohen, S. Folkman, R. L. Kahn, and S. L. Syme. 1994. "Socioeconomic Status and Health: The Challenge of the Gradient." *American Psychologist* 49:15–24.

Ahrentzen, S., D. W. Levine, and W. Michelson. 1989. "Space, Time, and Activity in the Home: A Gender Analysis." *Journal of Environmental Psychology* 9:89–101.

Barnett, R. C., and J. S. Hyde. 2001. "Women, Men, Work, and Family: An Expansionist Theory." *American Psychologist* 56:781–796.

Berto, R. 2005. "Exposure to Restorative Environments Helps Restore Attentional Capacity." *Journal of Environmental Psychology* 25:249–259.

Bringslimark, T., T. Hartig, and G. Grindal Patil. 2007a. "The Psychological Benefits of Indoor Plants: A Critical Review of the Experimental Literature." Manuscript under review.

Bringslimark, T., T. Hartig, and G. Grindal Patil. 2007b. "Psychological Benefits of Indoor Plants in Workplaces: Putting Experimental Results into Context." *HortScience* 42(3): 581–587.

Bringslimark, T., T. Hartig, and G. Grindal Patil. Manuscript under review. "Adaptation to Windowlessness: Do Office Workers Compensate and Does it Work?" Manuscript in preparation.

Cherulnik, P. D. 1993. *Applications of Environment-Behavior Research: Case Studies and Analysis.* Cambridge: Cambridge University Press.

Evans, G. W., and D. Johnson. 2000. "Stress and Open-Office Noise." *Journal of Applied Psychology* 85:779–783.

Evans, G. W., N. M. Wells, and A. Moch. 2003. "Housing and Mental Health: A Review of the Evidence and a Methodological and Conceptual Critique." *Journal of Social Issues* 59:475–500.

Grahn, P., and A. Bengtsson. 2005. "Lagstifta om utevistelse för alla! Låt våra gamla få komma utomhus när de önskar!" In *Den omvända ålderspyramiden. Stiftelsen Vadstena Forum*, edited by G. Blücher and G. Graninger.

http://www.ep.liu.se/ea/is/2005/003/is003-contents.pdf.

Hartig, T. 2001. "Guest Editor's Introduction" (special issue on restorative environments). *Environment and Behavior* 33:475–479.

Hartig, T. 2004. "Restorative Environments." In *Encyclopedia of Applied Psychology*, edited by C. Spielberger, vol. 3, 273–279. San Diego: Academic Press.

Hartig, T. 2007a. "Congruence and Conflict Between Car Transportation and Psychological Restoration. In *Threats from Car Traffic to the Quality of Urban Life: Problems, Causes, and Solutions*, edited by T. Gärling and L. Steg. Amsterdam: Elsevier.

Hartig, T. 2007b. "Three Steps to Understanding Restorative Environments as Health Resources." In *Open Space: People Space*, edited by C. Ward Thompson and P. Travlou. London: Taylor & Francis.

Hartig, T., A. Böök, J. Garvill, T. Olsson, and T. Gärling. 1996. "Environmental Influences on Psychological Restoration." *Scandinavian Journal of Psychology* 37:378–393.

Hartig, T., G. W. Evans, L. J. Jamner, D. S. Davis, and T. Gärling. 2003a. "Tracking Restoration in Natural and Urban Field Settings." *Journal of Environmental Psychology* 23:109–123.

Hartig, T., and U. Fransson. In press. "Leisure Home Ownership, Access to Nature, and Health: A Longitudinal Study of Urban Residents in Sweden." *Environment and Planning*

Hartig, T., G. Johansson, and C. Kylin. 2003b. "Residence in the Social Ecology of Stress and Restoration." *Journal of Social Issues* 59:611–636.

Hartig, T., F. G. Kaiser, and P. A. Bowler. 2001. "Psychological Restoration in Nature as a Positive Motivation for Ecological Behavior." *Environment and Behavior* 33:590–607.

Hartig, T., F. G. Kaiser, and E. Strumse. 2007. "Psychological Restoration in Nature as a Source of Motivation for Ecological Behavior." *Enviornmental Conservation*

Hartig, T., M. Mang, and G. W. Evans. 1991. "Restorative Effects of Natural Environment Experiences." *Environment and Behavior* 23:3–26.

Hartig, T., and H. Staats. 2006. "The Need for Psychological Restoration as a Determinant of Environmental Preferences." *Journal of Environmental Psychology* 26:215–226.

Heerwagen, J. H., and B. Hase. 2001. "Building Biophilia: Connecting People to Nature." *Environmental Design + Construction*, March–April, 30–36.

Heerwagen, J. H., and G. H. Orians. 1986. "Adaptations to Windowlessness: A Study of the Use of Visual Décor in Windowed and Windowless Offices." *Environment and Behavior* 18:623–639.

Heerwagen, J. H., and G. H. Orians. 1993. "Humans, Habitats, and Aesthetics." In *The Biophilia Hypothesis*, edited by S. R. Kellert and E. O. Wilson, 138–172. Washington, DC: Island Press.

Hobbs, R. J., and J. A. Harris. 2001. "Restoration Ecology: Repairing the Earth's Ecosystems in the New Millennium." *Restoration Ecology* 9:239–246.

Hobbs, R. J., and D. A. Norton. 1996. "Towards a Conceptual Framework for Restoration Ecology." *Restoration Ecology* 4:93–110.

Joye, Y. 2006. "An Interdisciplinary Argument for Natural Morphologies in Architectural Design." *Environment and Planning B* 33:239–252.

Kaplan, R. 1993. "The Role of Nature in the Context of the Workplace." *Landscape and Urban Planning* 26:193–201.

Kaplan, R. 2001. "The Nature of the View from Home: Psychological Benefits." *Environment and Behavior* 33:507–542.

Kaplan, R., and S. Kaplan. 1989. *The Experience of Nature: A Psychological Perspective*. New York: Cambridge University Press.

Kaplan, S. 1995. "The Restorative Benefits of Nature: Toward an Integrative Framework. *Journal of Environmental Psychology* 15:169–182.

Kellert, S.R. 2005. *Building for Life: Designing and Understanding the Human-Nature Connection*. Washington, DC: Island Press.

Kellert, S. R., and E. O. Wilson. 1993. *The Biophilia Hypothesis*. Washington, DC: Island Press.

Kluckhohn, F. R. 1953. "Dominant and Variant Value Orientations." In *Personality in Nature, Society, and Culture*, 2nd ed., edited by C. Kluckhohn, H. A. Murray, and D. M. Schneider, 342–357. New York: Knopf.

Kluger, A. N. 1998. "Commute Variability and Strain." *Journal of Organizational Behavior* 19:147–165.

Kuo, F. E. 2001. "Coping with Poverty: Impacts of Environment and Attention in the Inner City." *Environment and Behavior* 33:5–34.

Kuo, F. E., and W. C. Sullivan. 2001. "Aggression and Violence in the Inner City: Effects of Environment via Mental Fatigue." *Environment and Behavior* 33:543–571.

Larsen, L., J. Adams, B. Deal, B-S Kweon, and E. Tyler. 1998. "Plants in the Workplace: The Effects of Plant Density on Productivity, Attitudes, and Perceptions." *Environment and Behavior* 30:261–281.

Linden, W., T. L. Earle, W. Gerin, and N. Christenfeld. 1997. "Physiological Stress Reactivity and Recovery: Conceptual Siblings Separated at Birth?" *Journal of Psychosomatic Research* 42:117–135.

Lundberg, U., B. Mårdberg, and M. Frankenhaeuser. 1994. "The Total Workload of Male and Female White Collar

Workers as Related to Age, Occupational Level, and Number of Children." *Scandinavian Journal of Psychology* 35:315–327.

Novaco, R. W., W. Kliewer, and A. Broquet. 1991. "Home Environmental Consequences of Commute Travel Impedance." *American Journal of Community Psychology* 19:881–909.

Olmsted, F. L. 1997. "Public Parks and the Enlargement of Towns." In *The Papers of Frederick Law Olmsted: Supplementary Series: Vol. . 1, Writings on Public Parks, Parkways and Park Systems*, edited by C. E. Beveridge and C. F. Hoffman. Baltimore: Johns Hopkins University Press. (Original work published 1870)

Ouellette, P., R. Kaplan, and S. Kaplan. 2005. "The Monastery as a Restorative Environment." *Journal of Environmental Psychology* 25:175–188.

Parsons, R., L. G. Tassinary, R. S. Ulrich, M. R. Hebl, and M. Grossman-Alexander. 1998. "The View from the Road: Implications for Stress Recovery and Immunization." *Journal of Environmental Psychology* 18:113–140.

Purcell, A. T., E. Peron, and R. Berto. "Why Do Preferences Differ Between Scene Types?" *Environment and Behavior* 33:93–106.

Rydenstam, K. 1992. "I tid och otid: En undersökning om kvinnors och mäns tidsanvändning 1990/91" [At all times: How women and men use their time 1990/91]. SCB Report No. 79. Stockholm: Statistiska centralbyrå.

Saegert, S., and G. W. Evans. 2003. "Poverty, Housing Niches, and Health in the United States." *Journal of Social Issues* 59:569–589.

Saegert, S., and G. Winkel. 1990. "Environmental Psychology." *Annual Review of Psychology* 41:441–477.

Schoggen, P. 1989. *Behavior Settings*. Stanford, CA: Stanford University Press.

Shibata, S., and N. Suzuki. 2001. "Effects of Indoor Foliage Plants on Subjects' Recovery from Mental Fatigue." *North American Journal of Psychology* 3:385–396.

Sommer, R. 1983. *Social Design: Creating Buildings with People in Mind.* Englewood Cliffs, NJ: Prentice-Hall.

Staats, H., and T. Hartig. 2004. "Alone or with a Friend: A Social Context for Psychological Restoration and Environmental Preferences." *Journal of Environmental Psychology* 24:199–211.

Staats, H., A. Kieviet, and T. Hartig. 2003. "Where to Recover from Attentional Fatigue: An Expectancy-Value Analysis of Environmental Preference." *Journal of Environmental Psychology* 23:147–157.

Stokols, D., and R. W. Novaco. 1981. "Transportation and Well-Being." In *Transportation and Behavior*, edited by I. Altman, J. F. Wohlwill, and P. Everett, 85–130. New York: Plenum.

Stokols, D., and S. A. Shumaker. 1981. "People in Places: A Transactional View of Settings." In *Cognition, Social Behavior, and the Environment*, edited by J. H. Harvey, 441–488. Hillsdale, NJ: Lawrence Erlbaum.

Ulrich, R. S. 1979. "Visual Landscapes and Psychological Well-Being." *Landscape Research* 4:17–23.

Ulrich, R. S. 1983. "Aesthetic and Affective Response to Natural Environment." In *Behavior and the Natural Environment*, edited by I. Altman and J. F. Wohlwill, 85–125. New York: Plenum.

Ulrich, R. S. 1984. "View Through a Window May Influence Recovery from Surgery." *Science* 224:420–421.

Ulrich, R. S. 1993. "Biophilia, Biophobia, and Natural Landscapes." In *The Biophilia Hypothesis*, edited by S. R. Kellert and E. O. Wilson, 73–137. Washington, DC: Island Press.

Ulrich, R. S., R. Simons, B. D. Losito, E. Fiorito, M. A. Miles, and M. Zelson. 1991. "Stress Recovery During Exposure to Natural and Urban Environments." *Journal of Environmental Psychology* 11:201–230.

United Nations. 2002. *World Urbanization Prospects: The 2001 Revision.* New York: Population Division, United Nations.

Van den Berg, A. E., T. Hartig, and H. Staats. 2007. "Preference for Nature in Urbanized Societies: Stress, Restoration, and the Pursuit of Sustainability." *Journal of Social Issues* 63:79–96.

Van den Berg, A. E., S. L. Koole, and N. Y. Van der Wulp. 2003. "Environmental Preference and Restoration: (How) Are They Related?" *Journal of Environmental Psychology* 23:135–146.

Wells, N. M., and G. W. Evans. 2003. "Nearby Nature: A Buffer of Life Stress Among Rural Children." *Environment and Behavior* 35:311–330.

Wilson, E. O. 1984. *Biophilia: The Human Bond with Other Species.* Cambridge, MA: Harvard University Press.

chapter 10

Healthy Planet, Healthy Children: Designing Nature into the Daily Spaces of Childhood

Robin C. Moore and Clare Cooper Marcus

Although the epidemics and infectious diseases targeted by public health agencies during the last 130 years have largely been eradicated in the Western, industrial world, preventable lifestyle diseases have replaced them. Postmodern childhood is facing entirely new health threats resulting from rapid, massive cultural changes, including the impacts of new technologies on behavior. More and more of children's time is being "pulled" indoors away from nature by homework, video and computer screens, parental anxiety about stranger danger, and the dangers of automobile traffic (Jago et al. 2005). Richard Louv's book *Last Child in the Woods* has helped to focus public attention on the possible negative consequences for childhood health of these new risk factors. To protect children and support healthy lifestyles, new forms of "inoculation" are required, including changes to the built environments of children's daily lives.

Stimulated and emboldened by the many-layered, wide-ranging contents of *Children and Nature* (Kahn and Kellert 2002) and the empowering thrust of the biophilic building design symposium from which the present book derives, this chapter presents examples of designed environments that support or have the potential to support children's daily outdoor contact with nature and thus ensure the biophilic evolution of our planet and its human citizens. This chapter draws on the latest research findings, which suggest that a healthy, therapeutic effect is experienced by children who are directly exposed to nature (Wells and Evans 2003; Wells 2000; Kuo et al. 1998) and explores the role of physical design in improving the quantity and quality of exposure to nature by integrating it into the built environment. The majority of children worldwide live in urban environments, approximately half of them in urban centers of less than 500,000 population (Satterthwaite 2006). Thus our focus is the everyday life of urban children and concern for the quality of the environments where they spend most of their time, where "biophilic design" (supporting and stimulating children's biophilia) has most potency, where access to nature can be guided by design policy in childcare centers, schools, residential neighborhoods, and community facilities such as parks, museums, zoos, and botanical gardens. These topics will be addressed within the scope of this chapter because the fact is that they are not receiving adequate attention in current urban design practice.

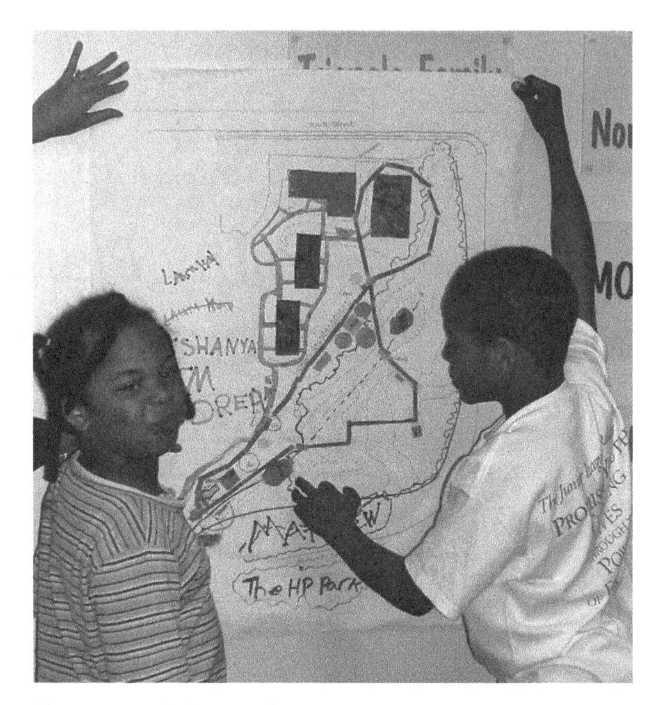

Figure 10-1: During a public housing community design workshop, these resident children are presenting their design proposals to improve the shared open space around their homes.

SUPPORTING A NEW BIOPHILIC CULTURE BY DESIGN

It is evident that we are at a turning point in history where opportunities for children to explore the natural world, until recently taken for granted, must now be intentionally created (Louv 2005; Rivkin 1995). To some this may seem a contradiction. How can the qualities of naturally occurring phenomena be deliberately recreated? The fact is that there is no other choice but to fully engage the urban planning, landscape architecture, and architecture professions in creating new, nature-based urban development policies to help ameliorate the new lifestyle health issues. On the other hand, solutions cannot be imposed but must evolve through community-based processes to engage stakeholders and users (including children) in creating design solutions (Cele 2006). Middle-age children (definitions of outer limits vary, but roughly between 6 and 12) are skilled and capable of evaluating their surroundings and explaining their likes, dislikes, fears, and perceptions of territorial barriers (Moore 1980)—and to make design proposals to improve their surroundings (see Figure 10-1).

Biophilic design for children is supported by precedents (case study designs that have withstood the test of time) that may inspire community action to help a new biophilic culture to take root. Our hope is that these examples will support the creation of policies to support more inclusive, healthy lifestyles. Compelling examples are needed to inform parents, teachers, early childhood professionals, school officials, neighborhood developers, and all those who want to advance the state of the art and capture the market represented by families seeking healthy, sustainable settings for their children. The selected design precedents cover a range of scales and contexts that reflect a variety of needs across the childhood age span. The examples also address issues of family characteristics and demographics and illustrate the constraints and opportunities for designed natural systems in a variety of urban contexts.

For biophilic design to be fully effective, it should extend beyond buildings into what Danish urban designer Jan Gehl has called the "life between buildings" (Gehl 2003), to embrace the outdoor habitat of our most important citizens: children. Outdoors is where immersion in nature is more feasible, where young bodies and minds can be engaged with peers in health-sustaining activities with their surroundings. This being so, it is surprising that recent sustainable design literature (Beatley 2000; Hough 1990; Thomas 2003) does not emphasize children as arguably the most important users of sustainable, "green" urban development.

A fundamental assumption of this chapter is that children are born as "biophilic beings," expressed in their intrinsic curiosity to explore and learn from the natural world without fear and intimidation (Kellert 1993). Based on interviews with environmental activists in Kentucky in the United States, and Oslo and Trondheim in Norway, Chawla (2006) presents a compelling theoretical framework and research-based statement addressing the critical role of childhood experience of nature in explaining adult environmental stewardship later in life. Wells and Lekies (2006), interviewed 2,000 adults across the United States to present further convincing evidence supporting the strong connection between environmentalism and childhood experience of nearby nature—especially if "wild." Effective biophilic design must integrate two domains of health: children and planet. Children must spend sufficient time in naturally rich, healthy environments for biophilia to be instilled as a lifelong affect which, in turn, will create a sufficiently large majority of biophilic citizens who love the world so strongly as to become adult environmentalists doing everything in their power to combat global warming and associated environmental issues (Chawla 2006) (see Figure 10-2).

Figure 10-2: A local nature reserve or botanical garden can offer rich opportunities for adults and children to share nature together. Knowledgeable, attentive adults can help children expand their awareness and appreciation of the beauty of nature.

Many barriers presently limit children's access to nature, which may prevent them from growing up with love and respect for the planet and a passion to protect it (Crain 2003). These barriers include the lack of direct experience of natural processes and materials in early childhood when sensory impact is the primary mode of learning; the negative messages from adults who have already lost their biophilic feeling for nature; the lack of use of living environments in schools where children receive primary education at a stage of development when minds and bodies are open to all that the world has to offer and where the seeds of understanding about how the world works are sown; the lack of rich, diverse, accessible sustainable landscapes in the residential districts where children live; and the lack of independent mobility and rich environmental experiences at a neighborhood level.

Currently, built environments often present barriers to children's independent mobility and therefore their experience of nature. To increase the "activity friendliness" of urban neighborhoods for children (de Vries et al. 2007), substantial structural urban design issues must be overcome such as traffic and road/sidewalk configuration, school and park planning, location of shared spaces in residential neighborhoods, location of walk/bike/skate/ski trails, residential density and site planning, and urban planning issues such as increasing walking for young people by ensuring recreation destinations close to home (Frank et al. 2007; Mackett et al. 2004).

In addition to ensuring that children's intrinsic biophilia is activated, developed and supported strongly enough to extend into adulthood, biophilic design simultaneously addresses children's health, a need most obviously expressed by burgeoning sedentary lifestyle trends, resulting in an obesity crisis for children and adults. The latter may be the most visible and possibly the most serious manifestation of the negative health impact of children's lifestyle changes in the last three decades or so, but it is not the only consequence—as addressed below.

CHILDHOOD LIFESTYLE HEALTH THREATS

Combating Sedentary Behavior

Worrying, negative health changes are affecting the physical, mental, and social functioning of children across the Western, industrialized world, changes so severe that the steady rise in life expectancy during the past two centuries may soon come to an end. A recent study of the effect of obesity on longevity in the United States (Olshansky et al. 2005) suggests that a growing proportion of children born today may die before their parents. In the United States, approximately 18 percent of children under 19 years old are overweight or at risk of being overweight (CDC 2007). These negative lifestyle conditions are even beginning to impact early childhood. In the United States, more than 10 percent of two- to five-year-olds are obese and more than 20 percent are overweight or at risk of being overweight (Ogden et al. 2002). The situation in some southern European countries is even worse. In Spain 13.9 percent of individuals aged two to twenty-four are obese and 26.3 percent are overweight (EEHC 2005).

Levels of movement and energy expenditure necessary for healthy physical development are not feasible when limited to indoor environments. Being outdoors is the best predictor of children's physical activity (Sallis, Prochaska, and Taylor 2000). However, today's children are not getting outdoors enough. This reduction in "free range childhoods" is a major unhealthy lifestyle factor. Although empirical data is lacking on this issue, compelling anecdotal information from concerned professionals, parents, and cultural commentators has accumulated, most recently contributed to by Pyle (1993; see also Chapter 12, which stresses the critical experiential loss resulting from reduced free range access to natural settings), adding force to Richard Louv's compilation of evidence (Louv 2005). Research conducted in the 1970s and 1980s provides substantial, evidence-based benchmarks of children's "range behavior" from an era when it was internally driven by children's maturity levels rather than external constraints of the built environment and adult control (Moore 1986a; Moore, 1980; Moore and Young 1978; Hart 1979).

The Threat of Automotive Traffic

Traffic danger exacerbated by inappropriate street design is the most obvious, measurable factor inhibiting children's outdoor behavior. Pedestrian-friendly residential street design has a long history stretching back to the 1875 layout for Bedford Park, Chiswick, London (Southworth and Ben-Joseph 2003). Innovative residential street designs emphasizing pedestrians and cyclists, including children (Eubank-Ahrens 1980; Francis 1980; Moore 1980), continued to evolve on both sides of the Atlantic (Southworth and Ben-Joseph 2003; Vernez-Moudon 1987; Appleyard 1980; Engwicht 1999). These well-documented precedents have yet to be fully embraced in the United States even though they are safer (Pucher and Dijkstra 2003). But even now, the latest European thinking on residential street design surprisingly underplays children's needs (HMSO 2007). Over the last two decades, children have been driven from residential streets by massive increases in traffic. Have children also disappeared from adult consciousness? They should still be considered the most important users of neighborhood streets (Moore 1991). When encouraged, they will express perceptions and opinions (Cele 2006) that are useful to adult policy makers who are willing to listen.

It is interesting to note that countries such as Denmark, Germany, the Netherlands, and Sweden, where higher levels of functional urban bicycle use are publicly visible, exhibit markedly lower rates of childhood obesity than the United States (Rigby and James 2003). Citizens of all ages can move around freely and safely without polluting the air because of the high-quality pedestrian/bicycle infrastructure designed into the urban fabric—indicating close collaboration between traffic engineers and urban designers (Figure 10-3). The pressing issue of children's independent mobility could be solved if traffic engineers, residential developers, and urban designers collaborated on child-friendly street design. Paradoxically, it has become increasingly difficult to support the argument in terms of child pedestrian traffic injury and death, because for years, child pedestrians have been disappearing from city streets perceived as dangerous (Hillman, Adams, and Whitelegg 1990). Alternative designs that would bring them back are needed.

Vehicle exhaust is a direct health threat. Although we were unable to identify the relative asthma rates for the countries cited, the Atlanta Summer Olympic Games study demonstrates the relationship between vehicle exhaust and childhood asthma in the United States. During the 17-day Olympic event, peak weekday traffic counts dropped 22.5 percent, peak daily ozone levels dropped 27.9 percent and asthma acute-care events in children assessed from four sources fell between 44.1 percent and 11.1 percent, with the highest level being statistically significant (Friedman et al. 2001). The effect of vehicle exhaust is also an indirect threat by keeping childcare center children indoors on "ozone alert" days.

Figure 10-3: Traffic-free urban trails and greenways expose children to nature and help them learn the joy of bicycle riding at an early age.

Impact on Cognitive Development

In an interview with the *Guardian* newspaper (Crace 2006), psychologist Michael Shayer reported the findings of a study sponsored by the Economic and Social Research Council (ESRC) of more than 10,000 11- to 12-year-old British children. The principal finding was that UK children have fallen two to three years behind in cognitive and conceptual development from where they were 15 years ago. When pushed to explain these findings, Shayer said, "The most likely reasons are the lack of experiential play . . . and the growth of a video-game, TV culture. Both take away the kind of hands-on play that allows kids to experience how the world works in practice and to make informed judgments about abstract concepts." The "rediscovery" of the importance of play in promoting children's health and positive parent-child relations is further supported by the American Academy of Pediatrics (Ginsburg 2007)—although, unfortunately, they do not mention the importance of *outdoor* play. Shayer also does not tie play to the outdoors in his speculations; however, a longitudinal study by Wells (2000) demonstrates a statistically significant correlation between nature and cognitive functioning of a group of low-income children when they moved to "greener" homes (measured by views from windows). If natural scenes viewed from indoors can have a measurable effect, imagine the possible impact of hands-on, outdoor immersion in nature.

Attention Functioning

Since being officially designated by the American Psychiatric Association in 1980, ADD (Attention Deficit Disorder) and ADHD (when "hyperactivity" is also exhibited) have become a hotly debated health issue (DeGrandpre 2001; Diller 1998). Lacking an authoritative, valid, reliable medical diagnosis, ADD/ADHD is typically "diagnosed" using behavioral criteria, some of which bear close resemblance to behaviors we might expect from normally active kids (Eberstadt 1999) cooped up in classrooms, acting as if they were in the woods. The most frightening fact related to ADD/ADHD is that an estimated nearly four million children are daily administered methylphenidate, a psychotropic drug (brand name *Ritalin*, similar in chemical composition to cocaine (http://learn.genetics.utah.edu/units/addiction/issues/ritalin.cfm) to control ADD/ADHD symptoms. The treatment is so popular in the United States that an estimated 80–90 percent of the world's production and consumption of Ritalin occurs there according to Eberstadt (1999), who cites estimates of production increases of 700 percent since 1990 and a doubling of consumption since 2000. Is there any more powerful statistic that underscores the distorted, misguided way we are beginning to regard childhood?

Outdoors as a Protective Shield for Mental, Social, and Physical Health

On the positive side, mounting evidence suggests that being outdoors in natural surroundings might be viewed as a "preventive treatment" for healthy attention functioning. Empirical studies are beginning to show statistically significant associations between nature (as little as trees seen through apartment windows) and improved attention functioning (Faber Taylor et al. 1998). Even small amounts of nature have been shown to exert a measurable, positive effect on children's attention functioning (Grahn et al. 1997).

Wide-ranging, independent behavior away from adult control can also have a positive social impact on children. Under these circumstances, they are afforded more opportunities for cooperative group play. Outdoors, children have more opportunities to collaborate with each other, whether to organize informal games, build a clubhouse, or go exploring without any particular goal in mind (Moore 1986a). Because such behavior is based on friendship and joint action to carry out projects, it builds democratic skills, facilitates cooperation and collective effort, and can help overcome prejudice against other children with varied backgrounds. Self-directed groups of children playing outdoors together build their own cohesive society and are better able to acquire self-reliance to overcome the challenges that life brings (see Figure 10-4).

Physical and social health and outdoor experiences also strengthen psychological health. A study by Wells and Evans (2003) suggests that nature nearby children's homes might buffer or moderate the effects of stressful life events on children's well-being—even among rural

Figure 10-4: Nature provides children with an inexhaustible supply of renewable play materials, motivating them to think independently, work together democratically to solve problems, and carry out self-initiated projects, with a sense of pride in their accomplishments.

children. A child with trustworthy friends, shared experiences in special places, and heightened self-esteem resulting from territorial control is more likely to maintain good mental health. Grahn et al. (1997) used standardized child development measures to compare the impact of outdoor environments on children in two typical Swedish nursery schools. Both had conventionally equipped outdoor environments but in one school, children also played in a lush woodland where they could spend outdoor time. Developmental measures of these latter children were remarkably different. In addition to improved attention functioning (supporting the later findings of Faber Taylor et al. 1998), Grahn and his team found that the children exposed to a more natural outdoor environment exhibited lower sickness rates (presumably because children get sick by exposure to each other indoors) as well as more advanced gross motor development, improved fitness, and increased imaginative and social play. At the neighborhood level, recent research indicates that in higher density areas, increased amounts of vegetation surrounding a child's residence protects against being overweight (Liu et al. 2007). Could it be that greener neighborhoods are more attractive for children to spend time outdoors?

These rigorous scientific research findings confirm the positive consequences that can accrue from outdoor play and direct experience of nature in terms of mental and physical health. Mounting evidence supports the notion that exposure to nature could be regarded as an essential childhood preventive health measure or "buffer effect," as discussed by Wells and Evans (2003, 315), who suggest that "environmental characteristics [such as nearby nature] may function as buffers or moderators of adverse conditions, serving as protective factors that contribute to resilience among children." (See Figure 10-5 in color insert.)

Boosting the Immune System

A further benefit of interaction with the outdoor natural environment is its association with the development of the human immune system. Research findings are beginning to demonstrate that the ubiquitous use of "germ-fighting" chemicals at home and in other environments used by children may have negative consequences, leading pediatric professionals to hypothesize that children are growing up with inadequately boosted immune systems. This may partly explain the dramatic growth of childhood asthma and other allergic ailments (Check 2004). The growing dependence on and easy availability of antibiotics may be part of the problem. Decreased immune stimulation ("training of immune system") through improved hygiene, fewer infections, fewer parasite infestations, et cetera, has resulted in the "hygiene" or "jungle" hypothesis suggesting that an overemphasis on hygiene may have reached a point of diminishing returns (Ring 2005). It is possible that exposure to nature, which in essence is nonsterile, may be a beneficial boost to a child's immune system, providing extra protection against illness.

How Does Nature Have an Effect?

The apparent health connection with outdoor nature (even in small doses) prompts speculation about possible explanations, ranging from the inherited preference for the "fractal array" of nature (see Chapter 1) and speculative predictions about children's relationships with nature based on evolutionary biology theories (Heerwagen and Orians 2002). Interpretation based on the biophilia hypothesis suggests that children are drawn to the natural outdoors because it is pleasurable and gives them a sense of well-being, expansive freedom, and agency or control over events (at the same time supporting health-enhancing, preventive behaviors). For children to reap the full benefit of being outdoors, opportunities for outdoor engagement with nature must be available as part of daily life, integrated with children's emerging developmental needs. This is especially true of very young children because their neurological and physical development is so rapid in the early years of life.

Out to Play

In middle childhood, schools and neighborhoods (containing the pathways and place destinations of children's home-based territories) must afford children sufficient daily physical activity for good health; they are therefore crucial targets for planning and design policy (Moore 1986a) (see Figure 10-6). Experientially rich territories can motivate the maturing child to get out and about, to explore and develop as a whole person, moderated by variables such as urban context, building density, and parental values and perceptions of safety; street traffic;

Figure 10-6: Children enjoying early morning exercise on the "peripheral trail" through the longleaf pine forest of Blanchie Carter Discovery Park, Southern Pines Primary School, Southern Pines, North Carolina. The children are members of the Walking (and running) Club, led by the school nurse every morning before school.

availability of playmate siblings and peers; and locations of schools, parks, open spaces, shops, and other local amenities relevant to children's interests. The cure for the lifestyle maladies of contemporary childhood seems glaringly obvious and simple: outdoor play in nature. Although this is easier said than done, great potential exists for counteracting sedentary lifestyle trends and the negative health consequences of inadequate time outdoors exposed to nature by reaffirming the benefits through empirical research and design based on the findings.

PROGRESSING AN INTERDISCIPLINARY, ACTION-RESEARCH STRATEGY

Environment and behavior (E&B) research has a 40-year track record and a developed repertoire of methodologies to study the sedentary lifestyle issue and help build the evidence base necessary to develop design solutions. Children's environments research, a subfield of E&B, has developed a substantial conceptual framework and methods that can be applied to this effort. Theories of territoriality, home range development, behavior setting, and affordance, currently applied by leading researchers, continue to offer potential for generating useful knowledge. Methods of direct observation of behavior and objective measurement of physical activity, combined with qualitative, child-friendly methods (drawings, child-taken photographs, journals, semi-structured interviews, child-led safaris), are appropriate data-gathering tools to measure children's behavior and perceptions. Multimethod quantitative/qualitative exploratory research offers the most potential for identifying relevant variables and measures. However, additional work is required to develop valid, reliable measures of the physical environment at a level of differentiation useful for design.

Action research is a viable strategy to adopt in the face of the tremendous need to rapidly generate new knowledge to serve as the evidence base for new designs. Correlation research already under way is generating an understanding of key associations to improve design decision-making. However, new, radically different designed environments with increased "ecological validity"[1] must be built and tested to assess their support of healthy lifestyles for children. Innovative models already exist on the ground (presented later). They represent key case study research opportunities for developing an understanding of early and middle childhood behavior and physical designs required to counteract unhealthy lifestyle trends.

LINKING SUSTAINABLE DESIGN AND HEALTHY CHILD DEVELOPMENT

Sustainable design has made tremendous technical strides in the design of buildings but less so in site design and the broader linking of urban planning to its ecological context so that the natural systems of the region become a daily experiential component of residential life and thus local culture. Until sustainable development is considered as a culture-building process, success will be limited. In this regard, the biophilic design of children's outdoor environments could provide a means for integrating technical and cultural domains through play, learning, and educational processes.

Many of the precedents to be discussed below may seem straightforward from a technical design perspective; however, they challenge the conventional wisdom of accepted practice relating to children's environments. Implicitly, they express a progressive education philosophy building on the traditions of Dewey, Montessori, Froebel, and others. Sometimes they contradict health and safety standards based on the conventional epidemiological (toxic environment) paradigm that overlooks the positive health-enhancing effect that "exposure" to the environment can have for children (Frumkin 2001).[2] They may also raise issues of liability in the conservative arena of risk management, reinforced by the lack of research evidence supporting the safety of such environments.[3] They will challenge entrenched attitudes about the scale of spending required to improve the biophilic quality of children's environments.

The precedents are "outdoors" because that is where children need to be to fully experience nature and benefit from its preventive health effects. New architectural forms are needed that emphasize continuous indoor-outdoor daily contact with natural systems. This is particularly true of cold climates, where glazed outdoor-indoor spaces would allow children daily interaction with plants in schools and childcare centers—as in a botanical glasshouse. A few precedents already exist (see Figures 10-7 and 10-8).

Figure 10-8: "Play partners" in the greenhouse engage children in learning about fascinating species such as the "sensitive plant." Glazed architecture can provide rich settings to serve children in child development centers and schools.

INSTITUTIONALIZED CHILDHOOD: THE POTENTIAL OF A NEW CULTURAL REALITY

The majority of young children are now growing up in institutions. Almost three-quarters of preschool children with working parents today spend part of each weekday in some form of childcare arrangement (Capizzano et al. 2000). The new reality of children as young as three months old spending long hours in childcare centers has arrived with little questioning of the possible developmental consequences of such a sudden, radical change in early childhood environments. Young children are spending the majority of their time in a new type of family with biologically unrelated adults and similarly aged children in new, nondomestic architectural forms. This is not necessarily a negative situation for child development. Indeed, research has identified positive benefits (Palacio-Quintin 2000), especially for children from socially deprived environments (Garces et al. 2002). The childcare center may be regarded as a new form of community care. However, with exceptions, typically little attention is given to the learning potential of the physical environment—both indoors and outdoors.

Figure 10-7: The Greenhouse at the Hammill Family Play Zoo, Brookfield Zoo, Brookfield, Illinois, provides a year-round setting for children and families to experience a rich variety of plants—including a banana tree. Each year the fruit is harvested by the children, who join the "banana parade" to feed them to the gorillas.

Early childhood architecture, including landscape design, could be celebrated as a subfield of the design professions with extraordinary potential for positively influencing environmental engagement and child development. And yet, childcare center buildings not only rarely match this promise but barely meet basic functional requirements such as providing floor level windows, interior daylight penetration, and ample transitional settings between indoors and outdoors. Outdoors, conventional playground equipment is typically provided rather than a dynamic, natural learning environment, which through play processes could offer new experiences each day instead of the repetition of static settings.

EARLY CHILDHOOD: WELCOME TO PLANET EARTH

For children, the "sedentary lifestyle" crisis means lack of opportunities for movement and play (Burdette and Whitaker 2005; Pellegrini and Smith 1998). In this regard, childcare centers offer an enormous opportunity for raising children in a "preventive environment" designed to support active lifestyles and healthy nutritional habits, connecting children and nature through design, beginning in the first year of life. When physical activity is emphasized in the preschool years, research suggests that it will track throughout childhood (Moore 2003) (see Figure 10-9).

Imagine designing an outdoor environment where a child's first birthday is not only a celebration of an individual's accomplishments in the 12 months since birth but also a celebration of the first steps of sensory integration with the world that will be the child's home for the rest of her life. Childcare centers can initiate cultural transformation which, while focused on the future, also must echo the history of our human ancestors from whom we have inherited our biophilia—and our responsibility to transmit it to future generations. From this perspective, the term "childcare center" hardly conveys the larger vision of childhood, community, and planet. "Child Development Center" (already used by some centers) would be an improvement, with "Earth Education" as a progressive extension of the center role. (See Figure 10-10 in color insert.)

Figure 10-9: A group of toddlers play with fallen leaves, experience their sensory properties, and explore their behavior on the curved surface of a hollow log—a type of activity that educational psychologist Michael Shayer (Crace 2006) suggests can boost cognitive development.

In 1992, the first author was asked to design an "infant garden" in a childcare center that served families of staff and faculty at North Carolina State University. At that time, the importance of contact with the natural world was hardly mentioned in the literature apart from the risk of insect stings and injury from poisonous plants. Then, as now, very little *design* research literature was available (Striniste and Moore 1989) along with limited practice-based texts. The second edition of Greenman's (2005) *Caring Spaces, Learning Places*, offers the most recent design advice on outdoor environments for infants and toddlers.

The lack of research models of best practice eventually resulted in the creation of a model site at a child development center located near North Carolina State University. Designed by the first author and constructed with two colleagues (then students in landscape architecture and horticultural sciences). The renovated site was completed in 1997 and has since served as a research site (Cosco 2006). At this center (and at other local centers where results from the first site were subsequently applied), infants and toddlers spend more time (usually more than an hour) outdoors each day in shady, diverse environments, immersed in natural settings in daily contact with plants and the animal life that they support. Preambulatory children (less than a year old) are commonly observed reaching out, grasping, touching, and smelling the variety of reachable plants.

A study by Yarrow, Rubinstein, and Pedersen (1975) observed that from birth children's attention is directed towards responsive environments, especially those that are diverse and complex (Figure 10-11). Once children begin to walk, their range of attention can rapidly expand to embrace the natural world, if provided. Yarrow et al.'s experimental laboratory findings are reflected in observations at two of the Natural Learning Initiative's (NLI)[4] naturalized research sites (including the one discussed above), where animals that attract attention (insects, amphibians, and birds) daily engage children's fascinated attention (Kaplan and Kaplan 1989).[5] Evi-

Figure 10-11: This very young child is fascinated by the fragrance of the sprig of rosemary he has picked from an adjacent planter. The smooth log provides a clean work surface above the surrounding sandy ground to support his exploration. Notice the fallen leaf clutched in his left hand as a prized possession.

dence of biophilia is readily observable, even by children under two—if their environment is designed to afford child-nature contact.

However, such affordances of nature depend on the natural diversity of children's immediate surroundings. A baseline assessment of outdoor quality in North Carolina childcare centers (Cosco and Moore unpublished report) showed that on average they contained three times as many manufactured components as natural components (mainly individual shade trees, grass, and woodchip safety surfaces). Field verification of these findings reinforced the conclusion that lush outdoor childcare environments are exceedingly rare.[6] As North Carolina is considered a progressive state in terms of childcare (at the time of writing, a statewide Committee on Outdoor Learning Environments is in session), it may be fairly assumed that other regions of the United States are certainly no better than North Carolina in the naturalized quality of their outdoor environments.

DESIGN FOR PHYSICAL HEALTH

Cosco (2006) conducted a comparative empirical study of three preschool (three- to five-year-olds) outdoor designs: one containing mostly manufactured equipment, a second containing a *mix* of natural and manufactured components, and a third containing manufactured equipment and natural areas segregated from each other. The second preschool play area supported higher levels of physical-activity play than the other two. Cosco concluded that its relatively dense mix of behavior settings (one of which was a broad, curvy, hard-surface, wheeled toy trail) and the number of children playing together at a given time, stimulated more social interaction, which, in turn, led to more active play than did the other two sites. She identified "*setting compactness* (higher numbers of children sharing multiple activities—in this case also surrounded by plants and wheeled toys)" as an attribute that may help explain the higher levels of activity (Cosco 2006, 123). This attribute is further linked to the more general phenomenon of "*additive effect* [our emphasis] of the layout of the site and its attributes (objects and events) on children's activities" (Cosco 2006, 120), explained by affordance theory (Gibson 2002). The specific role of vegetation integrated into setting design can be viewed as part of the additive effect or "buffer" (Wells and Evans 2003), acting as a crucial moderator in children's settings, positively affecting both the diversity, duration, and impact of outdoor play (Grahn et al. 1997). Building on these pioneering scientific studies, NLI is presently engaged in a multi-site study to confirm additive effect variables in outdoor preschool play areas that motivate or afford higher levels of physical activity and other types of play.

From a policy perspective, the greening of child development centers would seem a rather simple step. Instead of investing scarce financial resources exclusively in manufactured equipment and mulch (Cosco and Moore unpublished report), funds could be spent on relatively inexpensive trees, shrubs, perennial plants, and natural objects such as rocks and salvaged tree limbs. The play and educational value of these settings far exceed, dollar for dollar, settings such as climbing structures that lose attraction for some children if they must use them every day, year round (Moore and Wong

1997). This is not to deny that particular types of manufactured items have important functions. Indeed, items designed to support dramatic play, such as playhouses and various types of vehicles (trains, fire engines, and trucks), retain their attractiveness, especially when surrounded by pickable ingredients that "hunter-gatherer" children can use in dramatic play scenarios (see Figures 10-12 and 10-13).

To succeed, a greening strategy must engage the educational staff. However, many early childhood educators are not trained to work with children in outdoor environments. In response, some creative centers have hired a gardener as an assistant "outdoor teacher" to rectify this lack of expertise. However, as long as outdoor areas are labeled as "playgrounds" and are not seen as an integral part of the educational environment for both playing and learning, then the introduction of nature-play, will continue to be a challenging goal.

Maximum Exposure to Nature: Outdoors All Day

At "outdoors-in-all-weather nursery schools" and "forest kindergartens," children stay outdoors all day in all seasons. These alternative models started in Denmark in the 1990s and soon spread to the rest of Scandinavia

Figure 10-12: Vines and climbing plants can transform an otherwise bland chain-link fence to become, in this case, a cascade of creamy blossoms ready for early spring harvest by hunter-gatherer children—to be used as a "pizza" ingredient in the restaurant car of the nearby play train.

Figure 10-13: Naturalized play train chuffing through a forest

and Germany. Although no English-language comprehensive study of forest kindergartens has been identified, Keller (2006) lists four basic principles (translated from the German) that sum up the approach:

1. Nature, with its vast sources for play, provides space for the emergence of a child's fantasies, curiosity and creativity.
2. Direct contact with nature allows the minds of children to develop a sensitive appreciation for the earth.
3. The forest provides an ideal place for children to move freely about, thereby developing trust and gaining self-confidence.
4. In free play, above all, but also through daily routines, children gain competence in social relationships and in resolving conflicts. (http://www.whatcomwatch.org/php/WW_open.php?id=718)

There are now more than 500 forest kindergartens in Germany alone (Keller 2006). Forest kindergartens take the concept of education outdoors to its logical limit. In some models, the kindergarten consists of a small, one- or two-room, building housing an administrative office, storage for accoutrements and supplies for forest adventures, not always including a toilet other than the woods. Children meet there at the beginning of the day to collectively decide on a plan for the day (or half day, depending on age), assemble the gear needed, load it into a cart and backpacks, and take off into the forest to discover whatever befalls the group.

In 2005, the first author visited a forest kindergarten in a nature preserve in Munich, Germany. (See Figures 10-14 and 10-15 in color insert.) Upon approaching the site (a 10-minute walk into the forest), the quality of the atmosphere and the body language of the three- to five-year-olds were immediately striking. The group of 15 or so children were busily engaged in free-form activity in a clearing adjacent to the base building (two wagons constructed of timber), in the buffer of surrounding woodland, and down in the nearby creek. There, a five-year-old girl was sitting on a narrow sandbank surrounded by water, dabbling her feet in the flowing water, gently singing to herself. Certainly the teachers had an eye on her, but from a long distance. The girl was lost in a personal reverie for 15 minutes or more. Surely, such "spots of time" are never forgotten (Chawla 2002, quoting Wordsworth).

In the United States, the nearest equivalent to the forest kindergartens is the growth of preschools located in nature centers. The Nature Preschool at the Schlitz Audubon Nature Center near Milwaukee, Wisconsin (www.schlitzauduboncenter.com), is one of a small but growing number of nature-based preschools in the nation committed to both environmental education and active learning. The broad curriculum is based on seasonal changes and includes art, music, perceptual and cognitive skill development, large and small motor skill development, natural science exploration, and daily outdoor discovery in the center's 185 acres of diverse habitats. The children are able to experience the freedom of a seemingly limitless natural world. Playing and learning adventures occur throughout the center's prairie, forests, ponds, and marshes. The natural world is used as both theme and material in the education of the whole child. The stated goal is to develop the child's ability to work independently and cooperatively, to act in a caring and responsible way toward their environment and others, and to foster a love of nature.

Children ages three to five in the Audubon Nature Preschool (www.audubonnaturalist.org/cgi-bin/mesh/education/nature_preschool), located in the Edwin Way Teale Learning Center at the Woodend Sanctuary, Chevy Chase, Maryland, roam a 40-acre nature sanctuary. There they explore the wonders of the natural world through a balance of self-directed and teacher-directed activities in ecologically diverse aquatic, forest, and meadow habitats.

The Four Seasons Kindergarten in Ringe, Denmark (www.kompan.com/sw23720.asp), is a small nature-based early childhood facility. Constructed in 1997, it serves 30 three- to six-year-olds who are children of employees of Kompan, a leading international manufacturer of playground structures. Indoor facilities are provided by a 212 sq m "house." However, according to the kindergarten website, the children spend 80 percent of their time outdoors in a 3,000 sq m landscaped play garden. Each day, the children participate in tasks around the house, garden, or hen coop together with the five caregivers. They sit by the bonfire; draw on the veranda; or build with real hammers, nails, and saws. Gardening and cooking are part of the daily life of the kindergarten undertaken by caregivers and children together. When parents pick up the children in the afternoon they are dirty—from playing outdoors, tired—from playing outdoors, and happy—from playing outdoors.

The nature preschools of the United States and the Scandinavian/German forest kindergartens offer substantial models of nature-based early childhood, which need to be within reach of all communities to inspire progress towards full immersion of children in nature. Those seeking to promote nature pedagogy need to join forces with early childhood educators to develop a strategy and action plan to green the nation's childcare centers. This means not only buildings and outdoor spaces designed to satisfy LEED[7] standards and user criteria but also locations adjacent or within open spaces, forest preserves, parks, and greenways. The latter provide two-way access for walking and biking—for dropping off and picking up children as well as for exploring away from the center. Furthermore, both childcare centers and schools need to consider "green design" from the perspective of children's own need to explore and discover the natural world through play. Children themselves can contribute ideas by participating in the design process. Adult opinions vary regarding at what age children are sufficiently mature for this role. In executing NLI design assistance projects, we have found that by the age of four children can contribute worthwhile ideas and/or voice pros and cons of design proposals by other participants in the process.

Sun Exposure: A Word of Caution

Increasing, scattered evidence suggests that being outdoors for relatively long periods each day is beneficial to the health of the majority of young children (Fjørtoft 2001; Grahn et al. 1997). However, overexposure to direct sunlight can be a substantial health risk (Geller 2006), particularly in the middle part of the year. Boldemann et al. (2006, 306) stress, "Overexposure to ultraviolet radiation from the sun, particularly in childhood, is estimated to cause 80–90 percent of all skin cancers in Western societies." . . . sunburn is particularly hazardous to young children, as the skin does not "forget" the damage; however, sun exposure can be counteracted by design. Boldermann and colleagues showed, as we might expect, that reduced levels of sun exposure were associated with the presence of trees and shrubs in child development center play areas.

RETHINKING SCHOOL SITES

At five years old, school attendance in the United States is mandatory. By definition, school buildings and grounds should play a crucial role in biophilic design strategy. Fundamental to this notion is the concept of the elementary school as a center of neighborhood life, close enough to the majority of homes that children can make the trip back and forth on foot or by bicycle. The school grounds should serve as a space for learning and for children's play before and after school (Moore and Wong 1997; HMSO 2006; Beaumont and Pianca 2002). However, several barriers—longstanding and recent—constrain this objective. For decades, racial integration policies and the development of "magnet schools" in the United States have resulted in children being bussed to schools in locations outside their own residential neighborhoods, which means that school neighborhood friendship networks cannot be formed. More recently, in spite of research supporting the benefits of small neighborhood schools (Slate and Jones 2005; McRobbie 2001), in the name of economic efficiency elementary school sites have been dislodged from their walkable base in the neighborhood (Beaumont and Pianca 2002) and combined with middle schools on larger school campuses, increasing the school "carbon footprint."

Above all, schools should be safe and healthy environments for children, indoors and outdoors (Frumkin, Geller, and Rubin 2006). However, even though school buildings have moved to the forefront of "green building" design, the thinking about outdoor spaces remains unchanged from the perspective of users (especially children) and their educational potential as diverse, green habitats. School outdoor areas are still designated as "recess playgrounds," where children are expected to expend energy before going inside for academic work. Countering this view, the Toronto District School Board (http://ecoschools.tdsb.on.ca) regards schools and their grounds as eco-educational resources, as health-promoting outdoor environments, and as places for children's creative engagement with nature (Bell and Dyment 2006; Dyment 2005). However, this view is missing from the LEED approach, which focuses almost exclusively on building design, mainly from a technical costs and benefits point of view (Kats 2006). Attention to outdoor design is missing from the equation; even sustainable site-related issues such as stormwater management (and their educational potential) are overlooked.

Particularly alarming, and underreported, is the fact that an increasing number of school districts are curtailing or eliminating recess because it takes time away from academic studies (http://www.ipausa.org/recess-research.htm). This policy not only inhibits healthy child development but also is against international law in all UN member countries (except the United States, which has not ratified the Convention on the Rights of the Child). According to one news report, "As many as 4 out of 10 schools nationwide, and 80 percent of the schools in Chicago, have decided there is no time for recess. Instead of romping in playgrounds, kids are being channeled into more classes in an effort to make their test scores rise on an ever-higher curve . . .

(Schudel 2001). This regrettable policy has been contested and surely will eventually need to be rescinded and replaced with the opposite strategy to move educational programs outdoors and at the same time create attractive, usable, safe outdoor spaces for after-school activities—close to home.

Green building design policies related to schools need to expand in two directions before the theme of "green playing and learning" addressed in this chapter is sufficiently covered. First, green building design policies need to give equal prominence to both interior spaces *and* school grounds; second, they need to give equal weight to the behavioral requirements of users as they do to green technology requirements. To achieve its purpose of conserving the planet for the enjoyment of future generations, sustainable development practice must fully activate an educational role—especially in the design of institutions (including their outdoor spaces) where young people could learn not only *about* the natural world but also *in* and *through* the natural world (Moore and Wong 1997) (see Figure 10-16).

Examples reflecting this view have existed on the ground for decades as a result of an international movement, including groups in the United States, pushing the potential of school outdoor environments as places for education and enjoyment. Research evidence strongly suggests positive outcomes for children attending schools with naturalized sites. The first author's 10-year documentation of the naturalization of the Washington Environmental Yard, an inner-city schoolground in California, in terms of its impacts on the educational program and the children's daily experiences of the natural world, is a rich source of the multiple "playing and learning" roles of natural communities (Wechsler et al. 2003; Zask et al. 2001; Moore and Wong 1997). The Washington Environmental Yard responded to a special set of circumstances, where the boundaries of the possible could be pushed substantially. One of the most important outcomes was the demonstration of the motivational power of education outdoors. Many of the classroom teachers extended the mandated state curriculum into a rich outdoor environment on the schoolground as well as into the surrounding neighborhood and learning sites in the broader community. Children with varied learning styles were motivated to become engaged in learning when confronted by multiple hands-on opportunities because they triggered excitement and provided memorable grounding for later, more cognitive phases in the learning process. Research findings from the Washington Environmental Yard indicate the powerful impact of the on-site, outdoor natural educational resources on children's long-term affective relationship to their school (Moore and Wong, 1997). For children engaged every day both during and after school hours, the natural richness of the school grounds provided a well-understood added value and sense of pride in their school.

Figure 10-16: A naturalized outdoor classroom can immerse children's learning processes in nature as well as reduce demands on interior, air-conditioned space. Roof covering is a translucent, waterproof, ultraviolet-light-resistant fabric. As they work, children can enjoy the experience of rain pouring down around them or the play of sunlight and shadow of foliage vibrating in the wind over their heads.

At another Berkeley school, Martin Luther King Jr. High, a team led by Alice Waters, a former Montessori teacher and well-known restaurateur, removed a huge area of asphalt and replaced it with a school garden. Healthy nutrition and meal preparation by the children using the produce from the garden focuses on explicit curriculum objectives and health outcomes (http://www.edibleschoolyard.org/about.html). Murphy's (2003) empirical investigation of the Edible Schoolyard demonstrated positive impacts across several dimensions, including academic achievement, psychosocial adjustment, understanding garden cycles and sustainable agriculture, ecoliteracy, and sense of place. In their investigation of the impact of adults' experience of plants and gardening as children, Lohr and Pearson-Mims (2005, 476) concluded that, "Childhood experiences with nature influence adult sensitiveness to trees and that influence is very strong." This suggests that hands-on gardening and engagement with plants at an early age in child development centers and schools (where the children are), may be a crucial strategy for building an ethic of caring and protection for the natural world.

An international movement to restore school grounds as educational resources has been under way for decades in North America, Europe, and other regions of the world (Moore 2006). The Coombes School, near Reading in southern England (www.thecoombes.com), provides an advanced public education model of outdoor learning (Jeffrey and Woods 2003; see Figures 10-17 and 10-18). By collaborating with the school community, the teaching staff, led by Susan Humphries, created an extraordinarily diverse system of natural settings on the school grounds. (See Figure 10-19 in color insert.) The Coombes provides a fully evolved example of best practices so progressive that the documented model has been translated into Swedish (Olsson 2002).

Figure 10-17: One of a multitude of seasonal curricular events at the Coombes School is harvesting and comparing the tastes of the many varieties of apples planted on the school grounds in the last 30 years.

Figure 10-18: Apples taken to the classroom become the subject of a classification and group analysis lesson—before being stewed.

The educational and health-promoting role of the designed landscape is supported by research (Titman 1994; Kirkby 1989) and by the work of the Boston Schoolyards Initiative (http://www.schoolyards.org/education.htm), the Learning Landscapes Alliance (http://thunder1.cudenver.edu/cye/lla/about.html), and the Evergreen Foundation (http://www.evergreen.ca/en/; Dyment 2005). These sources of evidence all point to the same conclusion: schoolgrounds can be designed as natural learning environments that offer educational value and broad learning opportunities (Wechsler et al. 2003; Moore and Wong 1997; Zask et al. 2001; Murphy 2003; Moore and Cosco 2007), especially for learners whose style is not well adapted to indoor learning environments (Moore and Wong 1997). These innovations in indoor/outdoor education design have been under way for decades, pushing against the deeply embedded assumption that mandated learning objectives can only be implemented indoors, and demonstrating that hands-on learning outdoors can be more effective than an exclusively pressure-cooker approach (Lieberman and Hoody 1998). Heeding these results, the Blanchie Carter Discovery Park (BCDP) at Southern Pines Primary School, Southern Pines, North Carolina, was founded by a parent group to increase children's creative outdoor play and learning options—socially and environmentally (see Figures 10-20 and 10-21 in color insert). Children participated in the process in many different ways (see Figures 10-22, 10-23, and 10-24).

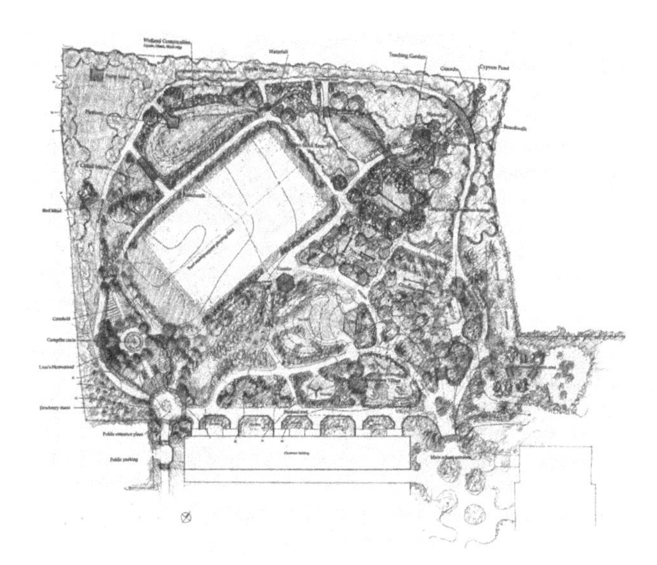

Figure 10-22: Blanchie Carter Discovery Park original master plan. The dual-use park was named after a former principal of Southern Pines Primary School and serves the school during school hours and the local community at other times. The multipurpose playing field is used by the junior soccer league. From the main school entrance (lower right), primary pathways distribute users to main settings. A peripheral trail provides travel around the entire circumference of the site and is used by the Walking Club every morning. Gazebos provide major landmarks for curricular and social activities. Groves of shade trees have grown up around the manufactured equipment settings. The labyrinth, added later, is located in the top left corner. A stream and wetland/pond have yet to be developed.

Figure 10-23: After a campout on the school grounds, Robin Moore facilitates an early morning Celtic tree blessing with the children.

The rise in schoolyard bullying (which could be interpreted as a symptom of underlying childhood social-psychological malaise) has yet to prompt a national movement to make schoolgrounds socially inhabitable. To do so will require the massive addition of natural resources on the nation's schoolgrounds and redirected outdoor education teacher training programs focused on schoolgrounds as educational and social resources (Wechsler et al. 2003; Moore and Wong 1997; Zask et al. 2001). Children and nature lobbies must convince local school boards to adopt biophilic policies to design schoolgrounds to support interdisciplinary environmental education—not only to meet criteria for sustainable development but also as places where children can learn to live together peacefully. Creation of BCDP had such a strong positive impact on the social relations between the children that the "time-out log" became a play object, as it was no longer needed for punishment.

SCHOOLGROUNDS AS NEIGHBORHOOD PARKS

Although the concept of the school park has been implemented in many municipalities, research has been limited to a number of case studies that indicate the potential of these sites as attractive places for children to interact with nature. Findings from a study of the model Washington Environmental Yard, using several meas-

Figure 10-24: Southern Pines Primary School hosts a red cockaded woodpecker workshop in the Blanchie Carter Discovery Park longleaf pine reserve.

Figure 10-25: The labyrinth at Blanchie Carter Discovery Park (built by two intern student playworkers from Leeds Metropolitan University, UK) is a place where children interact with nature and with each other; shown here during the early morning Walking Club.

ures, demonstrated children's strong affiliation with "biotic" elements such as ponds, streams (and all things aquatic), trees/shrubs, flowers, dirt, and sand compared to "abiotic" elements such as play equipment and asphalt (Moore 1986b). In a study of children's views of a schoolyard and other public places in Los Angeles, Loukaitou-Sideris (2003) found that nature-like elements (including grass, trees, and flowers) were the most frequently mentioned (42.9 percent) elements.

School parks, especially when located in older, denser, walkable urban neighborhoods, potentially offer significant exposure to nature for children—a function that is now more pressing for two reasons. First, the rising cost of urban land is making it more difficult for cities to acquire park sites; therefore, if school and park systems partner to combine capital and maintenance budgets, schoolgrounds/parks can be developed and maintained to a higher level of quality. Second, the rapid growth of families with both parents working has resulted in growing pressure for school sites to provide after-school programs for children. However, care must be taken to prevent such programs from becoming "school-after-school."

A key strategy is to make the outdoor environment so compelling that children will clamor to go outside (see Figure 10-25). In contrast to the rigid academic, indoor strictures of the school day, outdoors can provide diverse opportunities for group activity and creative expression in natural settings that equally attract children and program staff—who must be professionally trained to use the opportunities for after-school creative enjoyment.

However, such a profession does not exist in the United States. In other countries the field is well established under various titles including playworker (UK), social pedagog (Scandinavia), animator (France, Spain, and Latin America), and cultural worker or cultural animator (Germany). These professional groups are given teacher training at the college or university level to work in a broad variety of nonformal education community contexts. In spite of the lack of such professionals in the United States, our experience suggests that if the program environment is sufficiently conducive, creative community professionals will be motivated to become engaged because of the creative opportunities offered. A wonderful example of this was Project PLAE (Playing and Learning Adaptable Environments), held on the Washington Environmental Yard, where children of all abilities and ages participated in a summertime program of arts and environment workshops facilitated by local artists (Moore and Wong 1997, chap. 14).

NEIGHBORHOOD PARKS

In Moore's studies of urban childhood territories (Moore 1986a), neighborhood parks emerged as important places where children can escape from the restrictions of home, meet up with peers, have fun, enjoy nature, and learn about themselves and the world around them—especially the natural world. In drawings of their favorite places, natural elements (including parks) were the most frequently mentioned (Moore 1986a, 43). In her study of children's use of Los Angeles public spaces, including two parks, Loukaitou-Sideris (2003) found that of the elements most liked by the children a third (33.3%) were naturelike (grass, lake, trees, flowers, ducks, sand). A study by Milton, Cleveland, and Bennett-Gates (1995) shows how an urban park can offer a natural learning environment with unanticipated outcomes that included changed perceptions in students of themselves, of each other, of teachers, and of the park itself. The findings of Burgess, Harrison, and Limb's (1988) study of London parks strongly indicate the potential of local parks—especially where the "wild" landscape dominates—as attractive places for children and families to spend time together outdoors (see Figures 10-26, 10-27, and 10-28). These scattered findings suggest the need for increased research to build a solid field of literature to underpin the potential of parks as a crucial local resource for play, learning, and community development (Moore, 2003).

Figure 10-27: The broad, curvy pathways in Kids Together Park, surrounded by a rich landscape, stimulate children's active play, while accompanying parents can relax nearby.

Figure 10-26: Kids Together Park, Cary, North Carolina, demonstrates how manufactured play equipment, elegant arbors, and a natural landscape can be designed together to create a relaxed, intimate, comfortable place for users of all ages.

Figure 10-28: Large rocks designed into a park in Nantes, France, add a natural landscape challenge for children and an opportunity for interaction with caregivers.

COMMUNITY NATURE DESTINATIONS

In the last two decades on both sides of the Atlantic, several new models of community institutions have developed that have increased options for families seeking natural places to spend time together. New types of nonformal education institutions, including children's museums, children's zoos, children's gardens, and botanic gardens, together with established models such as adventure playgrounds and urban farms, offer extended opportunities for adventurous outdoor nature experiences and active living in the wider city environment, (see Figure 10-29). For most families, they serve as destinations beyond residential neighborhoods. Children must be taken by adults (parents, school staff, summer camp counselors), and in low-income neighborhoods may have no means of access.

Brevard Zoo in Melbourne, Florida, has partially solved this problem by building three public school classrooms, to immerse "at-risk" fifth-graders in the zoo as their learning environment (www.brevardzoo.org/education/zoo_school). Not surprisingly, the positive effect on some of the students in both academic achievement and personal growth has been remarkable (see Figure 10-30). However, the evidence is purely anecdotal (personal communication with the zoo director). In spite of the continuing investment in nonformal education environments, such as zoos, there is a dearth of research literature available that might offer stronger support for integrating formal and nonformal education systems. A hopeful sign is the Good to Grow initiative by the Association of Children's Museums (http://www.childrensmuseums.org/index.htm) to promote outdoor spaces in children's museums, which presently are found in ap-

Figure 10-29: Families enjoy nature together at Hammill Family Play Zoo stream, designed as a safe, secure setting for all ages.

Figure 10-30: Classrooms in the trees at Brevard Zoo, Melbourne, Florida, enable the curriculum for at-risk fifth-graders to be conducted at the zoo. (See www.brevardzoo.org/education/zoo_school.)

proximately a quarter of children's museums in the United States.

PROVIDING FOR CHILDREN'S NEEDS IN RESIDENTIAL ENVIRONMENTS: BEYOND PLAYGROUNDS

Since the early decades of the twentieth century, when municipalities first began to recognize the issue of children on busy streets, it has been assumed that city parks and playgrounds at regular intervals are the solution to the "problem" of children's play. One contemporary study of children's play concludes: "There is an uncritical and widely accepted belief among adults that children need places in which to play and that the playground is the space that best fulfills this need. An undercurrent of paternalistic concern (it's for the kids) and self-interest (it keeps them off-the-street [for which read 'my street'] sustains this commitment to neighborhood playgrounds" (McKendrick 1999, 5). Not only does the emphasis on the playground confine the legitimate (in adult eyes) locale of play to one particular setting, the ubiquitous, non-site-specific products of play equipment manufacturers dominate such settings, separating children from nature and the contextual landscape of their home region (Herrington 1999). It is argued by Woolley (2006) that entire urban open space systems have potential relevance for the independent movement of children around the city—if this potential was thought through from the beginning, as it was in some of the postwar British and Nordic New Towns (see below).

Another study recognizing children's need to have access to the wider urban landscape concludes: "Can enrichment of the small, local and generally confined spaces that are the playground, essential as that enrichment is, ever compensate for impoverishment of the broader environment that constitutes the child's more general universe and playscape?" (Cunningham and Jones 1999, 12). In spite of recent actions such as the *Childstreet* conferences and resulting Delft Manifesto on a Child-Friendly Urban Environment (www.urban.nl/childstreet2005/programme.htm), adult views that children's needs are best met by the provision of a specific, bounded, equipped play place persist. However, naturalistic studies of what children actually *prefer* reveal a marked preference for access to, and modification of, natural undesigned areas (Hart 1979; Moore 1986a). Even within the boundaries of a playground environment, marked differences exist between children's and adult's expectations. When asked by parents at Village Homes, Davis, California, to assist in the design of a playground for their children, and later to do the same at a local school, landscape architect Mark Francis discovered that " . . . children preferred challenging alternative and fantasy elements which incorporated loose parts and water and changed over time. Adults wanted more traditional play environments which are safe, neat, and fixed, with no water and clean edges" (Francis 1988, 69).

Despite its self-image as a child-oriented society, it is rare in the United States for a residential neighborhood to be designed with the needs of children—its least mobile and most vulnerable members—at the forefront of planning and policy decisions. A study conducted in 1976 of children's play in Oakland, California, concluded: "when it comes to the built environment of inner cities, children's needs are largely unrecognized and unmet or disregarded. . . . The constraints of the neighborhood environment can deprive children of a basic right of childhood—the right to experience and explore the world around them safely, spontaneously, and on their own terms" (Berg and Medrich 1980).

Thirty years later, little has changed. Despite the fact that studies over the last three decades have documented how children's use and enjoyment of their neighborhood has been severely curtailed (Gaster 1991; Lynch 1977; Hart 1986), there was virtually no change in public policy responding to this phenomenon until the rise of childhood obesity focused on lack of exercise as a partial explanation for this physical problem. Even this has not resulted in any radical call for change in how our neighborhoods are planned. Rather, emphasis has focused on the modification of existing streets for "walkability," the provision of sports programs, and programs to encourage walking to school such as the Walking School Bus. David Engwicht, its Australian inventor, suggests that its adoption around the world in programs organized by adults to accompany children to and from school is losing sight of the original intention: to support children's *independent* mobility (http://www.lesstraffic.com/index.htm). If children were genuinely involved in the planning, they would no doubt highlight the difference between going to school and returning home from school, a time for dawdling along the way to explore and play with friends.

A hopeful sign of a reactivated children's environments discourse is the range of recent publications pressing for an understanding of children's needs beyond "home, school, and playground" and for the right of children to have access to the whole urban environment. Recent books on this topic have emanated from northern Europe (*Children in the City* [Christensen and O'Brien 2003]); from Australia/New Zealand (*Creating Child Friendly Cities* [Gleeson and Sipe 2006]); from the UK (*Children and their Environments* [Spencer and Blades 2006]); and from an international group of authors (*Growing Up in an Urbanising World* [Chawla 2002a]).

PREFERRED PLAY ACTIVITIES: CHILDREN'S VIEWS

Only about half the days of the year are school days (even in year-round programs). The design of the neighborhood environment close to home is therefore crucial in terms of children's freedom to play outdoors with ready access to nature. Many studies have shown that the provision of equipped play areas or designed park space is not sufficient to meet children's needs for exploratory social and imaginative play (Van Andel 1990; Bjorkild-Chu 1977; Parkinson 1985; Moore, 1986a; Wheway and Millward 1997). Given the choice, children interact with all aspects of the neighborhood environment, and it is the relative diversity of such environments and the available access to them that are the most important factors for child development.

It is critical that residential neighborhoods and developments where children live have safe access to such diversity, especially so for girls, who after the age of eight or nine tend to have a significantly smaller home range than boys (Tranter and Doyle 1996; Moore 1986a). For children in industrially developed countries, the last few decades have seen a marked decrease in independent mobility. Studies in the UK (Hillman and Adams 1992), Australia (Tranter 1993), and the Netherlands (Van der Spek and Noyon, 1995) record steep declines in children's mobility and in the case of the Dutch study, a parallel decline in environmental awareness. In the UK in 1971, 80 percent of seven- and eight-year-olds were allowed to go to school without adult supervision. By 1990, this figure had dropped to 9 percent (Wheway and Millward 1997, 17). Mobility is not only important for a child's physical development, but it also is essential in promoting self-esteem, a sense of identity, and the capacity to take responsibility for oneself (Kegerreis 1993; Noschis 1992). Two elements fuel this change: "stranger danger," or parents' fears of child molestation, et cetera; and danger from traffic. Ironically, the traffic peak caused by parents dropping off and picking their children up from school is part of the traffic danger problem (Hillman 1991).

One of the impediments to the development of child-friendly neighborhoods may well be that the very qualities that are aesthetically pleasing to adults can be detrimental to children's needs. For example, a study of the effect of the physical environment on the play patterns of children in four Oakland, California, neighborhoods found that in the neighborhood with the lowest density and the hilly verdant terrain favored by upper-middle-income home buyers, children felt painfully isolated from each other and lacked access to places for spontaneous, unplanned play. In contrast, children living in more urban, higher-density (and flatter) neighborhoods tended to have a greater range and autonomy; friendship patterns were more casual, less structured, and tended to involve a greater age range.

Although children in all four neighborhoods had some access to parks and school playgrounds, many did not consider these "their own" and sought out unplanned, undeveloped open space. "These unplanned areas, which often were nothing more than a vacant lot or a garbage-strewn stream, met certain needs that developed play space could not. At the very least they offered privacy—for these were places where often no one *but* a child could go or would want to go. This should not be surprising, for it reflects children's desires to have something that is theirs, at a time when virtually everything else—houses, shops, streets, public transportation—is built for or 'belongs' to grownups"(Berg and Medrich, 340).

A study of children's play in two rural Welsh communities recorded that woodland featured prominently in children's accounts of favorite places to play. A wooded area provided a place to explore and also facilitated imaginative play, providing raw materials such as branches, bark, sticks, and leaves that triggered creativity. "Indeed, the imagery drawn upon in many games outside in the woods, in community spaces, or private spaces within their homes, drew heavily on this setting" (Maxey 1999, 22). A 10-year-old girl, when asked why she liked the woods best, responded, "Because there is lots to do, we can hide and build dens, we have a swing. . . . I like to see the animals collect things and . . . well, we just do what we want, we don't have to [pause] you know, do what we're told" (Maxey 1999).

Nature not only comprises green growing things but also two other elements that are significant to children: water and animals. In a Danish study, 88 children living in settings ranging from cities to villages were given cameras and asked to take pictures for one week of what they were doing and what was meaningful to them (Rasmussen and Smidt 2003). As well as elements of green nature (trees, shrubs, flowers, sand dunes, etc.) and places where they played (mounds, dens, campfire sites, tree swings, etc.), animals featured prominently—both those kept at home or school (mice, guinea pigs, rabbits) and those known in the neighborhood (cats, dogs, chickens, ducks, birds, horses). While urban or suburban green spaces may not be appropriate for farm animals, through the deliberate creation of habitat the presence of birds, insects, and small mammals can be guaranteed. Inclusion of creeks, natural or man-made ponds, wetlands, et cetera, can encourage habitation by fish and amphibians.

Besides elements of green nature, another natural element that is particularly attractive to children is water—whether standing in a pond, lake, marsh, or retention pond; or flowing in a creek, river, gutter, et cetera. As well as being a natural element that children find endlessly fascinating to touch, explore, float things on, et cetera, it also of course attracts wildlife. Wildlife corridors and greenways that are also creek valleys can influence the basic structure in neighborhood design, improve wildlife value (Hellmund and Smith 2006) and sustainability, and provide children with the added attraction of water in a near-home environment (Arendt 1996). But there are other ways in which water can be found in near-home play locations. As Google will tell you, the San Francisco Bay Area and Seattle's King County appear to be the national leaders in the "daylighting" of creeks buried in pipes decades ago. Children are major beneficiaries of these initiatives, especially when in public parks and schoolgrounds; for example, Blackberry Creek, Thousand Oaks School, Berkeley, was daylighted in 1995 (initiated on the ground in 1971 with a small, artificial ground level creek built by the first author and UC–Berkeley students to "mark" the creek hidden underground). In 2005, the living creek and its educational use by the school was appraised positively by Gerson, Wardani, and Niazi (2005).

Innovations in stormwater management are creating other opportunities for children to find water for play close to home (Jencks 2007). On residential blocks that are part of the Green Street program in Portland, Oregon, one parking lane is converted to a bioswale, with stormwater passing through an area of native plants and rock berms bringing nature into the neighborhood. Neighbors have to apply to be part of the program and to maintain the swale. At High Point, a Hope VI public housing scheme in Seattle, a complete retrofit included a 34-block water retention system with porous concrete, trees, and wide strips with native planting between the sidewalk and the parking lane, all draining to an on-site retention pond. Even in highly urbanized neighborhoods, creative infrastructure solutions can provide elements for children's water play. In the German city of Freiburg in Bresgau, the "baechle" or "little streams" provide small water courses where children float paper boats beside city streets (Lennard and Lennard 1992).

In a growing number of participatory studies, when children are asked what might be done to improve the environment of their neighborhood, they have many perceptive and practical comments including calming traffic, improving maintenance, creating places for different age groups from toddlers to teens, and providing more natural amenities, particularly trees (Chawla and Malone 2003; O'Brien 2003; Morrow 2003). In a study researching 12- to 15-year-old children's subjective experience of two neighborhoods in a town 30 miles from London, several children described the lack of wild places where they could play and make dens (Morrow 2003). A 12-year-old boy mentioned he didn't like the sprawling suburban neighborhood where he lived, "cos it's so built up, there's not much to do and like, where my sister lives, she lives in [another town], and just across the road there's a big forest, and my brother likes to go over there with their dog, and they'd be out for hours and hours, and that's what I like when I go there." (Morrow 2003, 170).

Another boy in the same neighborhood, Bart, age 13, described how a local park the children had dubbed "Motorway Field" could be improved: "Motorway Field is like a long strip, and at the end, there is this round bit. There's a few trees there, but it'd be nicer if . . . they planted more trees there, so it was like a little mini-forest where people can build dens, that won't be kicked in and stuff, so there's more variety of things to do" (Morrow 2003, 174). All the children interviewed described "not having enough to do" in terms of appropriate facilities, activities, and places to go. In this study and others (for example, Percy-Smith 2002), it is ironic that when children's views about their neighborhoods in the inner city and in a more affluent suburban location are compared, it is children in the latter who are more likely to find their environment "boring."

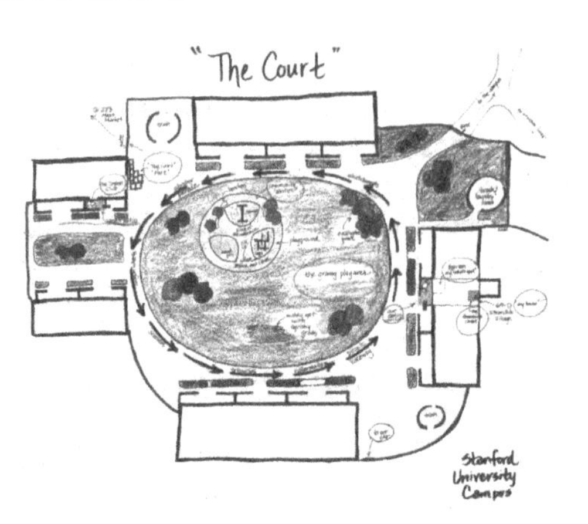

Figure 10-31: Shared outdoor space at Stanford University married students' housing, recalled as her favorite childhood place by a Berkeley architecture student.

PREFERRED PLAY ACTIVITIES: ADULT RECOLLECTIONS

When adults recall their favorite places of childhood, the great majority are outdoor locations (Cobb 1977; Cooper Marcus 1978; Chawla 1986; Louv 2005) and very often involve natural features (trees, streams, bushes, rocks, sand, woodland), and even in very urban settings, play with natural "loose parts" (leaves, seeds, twigs) is a most fondly remembered episode (see Figure 10-31).

In the middle years of childhood (about age 6 to 12), finding or creating special places in the landscape appears to be a common experience for children of all cultures. The power of the memory of such places in adulthood suggests that they play a unique and powerful role in the shaping of the self (Sobel 1990). In analyzing the special-place experiences of more than 100 adults and 200 children, Sobel noted the following recurrent descriptors: special places are found or created by children on their own; they represent an organized world for the child; they are secret, safe, and owned by their creators; in turn, such places empower their builders.

It is essential that we leave wild or semiwild places in our residential areas where such child-created spaces can naturally occur. Providing a playground, buying a playhouse, or building a tree-fort *for* your child just doesn't measure up. Not only does a child-created or found place contribute to a child's sense of autonomy and independence, recollections from adulthood indicate that they also provide a sense of solace in difficult times. In a paper discussing environmental autobiographies collected at three universities in the United States and Australia, Dovey includes a number of quotations that illustrate this point (Dovey 1990). One person recalled: "The willow tree in our backyard was our favorite thing from about four until it was cut down when I was eight. . . . It was the center of my childhood fantasies. The branches served as whips for horses, swords for duels, hair for mermaids. . . . When I was angry or upset I used to sit far above the world swaying in the breeze in the comfortable curve of its topmost branches." Another wrote: "One of my favorite places to go and tell my worries to was the big apple tree in my backyard. It was my refuge and for once I was able to talk and have someone that would always agree with what I was saying."

NEW BIOPHILIC FORMS OF RESIDENTIAL NEIGHBORHOOD

Residential neighborhoods designed on biophilic principles need a fine-grained integration of nature into children's everyday lives. Neighborhood nature can be integrated into private spaces around homes (some large enough for food production and biodiversity), and flow out into the public realm of residential streets, local commercial areas, neighborhood parks, schoolgrounds, open spaces, greenways, protected reserves, urban stream corridors, and "leftover" unbuildable wild spaces.[8] In the semipublic realm, community gardens, the grounds of childcare centers, gardens attached to community facilities such as health care facilities, libraries, recreation centers, and college campuses can be designed to offer contact with nature for children (Moore 1986a).

Levels of access depend on stage of maturity and degree of independent mobility (which is constantly changing across child populations); ideally, as many opportunities as possible for daily exposure to nature should be made available within the bounds of residential neighborhoods. Four models of child-friendly layouts, together with case studies, are discussed below.

1. CLUSTERED HOUSING AND SHARED OUTDOOR SPACE

A special case is made here for shared outdoor spaces within housing areas—a form that offers particular opportunities for exposure to nature and for children's independent mobility. We are speaking here of a particular form of outdoor space within a cluster of residential buildings (single-family homes, row houses, walk-up apartments, lofts, etc.) directly accessible to the residents of those buildings without crossing a street. Such spaces are neither private (like backyards or balconies) nor fully public (such as streets or parks) but something in between. Immediate residents share these spaces and either participate in their maintenance or pay a fee for the upkeep (usually the latter). Historic precedents of this form of a cluster of buildings enclosing an area of shared outdoor space include the monastic cloister garden; Oxford and Cambridge college quadrangles; 1920s California bungalow courts; 1960s Planned Unit Developments; and historic gardens and squares of cities in the United States and the UK such as London, Edinburgh, Baltimore, Boston, and New York.

Contemporary forms can be found in many medium-density housing developments (in both urban and suburban locations), as well as in cohousing and ecovillage developments (Bang 2005). In all such schemes, traffic and parking (in the form of garages or grouped parking lots) is kept to the periphery, and the living spaces of the surrounding dwellings face into the green heart of the block (see Figure 10–32). Private outdoor spaces in the form of backyards or patios provide a buffer between private and shared space, and a gate or break in a hedge or planting permits easy access from one to the other. Providing the space alone is not enough. Care must be taken in detailing circulation, planting, and furnishings so that the shared space includes pathways, open lawns for active play or sunbathing, shaded seating clusters for social meetings, play areas for younger and older children, and areas of shrubbery and unkempt areas where children can explore and make dens, et cetera. Space permitting, vegetable garden plots may be included for those lacking sufficient private space around the home (as in most cohousing communities).

The chief beneficiaries of shared open space are children. Systematic observational studies reveal that where the residences around such space are for families, more than 80 percent of the users of the outdoors are likely to be children (Cooper Marcus 1974; Cooper

Figure 10-32: Shared green space surrounded by row houses and apartments provides ample opportunities for children's nature contact. Cohousing, Wageningen, the Netherlands.

Marcus 1993; Moore and Young 1978; Cooper Marcus and Sarkissian 1986). In summary, the advantages of such space include:

1. Providing green views from home, which have been associated with positive psychological benefits (Ulrich 1999).
2. Offering children a traffic-free play area within sight and calling distance of home (Cooper Marcus 1974).
3. Reducing the anxiety of parents so they are more likely to let their children out to play in such spaces (compared to neighborhood streets or parks) since two of the greatest parental fears are eliminated: traffic and "stranger-danger" (Cooper Marcus 1974).
4. Facilitating spontaneous play between friends living nearby during brief periods (before the evening meal or after homework), when trips farther away from home are unlikely.
5. Including planting designs that can provide diverse wildlife habitats for birds, insects, small mammals, and amphibia, thus enriching the nature experience of both children and adults.
6. Strengthening a sense of community, ownership, and caring often lacking in contemporary urban/suburban neighborhoods (Cooper Marcus 2003).

Shared open space provides a vehicle for community development and the building of social capital beyond the nuclear family at a level less than the unfeasible prospect of a whole neighborhood. While the direct benefits to children are rather obvious, there are indirect benefits, which include use by older residents (particularly those who may live alone and/or do not own a car) who offer potential intergenerational social relationships with resident children. Provision of shared outdoor space serving housing for both families with children and older adults, for example, in assisted living, if carefully designed with their disparate and shared needs in mind, could be well-accepted and appreciated. An example of this approach is the Village of Woodsong, Shallotte, North Carolina, a traditional walking neighborhood "designed for tending to the basic rites of life" (www.villageofwoodsong.com/inde). A village center, mixed housing types, narrow streets, a park specifically designed for children, a woodland trail connection to the local elementary school, continuing care residences, a range of outdoor spaces, and natural areas provide for socializing, working, shopping and recreation within walking/biking distance. Indirect, angled alleyways are designed as secondary "secret" play routes for children. Collectively, the easily accessible, shared spaces of Woodsong are aimed at village-wide social integration. The development is still under construction so it is still too early to know if this design objective has been met.

Where shared outdoor spaces have been designed into family housing developments they have often been remarkably successful, especially in providing for safe play close to home, and in facilitating a sense of community. The examples below illustrate these points in a variety of forms.

Completed in 1964, St. Francis Square was the first of many similar medium-density garden-apartment schemes built in San Francisco during the era of urban renewal. The client for the 299-unit project (the Pension Fund of the ILWU) challenged the designers (Robert Marquis, Claude Stoller, and Lawrence Halprin) to create a safe, green, quiet community that would provide an option for middle-income families wanting to raise their children in the city. Built as a co-op, St. Francis Square occupies an 8.2-acre, three-block site in the city's Western Addition, and it has an overall density of 36.5 units per acre. Its design is based on a pedestrian-oriented site plan, with parking on the periphery and three-story apartment buildings facing onto three landscaped interior courtyards (see Figure 10–33).

The shared outdoor space, which is owned and maintained by the co-op, is critical to this community. Its trees screen the view of nearby apartments, reducing perceived density, and its grassy slopes, pathways, and play equipment provide attractive places for children's play. Sitting outside with a small child, or walking home from a parked vehicle (or from one of the three shared laundries), adult residents frequently stop to chat with one another. The courtyards at St. Francis Square are, in effect, the family backyard writ large. Behavior mapping data gathered in 1969 showed an overall child-to-adult ratio across the site of 7:3 and in the courtyards, a ratio of between 5:1 and 7:1 (see Figure 10-34). If these

Figure 10-33: St. Francis Square, San Francisco, is a successful inner-city, medium-high-density housing neighborhood for families with children. Parking is located on the periphery of the site and dwellings face onto three landscaped interior courtyards.

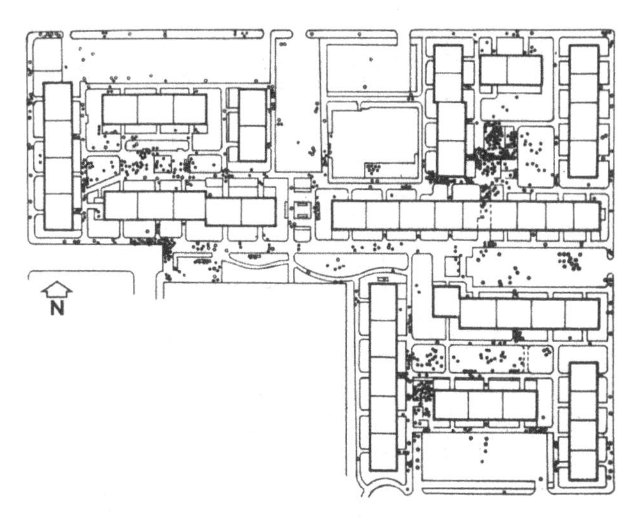

Figure 10-34: Aggregate map of people seen outdoors, 8 a.m. to 8 p.m., St. Francis Square, San Francisco. Solid black dots represent children; open circles represent adults. Observations were conducted on one weekday morning, one weekday afternoon, and one weekend morning, in June 1969 . The proportion of children to adults in the shared interior courtyards is between 5:1 and 7:1.

spaces were public parks, parents would likely not allow their children to play there alone, and residents would be less likely to help maintain the courtyards, question strangers, or help neighbors in need.

The findings of a postoccupancy evaluation of St. Francis Square conducted by the second author in 1969–1970 were confirmed and expanded by a further year of observation when she lived there with her family (1973–1974).[9] Numerous site visits since the original study, plus conversations with the current management, confirm the basic findings of almost 30 years ago about why the shared outdoor space at St. Francis Square is highly valued and well-used by residents: (1) narrow entries between buildings clearly mark the passage from the public space of street and sidewalk to the shared space; (2) the size of the courtyards (c. 150 × 150 ft.) and the ratio of the height of adjacent buildings to the distance between them (c. 1:6) gives them a human scale; (3) the courtyards are bounded by the units they serve, and almost all units have views into the outdoor space (facilitating child supervision); (4) attention and financial resources were focused on the quality of the courtyard landscaping; (5) fences provide a clear distinction between private outdoor patios and the shared space of the courtyards; and (6) easy access is provided from apartments and patios to the courtyards (see Figure 10-35 in color insert).

Southside Park is a 25-unit urban infill cohousing development in inner-city Sacramento, California, designed by Mogavero Notestine and Associates in consultation with the 67 residents (40 adults and 27 children). Completed in 1993, it contains 14 market-rate, 6 moderate-income, and 5 low-income condominiums. The site plan was inserted into Sacramento's existing street grid, with most of the houses clustered around an interior green (see Figure 10-36). The remaining houses (two rehabbed Victorians and several new units) were arranged in a smaller cluster across an alley. Front porches mark house entries from the street, while back porches and patios look out onto the common green (see Figure 10-37 in color insert). Residents eat meals together several times a week in the 2,500 sq. ft. common house.

Informal observations conducted during several visits confirm what residents and designers hoped for. Children play on the common lawns, pathways, and in the play-equipment area; adults meet and converse while outdoors with their children, using the common laundry, working in the raised garden beds, walking

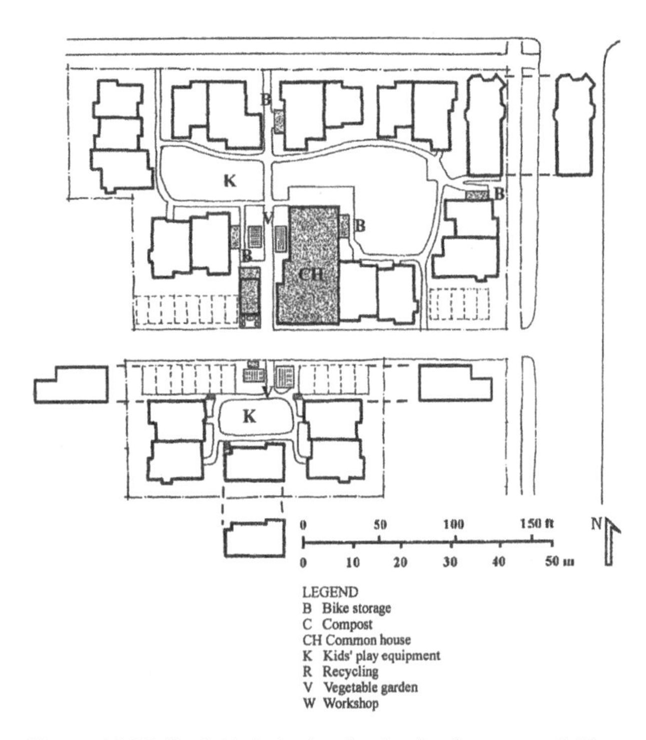

Figure 10-36: Southside Park cohousing site plan; Sacramento, California, designed by Mogavero Notestine and Associates

back and forth to their cars, or congregating at the common house. As at St. Francis Square, the sense of community and the range of children's outdoor play opportunities at Southside Park are supported by a layout that controls traffic flow and offers a central pedestrian green. Interestingly, the street-facing porches at Southside Park are used by residents for privacy, since the shared outdoor space on the interior of the block is such a social space. Cohousing, originally a grassroots phenomenon instigated by groups of individuals seeking a more neighborly and child-friendly lifestyle, has now been adopted into the mainstream and is delivered through top-down as well as bottom-up processes in northern Europe and, to some extent, in North America (Williams 2005).

While St. Francis Square and Southside Park were purpose built, it is possible to modify an existing urban block so that the interior becomes a shared green space. The Meadows occupies a city block in Berkeley, California. From 1963 to 1973, a lecturer in real estate at the University of California acquired 27 properties around a block, most of which were single-family residences built between 1900 and 1920. In 1971, in a conscious experiment to create a unique residential environment, he began removing backyard fences on the interior of the block as well as unused garages, extraneous outbuildings, and paved areas, replacing them with grass, flowers, shrubs, trees, and walkways to create a parklike shared space. The residents, who were all his tenants at the time, retained semiprivate patios, lawns, or planted areas close to their dwellings. The block was named The Meadows by its residents.

A study by Cavanna (1974) compared this block with an adjacent control block with regular fenced backyards using a questionnaire, behavior traces survey, and a systematic record of outdoor activities. In contrast to residents of the block where the fences had not been removed, residents of The Meadows had more social contacts (see Figure 10-38), felt safer in the areas around the houses, had a higher opinion of their neighborhood, spent more time outdoors at the back of the house, and considered their backyard environment to be more open, attractive, and better maintained. While this study was conducted almost 30 years ago, recent visits to this block revealed that the backyard fences have not been replaced, even though most dwellings are now owner-occupied (see Figure 10-39

Figure 10-38: The Meadows, Berkeley, California. Composite of the total number of social contacts.

in color insert). The central open space has mature trees, areas of grass, shrubbery, vegetable gardens, and a sand box, and is well used for children's play, studying, sunbathing, barbecues, basketball, and gardening. Residents maintain their own private (but unfenced) yards and patios, as well as adjacent portions of the shared outdoor space.

Although one might assume The Meadows to be a unique innovation, many similar historic examples exist. In Boston's South End, for example, Montgomery Park comprises one-third of an acre entirely enclosed by 36 brick row houses. Established as a formal garden by the original builder of the houses in 1865, by the mid-twentieth century it had become run down, and the shared space had been virtually abandoned. From the 1970s on, however, a new group of residents removed debris, improved drainage, planted a lawn and perennial borders, took down fences, lobbied to have phone lines buried, removed a service road that circles the park, and restricted access from adjacent streets by installing locked gates. By the 1990s, the orientation of most of the buildings was toward the back, with a brick pathway delineating the border between private backyards and shared space. The lush interior of the block is now equipped with movable garden furniture and is used for informal dining, children's play, annual potlucks, weddings, birthday parties, and garden tours (Morris 2001).

A recent article in the *Atlantic* monthly surveyed how variations on The Meadows and Montgomery Park may provide ways of redesigning conventional suburban blocks where the residents—especially those with children—are looking for more neighborly lifestyles, and for settings for play that are safer and more stimulating than conventional sidewalks (Drayton 2000).

To achieve the successful outcomes of the examples described above requires a carefully considered layout with regard to traffic flow, pedestrian circulation, and the location of shared open space, as well as attention to design details. The lack of such attention rendered the shared space in many postwar public-housing projects and the suburban Planned Unit Developments of the 1960s nonfunctional. Unfortunately, those who criticized such spaces for being poorly maintained no-man's-lands assumed (wrongly) that they could never work (Coleman 1985). There is ample evidence that the outdoor activity of resident children and adults, and a related sense of community, can be increased by careful attention to design. Not only do housing schemes with shared outdoor space *work*, people who can choose where and how to live actively seek them out. For example, of the hundred or so cohousing communities in North America completed or in the planning stage, all feature site plans where units face onto shared outdoor space as defined above.

Further evidence for the success of schemes with interior block green space has been compiled by Community Greens: Shared Parks in Urban Blocks, a nonprofit initiative based in Arlington, Virginia (www.communitygreens.org). Community Greens notes that homes in developments that abut shared outdoor space sell, generally, at prices 5–15 percent higher than the competition and the sales rate is also faster—two factors that benefit home builders' bottom lines. One developer in the northwest, Jim Soules of the Cottage Company, specializes in cottage homes that surround such shared green space. In fact, Soules will *only* develop these kinds of projects. Says Soules, "I will never build another project without a community green. Residents open their door to a private park . . . it's an emotional experience. That is what people are interested in" (Kate Herron, personal communication to Cooper Marcus, 2007). The communities in which Jim Soules operates have adopted a "cottage housing code" which allows small homes of about 1,000 sq. ft. to be built in neighborhoods of typically larger homes, providing that the development includes a community green, at least 50 percent of the homes abut the green, no home is more than 60 feet from the green, a minimum of 400 sq. ft. per dwelling unit of open space is provided, and the green is encompassed by houses on at least two sides.

In 2007, the City of Portland Bureau of Planning sponsored a design competition for "Family-Friendly Courtyard Housing," because they saw the need for fostering higher-density housing configurations that provide quality living environments for families with children. The competition guidelines state: "Common higher-density ownership housing types, such as small-lot row houses and detached houses, do not allow for outdoor spaces of sufficient size to serve the needs of families with children. Housing oriented to shared courtyards present opportunities for large useable, outdoor spaces that are not possible in the form of private yards at higher densities."

There are demographic, economic, and psychological reasons why residential layouts that balance vehicular needs, pedestrian use, and shared outdoor space are particularly appropriate at this time. With increasing numbers of families where both parents are employed, safe, communal play space *right outside* the house is especially useful (see Figure 10-40). Gone are the days, for most families, when the mother was home all day to walk or drive children to a nearby park. The potential sociability of a traffic-free, green area at the heart of a community is also appealing to the increasing number of single-person households (both young and elderly).

Because shared spaces are in protected locations and used by residents, a crucial point is that they can be managed to a higher level of natural diversity and aesthetic enhancement than more public spaces. As residents control shared space management, it means that functions of the space can be adjusted to match user needs as they change. The residents of the St. Francis Square co-op, for example, have made numerous changes to their shared spaces over the past 40 years. With a reduction in the number of children living there (the original families who raised their children there have no desire to move away), a play equipment area was recently removed and replaced by a small Japanese garden created and constructed by residents. Outside of private dwellings and their associated private outdoor space, there are relatively few opportunities for small groups to have the same sense of accomplishment through hands-on manipulation of the local environment. The social benefits of greening activities have been well documented over several decades (Plas and Lewis 1996). Evidence from interviews in communities with shared outdoor space indicates that such "working together" provides a profound sense of shared responsibility and community (Cooper 1970, 1971; Cooper Marcus 2003).

Figure 10-40: Shared greenspace, where children can play and adults meet, is increasingly important at a time when both parents may be working or a single parent is raising children alone. Co-op housing, False Creek neighborhood, Vancouver, British Columbia, Canada.

Resistance to the Provision of Shared Outdoor Space

Shared outdoor space in clustered housing can be found functioning successfully for both adults and children in everything from urban cohousing retrofits, to new urban and suburban affordable housing, to urban loft schemes, to sought-after bungalow courts dating from the 1920s. If the provision of shared outdoor space in clustered housing makes so much sense in terms of children's needs, what are the impediments to its more widespread adoption, particularly in new suburban developments? The opposition comes largely from the proponents of new urbanism who emphasize the importance of a return to the grid, and green space being provided almost exclusively in public parks and squares. New urbanist thinking places the aesthetics of the streetscape as a very high priority. Hence, parking is most often provided in rear-access alleys or in the interior of the block. While there is an urban form designated as the "square block" in new urbanist literature (Steutville and Langdon 2003, 1–11), and this could potentially result in the kind of clustered housing described above, the insistence that parking be provided off-street frequently results in this interior open space being filled with cars. For example, at Britton Courts, a new urbanist development of affordable housing for families in San Francisco, the interior of the block is filled with parking and is designated as a "Parking/Play Court." It is sad indeed when the needs of the car and the aesthetics of the streetscape take precedence over the needs of children. Although there are examples of small new urbanist courtyard schemes with interior hardscape, the development of neighborhoods such as St. Francis Square, Southside Park, or The Meadows with spacious areas of shared green space on the interior of the block would be virtually impossible under current new urbanist form–based codes.

Add to this the unsubstantiated statements such as that by a leading new urbanist proponent that "shared outdoor space at the back never, ever works" (Duany 2001), and the future of this form of housing is in jeopardy. For example, the site plan for an affordable housing scheme in Windsor, California, incorporated shared green outdoor space and was welcomed by its client, who had previously noted the success of Cherry Hill (discussed below). However, the City Planning Commission, citing new urbanism principles, insisted that the site must have a through street, that shared outdoor space "doesn't work," and that housing clustered around such a space creates "a ghetto" (Durrett, personal communication, 2005). Such misunderstandings of the social implications of site planning are disturbing, particularly in a lower-income setting where residents may not be able to sustain wider social networks or take their children to areas of public recreation or to natural settings such as nature reserves and parks. There is much progress yet to be made in professional education to counteract the prevailing level of ignorance in these matters. Collectively, supporters of biophilia-based neighborhoods need to present arguments to the proponents of new urbanism that there are other important options for residential settings where children and families predominate besides the standard houses-facing-onto-streets.

2. CUL-DE-SACS AND GREENWAYS

Another way in which safe access to nature can be ensured in a residential neighborhood is to create a site layout where local streets end in cul-de-sacs that abut a greenway or local park. Children can then move safely to a green area from their homes without crossing a street. The greenway itself might be a pedestrian or cycling connection to a local school, shops, or larger park (see Figure 10-41).

A systematic observational and interview study of children's informal play on twelve housing estates in the UK (Wheway and Millward 1997), noted that the favorite activity was being "on the move"—walking, running, cycling, meeting others, stopping for a while, moving on. When asked about their favorite play spaces, children consistently referred to green open spaces (parks, fields). If there was a single tree or a copse of trees, these were very popular for climbing, swinging,

Figure 10-41: Provision for safe, hard-surface play on a cul-de-sac (foreground) and access into a semi-natural greenway with walking and bike paths to other neighborhoods and urban amenities (Davis, California). Note connectivity to another cul-de-sac across the greenway.

or just "hanging out." Green areas and trees were cited as favorite places by 73 percent of the children; equipped play areas by only 21 percent. Cause for hope is the similarity of these results to the field data gathered by the first author in three contrasting neighborhoods in 1977 (Moore 1986a). In spite of dramatic changes in lifestyles, children still are searching for the same natural outdoor spaces as a generation ago.

Wheway and Millward's (1997) findings, together with requests from a majority of parents for their children to be within sight and calling distance of home, prompted the authors to recommend traffic-calmed cul-de-sac site plans with footpath networks to open spaces and play areas, permitting children's access to as large an outdoor environment as possible. "The ideal estate [development] would be designed so that children would be able to move freely throughout the neighbourhood, able to enjoy a wide variety of social interactions and opportunities for physical, imaginative and creative play" (Wheway and Millward 1997, 60).

In a U.S. study of cul-de-sacs in four northern California towns, a rigorous statistical analysis revealed that children who live on cul-de-sacs play outside in their neighborhood more often than children who don't, and moving to a cul-de-sac is associated with an increase in children's outdoor play (Handy et al. 2007). An extension of this study interviewing parents (and some children) in a fifth town reported that the neighborhood is an important setting for play for all children, but that 75 percent of those living on cul-de-sacs reported being highly active versus 55 percent of children on through streets. Traffic and strangers were cited as concerns by parents on both cul-de-sacs and through streets, but traffic was less of a concern for parents on cul-de-sacs. Forty percent of parents on through streets expressed concern about traffic, whereas 100 percent of parents on cul-de-sacs said that what they liked most about their street was safety from traffic. Thirty-five percent of parents on through streets asked for infrastructure to decrease traffic speed, versus zero percent of those on cul-de-sacs (Handy et al. 2007).

The cul-de-sac and greenway approach to residential neighborhood planning had its beginning in the English Garden City movement. The largest twentieth-century application can be seen in the postwar era in the British New Towns (for example, Stevenage), in the suburbs around Stockholm, Sweden (e.g., Vällingby and Fårsta) and in Tapiola, the New Town outside Helsinki, Finland. In all cases, green fingers radiate out from town centers, permitting safe pedestrian and bicycle access from homes to school, after-school centers, play areas, shops, services, and the subway. Importantly, in terms of children, many green areas were neither developed nor designed (except for pathways), leaving broad expanses of natural landscape, woodland, forest, and rocks (in Sweden and Finland) as inviting areas for exploration and play (see Figure 10-42). In Sweden, this form of planning occurred both because access to nature is a highly regarded cultural value and because, in the immediate postwar years, there was a labor shortage in Sweden promoting planning policies that created child-friendly environments, which encouraged women to return to the labor force.[10]

Figure 10-42: Layout of a Stockholm suburb allows children to expand their territory naturally and continuously as they mature cognitively and become more skilled in negotiating their environment. Woodland greenway provides plentiful contact with nature.

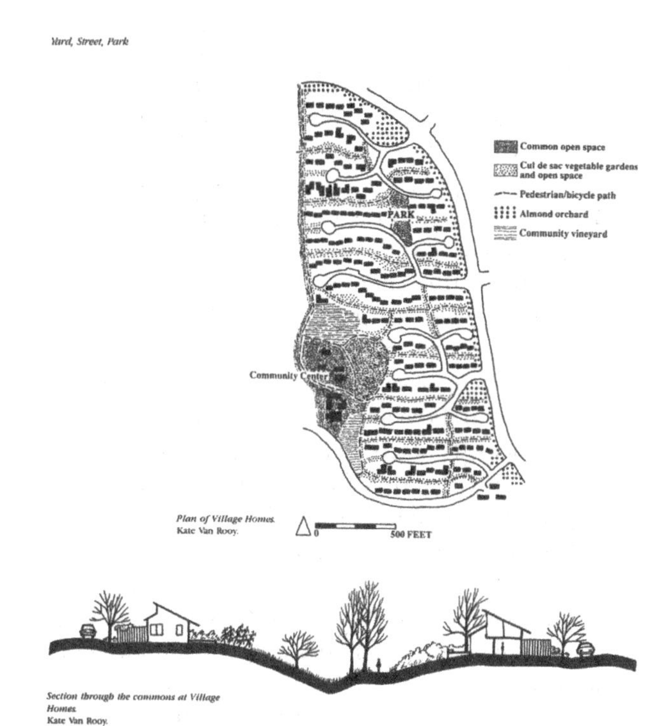

Figure 10-43: Layout of Village Homes, Davis

Access to a cul-de-sac *per se* does not necessarily guarantee access to nature; however, where the dead-end abuts a green area, or where the cul-de-sac itself loops around an area of greenery, safe access to nature is maximized. Two case studies described below illustrate this point.

Village Homes is a 244-unit neighborhood on a site of 60 acres completed in 1982 on the outskirts of the university town of Davis, California, 60 miles north of San Francisco (see Figure 10-43). Its designers, Michael and Judy Corbett, document how it began as a "hippie subdivision" derided by banks and the local real estate industry, but now has become the most desirable neighborhood in Davis (Corbett and Corbett 2000). Village Homes uses shared outdoor space as a successful aesthetic and social basis for neighborhood design. Individual houses are accessed from cul-de-sac streets with their backs facing onto pedestrian greenways, all leading to a central green. The long, narrow (23 ft.), tree-shaded, dead-end streets keep the neighborhood cooler in summer, save money on infrastructure, eliminate through traffic, and create quiet and safe spaces for children to play and neighbors to meet (see Figure 10-44). An extensive pedestrian common area at the heart of the neighborhood includes spaces for ball games and picnics, community-owned gardens, vineyards, and an orchard. Greenways provide access for bicycles and pedestrians traversing the neighborhood. Drainage swales instead of storm sewers collect storm water runoff in a system of linear wetlands, which greatly enhance the wildlife habitat and exploratory play opportunities between the backs of the houses and reduce summer irrigation costs by one-third. Neighborhood pathways follow the swales and connect to the main greenways (see Figure 10-45).

Figure 10-44: A shaded cul-de-sac with mature trees provides a setting for potential nature contact close to home (Village Homes, Davis).

Figure 10-45: Family out for a walk on one of the many greenways in Village Homes, Davis

This attractive environment, although accessible to outsiders, is definitely *not* a public park. Bounded by inward-facing residences, it provides a green heart to the neighborhood, a safe and interesting network of open spaces for children and adults. A study of neighboring revealed that residents report having three times more social contacts, and twice as many friends as residents of a nearby conventional control neighborhood (Lenz 1990, quoted in Francis 2003). With the recurring problem of children's diminishing independent mobility, Village Homes remains an outstanding example of territorial continuity, enabling each child to gradually expand her territory from private front yard or backyard (up to age 5 or so), to cul-de-sac street (age 5 to 7), to back swale pathway (age 8 to 10), to main greenway system (age 10 upwards), and from walking to bicycle (variable ages; see Figure 10-52). Not only does Village Homes offer each child a hierarchical movement system that affords independent mobility from an early age, but as an extra bonus it offers experience of a rich, diverse landscape along the way.

A systematic observation study of children's use of communal open space at Village Homes conducted in 1981 (see Figure 10-46) revealed that the great majority of activity (65 percent) occurred in green open spaces (bike paths, green belts, drainage swales, turf areas). The second most frequently used area was street space (20 percent), the quiet, shaded cul-de-sacs with slow-moving traffic (Francis 1984–85). In a later account of Village Homes, Francis notes: "What is unique about Village Homes from a child's perspective is the diversity of places provided, from streets to play areas to natural areas, and the almost seamless access provided to these places" (Francis 2003, 56).

The importance of Village Homes for children is illustrated by recollections of Christopher Corbett, son of the developers, who grew up there: "Growing up in Village Homes gave me a sense of freedom and safety that would be difficult to find in the usual urban neighborhood. The orchards, swimming pool, parks, gardens, and greenbelts within Village Homes offered many stimulating, exciting, joyful places for

Figure 10-46: Behavior mapping study of children's use of outdoor space in Village Homes, Davis, conducted by Mark Francis in 1981

me to play with my friends" (Corbett and Corbett 2000, 21).

Interestingly, in Francis's 1981 study, when children were asked to describe their favorite places at Village Homes, the most sacred were wild or unfinished places such as building sites and places with names such as "willow pond" and "clover patch." "These findings argue for neighborhood design that retains open space in its natural state, which children can manipulate to suit their own needs" (Francis 1984–85, 37; see Figure 10-47). Only 13 percent of the children observed by Francis were seen at an amenity at Village Homes specifically designed as a playground. When a team of Berkeley graduate students interviewed residents of Village Homes and two nearby subdivisions with similar layouts also in Davis, people named the cul-de-sacs and greenways as their favorite aspects of the neighborhood environment (University of California 2003). (See Figure 10-48 in color insert.)

Figure 10-47: Child playing in a drainage swale in the early days of Village Homes. Design enabled daily contact with nature from the beginning.

Citywide Greenway Networks

While Village Homes' network of cul-de-sacs linked to greenways is widely cited, and rightly so (see Figure 10-49), it is not unique. At DC Ranch, a master planned community near Phoenix opened in 1997, a common open space is located at the end of each cul-de-sac in lieu of the pie-shaped house lots that typically terminate such streets. The common open spaces link to a 13-mile system of paths and natural preserves with pedestrian underpasses providing safe passage for children under major streets (Gause 2002, 64). Reston, Virginia, at its inception in 1962 the largest new town in the United States, includes over 55 miles of trails with footbridges over vehicular streets, linking residential streets with each other and to extensive nature preserves (Gause 2002, 182).

Stream valleys, drainage swales, and ribbons of natural landscape with pedestrian and bike trails form the open space frameworks of these and a number of other successful U.S. new towns and master-planned communities created in the last 40 years. These include The Woodlands, near Houston; Columbia, Maryland; New Albany and Easton, Ohio; and Bonita Bay, Florida. The early-eighties planning of the last, for example, included the natural systems analysis of the site, preservation of

Figure 10-49: One of many greenways crisscrossing the city of Davis, California. "Elk" sign at right indicates path leading to a cul-de-sac that abuts the greenway.

hundreds of acres of wildlife habitat and innumerable small ponds and lakes, and the creation of 12 miles of bicycling and walking paths crisscrossing the community between a street system ending in cul-de-sacs, and leading to waterfront parks, playgrounds, hiking trails, boardwalks, et cetera (Gause 2002). An integrated system of environmental management includes applying xeriscape principles, restricting pesticide application, leaving snag trees undisturbed to provide habitat, and planting native grasses. A 50 percent increase in listed species was revealed by a wildlife survey conducted both before and seven years after the construction of a golf course. Although this is an upscale community and no information exists on children's use of the outdoor areas, the sensitive physical planning and environmental management offers a landscape model where child-nature contact could potentially be optimized.

Compared to contemporary towns and development based on new urbanist principles, these earlier planned communities used ecological planning and design as major determinants in creating land use patterns and street systems. Stream corridors and sensitive conservation areas were preserved; street patterns, often winding, with cul-de-sacs, were determined by natural elements of the site. In contrast, many of the early and influential new urbanist developments, such as Kentlands, Harbor Town, and Celebration, while certainly respecting the natural qualities of their sites, employ an apparent one-size-fits-all street pattern of bent grids and axials, eschewing cul-de-sac-greenway combinations.

A variation on providing nearby nature in the interior of an urban block, or a cul-de-sac abutting a greenway, is the provision of a green area at the center of a cul-de-sac. While not ideal, since a roadway separates houses from the natural area, with traffic-calming measures (narrow approach road, bulb-outs, speed bumps) the potential for accidents is minimized. If planted in a naturalistic fashion and ideally including at least one mature tree, such an area can provide nature contact very close to home.

One community designed in this fashion is Cherry Hill, a 29-unit development of townhouses for low- and moderate-income families with children in Petaluma, California, a small town north of San Francisco. The first residents moved into the project, built by the nonprofit Burbank Housing Development Corporation, in January 1992. The site was planned as a safe environment for the many children expected to live there. The project manager had read about the *woonerf* (a Dutch term roughly translated as "residential precinct," or "home zone" in the UK) used to calm traffic in northern Europe, and asked the designers to pursue the idea. They created a site plan with a narrow (22-ft.), one-way loop access road around a central green—in effect, a very large cul-de-sac (see Figures 10-50 and 10-51). Four paved courtyards off the loop permit cars to drive up to each house and provide hard-surface play areas. As in European examples, pedestrians and vehicles at Cherry Hill appear to coexist safely without sidewalks, since a narrow roadway, speed bumps, and the dead-end nature of the street pattern regulate the speed of cars. Unlike in neighborhoods with standard street grid patterns (such as those promoted by the new urbanists), no cars enter Cherry Hill except those belonging to residents or known visitors.

The success of these design decisions was confirmed by a study conducted by architecture graduate students in April 1993 under the direction of the second author (Cooper Marcus 1993). Interviews were administered to 17 of the 29 households, and 7.5 hours of behavioral observation were conducted in the shared outdoor spaces. Eighty-eight percent of the interviewed sample

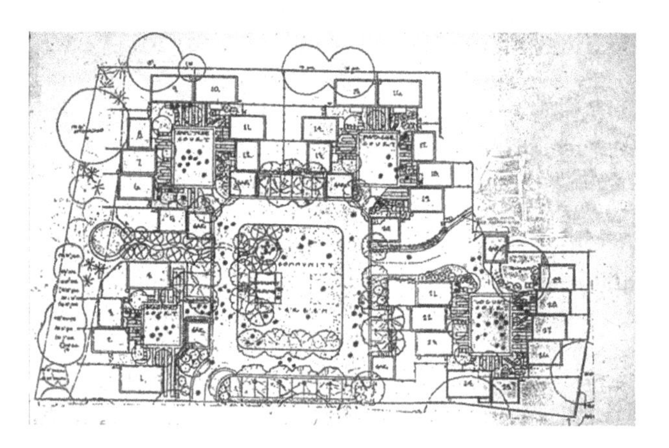

Figure 10-50: Cherry Hill, Petaluma, California, behavior map. Use of shared outdoor spaces by children and teens (aggregate of 7½ hours of observation, April 17–23, 1993).

Figure 10-51: Central greenspace at Cherry Hill, Petaluma, California, affordable family housing

socialized with other families in their immediate courtyard and almost two-thirds with families elsewhere in Cherry Hill. Eighty-eight percent reported they would recognize a stranger walking in Cherry Hill. Two-thirds were very satisfied with the site plan, citing safety for children, convenience, and feelings of intimacy and community as major reasons. Seventy-one percent rated a sense of community as "strong" or "very strong."

Behavior mapping of outdoor activities in 1993 revealed a heavy use by children, both of the traffic-calmed streets and of the central green area. (See Figure 10-50.) During daylight, nonschool hours, children were observed engaging in such activities as inline skating, rolling on a grassy slope, going around the loop on scooters, watching adults working on cars, clustering around an ice cream truck, collecting leaves, and digging for worms. Two sections of the roadway have been formally designated for games—four-square and basketball. It is reasonable to assume that most of these children's play activities could not be accommodated in a standard grid-pattern neighborhood with through traffic and no shared outdoor space. Significantly, half the parents said their children watched less TV since moving to Cherry Hill. The other half said they had no TV, or that their children watched about as much as before. Since being outdoors is a major correlate of children's physical activity (Sallis, Prochaska, and Taylor 2000), we may assume above average levels of physical activity in Cherry Hill children.

In Northpark, one of the newest "villages" in southern California's vast Irvine Ranch community, cul-de-sacs include landscaped islands (though not as large as those at Cherry Hill) breaking up the usual sea of asphalt. Instead of being terminated with a house, each cul-de-sac is linked by a pathway to sidewalks on adjacent streets, thus creating a child-pedestrian friendly network (Gause 2002, 105). While a small landscaped island in a cul-de-sac may seem a minute detail at the scale of a whole planned community, it can create the opportunity for nature contact close to home at the scale of a small child. Recalling favorite childhood places, a number of design students remembered significant features, especially trees, in just such spaces (Cooper Marcus 1978).

While it is clear that cul-de-sacs provide a safe and accessible locale for children's play close to home, a movement encompassing new urbanists, traffic engineers, planners, and some municipalities is successfully lobbying for eliminating their presence in new developments despite their popularity with home buyers (Efrati 2006). In a *Wall Street Journal* article titled "The Suburbs Under Siege," Amir Efrati notes: "Thanks to a growing chorus of critics, ranging from city planners and traffic engineers to snowplow drivers, hundreds of local governments from San Luis Obispo, Calif., to Charlotte, N.C., have passed zoning ordinances to limit cul-de-sacs or even ban them in the future. In Oregon, about ninety percent of the state's 241 cities have changed their laws to limit cul-de-sacs, while 40 small municipalities outside Philadelphia have adopted restrictions or an outright ban." Opponents argue that cul-de-sacs exacerbate traffic on nearby collector streets and that reimposing the grid redistributes traffic and encourages people to walk and not get into their car for every errand. This ignores the fact that when the dead-ends of cul-de-sacs are connected by walking and bike paths (forming a pedestrian "grid"), people are probably *more* likely to walk, as in Davis, California, though as yet there is only anecdotal evidence to support this (see Figure 10-52). In a study looking at cul-de-sacs in four northern California towns, data reveal that there is little difference between the proportion of people walking who live on unconnected cul-de-sacs as compared with those living on through streets (Handy et al. 2007).

Figure 10-52: A teen on in-line skates at Village Homes can expand her territory to virtually the whole of Davis, California, via an interconnected network of greenways. Note bridge crossing the drainage swale, a favorite play space of younger children.

The arguments for eliminating cul-de-sacs have everything to do with traffic engineering and with new urbanist arguments for erasing anything that resembles the conventional suburban layouts of the 1950s through 1980s. However, the free market tells another tale. As already mentioned, homes on cul-de-sacs tend to sell faster than other homes, and often command a higher price. Let us hope more municipalities follow the lead of Rock Hill, South Carolina, which changed its rules in 2007 banning cul-de-sacs, "requiring developers to cut pedestrian paths through their bulb-like tips to connect them to other sidewalks and allow people to walk through neighborhoods unimpeded" (Efrati 2006).

Rarely are the needs of children addressed by the proponents of new urbanism beyond the provision of neighborhood parks or playgrounds at regular intervals. While these amenities are certainly important, in the current atmosphere of parents' fears of traffic and stranger-danger, parks are not a viable alternative to outdoor play space for young children within sight of home.

Benefits of Greenways

A key component of a neighborhood well suited to the needs of children is one where it is easier (and safer) to walk or cycle than to drive. Greenways permitting movement between and through residential neighborhoods provide one such solution. If we can designate land as a wildlife corridor for the free movement of large mammals such as mountain lions, can we not equally regard our children as a precious species and provide for their safe movement through our increasingly hazardous environments? To summarize, some of the potential benefits of greenways for children and families include the following:

Greenways Are Accessible to Many

A greenway potentially provides a higher degree of nature contact than a traditional square park because of its linearity and high ratio of edge to area. After studying the use of local and long-distance green trails, Gobster (1995) recommends the creation of fine-grained networks of "mini-greenways" and "ribbons of nature" within urban environments. Having surveyed nearly 3,000 users of 13 greenway trails in metropolitan Chicago, he found that people using local trails (where the majority of users lived within five miles) used them more frequently to make shorter trips, including commuting, than those using regional or state trails located further from home. Hellmund and Smith suggest an upper limit of one mile from the farthest residence—two miles between trails—as an appropriate goal for a fine-grained network of local greenways (Hellmund and Smith 2006, 191).

A long-range, visionary project in Los Angeles involves converting the Los Angeles River (at present, mostly culverted) into a national urban wildlife refuge, bringing nature close to a large number of low-income families. This is part of an even more far-reaching project (which may take half a century or more) aimed at bringing wild nature within a quarter mile of every child in Los Angeles (Hester 2007).

Greenways Can Provide Walking and Cycling Linkage to Other Outdoor Spaces

In a definitive study of greenways, Hellmund and Smith (2006) recommend they should be combined with neighborhood-scale "minigreenways" and "pocket parks" to provide green space at multiple scales. An example of this can be found in the greenways created under the tracks of the Bay Area Rapid Transit system in Albany, California.

Greenways Can Be Used as Outdoor Classrooms

Schools could bring students to greenways to study local flora, monitor water quality, interview greenway users, et cetera. In the West Philadelphia Landscape Project, middle school and university students studying the Mill Creek neighborhood discovered a long-buried stream as the cause of flooding and subsidence. Having attracted the attention of the Philadelphia Water Department, a stormwater detention facility incorporating a wetland, water garden, and outdoor classroom was created in a vacant lot next to the school (see http://web.mit.edu/wplp/home.htm). Projects such as this are critical in raising the awareness of youth with regard to local ecology. This awareness can have long-term implications. David Sobel conducted a study of environmentalists to discover what in their past inspired them to care about the environment. The two main reasons were "many hours spent outdoors in a keenly-remembered wild or semiwild place in childhood or adolescence, and an adult who taught respect for nature" (cited in O'Shaughnessy 2000, 123).

Potential Conflicts Between Providing Child-Friendly and Wildlife-Friendly Green Spaces in Cities

Landscape ecologists generally refer to two basic types of green habitats in cities: patches and corridors. The patch is a relatively homogeneous nonlinear area that differs from its surroundings (Hellmund and Smith 2006, 46). The analogy in terms of site planning for human use would be what we have termed here shared outdoor space. A corridor is a strip of land of a particular type that differs from the adjacent land on either side, especially valued as a conduit for wildlife movement (Hellmund and Smith 2006, 46). Social planning analogies would include greenways, linear parks, riparian trails, et cetera, providing for human movement and connecting different neighborhoods.

The conflict between the design of green spaces for children's use and as wildlife habitats includes the following:

1. In the case of shared outdoor space in clustered housing, to optimize use by children and to maximize the potential for parents seeing a green space as safe, it needs to have distinct edges and be visible from adjacent homes. This suggests a round, square or rectangular shape with no hidden corners. Richard Forman has proposed that as a natural habitat an "ecologically optimized" patch should have enough roundness to ensure an interior habitat but with tentacle-like corridors extending out to facilitate plant and animal movement in and out of the patch (Hellmund and Smith 2006, 57). This shape would probably reduce its potential as a child-friendly landscape (unless the out-of-sight tentacles were closed to human use) as parents of young children might fear they had wandered away.

2. The vertical structure of the edge of a patch or a corridor with a variety of heights of shrubs, low trees, and high canopies is very important to birds and other wildlife. However, the understory may block the views from houses into the green space, making parents reluctant to let their children play there alone. An edge with no understory vegetation may increase its use by children but create a less-than-perfect wildlife habitat.

Possible Negative Unintended Consequence of Greenway Provision

Views to greenery are highly valued and may translate into higher house prices. A study of a master planned community near Seattle recorded that adult residents highly valued views to (but not necessarily use of) greenways adjacent to their homes (Kearney 2006). Creation of a greenway in an existing urban fabric can result in gentrification. For example, property values along the as yet incomplete Rose Kennedy Greenway (part of Boston's Big Dig project) increased 79 percent from 1988 to 2004, compared to a 41 percent increase citywide (Hellmund and Smith 2006, 163). In cases like these, it is possible that lower-income families with the least resources to drive to natural areas for recreation are also the least likely to be able to afford to live close to existing or newly created greenways.

Urban Promenades: An Alternative Model

While greenways passing through natural or barely altered landscapes provide good potential settings encouraging nature contact, in terms of nature access along urban pathway systems, the *urban promenade* is another but rare model. Setagaya Ward, Tokyo, known for many urban design innovations (see Useful Websites below), contains two well-known examples: the Kitazawagawa River Nature Path and the Yoga Promenade. The former (see Figure 10-53) is a broad curving pathway several blocks long, lined with cherry trees, which follows one side of a reconstructed urban stream brought to grade level and fed by the clean effluent of a local sewage treatment plant. The warm water and its high nutrient content ensures vigorous plant growth along the stream channel. The Yoga Promenade, designed by Group Zo, is a longer urban pedestrian pathway built to connect a subway station to the Setagaya Art Museum. While not as verdant as the Kitazawagawa River Nature Path, it offers many vegetated segments, an urban stream popular with local children, and a variety of aesthetic features and enhancements such as uniquely designed playful "lounging" street furniture, pavers inset with poems, and planting with poetic inscriptions. The Yoga Promenade is integrated into the urban fabric in a way that offers accessibility, including for children riding bicycles, to many community facilities along the way in addition to the art museum (see Figure 10-54).

3. ALLEYS

Residents in some inner-city neighborhoods are beginning to view converted back alleys as another potential site for children's play and nature contact. For example, in 2003, a group of residents in Baltimore's Patterson Park neighborhood, along with Community Greens and other NGOs, began the process of turning underutilized city property—the littered, neglected alleys behind their homes—into safe places for children to play and adults to unwind. After a lengthy process of resident envisioning, petition signing, legal maneuvering, and fundraising, two pieces of legislation were created that give city residents the option to gate and green their alleyways. The first piece of enabling legislation, passed in 2004, changed the city charter of Baltimore, empowering the city to gate a right-of-way (e.g., an alley) and lease it to abutting homeowners. The second piece of legislation, a 2007 ordinance, outlines many provisions and requirements of gating and greening. The ordinance stipulates that 80 percent of homeowners living on a block must agree and sign a petition to gate and green an alleyway if existing traffic is not im-

Figure 10-53: Kitazawagawa River Nature Path, Setagaya Ward, Tokyo, follows a reconstructed urban stream brought to grade level and fed by the clean effluent of a local sewage treatment plant. The warm water and its high nutrient content ensures vigorous plant growth along the stream channel.

Figure 10-54: The Yoga Promenade, an urban pedestrian pathway built to connect a Tokyo subway station to the Setagaya Art Museum

peded. If existing traffic is impeded, then 100 percent of occupied homes need to sign the petition. (Abandoned homes are not counted in the "voting pool.")

Remarkable changes have occurred on the first block in Patterson Park where alley gating and beautification have taken place. Garbage pickup has moved to the front; crime and littering have been eliminated; and a garbage-strewn no-man's-land has been converted with planters, potted plants, benches, and a barbecue grill into a space well used by adults and children, which is viewed as an extension of everyone's home. (See Figure 10–55 in color insert.) (In this case, parking was accommodated on the street before the alley closure.)

While ungated alleys are not ideal settings for children's nature contact, in existing high-density urban settings, evidence shows that they are often used by children in creative ways (Moore and Young 1978; Moore 1986a). In more suburban settings, proponents of new urbanism promote the use of alleys to allow houses to be sited closer together and to ensure that curb cuts and garages do not mar the streetscape. While some such alleys (e.g., at Celebration, Florida) *do* contain green elements and possibly function as casual play areas, others are designed as stark utilitarian spaces. The advertising literature of some developers espousing new urbanism in Santa Fe, for example, refers to alleys as places for children to play. It is hardly credible that a setting for cars, trash cans, recycling bins, and power lines somehow serves children's healthy development. Common sense suggests that children who grow up amid natural settings (such as the creeks, fruit trees, wildlife, and gardens of Village Homes) will be rewarded with more nature contact and more positive health outcomes. One has to wonder if residents living on suburban alleys in new urbanism–inspired neighborhoods will, some time in the future, turn to the solution of gating and greening these spaces as have the residents of the Baltimore neighborhoods cited above.

4. *WOONERVEN* AND HOME ZONES

Studies by Moore (1991) and others demonstrate the historic importance of streets for children's social life. Reflecting this fact, a fourth model of neighborhood design promoting children's safe outdoor play and potential nature-contact is one that had its birth in northwest Europe. The *woonerf* or "residential precinct" was first developed in the Netherlands to curb speeding traffic on inner-city, grid-pattern streets. The street is transformed by means of speed bumps, bulb-outs, planters, trees, benches, play spaces, et cetera, into a space for pedestrians where local traffic has access at only very low speeds. Pedestrians and cars share the paved space of the street (with no specific sidewalks), with pedestrians having legal priority. Entrances to the shared zone are clearly marked; through traffic is discouraged, while residents have auto access to the front of dwellings (Pressman 1991). (See Figure 10–56 in color insert.)

The success of the first *woonerf* schemes in Delft triggered the spread of this urban form to other Dutch cities, then to suburban Dutch neighborhoods. The shared street concept became accepted and established through guidelines and regulations in the Netherlands and Germany (1976); England, Denmark, and Sweden (1977); France and Japan (1979); Israel (1981); and Switzerland (1982) (Ben-Joseph 1995). Studies and surveys of shared streets in Europe, Japan, Australia, and Israel have found reductions in traffic accidents, increased social interaction and play, and a high degree of satisfaction by the residents. A carefully observed study of activity on two streets in a mixed-use, high-density, inner-city neighborhood in Hannover, Germany, before and after conversion to *woonerven*, documented an increase in children's outdoor play after the conversion (Eubank-Ahrens 1991). While this does not necessarily translate into nature contact, inclusion of trees with an understory of bushes or planting beds for residents to maintain, would provide nature contact during the increased outdoor play.

A recent rebirth of the *woonerf* movement in the UK, where these play-streets are called home zones, is generating a number of child-friendly models in urban neighborhoods. A recent study of such UK developments discusses the evolution of street life as explored by Levitas (1986), and how the use of streets quite closely reflects the values and priorities of society. "She highlights the dominant view that streets have become seen as links rather than a locus, and that increasingly the street is recognized for its transit capabilities rather than its ability to provide for a range of rich and diverse human behavior" (Levitas 1986, 232, quoted in Biddulph 2003, 218).

There are no fully developed "home zone" examples in the United States, primarily due to opposition from traffic engineers, road-building companies, and fire and police departments. The principal impediment is the fact that the Institute of Traffic Engineers has never adopted the concept. As one expert remarked: "As long as they do not back it up or publish suggested guidelines, public officials (and especially the city's legal department) will not endorse it. . . . Most countries in Europe and Asia have adopted guidelines for the design and construction of such spaces" (Ben-Joseph, personal communication, 2007). However, this is a model that should be still considered as an ideal way of creating safe outdoor play close to home in built-up neighborhoods where there is no possibility of creating inner-block green space, and where there are no alleys to gate and convert.

Ben Joseph considers that the concept holds true, even for new urbanist developments that advocate interconnected street networks. "Increased accessibility on all streets raises the likelihood of cut-through traffic and of speeds inappropriate to residential neighborhoods—the original impetus for abandoning the grid . . . more than sixty years ago. Shared streets in a connected system can eliminate the deficiencies of the grid. Speed will be reduced and through traffic by non-residents discouraged, yet connective factors . . . will be much more numerous than in the typical hierarchical, disconnected street system" (Ben-Joseph 1995, 512).

LEED NEIGHBORHOOD DEVELOPMENT AND ACCESS TO NATURE

The Natural Resources Defense Council has partnered with the U.S. Green Building Council and the Congress for New Urbanism to certify exemplary neighborhood development through the LEED for Neighborhood Development rating system. Pilot projects are being reviewed in 2007 to test the strengths and weaknesses of the rating system.

A few features covered in the rating system mirror recommendations made in this chapter. For example, the possibility of creating a *woonerf* is mentioned, and under "Street Network," where cul-de-sacs are created, at least 50 percent are required to have through-connections for pedestrians and cyclists. While habitat conservation, restoration and management are covered in detail, nothing is specifically mentioned regarding human access to these areas. Parks, squares, plazas, et cetera, are mentioned in terms of their required size and dimensions, and their proximity to the project being reviewed, but there is no discussion of the quality of their design or any required components. Such a rating system for public green spaces may well result in the same minimally detailed and furnished flat green parks seen in many contemporary new urbanist developments that offer little in the way of nature contact for either adults or children. There is no mention or recognition in this rating system of the value of shared outdoor space, as defined above.

To gain credit for "School Proximity," the project being reviewed needs to be located or designed so that at least 50 percent of dwelling units are within a half-mile walking distance to the school, but the rating system does not help to encourage nature-contact in school grounds, on the walk to and from school (e.g., via a greenway), at a childcare center, or in any other setting where children may spend part of their day.

The LEED for Neighborhood Development Public Health Report (www.greenbuildingcouncil.org) provides an excellent overview of the current debate and research regarding public health and planning, focussing largely on cardiovascular health and air pollution, traffic accidents, physical activity and urban form, but (sadly) has very limited acknowledgment of children's needs and no mention of the psychological or spiritual value of access to nature.

It is essential that those of us committed to the principles of biophilic planning and design become vocal during the 2008–2009 public comment period for the postpilot version of the LEED for Neighborhood Development, so that access to nature becomes an intrinsic component of those neighborhoods rated as exemplary through this review process.

CONCLUSIONS

There are clearly many socioeconomic and locational factors affecting whether natural areas are sought out and used by children. For example, in a small, relatively safe California city, 14- to 18-year-olds sought out natural areas where they could be alone or with friends in an informal, unsupervised way (Eubanks Owens 1988). In a crime-ridden area of Los Angeles, however, children ages 9 to 11 almost unanimously rejected parks and other public spaces as the domain of gangs and "bad people," despite the fact these areas were created (by adult decision-makers) for their recreation (Buss 1995). But it is important to note that in the extensive international research of the Growing Up in Cities project, when children's views were solicited concerning how their environment functioned for them in 8 urban communities in the 1970s, and 16 researched since 1995, "safe, clean green spaces with trees, whether formed or wild, extensive or small . . ." was one of nine positive indicators of community quality from the children's perspective (Chawla and Malone 2003).

This, together with the studies of different models of child-friendly neighborhoods discussed above, and the repeated negative indicators cited by children (heavy traffic, violence, bullies, gangs, litter, pollution and lack of places to play and meet friends) provide us with some parameters for planning children's access to nature in residential settings:

- A natural or quasi-natural area needs to be created or protected within sight and calling distance of a majority of homes occupied by families with children. This area needs to facilitate as many varieties of play as possible, from spaces where digging in, or molding, dirt or sand is encouraged (preferably with access to water nearby); to semiwild areas where dens might flourish; planting beds for gardening; trees for climbing; paths for wheeled toys; equipment for swinging, sliding, climbing, et cetera (see Figure 10-57).
- Access from homes needs to be safe, not requiring a street crossing wherever possible. Studies of children's perceptions of their own neighborhoods repeatedly cite the problem of traffic in limiting mobility and access to places they want to go (Hillman 1993; Davis and Jones 1996, 1997; Morrow 2003; O'Brien 2003; Wheway and Millward 1997).
- The space needs to be well-maintained (no litter, no pollution) without the removal of those "loose parts" valuable in creative play.
- Adults and children alike need to understand that this is a legitimate locale for children's play.

Figure 10-57: Shared outdoor space must provide for as great a variety of children's activities as possible, from digging in sand or dirt, to climbing or riding a bike, to creating a den under a bush or examining insects or leaves (St. Francis Square, San Francisco).

- The space needs to provide for all age groups, from toddlers to late-teens, without any one group or gender dominating the use of the space and intimidating others.[11]
- Where possible, the space needs to link via a greenway path and bikeway system to other natural areas and to schools, local shops, library, et cetera.

SUMMARY

Children's lack of safe access to wild or semiwild nature does not bode well in terms of inspiring and motivating a coming generation of environmental stewards. Urban environmental design and landscape architecture partnered with allied fields (public health, urban planning, parks and recreation, horticulture) have a crucial role to play in alerting society to this concern, as well as turning the tide on sedentary indoor lifestyle trends and the negative health consequences, beginning in the first year of life.

Crucial policy areas, new urban forms, and innovative settings and components need to be developed and tested. They include school locations, neighborhood pathway networks, their application to trips to and from school, shared open space in residential areas, housing patterns with child-friendly outdoors including child-friendly streets, neighborhood parks and local open space, nonformal education facilities such as botanical gardens, greenways, and urban trails and urban promenades. Empirical research is urgently required to fully understand the environment-behavior relations in these settings to inform responsible practice—as well as to provide evidence to counter some of the still unsupported claims of new urbanism. Most important, the quality of the place where the majority of U.S. children spend their early childhood—childcare centers—must become a central focus of biophilic design.

Although the task will never be complete because of the dynamic nature of postmodern culture, in the last several decades, well-developed, evidence-based precedents have accumulated that offer best practice guidance for the design of children's everyday environments. Such models can help us imagine what an urban environment would be like if it were designed to fully support the biophilic development of children and thus the future health of our planet, our place in the universe.

ENDNOTES

1. By ecological validity we mean that the environments being investigated must exhibit sufficient physical diversity that a broad range of human response is elicited. Unless research includes new types of environments attempting to respond to the health crisis through design innovation, the knowledge base will remain static and narrow.
2. An example in the first author's experience was the mandated removal by the local public health department of a beautiful bog garden constructed as a preschool play and learning setting in a childcare center preschool area, on the grounds that it could "harbor vermin." The only animals observed there, however, were dragonflies, other flying insects, and birds.
3. Safety regulations for children's play environments until now have been driven by data gathered in hospital emergency rooms (NEISS–National Electronic Injury Surveillance System) and product-related data gathered by the U.S. Consumer Product Safety Commission (CPSC). Understandably, the role of government is to protect citizens from harmful products, which become the focus of attention when unintended injury rates get public attention (as was the case with public playgrounds in the 1970s and 1980s). The problem is that we assume that environments that don't show up in the statistics are safe. Furthermore, legal liability has distorted our view of children's environments. Perceived safety and liability tend to be the central focus in decisions about provision, instead of play value and developmental outcomes. See Moore (2006) for extended discussion of this issue.
4. The Natural Learning Initiative (NLI) is a research, design assistance, and professional development unit of the College of Design, North Carolina State University. NLI was founded in 2000 with the purpose of promoting the importance of the natural environment in the daily experience of all children, through environmental design, research, education, and dissemination of information. For more information, visit www.naturalearning.org.
5. Since most of these animals are far more agile than children under two years old, there is very little chance that they can be touched, let alone caught or mouthed. By age three to

five, preschool children have learned where small animals live and enjoy hunting and catching them. For this age, teachers are crucial role models with the task of facilitating respectful, caring behavior toward animals (Myers 2007).

6. Completed survey responses (n=326) were received from approximately 10% of the licensed childcare centers in North Carolina. Based on the results, the 25 highest-scoring centers were visited in the field. Of these, only three or four could be labeled as exemplary outdoor environments according to the quantity and quality of natural settings.
7. LEED (Leadership in Energy and Environmental Design) is part of the U.S. Green Building Council (USGBC), a 501(c)(3) nonprofit composed of more than 10,000 member organizations from the building industry united by a common purpose: to transform the building marketplace to sustainability. The Green Building Rating System(tm) is the nationally accepted LEED benchmark for the design, construction, and operation of high-performance green buildings.
8. Such spaces are typically unprotected and vulnerable to development as urban land prices rise. Hence the concept of "designed" vacant lots as part of the local public open space system.
9. There are now many fewer children living at St. Francis Square, as parents who raised their children there in the 1970s and 1980s have opted to stay on as "empty nesters."
10. Meanwhile in the United States, the opposite was happening as women were encouraged to stay home, opening up jobs for returning veterans and becoming full-time housewife-chauffeurs in the sprawling suburbs spawned by low-interest mortgages, freeway construction, and increased car ownership.
11. In an Australian participatory action research project where high school students engaged in redesigning an unused outdoor space, they complained that planners and city officials always think about facilities for toddlers—that is, playgrounds—and rarely for older children and teens (Chawla and Malone 2003, 129–134).

REFERENCES

Appleyard, D. 1980. *Livable Streets*. Berkeley: University of California Press.

Arendt, R. 1996. *Conservation Design for Subdivisions*. Washington, DC: Island Press.

Bang, J. 2005. *Ecovillages: A Practical Guide to Sustainable Communities*. Edinburgh: Floris Books.

Beatley, T. 2000. *Green Urbanism: Learning from European Cities*. Washington, DC: Island Press.

Beaumont, Constance, and Elizabeth Pianca. 2002. "Why Johnny Can't Walk to School: Historic Neighborhood Schools in the Age of Sprawl." Washington, DC: National Trust for Historic Preservation.

Bell, Anne C., and Janet E. Dyment. 2006. "Grounds for Action: Promoting Physical Activity Through School Ground Greening in Canada." Toronto: Evergreen. http://www.evergreen.ca/en/lg/pdf/PHACreport.pdf.

Ben-Joseph, E. 1995. "Changing the Residential Street Scene: Adapting the Shared Street *(Woonerf)* Concept to the Suburban Environment." *Journal of the American Planning Association* 61(4): 504–515.

———. 2007. Personal communication.

Berg, M., and E. Medrich. 1980. "Children in Four Neighborhoods: The Physical Environment and Its Effect on Play and Play Patterns." *Environment and Behavior* 12(3): 320–348.

Biddulph, M. 2003. "Towards Successful Home Zones in the UK." *Journal of Urban Design* 4(3): 217–241.

Bjorkild-Chu, P. 1977. "Children's Outdoor Environment. Summary." *Man-Environment Systems* July/Sept. 250–251.

Boldemann, C., M. Blennow, H. Dal, F. Mårtensson, A. Raustorp, K. Yuen, and U. Wester. 2006. "Impact of Pre-School Environment upon Children's Physical Activity and Sun Exposure." *American Journal of Preventive Medicine* 42:301–308.

Burdette, Hillary, and Robert Whitaker. 2005. "Resurrecting Free Play in Young Children: Looking Beyond Fitness and to Attention, Affiliation, and Affect." *Archives of Pediatrics & Adolescent Medicine* 159(1): 46–50.

Burgess, J., C. Harrison, and M. Limb. 1988. "People, Parks and the Urban Green: A Study of Popular Meanings and Values for Open Spaces in the City." *Urban Studies* 25(6): 455–473.

Buss, S. 1995. "Urban Los Angeles from Young People's Angle of Vision." *Children's Environments* 12(3): 340–351.

Capizzano, J., G. Adams, and F. Sonenstein. 2000. *Child Care Arrangements for Children Under Five: Variation Across States*. Washington, DC: The Urban Institute.

Cavanna, R. C. 1974. Backyard options in residential neighborhoods. Master's thesis, University of California–Berkeley.

Cele, S. 2006. *Communicating Place: Methods for Understanding Children's Experience of Place*. Stockholm: Stockholm University.

Centers for Disease Control and Prevention (CDC). National Center for Health Statistics. 2007. "Prevalence of Overweight Among Children and Adolescents: United States, 2003–2004." http://www.cdc.gov/nchs/products/pubs/pubd/hestats/overweight/overwght_child_03.htm.

Chawla, L. 1986. "The Ecology of Environmental Memory." *Children's Environments Quarterly*, Winter, 34–42.

———, ed. 2002a. *Growing Up in an Urbanising World*. Paris: UNESCO/Earthscan.

———. 2002b. "Spots of Time: Manifold Ways of Being in Nature in Childhood." In *Children and Nature*, edited by J. Kahn and S. Kellert. Cambridge, MA: MIT Press.

———. 2006. "Learning to Love the Natural World Enough to Protect It." *Barn* 2:57–78.

Chawla, L., and K. Malone. 2003. "Neighborhood Quality in Children's Eyes." In *Children in the City: Home, Neighborhood and Community*, edited by P. Christensen and M. O'Brien. London: RoutledgeFalmer.

Check, Erika. 2004. "Link from Hygiene to Allergies Gains Support." *Nature* 25(428): 354.

Christensen, P., and M. O'Brien, eds. 2003. *Children in the City: Home, Neigbourhood and Community*. London: Routledge Falmer.

Cobb, Edith. 1977. *The Ecology of Imagination in Childhood*. Dallas: Spring Publications.

Coleman, A. 1985. *Utopia on Trial: Visions and Reality in Planned Housing*. London: H. Shipman.

Cooper, C. 1970. *Resident Attitudes Towards the Environment at St. Francis Square, San Francisco: Summary of the Initial Findings*. Berkeley: University of California–Berkeley. Institute of Urban and Regional Development.

———. 1971. "St. Francis Square: Attitudes of Its Residents. *AIA Journal* 56:22–27.

Cooper Marcus, C. 1974. "Children's Play Behavior in a Low-Rise, Inner City Housing Development." In *Childhood City*, edited by R. Moore and D. Carson. Stroudsberg, PA: Dowden, Hutchinson, and Ross.

———. 1978. "Remembrances of Landscapes Past." *Landscape* 22(3).

———. 1993. "Postoccupancy Evaluation of Cherry Hill, Petaluma, CA." Berkeley: University of California–Berkeley.

———. 2003. "Shared Outdoor Space and Community Life." *Places: Quarterly Journal of Environmental Design* 15(2): 32–34.

Cooper Marcus, C., and W. Sarkissian. 1986. *Housing As If People Mattered: Site Guidelines for Medium-Density Family Housing*. Berkeley: University of California Press.

Corbett, J., and M. Corbett. 2000. *Designing Sustainable Communities: Learning from Village Homes*. Washington, DC: Island Press.

Cosco, N. 2006. Motivation to move: physical activity affordances in preschool play areas. Doctoral thesis, School of Landscape Architecture, Edinburgh College of Art, Heriot Watt University, Edinburgh.

Cosco, N., and R. Moore. Baseline survey of environmental conditions of outdoor play areas in North Carolina child-care centers. North Carolina State University, College of Design.

Crace, John. 2006. "Children Are Less Able Than They Used to Be." *Guardian*, January 24.

Crain, W. 2003. *Reclaiming Childhood: Letting Children Be Children in Our Achievement-Oriented Society*. New York: Henry Holt.

Cunningham, C., and M. Jones. 1999. "The Playground: A Confession of Failure?" *Built Environment* 2(1): 11–17.

Davis, A., and L. Jones. 1996. "Children in the Urban Environment: An Issue for the New Public Health Agenda." *Health and Place* 2(2): 107–115.

———. 1997. "Whose Neighborhood? Whose Quality of Life? Developing a New Agenda for Children's Health in Urban Settings." *Health Education Journal* 56:350–363.

DeGrandpre, Richard. 2001. *Ritalin Nation: Rapid-Fire Culture and the Transformation of Human Consciousness*. Boston, MA: W.W. Norton.

Diller, Lawrence. 1998. *Running on Ritalin: A Physician Reflects on Children, Society and Performance in a Pill*. New York: Bantam Books.

Dovey, K. 1990. "Refuge and Imagination: Places of Peace in Childhood." *Children's Environments Quarterly* 7(4): 13–17.

Drayton, W. 2000. "Secret Gardens: How to Turn Patchwork Urban Backyards into Neighborly Communal Parks." *Atlantic Monthly* 285(6).

Duany, A. 2001. Keynote address, Pacific Coast Builders' Conference, San Francisco.

Durrett, C. Personal communication 2005.

Dyment, Janet. 2005. "Gaining Ground: The Power and Potential of School Ground Greening in the Toronto District School Board." http://evergreen.ca/en/lg/gaining_ground.pdf.

Eberstadt, Mary. 1999. "Why Ritalin Rules." *Policy Review* 94, April–May.

Efrati, A. 2006. "The Suburbs Under Siege: Homeowners Love Cul-de-Sacs, Planners Say They're Perils; Taking Sides in Minnesota." *Wall Street Journal Online*, June 2.

Engwicht, D. 1999. *Street Reclaiming: Creating Livable Streets and Vibrant Communities*. Gabriola Island, BC: New Society Publishers.

Eubank-Ahrens, B. 1991. "A Closer Look at the Users of *Woonerven*." In *Public Streets for Public Use*, edited by A. Vernez-Moudon. New York: Columbia University Press.

Eubanks Owens, P. 1988. "Natural Landscapes, Gathering Places, and Prospect Refuges: Characteristics of Outdoor Places Valued by Teens." *Children's Environments Quarterly* 5(2): 17–24.

European Environment and Health Committee (EEHC). 2005. "Spain: The NAOS Strategy: A Strategy for Nutrition, Physical Activity and Obesity Prevention." http://www.euro.who.int/eehc/implementation/20051129_2.

Faber Taylor, A. , A. Wiley, F. Kuo, and W. Sullivan. 1998. "Growing Up in the Inner City: Green Spaces as Places to Grow." *Environment and Behavior* 30:3–27.

Fjørtoft, I. 2001. "The Natural Environment as a Playground for Children: The Impact of Outdoor Play Activities in Pre-Primary School Children." *Early Childhood Education Journal* 29(2): 111–117.

Francis, M. 1984–85. "Children's Use of Open Space in Village Homes." *Children's Environments Quarterly* 1(4): 36–38.

———. 1988. "Negotiating Between Children and Adult Design Values in Open Space Projects." *Design Studies* 9(2): 67–75.

———. 1991. "The Making of Democratic Streets." In *Public Streets for Public Use*, edited by A. Vernez-Moudon. New York: Columbia University Press.

———. 2003. *Village Homes: A Community by Design*. Washington DC: Island Press.

Frank, L., J. Kerr, J. Chapman, and J. Sallis. 2007. "Urban Form Relationships with Walk Trip Frequency and Distance Among Youth." *Amercan Journal of Health Promotion* 21(4 Supplement): 305–311.

Friedman, M., K. Powell, L. Hutwagner, L. Graham, and W. Teague. 2001. "Impact of Changes in Transportation and Commuting Behaviors During the 1996 Summer Olympic Games in Atlanta on Air Quality and Childhood Asthma." *Journal of the American Medical Association* 285(7): 897–905.

Frumkin, H. 2001. "Beyond Toxicity: Human Health and the Natural Environment." *American Journal of Preventive Medicine* 20(3): 234–240.

Frumkin, H., R. Geller, and I. L. Rubin, eds. 2006. *Safe and Healthy School Environments*. New York: Oxford University Press.

Garces, E., D. Thomas, and J. Currie. 2002. "Longer-Term Effects of Head Start." *American Economic Review* 9(4): 999–1012.

Gaster, S. 1991. "Urban Children's Access to Their Neighborhood: Changes over Three Generations." *Environment and Behavior* 23(1): 70–85.

Gause, J., ed. 2002. *Great Planned Communities*. Washington, DC: Urban Land Institute.

Gehl, J. 2003. *Life Between Buildings: Using Public Space*. Copenhagen: Danish Architectural Press.

Geller, R. 2006. "Safety in the Sun." In *Safe and Healthy School Environments*, edited by H. Frumkin, R. Geller, and I. L. Rubin. New York: Oxford University Press.

Gerson, S., J. Wardani, and S. Niazi. 2005. "Blackberry Creek Daylighting Project, Berkeley: Ten-Year Post-Project Appraisal." Water Resources Center Archives, Restoration of Rivers and Streams (University of California, Multi-Campus Research Unit). http://repositories.cdlib.org/cgi/viewcontent.cgi?article=1056&context=wrca.

Gibson, Eleanor. 2002. *Perceiving the Affordances: A Portrait of Two Psychologists*. Mahwah, NJ: Lawrence Erlbaum.

Ginsburg, K., and the Committee on Communications and Committee on Psychosocial Aspects of Child and Family Health. 2007. "The Importance of Play in Promoting Healthy Child Development and Maintaining Strong Parent-Child Bonds." *Pediatrics* 119(1).

Gleeson, B., and N. Sipe. 2006. *Creating Child Friendly Cities: Reinstating Kids in the City*. Abingdon, UK: Routledge.

Gobster, P. 1995. "Perception and Use of a Metropolitan Greenway System for Recreation." *Landscape and Urban Planning* 33:401–413.

Grahn, P. , F. Mårtensson, B. Lindblad, P. Nilsson, and A. Ekman. 1997. "Ute på Dagis" (Out in the preschool). *Stad and Land* 145.

Greenman, Jim. 2005. *Caring Spaces, Learning Places: Children's Environments That Work*. 2nd ed. Redmond, WA: Exchange Press.

Handy, S., S. Sommer, J. Ogilvie, X. Cao, and P. Mokhtarian. 2007. "Cul-de-Sacs and Children's Outdoor Play: Quantitative and Qualitative Evidence." Presented at Active Living Research Conference, San Diego.

Hart, R. 1979. *Children's Experience of Place: A Developmental Study*. New York: Irvington Press.

———. 1986. *The Changing City of Childhood*. New York: The City College Workshop Center.

Heerwagen, J., and G. Orians. 2002. "The Ecological World of Children." In *Children and Nature: Psychological, Sociocultural, and Evolutionary Investigations*, edited by P. Kahn and S. Kellert. Cambridge, MA: MIT Press.

Hellmund, P., and D. Smith. 2006. *Designing Greenways: Sustainable Landscapes for Nature and People*. Washington, DC: Island Press.

Her Majesty's Stationery Office (HMSO). 2006. *Learning Outside the Classroom Manifesto*. Nottingham: Department for Education and Skills.

———. 2007. *Manual for Streets*. London: Thomas Telford Publishing/Her Majesty's Stationary Office.

Herrington, S. 1999. "Playgrounds as Community Landscapes." *Built Environment* 25(1): 25–34.

Hester, R. 2007. "Ecological Democracy." Public lecture, Department of Landscape Architecture and Environmental Planning, University of California–Berkeley, April 2.

Hillman, M. 1993. *Children, Transport and the Quality of Life*. London: Policy Studies Institute.

Hillman, M., and J. Adams. 1992. "Children's Freedom and Safety." *Children's Environments Quarterly* 9(2): 10–22.

Hillman, M., J. Adams, and J. Whitelegg. 1990. *One False Move: A Study of Children's Independent Mobility*. London: PSI Press.

Hough, Michael. 1990. *Out of Place: Restoring Identity to the Regional Landscape*. New Haven, CT: Yale University Press.

Jago, R., T. Baranowski, J. Baranowski, D. Thompson, and K. Greaves. 2005. "BMI from 3–6 Years of Age Is Predicted by TV Viewing and Physical Activity, Not Diet." *International Journal of Obesity* 29(6): 557–564.

Jeffrey, Bob, and Peter Woods. 2003. *The Creative School: A Framework for Success, Quality and Effectiveness*. New York: RoutledgeFalmer.

Jencks, R. 2007. "Stormwater management". Public lecture, Department of Landscape Architecture and Environmental Planning, University of California-Berkeley, March 14.

Kahn, P., and S. Kellert, eds. 2002. *Children and Nature: Psychological, Sociocultural, and Evolutionary Investigations*. Cambridge, MA: MIT Press.

Kaplan, Rachel, and Stephen Kaplan. 1989. *The Experience of Nature: A Psychological Perspective*. Cambridge: Cambridge University Press.

Kats, G. 2006. *Greening America's Schools: Costs and Benefits*. Washington, DC: Capital E.

Kearney, A. 2006. "Residential Development Patterns and Neighborhood Satisfaction: Impacts of Density and Nearby Nature." *Environment and Behavior* 38(1): 112–139.

Kegerreis, S. 1993. "Independent Mobility and Children's Mental and Emotional Development." In *Children, Transport and the Quality of Life*, edited by M. Hillman. London: Policy Studies Institute.

Keller, R. 2006. "Forest Kindergartens in Whatcom?" *Whatcom Watch Online*, July 2006. <http://www.whatcomwatch.org/php/WW_open.php?id=718>

Kellert, S. 1993. "The Biological Basis for Human Values in Nature." In *The Biophilia Hypothesis*, edited by S. Kellert and E. O. Wilson. Washington, DC: Island Press.

Kirkby, Mary Ann. 1989. "Nature as Refuge in Children's Environments." *Children's Environments Quarterly* 6(1): 7–12.

Kuo, F., M. Bacaicoa, and W. Sullivan. 1998. "Transforming Inner-City Landscapes: Trees, Sense of Safety, and Preference." *Environment and Behavior* 30:28–59.

Lennard, H. L., and S. Crowhurst Lennard. 1992. "Children in Public Places: Some Lessons from European Cities." *Children's Environments Quarterly* 9(2): 37–47.

Lenz, T. 1990. A Post-Occupancy Evaluation of Village Homes, Davis, California. Master's thesis, Technical University of Munich.

Levitas, G. 1978. "Anthropology and Sociology of Streets." In *On Streets*, edited by S. Anderson. Cambridge, MA: MIT Press.

Lieberman, G., and L. Hoody. 1998. *Closing the Achievement Gap: Using the Environment as an Integrating Context for Learning*. Poway, CA: State Education and Environment Roundtable, Science Wizards.

Liu, G., J. Wilson, R. Qi, and J. Ying. 2007. "Green Neighborhoods, Food Retail and Childhood Overweight: Differences by Population Density." *American Journal of Health Promotion* 21(4 Supplement): 317–325.

Lohr, V. I., and C. H. Pearson-Mims. 2005. "Children's Active and Passive Interactions with Plants Influence Their Attitudes and Actions Toward Trees and Gardening as Adults." *HortTechnology* 15(3): 472–476.

Loukaitou-Sideris, A. 2003. "Children's Common Ground." *APA Journal* 69(2): 130–143.

Louv, R. 2005. *Last Child in the Woods: Saving Our Children from Nature-Deficit Disorder*. Chapel Hill, NC: Algonquin Books.

Lynch, K. 1977. *Growing Up in Cities*. Cambridge, MA: MIT Press.

Mackett, R., L. Lucas, J. Paskins, and J. Turbin. 2004. "The Therapeutic Value of Children's Everyday Travel." *Transportation Research Part A* 39:205–219.

Matthews, M.H. 1987. "Gender, home range and environmental cognition". *Transactions of the Institute of British Geographers* 12(1): 43–56.

Maxey, I. 1999. "Playgrounds: From Oppressive Spaces to Sustainable Places?" *Built Environment* 2(1): 18–24.

McKendrick, J. 1999. "Playgrounds in the Built Environment." *Built Environment* 2(1): 5–10.

McRobbie, Joan. 2001. *Are Small Schools Better? School Size Considerations for Safety and Learning*. San Francisco: WestEd Policy Brief.

Milton, B., E. Cleveland, and D. Bennett-Gates. 1995. "Changing Perceptions of Nature, Self, and Others: A Report on a Park/School Program." *Journal of Environmental Education* 26(3): 32–39.

Moore, L., A. Di Gao, L. Bradlee, A. Cupples, A. Sundarajan-Ramamurti, M. Proctor, M. Hood, M. Singer, and C. Ellison. 2003. "Does Early Physical Activity Predict Body Fat Change Throughout Childhood?" *Preventive Medicine* 37:10–17.

Moore, R. 1980. "Collaborating with Young People to Assess Their Landscape Values." *Ekistics* 281:128–135.

———. 1986a. *Childhood's Domain: Play and Place in Child Development*. Dover, NH: Croom Helm.

———. 1986b. "The Power of Nature: Orientations of Girls and Boys Toward Biotic and Abiotic Settings on a Reconstructed Schoolyard." *Children's Environments Quarterly* 3(3): 52–69.

———. 1991. "Streets as Playgrounds." In *Public Streets for Public Use*, edited by A. Vernez-Moudon. New York: Columbia University Press.

———. 2003. "How Cities Use Parks to Help Children Learn." In *City Parks Forum Briefing Papers*. Chicago: American Planning Association.

———. 2006. "Playgrounds: A 150-Year-Old Model." In *Safe and Healthy School Environments*, edited by H. Frumkin, R. Geller, and I. Rubin. New York: Oxford University Press.

Moore, R. and N. Cosco. 2007. "Greening Montessori School Grounds by Design." *North American Montessori Teachers Association (NAMTA) Journal* 32(1): 128–151.

Moore, R., and H. Wong. 1997. *Natural Learning: Creating Environments for Rediscovering Nature's Way of Teaching*. Berkeley, CA: MIG Communications.

Moore, R., and D. Young. 1978. "Childhood Outdoors: Toward a Social Ecology of the Landscape." In *Human Behavior and Environment*, edited by I. Altman and J. Wohlwill. New York: Plenum Press.

Morris, A. 2001. Montgomery Park. Washington DC: Community Greens.

Morrow, V. 2003. "Improving the Neighborhood for Children: Possibilities and Limitations of 'Social Capital' Discourses." In *Children in the City: Home, Neighborhood and Community*, edited by P. Christensen and M. O'Brien. London: RoutledgeFalmer.

Murphy, Michael. 2003. *Education for Sustainability: Findings from the Evaluation Study of The Edible Schoolyard*. Berkeley, CA: Center for Ecoliteracy.

Myers, G. 2007. *The Significance of Children and Animals: Social Development and Our Connections to Other Species*. 2nd ed. West Layfayette, IN: Purdue University Press.

Noschis, K. 1992. "Child Development Theory and Planning for Neighborhood Play." *Children's Environments* 9(2): 3–9.

O'Brien, M. 2003. "Regenerating Children's Neighbourhoods: What Do Children Want?" In *Children in the City: Home, Neighborhood and Community*, edited by P. Christensen and M. O'Brien. London: RoutledgeFalmer.

Ogden, C., Fregal, K., Carroll, M., Johnson, C. 2002. "Prevalence and Trends in Overweight Among US Children and Adolescents, 1999–2000." *Journal of the American Medical Association* 288(14): 1728–1732.

Olshansky, S. J., D. Passaro, R. Hershow, J. Layden, B. Carnes, J. Brody, L. Hayflick, R. Butler, D. Allison, and D. Ludwig. 2005. "A Potential Decline in Life Expectancy in the United States in the 21st Century." *New England Journal of Medicine* 352(11): 1138–1145.

Olsson, Titti. 2002. *Skolgården Som Klassrum*. Lund: Runa Förlag.

O'Shaughnessy, M. 2000. "The Child and the Natural Environment." *North American Montessori Teachers Association Journal* 25:119–144.

Palacio-Quintin, E. 2000. "The Impact of Day Care on Child Development." *ISUMA: Canadian Journal of Policy Research* 1(2).

Parkinson, C.E. 1987. Children's Range Behavior: A review of studies of the ways in which children venture from their homes during the middle years of childhood. Birmingham, UK: Play Board.

Pellegrini, Anthony D., and P. K. Smith. 1998. "Physical Activity Play: The Nature and Function of a Neglected Aspect of Play." *Child Development* 69:577–598.

Percy-Smith, B. 2002. "Contested Worlds: Constraints and Opportunities in City and Suburban Environments in an English Midlands City." In *Growing Up in an Urbanising World*, edited by L. Chawla. London: Earthscan.

Plas, J. M., and S. E. Lewis. 1996. "Environmental Factors and Sense of Community in a Planned Town." *American Journal of Community Psychology* 24(1): 109–43.

Pressman, N. 1991. "The European Experience." In *Public Streets for Public Use*, edited by A. Vernez-Moudon. New York: Columbia University Press.

Pucher, J., and L. Dijkstra. 2003. "Promoting Safe Walking and Cycling to Improve Public Health: Lessons from the Netherlands and Germany." *American Journal of Public Health* 93(9): 1509–1516.

Pyle, R. 1993. *The Thunder Tree: Lessons from an Urban Wildland*. New York: Houghton Mifflin.

Rasmussen, K., and S. Smidt. 2003. "Children in the Neighborhood, the Neighborhood in the Children." In *Children in the City: Home, Neighborhood and Community*, edited by P. Christensen and M. O'Brien. London: RoutledgeFalmer.

Rigby, N., and P. James. 2003. *Obesity in Europe*. London: International Obesity Task Force.

Ring, J. 2005. *Allergy in Practice*. New York: Springer.

Rivkin, Mary. 1995. *The Great Outdoors: Restoring Children's Right to Play Outside*. Washington, DC: National Association for the Education of Young Children.

Sallis, J., J. Prochaska, and W. Taylor. 2000. "A Review of Correlates of Physical Activity of Children and Adolescents." *Medicine Science of Sports Exercise* 32(5): 963–975.

Satterthwaite, D. 2006. *Outside the Large Cities: The Demographic Importance of Small Urban Centres and Large Villages in Africa, Asia and Latin America*. London: International Institute for Environment and Development.

Schudel, M. 2001. "No-Fun Zones: Schools Take a Recess Timeout—Commentary." *Technos*, winter.

Slate, John, and Craig Jones. 2005. *Effects of School Size: A Review of the Literature with Recommendations*. Aiken: University of South Carolina at Aiken, National Clearing House for Educational Facilities.

Sobel, D. 1990. "A Place in the World: Adult Memories of

Childhood's Special Places." *Children's Environments Quarterly* 7(4): 13–17.

Southworth, M., and E. Ben-Joseph. 2003. *Streets and the Shaping of Towns and Cities*. Washington, DC: Island Press.

Spencer, C., and M. Blades, eds. 2006. *Children and Their Environments: Learning, Using and Designing Spaces*. Cambridge: Cambridge University Press.

Steutville, R., and P. Langdon. 2003. *New Urbanism: Comprehensive Report and Best Practices Guide*. Ithaca, NY: New Urban Publications.

Striniste, N., and R. C. Moore. 1989. "Early Childhood Outdoors: A Literature Review Related to the Design of Childcare Environments." *Children's Environments Quarterly* 6(4): 25–31.

Thomas, R., ed. 2003. *Sustainable Urban Design: An Environmental Approach*. London: Taylor and Francis.

Titman, W. 1994. *Special Places, Special People: The Hidden Curriculum of School Grounds*. Surrey, UK: World Wide Fund for Nature.

Tranter, P. 1993. *Children's Mobility in Canberra: Confinement or Independence?* Canberra: University College, UNSW.

Tranter, P., and J. Doyle. 1996. "Reclaiming the Residential Street as Playspace." *International Play Journal* 4:81–97.

Ulrich, R. 1999. "Effects of Gardens on Health Outcomes: Theory and Research." In *Healing Gardens: Therapeutic Benefits and Design Recommendations*, edited by C. Cooper Marcus and M. Barnes. New York: Wiley.

University of California, Department of City Planning. 2003. Unpublished paper for ID241.

Van Andel, J. 1990. "Places Children Like, Dislike, and Fear." Children's Environment Quarterly 7(5): 24–31.

Van der Speck, M. and R. Noyon. 1997. "Children's Freedom of Movement in the Streets." In *Growing Up in a Changing Urban Landscape*, edited by R. Camstra. Assen: Van Gorcum.

———. 1995. "Children's freedom of movement in the streets." Paper presented at the International Conference on Building Identities. Gender Perspective on Children and Urban Space, Amsterdam, The Netherlands, 11–13 April 1995.

Vernez-Moudon, A., ed. 1991. *Public Streets for Public Use*. New York: Columbia University Press.

de Vries, S., I. Bakker, W. van Mechelen, and M. Hopmann-Rock. 2007. "Determinants of Activity-Friendly Neighborhoods for Children: Results from the Space Study." *American Journal of Health Promotion* 21(4): 312–316.

Wechsler, H., R. Devereaux, M. Davis, and J. Collins. 2003. "Using the School Environment to Promote Physical Activity and Healthy Eating." *Preventive Medicine* 31(2): 121–137.

Wells, N. M. 2000. "At Home with Nature. Effects of 'Greenness' on Children's Cognitive Functioning." *Environment and Behavior* 32(6): 775–795.

Wells, N. M., and W. Evans. 2003. "Nearby Nature: A Buffer of Life Stress Among Rural Children." *Environment and Behavior* 35(3): 311–330.

Wells, N. M., and K. Lekies. 2006. "Nature and the Life Course: Pathways from Childhood Nature Experiences to Adult Environmentalism." *Children, Youth and Environments* 16(1): 1–24.

Wheway, R., and A. Millward. 1997. *Child's Play: Facilitating Play on Housing Estates*. Coventry, UK: Chartered Institute of Housing.

Williams, J. 2005. "Designing Neighborhoods for Social Interaction: The Case of Cohousing." *Journal of Urban Design* 10(2): 195–227.

Woolley, H. 2006. "Freedom in the City: Contemporary Issues and Policy Influences on Children and Young People's Use of Public Open Space in England." *Children's Geographies* 4(1): 45–59.

Yarrow, Leon, Judith Rubinstein, and Frank Pedersen. 1975. *Infant and Environment: Early Cognitive and Motivational Development*. New York: Wiley.

Zask, A., E. van Beurden, L. Barnett, L. Brooks, and U. Dietrich. 2001. "Active School Playgrounds—Myth or Reality? Results of the 'Move It or Groove It' Project." *Preventive Medicine* 33:402–408.

Useful Websites

www.unesco.org/most/growing.htm for updated project reports on Growing Up in Cities

www.unicef-icdc.org for information on the Child Friendly Cities Programme of UNICEF, which focuses on monitoring and implementing the rights of children in urban areas as stipulated in the UN Convention on the Rights of the Child

www.communitygreens.org for information on historic and contemporary examples of shared green space in the interiors of urban blocks

www.plangreen.net for information on compact, mixed-use, green development that integrates native ecosystems

http://www.city.setagaya.tokyo.jp/topics/bunkoku/outline/guide002.html for information on urban landscape and design innovations in Setagaya Ward, Tokyo

www.homezones.org for information on UK examples of shared streets (*woonerven*)

http://www.homezones.org.uk/public/downloads/Tim_Gill_Childstreet_Paper.pdf for a copy of the paper by Tim Gill on Home Zones

http://www.planning.org/cpf/ for information on the City Parks Forum

http://www.naturalearning.org/helpchildrenlearn.html for access to the City Parks Forum publication "Urban Parks Help Children Learn," by Robin Moore

http://www.whatcomwatch.org/php/WW_open.php?id=718 for information about forest kindergartens

http://www.edibleschoolyard.org/about.html for information about the Edible Schoolyard at Martin Luther King Jr. Middle School, Berkeley, California

http://www.urban.nl/childstreet2005/ for diverse information about and from the Childstreet conference held in Delft, the Netherlands, August 2005

http://www.urban.nl/childstreet2005/downloads/delft_manifesto_draft.pdf for a copy of the Delft Manifesto on a Child Friendly Urban Environment–drafted on behalf of the Childstreet conference participants, August 23, 2005

http://www.europoint-bv.com/download/1163606010 for a copy of the Manifesto of the European Child Friendly Cities Network, from the Child in the City conference, Stuttgart, Germany, October 2006

http://www.farmgarden.org.uk/ for information about Urban Farms in the UK and Europe

http://efcf.vgc.be/index.html for information about city farms in Europe

http://www.greenteacher.com/ for information about *Green Teacher*, a magazine by and for educators to enhance environmental and global education across the curriculum at all grades

http://www.ecoliteracy.org/ Center for Ecoliteracy, to access writings of Fritof Capra and others

chapter 11

Children and the Success of Biophilic Design

Richard Louv

Winston Churchill was on target when he said that we shape our buildings and then they shape us. Most of all, our buildings and communities shape our children.

Nearly three decades ago, I visited the ecocommunity of Village Homes, the first fully solar-powered housing development in the United States, built in 1976 on 70 acres of tomato fields in the college town of Davis, California, by Judy and Michael Corbett. If such a thing as biophilic design existed then, Village Homes was it.

As Michael escorted me around this 200-home neighborhood, I was struck by the inside-out nature of the place: garages were tucked out of sight; homes pointed inward toward open green space, walkways, and bike paths. The community was infused with flowers and vegetable gardens. Grapevines on roofs thickened in the summer, providing shade, and thinned in the winter, letting the sun's rays through. (At least in the early years of Village Homes, residents produced nearly as much food as the original owner, a farmer.) Orchards surrounded this community.

"We've got a group of kids called 'the harvesters,'" the Corbetts' teenage daughter, Lisa, said. The orchards are set aside for the kids; we go out and pick the nuts and sell them at a farmers' market at the gazebo in the center of the village."

Many years later, when researching *Last Child in the Woods*, I called Michael and asked him if he had observed changes or unexpected behavior among the parents or young people who grew up at Village Homes.

"The parents loved it here because their kids were easy to watch; there was no through-traffic, so it was safe," he said. "The kids really got involved with the gardens and harvesting the fruit from the orchard. They developed a respect for where food came from. The junior high kids were particularly interested in gardening—they started gardening on their own. This was less true of the high school kids. Interesting—not once in twenty years have I seen the kids who live here throw a tomato or fruit at anyone else. Kids from outside Village Homes did it, but our kids chased them out."

From the beginning, the waiting list for prospective buyers was long. They were attracted to the development's efficiency; the heating bills of most Village Homes residents were a third to a half of those paid by residents in surrounding neighborhoods. And they were attracted to the intrinsic humanity of the design, which connected people to nature on a daily basis. Word spread; developers and architects from around the world visited Village Homes. And, as the years passed, similar ecocommunities started springing up across parts of Western Europe.

Yet, in the United States, to date, no developer has replicated the Corbetts' ecocommunity concept. Instead, a very different kind of community—an exocommunity—now dominates the suburban landscape. This is the kind of development that protects its exterior with a hard exoskeleton of exclusivity, gates, and walls. In the exocommunity, children grow up in covenant-controlled environments where design favors martially trimmed postage-stamp yards, and where community associations dictate the color of curtain liners and prevent families from planting gardens. The exocommunity effectively criminalizes natural play. Just try to put up a basketball hoop in some of these communities, let alone allow kids to build a fort or a tree house. A San Diego woman told me recently that her community association had recently outlawed chalk drawing on the sidewalks.

One wonders how children growing up in this culture of control will define freedom as adults. What happens when all the parts of childhood are soldered down, when the young no longer have the time or space to play in their family's garden, cycle home in the dark with the stars and moon illuminating their route, walk down through the woods to the river, lie on their backs on hot July days in the long grass—what happens when such experiences are virtually impossible in much of the built environment?

Theoretically, people are free to move to other neighborhoods. But exocommunities dominate the growing doughnuts of development surrounding most American cities. Not every exocommunity enforces every covenant restriction; nonetheless, the message gets through, and the medium is the community.

To every trend, however, there is a countertrend.

The moment may have come for an extension of the Corbetts' dream of ecocommunity—that is, an ethic that incorporates nature into the design of our homes, schools, and communities; reintroduces natural spontaneity; and builds the enthusiasm necessary to make that happen. Why now? For one, global warming concentrates the attention. Another reason is the growing body of knowledge concerning the health benefits of connectedness to nature. Although the marketing for Village Homes promoted energy savings and a sense of community, the hidden benefit—the real selling point for the twenty-first century—may have more to do with children and health than with heating bills.

First hypothesized by Harvard University scientist and Pulitzer Prize–winning author Edward O. Wilson, *biophilia* is most often described as our biologically based affinity for natural settings. For workplaces, schools, hospitals, and neighborhoods, biophilic design has emerged as a promising way to add value to the energy-centric concept of sustainable or green design. As a word and concept, *sustainability* is surely important, but it suggests stasis, bringing our environment up to par, as if we know what constitutes par. Many of us, particularly the young, hunger for a more powerful frame, one that suggests creativity.

In his book *Building for Life: Designing and Understanding the Human-Nature Connection*, Stephen Kellert uses the term *restorative environmental design*. Restorative environmental design, he says, "incorporates the complementary goals of minimizing harm and damage to natural systems and human health as well as enriching the human body, mind, and spirit." That may become the brand of choice. But for the purposes of this chapter, let's stick with the term *biophilic design*. At least it's short and specific. Beyond the name is the frame. Here is my suggested working framework: Sustainable or green design is essentially about conserving energy and leaving a small footprint on the earth; biophilic design is about conserving energy and producing human energy.

Some designers now consciously avoid the word *green*, partly because of the political baggage the word carries in some quarters. But there's a second, more important reason: Biophilic designers are taking the next step, inclusive of but beyond sustainability. They're describing a way to increase the productivity and creativity of the people who work or live in those buildings. When considering the impact of community design on child development, the difference between only saving nonrenewable energy and producing renewable human energy is no small distinction.

* * *

The social and cultural obstacles to good community design are closely related to the barriers that keep children from experiencing nature firsthand.

In the United States, parents cite a number of everyday reasons why their children spend less time in nature than they themselves did, including disappearing access to natural areas, competition from television and computers, dangerous traffic, more homework, and other time pressures. Most of all, parents cite fear of stranger-danger, as round-the-clock news coverage conditions them to believe in an epidemic of child-snatchings, despite evidence that the number has been falling for years. The reason for this dissonance is, primarily, television. Decades ago, George Gerbner, professor of communications and dean emeritus of the Annenberg School of Communication in Philadelphia, described what he called the "mean world syndrome," meaning that people who watch a lot of TV think the world is more dangerous than it actually is. Among the symptoms: a pervasive sense of insecurity and vulnerability.

"Twenty-five years ago, some people thought Gerbner's theory was hype and overstatement, but not today," says Frank Gilliam, founding director of the Center for Communications and Community at UCLA. "A few years ago, there was some evidence health was crowding out crime on the news, but we discovered that the coverage was all the bad things that can happen: poisonings, unsafe jungle gyms, coupled with crime."

As described earlier, television is not the only delivery system for the antichild message. Our institutions, urban/suburban design, and cultural attitudes unconsciously associate nature with doom, while disassociating the outdoors from joy and solitude. As a medium for the message of fear, the design of communities is both a symptom and a cause of the disconnection from nature. An unintended antinature message is even codified into the design and regulatory structures of many of our communities—effectively banning much of the kind of play that we enjoyed as children.

It should come as no surprise, then, that even as cyberspace expands, the physical landscape of childhood is shrinking. The Centers for Disease Control and Prevention (CDC) reports that, in a typical week, only 6 percent of children ages 9 to 13 play on their own. Studies by the National Sporting Goods Association (NSGA), a trade group, and American Sports Data, a research firm, show a dramatic decline in such outdoor activities as swimming and fishing in the past decade. Even bike riding is down 31 percent since 1995. A child is six times more likely to play a video game on a typical day than to ride a bike, according to surveys by the Kaiser Family Foundation and the CDC. When children do go outside, they're usually under adult surveillance, playing organized sports. However, some organized sports for children are showing signs of decline: Little League participation has fallen to 2.1 million children, down 14 percent from its peak in 1997; meanwhile, overall baseball playing, including pick-up games in the neighborhood, has declined nearly twice as fast, according to the NSGA surveys.

The childhood trend away from nature appears to be occurring most rapidly in English-speaking countries. In 1986, Robin Moore, a professor of landscape architecture at North Carolina State University, charted the shrinkage of natural play spaces in urban England, a transformation of the landscape of childhood that occurred within a space of 15 years. The growing child-nature gap also exists beyond English-speaking countries. In 2002, another British study discovered that the average eight-year-old was better able to identify characters from the Japanese card trading game Pokémon than native species in the community where they lived; pikachu, metapod, and wigglytuff were names more familiar to them than otter, beetle, and oak tree. Similarly, Japan's landscape of childhood, already downsized, grew smaller. For almost two decades, the well-known

Japanese photographer Keiki Haginoya photographed children's play in the cities of Japan. In recent years, "children have disappeared so rapidly from his viewfinder that he has had to bring this chapter of his work to an end," Moore reports.

The result, as I have named it, is *nature-deficit disorder*. This is not a known medical condition, but a useful phrase to consider the human costs of alienation from nature, among them diminished use of the senses, attention difficulties, and higher rates of physical and emotional illnesses.

* * *

The good news is that at the very moment that the bond between the young and the natural world is in danger of breaking, recent research links the mental and physical health of children and adults, as well as cognitive functioning and creativity, directly to nature experiences. This growing body of knowledge has profound implications for the design of future communities, and the liberties within them.

In hospitals, biophilic design is associated with faster recovery time, decreased use of strong painkillers, and less postoperative anxiety among patients recovering from open-heart surgery. In a study of four hospital gardens in the San Francisco Bay area, Clare Cooper Marcus and Marni Barnes learned that these spaces were highly used and valued for their restorative effects on patients, visitors, and staff. In the last decade, a growing movement has seen healing gardens created at many hospitals. More recently, the provision of these restorative nature spaces has become even more specialized, with patient-specific gardens being designed by landscape architects for patients with cancer; patients requiring physical rehabilitation; those with Alzheimer's disease and other forms of dementia; and people suffering from depression and burn-out syndrome. The research findings regarding stress and nature by Roger Ulrich, Terry Hartig, Stephen and Rachel Kaplan, and others have convinced the medical world that designed natural areas in healthcare settings are not just cosmetic niceties, but actually facilitate the healing of patients and the restoration of busy staff and worried visitors.

While news media remain enamored of electronic neural networks said to enhance human intelligence, an immediately available intelligence-enhancing environment already exists: immersion in nature also enhances our neural networks, along with our physical and psychological health. For several years, I worked with a council of neuroscientists concerned with children. When creating laboratory environments, they attempt to replicate the natural world; only within such environments can the impact on brain development of, say, toxic stress, be measured. Yet, when asked how the natural world itself affects brain development, they usually drew a blank. "How do you define nature?" they would ask.

We would hope that such limitations of imagination and science will soon lift. Neuroscientists now know that genetics play a role in the development of brain architecture, but not a final role.

"Our brains come with blueprints—that's our genetic inheritance. But situations and conditions determine how a child's brain architecture actually gets built," explains Jack Shonkoff, a Harvard professor. He is chair of the National Scientific Council on the Developing Child, which brings neuroscientists, developmental psychologists, and economists together to review new research on how early development actually unfolds. "A child's genes influence the initial blueprint, but are turned on and off by interactions with the environment—in the home, in the neighborhood, in the child's environment of relationships."

A logical next question, deserving more scientific attention, is: What role do nature and its expressions in the built environment play in brain development? Does biophilic design build better brains?

When outside in woods or fields or on water, children stretch all of their senses, something they do not do in front of a screen. Howard Gardner, a professor of education at Harvard University, developed his influential theory of multiple intelligences in 1983. Gardner argued that the traditional notion of intelligence, based on IQ testing, was far too limited; he instead proposed seven types of intelligences to account for a broader range of human potential in children and adults. A few years ago, he added an eighth intelligence: naturalist intelligence ("nature smart"), which can apply to every child.

More recently, a 2005 study by the California Department of Education found that students in schools with nature-immersion programs performed 27 percent better in science testing than kids in traditional classrooms. These students were also more likely to play cooperatively. At schools that employ outdoor classrooms, studies have shown substantial testing improvements, particularly in science.

These studies suggest that biophilic design changes in schools and homes do not have to be complex or expensive. Outdoor classrooms cost less than brick and mortar buildings. Design permitting, natural light is a cheap and renewable brain resource. Vivian Loftness, a professor at Carnegie Mellon School of Architecture, reports 20–26 percent higher test scores in classrooms with ample natural light. "Yet, we're sealing up our campuses, even as we speak," she says.

"Natural spaces and materials stimulate children's limitless imaginations and serve as the medium of inventiveness and creativity," says Moore, who is an international authority on the design of children's play and learning environments. For example, in Sweden, Australia, Canada, and the United States, studies of children in schoolyards with both green areas and manufactured play areas found that children engaged in more creative forms of play in the green areas. Swedish researchers compared children in two daycare settings: at one, the quiet play area was surrounded by tall buildings, with low plants and a brick path; the second was based on an "outdoors in all weather" theme and was set in an orchard surrounded by pasture and woods. Adjacent to the school was an overgrown garden with tall trees and rocks. The study revealed that children in the green daycare, who played outside every day, regardless of weather, had better motor coordination and more ability to concentrate.

While nature experience should not be seen as a panacea or a substitute for appropriate medication, it can help relieve the everyday pressures that may lead to childhood depression.

More than 100 studies of children and adults show that spending time in nature reduces stress. Research conducted by the Human-Environment Research Laboratory at the University of Illinois shows that contact with the natural world significantly reduces symptoms of Attention Deficit Disorder in children as young as age five. I am moved when I hear how parents notice significant changes in their hyperactive children's behavior when they take them hiking or encourage them to enjoy other nature-oriented outings. Camping programs, accustomed to facilitating emotional well-being since the early 1900s, increase self-esteem, especially for preteens. Children with disabilities also benefit. One study of 15 residential summer camp programs with specialized programs for children with disabilities—including learning disabilities, autism, sensory disabilities, moderate and severe cognitive disabilities, physical disabilities and traumatic brain injury—revealed that participating children demonstrated improved initiative and self-direction that transferred to their lives at home and in school.

As a species, we have known all of this, intuitively, for thousands of years. But only now in Western society is science beginning to fully appreciate the role of nature experience in child development. The time is ripe for Village Homes and developments like it to take the next step—not only beyond the exocommunity but also beyond the current definition of the ecocommunity.

* * *

What the Corbetts pioneered, or an evolution of their design, may finally fit the times. Such an approach would reflect the kind of thinking that has been emerging in public health circles for years, although not always consciously. To improve health by trying to change human behavior—preaching to people that they should change their habits—is not as efficient as using design to transform the environment in which people live. For example, the drop in traffic accident fatalities, relative to the population, has less to do with teaching people to drive better than it does with designing safer cars and better highways.

If biophilic design does produce the kind of growth and positive change in human beings that the early research suggests, we should move quickly as a society toward biophilic community design, not only to save us from global warming, but also to allow our children to develop optimally, and to countermand the growing nature deficit in their lives.

Imagining this future is one thing; making it happen is another. At a recent conference on biophilic design, Stephen Kellert, referring to the growing excitement about biophilic or restorative design, asked, "Where do we go from here?" One architect raised his hand and said, "Just let us do what we do." Fair enough. The pioneers should be encouraged to pioneer. But that approach will produce, at best, a few hundred or a few thousand biophilic homes for the wealthy. We need a movement to encourage biophilic design on a mass level. Focusing on children's psychological and physical health, and their learning abilities, offers a doorway to the popularizing of biophilic design.

The issue of the disconnection of children from nature has a peculiar effect on people from all walks of life. Some U.S. developers have expressed keen interest in applying some of the same principles that, in the case of Village Homes, seemed so troublesome to replicate. The *Sacramento Bee* reported in July 2006 that Sacramento's biggest developer, Angelo Tsakopoulos and his daughter Eleni Tsakopoulos-Kounalakis, who together run AKT Development, "have become enthusiastic promoters" of the idea that residential developments can connect kids to nature. Such thinking is "really going to change how we build neighborhoods," said Tsakopoulos-Kounalakis.

As I reported in the March 2007 issue of *Orion* magazine, not long after the publication of *Last Child in the Woods* I received an e-mail from Derek Thomas, founder and chief operating officer of Newland Communities, the nation's largest privately owned residential development company. "I have been reading your new book and am profoundly disturbed by some of the information you present," he wrote. I must admit that discomforting developers gives me comfort; as a boy, I pulled out dozens—perhaps hundreds—of survey stakes to slow the bulldozers that were taking out the trees near *my* woods to make way for housing developments. But Thomas said he wanted to do something positive. He invited me to an envisioning session in Phoenix to "explore how Newland can improve or redefine our approach to open space preservation and the interaction between our homebuyers and nature."

I accepted his invitation to Phoenix. There, I offered my sermonette to a conference room filled with about 80 developers, builders, and real estate marketers. I told them that they were partially responsible for the disconnection, not only because of their destruction of habitat, but also because of the way they build developments and the covenants they place on these communities, restrictions that have virtually criminalized outdoor play—climbing trees, building forts, even chalk-drawing on sidewalks. I was ready to flee the lions. But then Thomas, a bearded man with an avuncular demeanor, stood up and said to the developers and real estate marketers: "I want you all to go into small groups and solve the problem: How are we going to build communities in the future that actually connect kids with nature?"

The room filled with noise and excitement. A half hour later, the groups reported their ideas. Some were practical: leave some land and native habitat in place (good place to start). Employ green design principles. Incorporate nature trails and natural waterways. Throw out the conventional covenants and restrictions that discourage or prohibit natural play; rewrite the rules to encourage it, to allow kids to build forts and tree houses or plant gardens. Create small on-site nature centers. "Kids could become guides, using cell phones, along nature trails that lead to schools at the edge of the development," someone suggested. The quality of their ideas mattered less than the fact that they had them. These were, after all, mainstream developers.

I'm realistic. I'm not holding my breath that we'll soon see such developments spreading across the landscape. Nor should such principles be exploited to encourage more sprawl. That won't do. Better to see vast decaying strip-mall biophilification, or the transformation of the ghost towns of the Great Plains states into biophilic ecovillages.

Beyond Village Homes and the green design of individual buildings in the United States, developers and builders can look to Western Europe for a whole-community approach. There, some of the newest neighborhoods are becoming more livable and loveable by protecting and regenerating nature. Timothy Beatley, in his book *Green Urbanism: Learning from European Cities*, describes Morra Park, an ecovillage in the city of Drachten in the Netherlands: it has a closed-loop canal system, in which stormwater runoff is moved by the power of an on-site windmill and circulated through a manufactured wetlands where reeds and other vegetation filter the water naturally, making it clean enough for residents to swim in. In a charming photograph of this village, a boy poles his crude raft along the stream. A similar Dutch development called Het Groene Dak (The Green Roof) incorporates a communal inner garden, "a wild, green, car-free area for children to play and residents to socialize," writes Beatley. At a similar, suburban ecovillage in Sweden, "large amounts of woodland and natural area have been left untouched." To minimize impact on nature, homes are built on pillars and designed "to look as though they had been lowered out of thin air."

Beatley describes an impressive array of European green-city designs—for example, cities with half the land area devoted to forest, green space, and agriculture; cities that have not only preserved nearby nature, but reclaimed some inner-city areas for woods, meadows, and streams. These neighborhoods are both denser and more livable than our own. Nature, even a suggestion of wildness, is within walking distance of most residences. In contrast to "the historic opposition of things urban and natural," he writes, green cities "are fundamentally embedded in a natural environment. They can, moreover, be re-envisioned to operate and function in natural ways—they can be restorative, renourishing and replenishing of nature and the human beings who are part of that nature."

An incentive to developers and builders is suggested by Louise Chawla, a professor in the College of Architecture and Planning at the University of Colorado, who served as international coordinator of UNESCO's Growing Up in Cities project. She is pushing for a Children and Nature Design Certification along the lines of the green industry's Leadership in Energy and Environmental Design (LEED) Green Building Rating System. LEED is the nationally accepted benchmark for the design, construction, and operation of high-performance green buildings. A certification program that would link children and nature to good building and community design would help move this cause forward.

Biophilic community design should not be building-bound, but should be seen in the larger context of the *zoopolis*—pronounced like *metropolis*—a word that Jennifer Wolch, a professor at the University of Southern California, and director of the Sustainable Cities Project, uses when she imagines areas in cities transformed into natural habitats through land planning, architectural design, and public education.

Such thinking by any name is not new, but newly remembered. Consider these wise words: "Any mind with sufficient imagination to grasp it must be stimulated by this conception of the city as one great social organism. Whole future welfare is in large part determined by the actions of people who comprise the organism today and, therefore, by the collective intelligence and will that control these actions. The stake is vast; the possibilities, splendid." That was said in 1916 by the legendary landscape designer Frederick Law Olmstead, who designed many of our nation's great urban parks, including New York City's Central Park. Olmstead was commissioned by nineteenth-century industrialists, who acknowledged the connection between nature and the health—and productivity—of their workers. It's hard to imagine any city establishing an urban park as splendid and humane as Central Park today. Or is it?

In my own community, champions of San Diego's urban canyons refer to them as the region's lungs, its bronchi—and they're not just talking about air quality. A San Diego Regional Canyonlands Park created from the hundreds of miles of connected and disconnected canyons would preserve natural habitat, but also promote human health and well-being.

Mike Stepner, principal of the Stepner Design Group and a professor at the New School of Architecture and Design in San Diego, argues that the city's natural canyons offer a unique opportunity to use biophilic design as a central organizing principle for the region's future: "I'm not only interested in preserving the canyons, but bringing their design forms, their spirit, up into the surrounding neighborhoods."

As part of creating such a park, Stepner believes in bringing the canyons to the neighborhoods, not just bringing the neighbors to the canyons. Urban planners and canyon protectors could move the conscious, organized extension of the look and feel of the canyons, along with native canyon botany, into surrounding neighborhoods. The spirit, forms, and life of the canyonlands would inform the architecture of new or renewed buildings and homes, along "boulevards, parks, plazas, and other found space that brings nature into the neighborhood and provides places for people to interact," says Stepner. Every one of these canyons is a potential outdoor classroom within walking distance of a public school.

In the past, we thought of nature as being out there, separate, beyond where we spend our daily lives. But in the twenty-first century, that approach will be seen as a quaint, destructive artifice.

Now, within such a zoopolis, homes and neighborhoods and schools and playgrounds would be designed and sustained using biophilic principles. Children, vulnerable to both the good and the bad of architecture and urban design, would stand to gain the most.

Educationally, socially, economically, spiritually, such a community would harness the power of place. Author and biologist Robert Michael Pyle coined the elegant phrase "the extinction of experience" to describe our lost connection to that power. "Place is what takes me out of myself, out of the limited scope of human activity, but this is not misanthropic," he writes. "A sense of place is a way of embracing humanity among all of its neighbors. It is an entry into the larger world."

Biophilic design, by any name, brings us home.

REFERENCES

Balmford, A., L. Clegg, T. Coulson, and J. Taylor. 2002. "Why conservationists Should Heed Pokémon." *Science* 295(5564): 2367–2367.

Beatley, Timothy. 2000. *Green Urbanism: Learning from European Cities.* Washington, DC: Island Press.

Cooper Marcus, Clare, and Marni Barnes. 1995. *Gardens in Healthcare Facilities: Uses, Therapeutic Benefits, and Design Recommendations.* Martinez, CA: Center for Health Design.

Corbett, Michael, Judy Corbett, and Robert L. Thayer. 2000. *Designing Sustainable Communities: Learning from Village Homes.* Washington, DC: Island Press.

"Effects of Outdoor Education Programs for Children in California." 2005. Palo Alto, CA: American Institutes for Research. http://www.sierraclub.org/youth/california/outdoorschool_finalreport.pdf.

Faber Taylor, Andrea, Frances E. Kuo, and William C. Sullivan. 2001. "Coping with ADD: The Surprising Connection to Green Play Settings," *Environment and Behavior* 33(1): 54–77.

Gardner, Howard. 2003. "Multiple Intelligences after Twenty Years." Paper presented at the American Educational Research Association, Chicago, Illinois, April 2003). © Howard Gardner: Harvard Graduate School of Education, Cambridge, MA.

Kellert, S. 1993. "Introduction." In *The Biophilia Hypothesis,* edited by S. R. Kellert and E. O. Wilson. Washington, DC: Island Press/Shearwater.

———. 2005. *Building for Life.* Washington, DC: Island Press.

Louv, Richard. 2005. *Last Child in the Woods: Saving Our Children from Nature-Deficit Disorder.* Chapel Hill, NY: Algonquin Books.

———. 2007. "Leave No Child Inside." *Orion:* 54–61.

Moore, Robin C. 1986. *Childhood's Domain: Play and Place in Child Development.* Dover, NH: Croom Helm.

———. 1997. "The Need for Nature: A Childhood Right." *Social Justice* 24(3): 203.

Vellinga, Mary Lynne. 2006. "Tuning In Call of the Wild: Developer Touts Book as a Guide to Get Kids Outdoors." *Sacramento Bee.* July 3, metro final ed., sec. A.

Wilson, Edward O. 1984. *Biophilia.* Cambridge, MA: Harvard University Press.

chapter 12

The Extinction of Natural Experience in the Built Environment

David Orr and Robert Michael Pyle

This is a tale of two journeys—one drawn large in miles and disturbing vistas, one sketched in smaller strokes from alleys, ditches, and vacant lots—converging in a mutual sense of what must be done if natural experience is to survive. Both of our lives' trajectories have been infinitely enriched by intimate exposure to the physical details of the natural world, human and otherwise. Yet these paths have also brought us both to an excruciating sense of the loss of natural texture in the occupied domain—the kindly gifts of geological, biological, and cultural development that make the world worth living in and make us worthy of living here. We share an extreme sense of privilege in having known the richness of nature as we have, and a dire concern for the social and biological aftermath of what more and more appears to be its precipitous decline. We call this ordeal of loss *the extinction of experience.*

The following intertwined considerations, in our alternating voices, tell how the leakage of experience from the collective lives of modern culture is happening, what it may mean for us all at both local and broader scales, and how it might be stanched. By weaving our essays together (D. O.'s warp in roman, R. P.'s woof in italics), we hope to take readers with us as we shift our focus from the motes in front of our noses to wide prospects of life and loss, and back again. For only when we both notice and understand the relationship of the particular grain to the overwhelming span of the big picture can we hope to come to terms with the degree of diminishment confronting human life today—and what we stand to lose, or gain, by how we respond. We invite readers to come along on both our journeys at once, adjusting their bifocals to the near view, then to the distant, and back again.

* * *

The manner in which we experience nature is changing. More and more, nature comes to us by indirection, packaged by mall designers, television producers, architects, developers, and urban planners, most of whom have regarded unmanaged contact between people and nature as inconvenient or unprofitable. The vendors of "virtual reality" would sell us even more cleverly seductive ways to experience a contrived nature. Even in national parks the experience of nature has been de-natured in order to accommodate the automobile and hordes of tourists, mediated by electronic devices. As the global economy has become more automated and centralized, few earn their livelihood any longer in direct contact with nature by farming, forestry, or fishing. In what is awkwardly called "the built environment" nature is held at bay by windows that do not open and sounds crafted to soothe our angst over the drone of HVAC systems providing filtered air. We have become mostly an indoor species spending upward of 95 percent of our time sealed away from anything remotely like authentic nature.

Strictly speaking, the experience of nature cannot be extinguished, since in the broadest sense there is nothing else. But what we traditionally think of and designate as "natural" experience—that is, contact with plants and animals other than our own species, and the landscapes that harbor them—can be radically diminished. Even in a sensory deprivation chamber, one would continue to experience what we call human nature, if in a bizarre way. In that condition we might even experience a vague yearning toward something better for which we have no name.

Along a continuum between, say, a bland and contrived nature to one that is totally capricious, *Homo sapiens* evolved somewhere in the middle in a world that was beautiful, dangerous, and mostly dependable. Lacking claws, speed, and strength, we were vulnerable to predators and natural disasters and such dangers are likely imprinted on us as well as the more benign feelings of biophilia. In time the mastery of nature became one of our defining traits, first through the ability to use fire and primitive tools, later in the arts of agriculture and industry, and now in technologies that extend into the far reaches of the atom, the gene, and space. But the mastery of nature is a mixed blessing, as we are coming to see, and far more complicated than it seems.

C. S. Lewis once observed that human mastery of nature really meant that some men mastered some aspects of nature in order to control other men. Mastery of nature, then, is thoroughly political. It is also paradoxical. As our power over nature has become more extensive and intrusive, nature in the large is becoming more volatile and in some ways less predictable. Look no farther than the connection between our control of nature powered by fossil fuels and the correlative changes in climate that include more severe storms, more extensive droughts and heat waves, rapidly changing ecosystems, and rising sea levels. In the emerging greenhouse world our experience of nature will change dramatically. Unless we choose otherwise and act very soon, our children will live between the extremes of a contrived, simulated nature sold to them as a commodity and one in the large that is far more menacing than humans have ever experienced.

The world of childhood, in particular, is changing dramatically. The great outdoor playground of fields, streams, woods, and shoreline is being degraded by heat waves, droughts, changing ecologies, and the loss of biological diversity. New diseases and water-borne viruses will progressively render nature threatening, not inviting. Curiosity will retreat indoors and the sense of wonder will shrivel in that confinement. To a degree, I suppose, biophilic design can compensate for the losses in direct contact with nature. But the great maternal bond between humans and the world of the Holocene will have been mutilated, if not broken entirely. In other words, if the cultivation of our innate affinity for life and lifelike processes is to be preserved, it must be preserved in the large—and that requires, among other things, the stabilization of greenhouse gases somewhere below the level at which planetary destabilization becomes a positive feedback system.

* * *

My concept of "the extinction of experience" arose out of necessity in the spring of 1975. I had been asked by biologist Charles Remington, at that time my major professor at Yale University, to substitute for a presentation he was to make at the American Association for the Advancement of Science in Boston. The topic of the symposium was

"Wildlife in the Year 2001." When I considered what to speak about, my mind drifted (as it often did) to the landscape of my own awakening to conservation. This was the High Line Canal, a century-old irrigation ditch that carried water from the South Platte River to the agricultural hinterlands of Denver, Colorado.

I grew up in the tract-house fringes of Aurora, a suburb on Denver's eastern edge. The moisture of the canal, and the corridor it provided between the mountains and the plains, made it a magnet for life. This is where I became acquainted with butterflies and eventually collected and learned 10 percent of the American fauna. My own subdivision, one of the first postwar tracts grafted onto the High Plains, was a built environment of no great distinction. It did possess several traits, however, that set it apart from many subdivisions of today: 1. The yards were generous, diversely planted, and largely unsprayed with chemicals. 2. The neighborhood parks were weedy and rank, not yet unburdened of the possibility of surprise among the roots and shade of surviving cottonwoods, nor regulated such that kids on the loose could be arrested and booked for digging a hole or building a fort (as recently happened on Jefferson County Open Space, not far away to the west of our dusty redoubt on the High Plains). 3. The edge of the densely built neighborhoods, and semiwild habitats such as the High Line Canal, lay nearby, in reach by foot or by bicycle. 4. Our blocks were populated by families who did not find it strange to accord their children the freedom of the day: once homework or yard chores were finished, it was "Bye, Mom, see you at dinner" for most of the kids.

I have asked dozens of audiences of adults whether they had this kind of freedom and these kinds of experiences in their childhood. The great majority of people involved in resource-related fields, as well as the arts and humanities, and many others who express concern for the environment, profess the vital place of wild play in their past. Common currency of these days out-of-doors included vacant lots, patches of woods, old fields, particular trees, rocks, and watercourses; building forts; catching crawdads, tadpoles, and insects; and just plain prowling without supervision. Nor was this strictly an American phenomenon. A German filmmaker, 40 years old, recently described his own childhood near Cologne to me, and I was struck by how much it was defined by the same elements. In a sad sequel, most of these people admit that they could no longer find their special places in anything like the condition in which they knew them as kids; and they confess the loss of the needful experience they provided in their own children's lives, and in their own stressful days as adults. Nor do their children and grandchildren enjoy the freedom to roam and explore that most of my generation took for granted. This strikes me as an enormous cultural change.

For my brother Tom and me, the escape hatch from our suburban grid lay in the green loops of the High Line Canal. There we would don the identities of Bill and Joe, a couple of guys on adventure. Sometimes the adventures were all too real: In The Thunder Tree, *I describe the massive hailstorm that pinned us down in a great hollow cottonwood, as cattle were killed in the adjacent field, their backs broken by hail the size of Rocky Ford canteloupes. For all we knew, the canal went all the way to Kansas, and our activities knew no limits but the ditchrider, the length of the day, and the amplitude of our own imaginations. I know that it was the canal that made me who I am.*

So when I came, some twenty years later, to consider the future of wildlife yet another twenty-five years on, I thought of what might no longer be commonly available to kids in 2001. By the time I'd left Aurora for college in 1965, much had already been lost from my home environs. No longer could a boy with a bug net walk out of Hoffman Heights on a morning in May with the expectation of finding Olympia marblewing butterflies. Most of the prairie dog colonies had gone the way of the bulldozer. Even the great cottonwoods of the canal were beginning to be cut, charged with the felony of being phreatophores—thieves of the water necessary for more toilets. I wondered whether this landscape would retain the capacity to entrance future generations of kids, as it had me—to move them toward a love of the land, and to care for it. And it occurred to me that if it could not, what would have happened could be called an extinction of experience. *So that's what I titled my talk in Boston. It became an essay in* Horticulture *magazine (Pyle, 1978), then a chapter in* The Thunder Tree *(Pyle, 1993). One way and another, this phrase defines what I have been engaged in ever since.*

The extinction of experience refers to the loss of common features in the everyday environment where most of the people live, often urban and suburban settings. Specifically, I mean the loss of such elements of diversity within any given individual's radius of reach. The radius of reach is much smaller for the very young, the very old, the poor, and the disabled. When local extinctions occur, the species or textures of life they represent might as well be gone altogether, in one important, existential sense.

I believe that the aggregate of such depletions amounts to an impoverishment of experiential learning, imprinting, and connecting, such that its victims become more and more disassociated from the particulars of nature and culture that give the world its savor. This inevitably leads to alienation, which in turn results in apathy, inaction, and further loss. Thus the tragic cycle of the extinction of experience, summed up in The Thunder Tree *like this: "So it goes, on and on, the extinction of experience sucking the life from the land, the intimacy from our connections. This is how the passing of otherwise common species from our immediate vicinities can be as significant as the total loss of rarities. People who care, conserve; people who don't know don't care. What is the extinction of a condor to a child who has never known a wren?"*

* * *

Having seen pictures of the devastation did not prepare me for the reality of New Orleans. Mile after mile of wrecked houses, demolished cars, piles of debris, twisted and downed trees, and dried mud everywhere. We stopped every so often to look closely into abandoned houses in the Ninth Ward and along the shore of Lake Pontchartrain: mud lines on the walls, overturned furniture, moldy clothes still hanging in closets, broken toys, a lens from a pair of glasses . . . once cherished and useful objects rendered into junk. Each house with a red circle painted on the front indicated results of the search for bodies. Some houses showed the signs of desperation: holes punched through ceilings as people tried to escape rising water. The smell of musty decay was everywhere, overlain with an oily stench. Despair hung like Spanish moss in the dank, hot July air.

Ninety miles to the south, the Louisiana delta is rapidly sinking below the rising waters of the Gulf. This is no "natural" process, but rather the result of decades of mismanagement of the lower Mississippi that became federal policy after the massive flood of 1927. Sediment that built the richest and most fecund wetlands in the world are now deposited off the continental shelf—part of an ill-conceived effort to tame the river. The result is that the remaining wetlands, starved for sediment, are both eroding and compacting, sinking below the water and perilously close to no return. Oil extraction has done most of the rest by cutting channels that crisscross the marshlands, allowing the intrusion of salt water and storm surges. Wakes from boats have widened the original channels, further unraveling the ecology of the region. The richest fishery in North America and a unique culture that once thrived in the delta are disappearing, and with it the buffer zone that protects New Orleans from hurricanes. "Every 2.7 miles of marsh grass," in Mike Tidwell's words, "absorbs one foot of a hurricane's storm surge" (2003, 57).

And the big hurricanes will come. Kerry Immanuel, an MIT scientist and once greenhouse skeptic, researched the connection between rising levels of greenhouse gases in the atmosphere, warmer sea temperatures, and the severity of storms. He's a skeptic no longer (Immanuel 2005). The hard evidence on this and other parts of climate science have moved beyond the point of legitimate dispute. Carbon dioxide, the prime greenhouse gas, is at the highest level in at least the last 650,000 years. CO_2 continues to accumulate by about 2.5+ parts per million per year, edging closer and closer to what some scientists believe is the threshold of runaway climate change. British scientist James Lovelock compares our situation to being on a boat upstream from Niagara Falls with the engines about to fail.

If this were not enough, the evidence now shows a strong likelihood that sea levels will rise more rapidly than previously thought. The third report of the Intergovernmental Panel on Climate Change (2001) predicted about one meter rise in the twenty-first century, but more recent evidence puts this figure at six to seven meters, the result of accelerated melting of the Greenland ice sheet and polar ice along with the thermal expansion of water (Kerr 2006).

* * *

On my way out of a shopping mall in Vancouver, Washington, just before closing time, I saw something that both shocked and thrilled me. I passed the mall's concrete, geometric water feature, and there I noticed a group of young teenage boys skipping flat rocks across the surface of the artificial pond. One got a three, the next a fiver. "Beat that," he shouted. "I heard ten is the most you can get," said another. Only the boys weren't flinging rocks—they were using quarters they had recovered from the pool. They'd plucked out the coins not to keep them, but to use them as skipping stones.

Conservation writers for decades have worried about the loss of everyday contact with nature. Ian McHarg's Design with Nature *(1992) predicted some of the problems involved with local diminishment, and in fact anticipated many of the issues of biophilic design. Since then, several writers have expressed concern over the loss of childhood's special places, freedoms, and common contact with the more-than-human encountered out-of-doors (Sobel 1993; Stafford 1986; Thomashow 1995; Nabhan and Trimble 1994; Kahn and Kellert 2002; Finch 2003). Only recently has this syndrome of loss and its consequences received a name and an encyclopedic examination, in Richard Louv's* Last Child in the Woods: Saving Our Children From Nature-Deficit Disorder *(2005). Between the outright extirpation of biological diversity in most people's vicinities, the blandishments of the virtual, electronic realm, the bogeyman phenomenon, our litigious society, and the busyness of modern children's lives, the extinction of experience is rapidly becoming the norm in contemporary culture. The lyric that says "See the children run, as the sun goes down, and we lie in fields of gold" (Sting, 1993) is becoming merely a sentimental and bucolic souvenir today.*

Yet the impulse remains for children to be like children—in search of adventure, out for fun beyond the scintillating screen. I learned this anew through that recent experience at the Vancouver Mall. I watched the boys skipping quarters for a while, then complimented them on their skill. "Did you learn that on a pond or crick near your home?" I asked.

"Nah," said one of the boys. "We don't have anywhere like that. We just come here." My heart was both broken to hear that fact stated outright, and thrilled to know that they'd found a way to exercise this child's right, this rite of growing up, in spite of the homogenization of their habitat. This says to me that all is not lost, that nature-deficit disorder may be reversible, if only our atavistic insistence on keeping our roots watered by wildness can be nurtured. The water feature of a shopping mall might stand in for now; but think how much more powerfully those lads might be affected by an actual pond in their 'hood?

Here, in my view, is where landscape and building architects, planners, managers, developers, and all those who ultimately control our townscapes come in. For antidotes to the extinction of experience do exist, even in landscapes traditionally thought of as relentlessly and irredeemably urban. And if we were to succeed in maintaining the link between the young and the land, the harvest for the future could be incalculable. If, on the other hand, the extinction of experience prevails to its obvious conclusion, the consequences for the culture will be baleful, and everlasting.

* * *

Nine hundred miles to the northeast of New Orleans as a sober crow would fly it, Massey Energy Inc., Arch Coal, and other companies are busy leveling the mountains of Appalachia to get at the upper seams of coal in what was once one of the most diverse and relatively undisturbed forests in the United States and one of the richest ecosystems anywhere. Throughout the coalfields of West Virginia and Kentucky, extractors have already leveled about 1.5 million acres and damaged a good bit more. Coal is washed on-site, leaving behind billions of gallons of dilute, asphalt-like gruel laced with toxic flocculants and heavy metals. An estimated 225 containment ponds are located over abandoned mines in West Virginia, held back from the communities below only by earthen dams prone to failure either by collapse or by draining down through the old mine tunnels that honeycomb the region. One did fail on October 11, 2000, in Martin County, Kentucky, when the slurry broke through a thin layer of shale, flowing into mines and out into hundreds of miles of streams and rivers. The result was the permanent destruction of waterways and property values of people living in the wake of an ongoing and mostly ignored disaster.

Such mayhem is typical of the coalfields. They comprise a third-world colony within the United States; a national sacrifice zone in which fairness, decency, and the rights of old and young alike are trampled on behalf of the national obsession with "cheap" electricity. For his role in trying to enforce even the flimsy laws that might have held Massey Energy slightly accountable for its flagrant and frequent malfeasance, a mine safety inspector was persecuted. The Bush administration failed to fire Jack Spadaro from the Interior Department, but eventually forced him to retire.

Jack is in the first plane to take off from Yeager Field in Charleston along with the chief attorney for the largest corporation in the world. Hume Davenport, founder of Southwings Inc., is the pilot of the four-seat Cessna. The ground recedes below us as we pass over Charleston and the Kanawha River lined with barges hauling coal to power plants along the Ohio River and points more distant. On the western horizon appears the John Amos plant. Owned by American Electric Power, this plant by one estimate releases more mercury to the environment than any other facility in the United States, as well as hundreds of tons of sulphur oxides, hydrogen sulfide, and CO_2. For a few minutes we can see the deep green of wrinkled Appalachian hills below, but very soon the first of the mountaintop removal sites appears, followed by another and then another. The pattern of ruin spreads out below us for many miles in all directions, as far as we can see. From 5,000 feet, trucks with 12-foot diameter tires and drag lines that could pick up two Greyhound buses at a single bite look like Tonka toys in a sandbox. What is left of Kayford Mountain comes into sight. It is surrounded by leveled mountains and a few still being leveled. "Overburden," the mining industry term for dismantled mountains, is dumped into valleys covering hundreds of miles of streams—an estimated 1,500 miles in the past 25 years. Many more miles will be buried if the coal companies have their way. Coal slurry ponds loom above houses, towns, and even elementary schools. When the earthen dams break on some dark rainy night, those below will have little if any warning before the deluge hits.

Jack Spadaro is our guide to the devastation. He is a heavyset, rumpled, and bearded man with a knack for describing outrageous things calmly and with clinical precision. A mining engineer by profession, he spent several frustrating decades trying to enforce the laws, such as they are, against an industry with friends in high places in Charleston and the District of Columbia. In a flat, unemotional monotone he describes what we are seeing below. Aside from the destruction of the Appalachian forest, the math is all wrong. The slopes are too steep, the impoundments too large, the angles of slope, dam weight, and proximity of houses and towns are the geometry of tragedies to come. He points out the Marsh Fork elementary school, situated close to a coal loading operation and below a huge impoundment back up the hollow. In the event of a dam failure, the evacuation plan calls for the principal to use a bullhorn to initiate the evacuation of the children ahead of the 50-ft. wall of slurry coming their way at maybe 60 miles an hour. If all works according to the official evacuation plan, they will have two minutes to get to safety, but there is no safe place for them to go. And so it is in the coal fields—ruin at a scale for which I have no adequate words; ecological devastation to the far horizon of topography and time. We say that we are fighting for democracy elsewhere, but no one in Washington or Charleston seems aware that we long ago deprived some of our own of the rights to life, liberty, and property.

Under the hot afternoon sun we board a 15-person van to drive out to the edge of the coal fields to see what this tragedy looks like on the ground. On the way to Kayford Mountain, we exit from the interstate at Sharon onto winding roads that lead to mining country. Trailer parks, evangelical churches, and truck repair shops line the road. Small, often lovingly tended houses intermix with others abandoned long ago when underground mining jobs disappeared. The two-lane paved road turns to gravel and climbs toward the top of the hollow and Kayford Mountain. Within a mile or two the first valley fill appears—a green V inserted between wooded hills. Reading the signs made by water coursing down its face, Jack Spadaro notes that this one will soon fail. Valley fills are mountains turned upside down: rocky mining debris and trees illegally buried along with more sinister things, many locals believe, brought in by unmarked trucks in the dead of night. Jack adds that some valley fills may contain as much as 500 million tons of blasted mountain and run for six miles. We ascend the slope toward Kayford, passing "no trespassing" signs around the gate to the mining operations.

Larry Gibson, a diminutive bulldog of a man fighting for his land, meets us at the summit, really a small peak on what was once a long ridge. His family has been on Kayford since the eighteenth century, operating a small coal mine. Larry is the proverbial David fighting Goliath, but he has no slingshot except that of moral authority spoken with a fierce, inborn eloquence. Those traits and the raw courage he shows every day have landed his picture in *Vanity Fair*, *National Geographic*, and other newsstand magazines. Larry's land remains intact so far because he made 40 acres of it into a park and has fought tooth and nail to save it from the lords of Massey Energy. They have leveled nearly everything around him and have punched holes underneath Kayford, because the mineral rights below and the ownership of the surface were long ago separated in a shameless scam perpetrated on illiterate and gullible mountain people.

Larry describes what has happened using a model of the area that comes apart more or less as the mountains around him have been actually dismantled. As he talks he takes the model apart piece by piece, leaving the top of Kayford like a knob sticking up amidst the encircling devastation. So warned, we walk down the country lane to witness the advancing ruin. Fifteen of us stand for half an hour on the edge of the abyss, watching giant bulldozers and trucks at work below us. Plumes of dust from the operations rise several thousand feet above us. The next set of explosive charges is ready to go on an area about the size of a football field. Every day some three million pounds of explosives are used in the eleven counties south of Charleston. This is a war zone. The mountains are the enemy, coal profits the spoils, and the people of these hills are collateral damage.

On the late afternoon drive back to Charleston we pass the coal-loading facilities along the Kanawha River. Mile after mile of barges are lined up to haul coal to hungry Ohio River power plants, the umbilical cord between mines, mountains, and us—the consumers of cheap electricity. Over dinner that night, two residents of Mingo County describe what it is like to live in the coalfields. Without forests to absorb rainwater, flash floods are a normal occurrence. A 3-inch rain can become a 10-foot wall of water cascading off the flattened mountains and down the hollows. The mining industry calls these "acts of God" and thoroughly bought public officials agree, leaving the victims with no recourse. Coal slurry contaminates groundwater, as do the chemicals used to make coal suitable for utilities. Well water becomes so acidic that it dissolves pipes and plumbing fixtures. Cancer rates are off the charts. Coal companies have always been major buyers of politicians, and Donald Blankenship, head of Massey Energy, is no exception, investing in precisely the kind of representatives he likes—the sort that can subordinate land and people to profit. His campaign to ravage the rest of West Virginia is perversely titled "For the Sake of the Kids."

Pauline and Carol from the town of Sylvester, both in their seventies, are known as the "dust busters" because they go around the town wiping surfaces covered with coal dust from a nearby loading facility with white cloths. At open hearings, they present these dustcloths to the irritated and unmovable servants of the people as evidence of foul air. Black lung disease and silicosis are now common among not only miners, but also among young and old who have never set foot in a mine. Pauline, a fiercely eloquent woman whose husband was wounded and captured by the Germans in the Battle of the Bulge in 1944, asks us, "Is this what he fought for?" The clock reads 9:30 p.m., and we quit for the day.

To destroy millions of acres of Appalachia in order to extract maybe twenty years of coal, while destroying the wonders of the mixed mesophytic forest of northern Appalachia along with habitat for dozens of endangered species, contaminating groundwater and rendering the land uninhabitable and unusable, is not just stupid—it is a derangement for which we as yet do not have adequate words, let alone the good sense and the laws to stop it. Unlike deep mining, mountaintop removal employs few workers. Glib talk of the economic potential of flatter places is just that: glib. Coal companies' efforts to plant grass and a few trees here and there are like putting lipstick on a corpse.

Virtually every competent, independent study of energy use in the past 30 years has concluded that we could cost-effectively halve our energy expenditure while strengthening our economy and standard of living, diminishing asthma and lung disease, and improving environmental quality through conservation and alternative sources. How far does the plume of heavy metals coming from coal-washing operations go down the Kanawha, Ohio, and Mississippi and into the drinking water of communities elsewhere? What other enterprises, based on the sustainable use of forests, nontimber products, eco-tourism, and human craft skills, might flourish in these hills? Why do the profits from coal mining leave the state? Why is so much of the land owned by absentee corporations like the Pocahontas Land Company? And ultimately, what is the true cost of "cheap" coal? Accounting for the costs of coal should consider the rising tide of damage and insurance claims attributable to climate change. Before long we will wish that we had not destroyed so much of the capacity of the Appalachian forests and soils to absorb the carbon that makes for bigger storms and more severe heat waves and droughts.

Nearly a thousand miles separate the coalfields of West Virginia from New Orleans and the Gulf Coast, yet they are a lot closer than that. The connection is carbon. Coal is mostly carbon and for every ton burned, 3.6 tons of CO2 eventually enters the atmosphere—raising global temperatures and warming oceans, thereby creating bigger storms, melting ice, and raising sea levels. For every ton of coal extracted from the mountains, perhaps a 100 tons of what is tellingly called "overburden" is dumped, burying streams and filling the valleys and hollows of West Virginia, Kentucky, and Tennessee. And between the hills of Appalachia and the sinking land of the Louisiana coast, tens of thousands of people living downwind from coal-fired power plants die prematurely each year from inhaling smoke laced with heavy metals that penetrate deeply into lungs.

Like all life forms, we search out great pools of carbon to perpetuate ourselves. It is our mismanagement of carbon that threatens the human future, and this is an old story. Humans have long fought for the control of carbon found in rich soils and deep forests and later in fossil fuels. Carbon exploitation is the root of all evil and original sin wrapped up together, leading quite possibly to death by heat for much of life on Earth, including ourselves. This is what James Lovelock calls the revenge of Gaia which, if it comes to pass, will be hell on Earth.

* * *

Carbon and oxygen, distilled through sunlight, collaborate to carry off the alchemy of photosynthesis, which makes the color green. We have come to call that which we deem environmentally acceptable, or progressive, "green." Thus comes carbon and its adaptive or maladaptive transformation to stand on either side of the balance of human survival. In the practice of biophilic design, we attempt to incorporate those elements that we consider to be green—both literally and metaphorically—into biophilic design.

Many attempts have been made to accommodate or at least give a nod toward the green in modern built environments, some of them elegant and successful. But many others fail for a variety of reasons that seem to go unnoticed by those in charge. Landscape designers go to great effort (and exact considerable spending by contractors) in order to insert green features that will quickly be trampled, littered, or otherwise ignored or disrespected. Or completely contrary nearby elements, such as noise, light, structure, or blight may vitiate the good intentions. I am no architect, merely an open-eyed naturalist. But certain strategies seem apparent to me for increasing the success and joyful use of green spaces within the built and managed footprint, while attracting and offering some of the diversity usually lost in dense cityscapes and sprawling suburbs, while offering a counterbalance to alienation in the bargain.

Here are a few ideas in that direction:

1. Incorporate building-side gardens and dooryards that invite diversity right into the precincts of the urban captive. Too often such spaces are wasted on strictly ornamental plants with few wildlife values. A hawthorn outside the window where cedar waxwings gather to strip the haws in winter, where multitudes of pollinators visit the thick white bloom in May, returns far greater dividends of diversity than a barren cypress. Good, free advice along these lines is available from the National Wildlife Federation's Backyard Habitat Program.

2. Don't ignore elements of diversity that harmlessly occupy buildings. At one university where I taught, handsome red-and-black box-elder bugs congregated in academic structures in winter as if they were native hibernal caves. Instead of using these benign and beautiful animals for education and healthy amusement, my building's managers contracted to spray them—thus threatening to introduce toxins into our workspace (until we stopped the ill-advised caper) even as the bugs were naturally ready to disperse in spring. Watching them do so might furnish a perfect transition from indoors to out as the season warms, yet such benisons of the built environment are seldom accepted.

3. Get rid of practices inimical to the retention of healthy experience in the building's immediate environs. Campuses are meant to be the very seat of tranquility in favor of scholarship; they are the Groves of Academe, after all. Leafy, mature neighborhoods with dappled sun and shadows offer a similar sense of bucolia for residents. Yet I have come to understand that nowhere is more likely to have its serenity shattered by leaf blowers than the cam-

pus or the leafy neighborhood. These screaming, stinking, two-stroke appurtenances have no place in built environments with any pretension of achieving the benefits of nature within the city wall. The degree to which they and other noisy devices are accepted, even (or especially) in self-consciously green settings, is outrageous, and utterly defeating to the purpose of place. After all, the theft of silence is part of the extinction of experience.

4. The same goes for toxins. A movement for healthy communities and schools is well under way in certain progressive cities, such as Seattle, where most biocides have been discontinued in parks and schoolyards. But the intentions of too many greenscapes are cruelly and dangerously perverted on a routine basis by the application of chemicals with known toxic impacts upon fish, amphibians, and people. One recent summer I witnessed a tractor-drawn, multi-head spray boom poisoning the immense lawns of the University of Maryland at College Park, where students walked barefooted, basked, and threw Frisbees, and soccer and cheerleader camps got underway. The quads of several campuses where I have taught have been similarly mistreated as soon as the season invited sprawling on the quad to study, snog, sun, or dream. Not only do these chemicals endanger the health of students and others, but they eliminate the variety of plants and insects that can make an unsprayed lawn interesting and far from barren. Roadside, trailside, railside, ditch and dike bank, and other such spray regimes all extinguish experience in our midst, unnecessarily and dangerously, for chemical company profit. Municipalities and landscapers that tout green values and then spray are guilty of hypocrisy in the first degree.

5. Eliminate the barbering and manicuring of open spaces in communities. Kids need real wilds near their homes, even if only in bits and scraps, and can make much of little. But the impulse to minimize the undergrowth and overgrowth, to pave the paths, and to turn every surviving glen into a bench spot or interpretive stop simply undermines the utility of open space for children. This applies to nature reserves as well. While strict preserves are necessary to conserve rarities, the imposition of fences, must-stay-on trails, and regulations render the wild spots off limits to exploring youth just as if the place in question were lost at the hands of developers. The built environment must find ways of incorporating unruly places full of water, weeds, and life, if it is to resist extinction and encourage kids and parents to connect with something beyond a TV or computer screen.

Children must have the leisure, encouragement, and security to explore such places, unsupervised except perhaps at a benign distance. Adults, in this hypertrophied commercial world, also need the ease and serenity that such kinds of places can provide.

6. Toward these ends, we must learn to venerate the vacant lot (Pyle 2002). Forest sociologist William Burch of Yale has reported that massive demolition of abandoned houses in inner Detroit has left some 60,000 lots in a feral state. While most consider this condition to be unsightly and undesirable, Burch (Grove et al. 1993) points out that it has created thousands of pocket-handkerchief urban forests of Ailanthus and other adaptable species. Few cities, in this overheated real estate climate, have the ability or courage to save vacant lots. But the default should be no new footprints on urban brownfields and other vacant lots until their potential for wildlife and kids has been fully considered. The fact is that nothing is less vacant to a curious kid than a vacant lot.

If we took the greening of the built environment seriously, we could easily enrich the radius of reach of the municipal citizen and the enburbed child. We could take a serious swipe at two of the most dangerous conditions of our time—the barrier between humans and the rest of nature, and the extinction of experience.

* * *

Biophilia is defined as the affinity for life and lifelike processes. From the work of E. O. Wilson, Stephen Kellert, and others, the word has emerged as a plausible description of an attachment to nature that is hardwired into us. Biophilia likely evolved during the early Holocene—a time mostly of climatic stability and ecological fecundity. Humans were few and nature was vast. Our affinity for life and lifelike things was honed over the ages in what was a paradise, even with predators. Try as we might we cannot recover the full sense of awe, wonder, and fear that our ancestors might have felt, but however rarified, we feel its tug even in the clutter and crowds of our human-dominated world. We fit here and we feel it in more ways than we can know.

But there are important differences in how one interprets biophilia. At one level, the evidence strongly indicates that we heal faster in the presence of nature, that we learn better and faster with natural lighting, and perhaps that we are more sociable in settings that reflect natural systems. This appears to be uniformly true across classes and cultures. It is simply how the bodies, minds, and emotions of *Homo sapiens* respond to environments with certain natural features and characteristics that resemble those that prevailed throughout most of its evolutionary career. But biophilia can also be interpreted as a desirable trait to be cultivated through education and experience. People with the right kinds of education and experience, accordingly, will be more likely to protect nature having come to bond with it and perhaps even to love it.

These two views, among others, are not mutually exclusive, but they represent very different emphases and possibilities. Assuming biophilia to be hardwired into us, but requiring neither affection on our part nor exposure to authentic nature, we might imagine that artificial settings reflecting natural features could elicit the same qualities of healing, learning, sociability, and comfort. Artificial but biophilicly designed settings might be thought to be as good as or better than the real thing. The second view, however, requires that we become attached to particular places, articulate about our feelings, and ecologically competent stewards. In the first view, people are passive, unconsciously manipulated by the designers' art in weaving together the strands of nature and human nature. The second requires an engaged citizenry who regard themselves as part of nature and take pleasure in its company. The first might, to a great extent, find their need for natural contact met in biophilicly designed buildings with plants, daylighting, and white sound and require no contact with nature on its terms. The second would find that kind of confinement, however artful, to be unfulfilling. Faced with the degradation of natural systems, the first would gladly retreat indoors, perhaps purchasing their biophilic needs in the form of carefully crafted "water parks," "wilderness theme parks," and even parlors that sell increasingly realistic virtual simulations of nature but more cheaply and without the bugs and dangers of the real thing. In such circumstances, the others would find themselves bereft and grieving.

I prefer to think of biophilia as joining both aspects; as a choice, not just an unconscious response; and as a higher and valuable, but fragile, human quality. It is clear that humans can lose their sense of attachment to nature, becoming indifferent and biophobic, whatever their unconscious response to it may be. In this interpretation, biophilia might be the basis for a broader and deeper human response to what E. O. Wilson has described as the "bottleneck," the convergence of climate change, species loss, ecological degradation, and human population growth in the coming decades. In such circumstances, biophilia—the expression of an articulate and competent affection for nature—will be essential to human survival.

Such a response, however, will become more difficult to nurture in the world we are making. Through the labors of a small army of climate scientists the picture coming into sharp focus indicates a vast change in the human condition. The media still refer to it as global warming, but it will be no such thing. It is, rather, global destabilization driven by higher temperatures, heat waves, droughts, bigger storms, changing ecosystems, novel diseases, famine, wars, economic chaos and political instability. Allowed to continue much longer, it may rapidly throw ecological and human systems everywhere into chaos.

Nature in such circumstances will become a lot less lovable. It will be regarded, rather, as hostile and alien, hence to be avoided. The nature of the Alabama coast that once sparked the excitement of a young boy nicknamed "Snake" Wilson will be far less likely to do so for other children in the century ahead. Much of that area will change radically as fire, heat waves, desiccation, bigger storms, and rising sea levels alter this place and other places beyond recognition. Children are more likely to retreat indoors, further compounding the problem that Richard Louv tellingly calls "nature deficit disorder."

We are making national and international sacrifice zones in places like the Gulf Coast and the coalfields of Appalachia. What will nature be to the children growing up in places made increasingly capricious and dangerous? We are also making the Earth itself a sacrifice zone—rendered less fecund and hospitable for subsequent generations. Other than as a science of manipulation and artificiality, will our innate sense of attachment to nature merely flicker for a while, atrophy, and finally gutter out?

* * *

We believe that biophilia, as a design art, means the careful integration of the full range of our senses with particular places and spaces. Between concept and application, however, lie many challenges. This practice obliges us to identify the integral (but often unnoticed) features of places that support health, well-being, and belonging. It asks us to weave these threads into a pattern of reengagement with wider nature rather than mere manipulation of human nature. Biophilia could become the basis for a science of design that broadly informs architecture and landscape architecture of schools, hospitals, commercial buildings, housing, parks, and countryside. But it is also possible that we will document the presence and power of biophilia in great detail just as the conditions that gave rise to it are lost.

Preservation of our innate affinity for life means preserving the conditions that allowed our kind to evolve and flourish, and the reverse is just as true. But to do so we will have to learn how to manage (or cooperate with) the big systems of Earth, notably the carbon cycle—and that will require monetizing, legislating, and ethicizing the use of carbon. And we will have to extend the idea of biophilia in time to ensure that our presumed "natural rights"—life, liberty, and property—come to include the right of every child to intimate experience with authentic nature. The children whose lives have been devastated by Katrina. and those who will be brutalized by larger storms to come; the children growing up in coalfields, or in third-world barrios; and the quarter of a billion children that the United Nations says are working as slaves in the global economy: few of these will ever hear the word biophilia, or experience the potential richness of the world. Even the children of privilege and relative security will have less and less to do with the actual, ecstatic, out-of-doors world. Unless, that is, the extinction of authentic experience comes to be seen as an intolerable cost to our descendants and to humanity. Only then might we act to avoid the worst of what could lie ahead.

REFERENCES

Finch, K. 2003. "Extinction of Experience: A Challenge to Nature Centers (Or, How Do You Make a Conservationist?)" including sidebars by G. T. Maupin. *Directions*, October–December, 2–10.

Grove, M., K. E. Vachta, M. H. McDunnough, and W. R. Burch. 1993. "The Urban Resources Initiative: Benefits from Forestry." In *Managing Urban and High-Use Recreation Settings*, edited by P. Gobster. General Technical Report NC-163. St. Paul, MN: U.S. Department of Agriculture, Forest Service, North Central Forest Experiment Station.

Immanuel, K. 2005. "Increasing Destructiveness of Tropical Cyclones over the Past Thirty Years." *Nature*, August 4, 686–688.

Kahn, P. H., Jr., and S. R. Kellert, eds. 2002. *Children and Nature: Psychological, Sociological, and Evolutionary Investigations*. Cambridge, MA: MIT Press.

Kerr, R. 2006. "A Worrying Trend of Less Ice, Higher Seas." *Science*, March 24, 1698–1701.

Louv, R. 2005. *Last Child in the Woods: Saving Our Children from Nature Deficit Disorder*. Chapel Hill, NC: Algonquin Books.

McHarg, I. 1992. *Design with Nature*, 25th anniversary ed. New York: Wiley. (First published 1969).

Nabhan, G. P., and S. Trimble. 1994. *The Geography of Childhood: Why Children Need Wild Places*. Boston: Beacon Press.

Pyle, R.M. 1978. "The Extinction of Experience." *Horticulture* 56(1): 64–67.

———. 1993. *The Thunder Tree: Lessons from an Urban Wildland.* Boston: Houghton Mifflin.

———. 2002. "Eden in a Vacant Lot: Special Places, Species, and Kids in the Neighborhood of Life." In *Children and Nature: Psychological, Sociological, and Evolutionary Investigations*, edited by P. H. Kahn Jr. and S. R. Kellert, 305–327. Cambridge, MA: MIT Press.

Sobel, D. 1993. *Children's Special Places: Exploring the Role of Forts, Dens, and Bush Houses in Middle Childhood.* Tucson, AZ: Zephyr Press.

Stafford, K. 1986. *Having Everything Right*. Lewiston, ID: Confluence Press.

Sting. "Fields of gold," 1993. *Ten Summoner's Tales*. A&M Records.

Thomashow, M. 1995. *Ecological Identity*. Cambridge, MA: MIT Press.

Tidwell, M. 2003. *Bayou Farewell.* New York: Vintage.

PART III

The Practice of Biophilic Design

chapter

13

Biophilia and Sensory Aesthetics

Judith H. Heerwagen and Bert Gregory

We know that people enjoy being outdoors. This interest and psychological need to connect with nature draws us to many different outdoor settings, such as urban parks, wilderness areas, backyard gardens, and seashores. We also know that people and organizations are willing to pay for natural views from homes and workplaces (Heerwagen 2006). And we are learning more every day about the benefits to health and well-being of a connection to nature in daily life, including hospitals, schools, offices, suburban neighborhoods, and high density urban housing (see Chapter 6 and others in this volume).

Is the need to connect with nature born of urban living or a deeper value, one that has unfolded throughout the course of human evolution? If biophilia has a genetic explanation, as asserted by E. O. Wilson (1984) and others (Heerwagen and Orians 1993; Orians and Heerwagen 1992; Iltis 1968), then access to nature is a basic human need, not a culturally determined preference. Of course, a genetic basis for biophilia does not mean that it has no cultural component. The ideas developed in this chapter link the creation of a unique, local sense of place to the cultural expression of biophilia. Iltis refers to this phenomenon as the "humane human environment" which is a "compromise between our genetic heritage, which we cannot deny except at great emotional and physical misery, and the fruits of an unbelievably varied civilization which we are loath to give up" (Iltis 1968, 117).

To investigate how biophilia can be expressed in building design, we need to understand what it is about nature that creates a sense of pleasure, well-being, and engagement with place. We know from a significant body of existing research that particular elements are important, namely water, large trees, flowers, and rich vegetation (Orians and Heerwagen 1992; Heerwagen

Figure 13-1: A walk in the woods reveals the movement of light and shadow—connecting us with nature and bringing its benefits home.

and Orians 1993; Ulrich 1993). We also know that certain spatial characteristics have strong appeal, such as views to the horizon (Appleton 1975), provision of refuge and protection (Appleton 1975), and a sense of enticement that provokes exploration (Hildebrand 2000).

At the same time, many of nature's sensory and compositional attributes have not yet been analyzed. This chapter will explore how design can evoke nature's qualities, relationships, and structures without direct replication. This approach is widely relevant, but especially to buildings in urban areas that lack natural amenities. The ideas are scalable and can be applied at the room, building, and neighborhood levels.

LET THERE BE LIGHT . . . AND SOUND, ODORS, COLORS, MOVEMENT, AND PATTERNS

Nature is rife with sensory richness and variety in patterns, textures, light, and colors. All organisms respond with genetically programmed reflexes to the diurnal and seasonal patterns of sunlight and climate. All organisms distinguish between food and nonfood, predators and prey, and safety versus danger. Why would humans be any different?

Throughout evolution, our ancestors needed to respond appropriately to environmental conditions, whether to seek comfort and shelter, eat, hunt or move safely through the environment. Variation in the color and form of plants signaled edibility or toxicity, and knowing the difference was a matter of survival. Being able to forecast the weather in the setting sun, clouds, and wind guided decisions about where to set up camp for the night. As Steven Pinker writes in *How the Mind Works* (1999), "The brain strives to put its owner in circumstances like those that caused its ancestors to reproduce." Our minds have evolved to perceive and seek out beneficial places and things, and likewise to avoid their opposites. Beneficial options include environments, living organisms, and natural processes that sustain life, especially water, light, and fire.

In this chapter, we consider how our inherent connection to nature can form the basis for a biophilic approach to design, derived from the qualities of natural settings that people find particularly appealing and aesthetically pleasing. We regard biophilia (which literally means "love of life") as key to creating places imbued with positive emotional experiences—enjoyment, pleasure, interest, fascination, and wonder—that are the precursors of human attachment to and caring for place.

We also explore the sensory richness and ambient variability that is abundant in nature and in many historic buildings, but meager in our modern built environments. We recognize that stimuli and events in nature are not always perceived in a positive way. Many aspects of nature, such as violent storms, decaying animals, dirty water, and dark places elicit dislike, anxiety, fear, and avoidance and fall into the category of "biophobia." Biophobic features, while not the subject of this chapter, could be employed to induce avoidance behavior regarding hazardous places or toxic substances. For instance, Mr. Yuk(tm) stickers are widely used to convey the message, "Don't ingest this."

Unlike the modern era of sight-dominant architecture (Pallasmaa 2005), natural environments demand the integration of information from all five senses. Pallasmaa describes the senses as "aggressively seeking" mechanisms rather than passive receptors, a description well suited to the idea that our sensory system evolved to process data from our surroundings. Pallasmaa argues that modern architecture's "occular bias" creates buildings with "a striking and memorable visual image" at the expense of other sense modalities. "In my view," he writes, "the task of architecture is 'to make visible how the world touches us,' as Merleau-Ponty said of the paintings of Cezanne" (Pallasmaa 2005, 56). A rich sensory environment surrounds us, not just with visual delight, but also with sounds, haptic sensations from the feel of wood or stone, and variations in temperature and light as we move through a space.

NATURAL AESTHETICS

Frank Lloyd Wright looked to the natural world for inspiration:

> A sense of the organic in Nature is indispensable to an architect; where can he develop this sense so surely as in this school? A knowledge of the relation of form to function is at the root of his practice; where can he find the pertinent object lessons Nature so readily furnishes? Where study the differentiations of form that go to determine character as he may study them in the trees? (Twombley 2003, 8)

Similarly, Louis I. Kahn's work has been compared to the aesthetics of a flower, an explanation for which is found in his words:

> When sight came, the first moment of sight was the realization of beauty. I don't mean beautiful, or very beautiful, or extremely beautiful. Just simply beauty itself, which is stronger than any adjectives that you might find to add to it. It is total harmony without knowing, without reservation, without criticism, without choice. It is a feeling of total harmony as though you were meeting your maker, the maker being that of nature, because nature is the maker of all that is made. You cannot design anything without nature helping you. (Lobell, 1979)

Nature clearly inspires many designs, yet it is less obvious which aspects of nature evoke inspiration and why. Perhaps the answer awaits us in a walk through the woods. Nicholas Humphrey (1980), in an article on the evolution of aesthetic sensitivity, advises designers:

> Go out to nature and learn from experience what natural structures men find beautiful, because it is among such structures that men's aesthetic sensitivity evolved. Then return to the drawing board and emulate these structures in the design of your city streets and buildings.

In the remainder of this chapter, we take Humphrey's advice and look at how architecture can evoke the qualities of nature through the use of light, air, materials, color, spatial definition, movement patterns, openings and enclosures, and connections to the outdoors. We begin with a discussion of seven attributes of nature that form the structure of our approach:

Sensory richness
Motion
Serendipity
Variations on a theme
Resilience
Sense of freeness
Prospect and refuge

We recognize that these attributes often overlap and that boundaries between them are often blended rather than discrete. Nonetheless, it is useful to discuss them as separate features for the sake of creating a new palette for design.

SENSORY RICHNESS

Natural environments have an abundance of odors, sounds, tastes, smells, haptic sensations, and visual patterns that fluctuate with time (daily and seasonal) and weather. This creates cyclical patterns, as well as irreg-

Figure 13-2: Outdoor environments reveal overlapping and complementary colors, sounds, scents, and seasonal variations.

ular variation, such as the change in ambient air quality after a storm. Seasonal variation in light, temperatures, and rainfall form the basis for major shifts in behavior and psychological states. In northern habitats, for example, reduced daylight in the winter is associated with increased levels of Seasonal Affective Disorder (SAD), characterized by "light hunger." Strategies to relieve symptoms include bright light therapy, better access to indoor daylight, and increased time outdoors (Heerwagen, 1990). (See Figure 13-3 in color insert.)

For much of human history, hunting and gathering groups have had seasonal campsites associated with the availability of resources. Agricultural societies, in turn, have celebrated the beginning and end of the growing season with communal events. And who hasn't enjoyed the first warm spring day when open windows and doors let the sounds and odors of nature sweep through the house! Alas, most of our buildings with their sealed envelopes do not allow our spirits this seasonal reprieve.

Diurnal changes also abound in nature. Variation in brightness, color of sunlight, the angle of light entering a space, the color of the sky, temperatures, air movement, and sounds vary from morning to night, starting with quietness in the morning, activity during the day, and quietness again at night as nature settles down.

The environment's sensory qualities also vary with local conditions. A walk through the woods readily reveals differences to the keen observer. The same species of flower or tree may exhibit different growth patterns and colors due to alterations in soil, light, and water conditions. Furthermore, the environment's overall sensory load changes across habitat types. Odors and colors can be very intense in a jungle, but muted in a prairie or desert landscape.

MOTION

Nature is always on the move. A walk in the woods reveals many kinds of activity—birds and small animals stirring, leaves rustling in the breeze, water gurgling in a stream, light and shade shifting as the tree canopy opens and closes to sunlight, early morning mist lifting, and clouds responding to air currents. Additionally, our own movement through the woods creates sensations that we would not experience as passive observers. As pleasant as the walk may be, the experience can change dramatically with a sudden storm. The soft breeze becomes a violent rush of air while the gentle movement of leaves gives way to erratic thrashing and bending of branches. Quietness is replaced with crashing thunder, and the soft interplay of light and shadow is replaced with gloom and bursts of lightening, creating a sense of peril that makes refuge all the more appealing, as Hildebrand discusses (see Chapter 16). (See Figure 13-5 in color insert.)

As this walk in the woods shows, all movement is not equally pleasing. Motion that ebbs and flows in a rhythmic way is soothing and pleasant, while sudden erratic activity is alarming. Psychiatrist Aaron Katcher refers

Figure 13-4: Nature exhibits different movements, from slow rustling to erratic thrashing, which elicit different human responses.

to motion perceived positively as "Heraclitean movement" and describes it as "always changing, yet remaining the same." A fish tank, for example, readily captures attention and ongoing fascination. Studies by Katcher show that not only do people enjoy watching the fish, they also experience reduced stress and improved relaxation in clinical settings (Katcher and Wilkins 1993), leading him to speculate that Heraclitean movement signals comfort and safety. By contrast, fast, erratic movement, such as that shown by schools of fish when a predator approaches, causes concern, arousal, and tension.

Biophilic design should also take into consideration the motion of people. The favorite human pastime of watching people move through an outdoor urban space can be explained, because the movement has the same rhythmic pattern as fish in benign environments and is every bit as fascinating. William Whyte's *Social Life of Small Urban Spaces* links people to spatial amenities and attributes in ways that are highly consistent with biophilic design; for example, he evaluates locations from the perspective of refuge, vegetation, water, food, and multiple view corridors. Even views of highway traffic from an elevated position create an appealing visual movement—a sensation that is probably not experienced from the perspective of the driver! In this case, distance matters a great deal.

SERENDIPITY

Ephemeral and unexpected encounters with animals, vegetation, and spatial features are common in natural environments. The sudden appearance and disappearance of a deer in the woods or of a rare bird elicit interest and even joy. An unusual rock outcropping, a ring of mushrooms, a snake underfoot, a rare wildflower all cause people to stop and look more closely. The switch in focus from the larger, ambient environment to a specific element often leads to closer inspection of immediate surroundings and discovery of other details that are missed when attention is diffuse. (See Figure 13-7 in color insert.)

Incentives to shift attention from the large to the small, from the wide environment to the close-at-hand, are design features in many Japanese gardens. The manipulation of perspective is often subtle, such as a change in the texture of the path or the sound of water flowing through a wooden trough. But the effect is predictable: people inevitably stop and search for the sound, or look down at their feet and, in doing so, discover another element that they would have missed otherwise.

In nature, there is much to discover by looking closer. A stone from standing height may look like a gray blob, but on closer inspection it reveals beautiful textures, colors and patterns. Who can walk along the water's edge without hunting for the perfect rock or

Figure 13-6: The unexpected sneaks up on you in nature, from small changes in texture or color to an animal drinking from a pool just around the bend.

skipping stone? No one needs to teach us to do this. It is intuitive and begins in childhood as soon as mobility is possible. A nature walk with a toddler is a rediscovery of small worlds and artifacts that we tend to overlook in our hurry to get from one place to another.

We see examples of serendipity in building ornamentation or in the rays of light that enter a building at a certain time of day, at a certain angle, illuminating surfaces and creating ephemeral patterns. We also experience serendipity in public places, such as the pig "footprints" in Seattle's Pike Place Market that meander outward from a beloved pig statue at the entry to the Market. The statue has become a favorite place for family photos because kids (and adults!) can sit on top of Rachel the Pig to pose.

A desire for serendipitous experience accounts for much behavior in urban settings, especially movement without a specific goal in mind (Gehl 2001). Browsing, wandering, and shopping all are such behaviors, and they are highly influenced by the inviting quality of the physical environment. Shopping mall design is a perfect case in point. The new "life style centers" feature water elements, village-type settings, pleasant landscaping with large trees, meandering pathways, multiple places to see and be seen, and a multiplicity of shops and restaurants.

VARIATIONS ON A THEME

Natural elements—trees, flowers, animals, shells—show both variation and similarity in form and appearance due to growth patterns. Humphrey refers to this phenomenon as "rhyming" and claims that it is the basis for aesthetic appreciation—a skill that evolved for classifying and understanding sensory experience, as well as the objects and features of the environment. He writes, "beautiful 'structures' in nature and art are those which facilitate the task of classification by presenting evidence of the taxonomic relationships between things in a way which is informative and easy to grasp" (Humphrey 1980, 63).

Similarities, not duplications, in patterns of various scales can also aid overall comprehension, especially in unfamiliar spaces. And because we find this type of patterning beautiful, it remains appealing with repeated experience. Smith (1980) discusses comparable concepts and suggests four genetically influenced "aesthetic programs": sense of pattern, appreciation of rhythm, recognition of balance, and sensitivity to harmonic relationships. He argues that pattern and rhyming are particularly strong in vernacular design styles. Over time, traditional crafts and materials create a rhythmic visual "growth" pattern that unfolds as one moves through the space. (See Figure 13-9 in color insert.)

Many aspects of the natural environment show fractal structuring, defined as self-similarity at different scales. Fractal structuring, found in trees, clouds, waves,

Figure 13-8: Rhythms and patterns create both balance and a sensitivity to environmental relationships.

snowflakes, coastlines, rock patterns, and rivers, has been referred to as "the fingerprints of nature," as well as the "new aesthetic" (Spehar et al. 2003). A rapidly expanding body of research is investigating the relationship between fractal dimensions and aesthetics in landscapes, skylines, pavement patterns, and urban environments (Hagerhall et al. 2004; Spehar et al. 2003; Mikiten et al. 2006). Growing evidence indicates that intermediate fractal structures (neither too simple nor complex) are generally preferred, whether in natural or built elements and scenes. Clouds, waves, parks with scattered trees, and many woody plants and trees are all intermediate fractal forms in nature (Spehar et al. 2003). Aesthetically pleasing buildings, such as the domes or "interior skies" of Persian architecture, demonstrate intermediate fractal patterning in the built environment (Sarhangi, 1999). Although the Persian domes are highly complex visually, they are not perceived as chaotic due to their high degree of patterning (or "patterned complexity," according to Hildebrand, 1999). Some researchers speculate that this visual system—showing a preference for fractal rather than Euclidean geometry—evolved in response to its prevalence in nature (Mikiten et al. 2006; Gilden et al. 1993.

Salangaros and Masden (see Chapter 5) argue that built environments with fractal features are "neurologically nourishing," because they reconnect humans with biologically preferred environments and natural elements. "Human beings connect physiologically and psychologically to structures embodying organized complexity more strongly than to environments that are either too plain or which present disorganized complexity." (Salangaros and Masden).

RESILIENCE

Many natural systems and species show a high degree of persistence in the presence of perturbations and disturbances. Krebs (1985) describes this resilience as a community-level concept concerned with how much disruption a group can tolerate before it shifts to another configuration. At the same time, it is important to recognize that some level of perturbation is actually desirable to prevent one species from eliminating all others.

Resilience is also affected by the web of relationships that connect the composition of species within an ecological community. Waste from one animal becomes food for another; unused space becomes a niche for a newcomer; decaying trees become resources and living spaces for a variety of plants and animals. Deaths and births, as well as migrations in and out of a community, maintain the overall species composition of a particular community. (See Figure 13-11 in color insert.)

The built world, however, is less resilient to natural forces and stresses. It is worth exploring how buildings

Figure 13-10: Living things create a web of resilient connections—where waste becomes resource, empty space becomes habitat.

could be more resilient, perhaps through characteristics of nature, such as the ability to bend in the wind. In this regard, biophilia and biomimicry begin to overlap by integrating natural adaptations with aesthetic appeal.

SENSE OF FREENESS

Natural environments offer many choices and opportunities to support behavioral and emotional needs, and rarely funnel behavior in a particular direction. The lack of boundaries between spaces expands sensory awareness, and creates both a psychological and physical sense of freedom. Even where boundaries do exist, such as water or mountains, they are permeable and can be crossed, although perhaps requiring special supports.

Figure 13-12: In nature, boundaries are permeable and movement is free.

Barriers and blockades that reduce sensory connections inside the building, as well as between the building and the outdoors, hinder a sense of freeness in today's built environment. How can we create a sense of freeness when concerns with security and safety abound, keeping doors locked, windows shuttered, and walls everywhere. Appleton (1975) and Hildebrand (see Chapter 16) show that freeness can be evoked even with modest interior openings, multiple view corridors, and the opening up of interior and exterior vistas as one moves through space. (See Figure 13-13 in color insert.)

In human experience, control of one's destiny applies not only to choosing life's path, but also to interacting with the physical environment. In the most elemental terms, being locked in a room without a key is not being free. An open door allows for escape. Perhaps in response to primal conditions, an open door to the outside signifies fresh air and the ability to move towards the sun. Additional options for movement amplify choice, and thereby, a sense of freeness. When paths are linked directly to the outdoors, such as to an exit, deck, or patio, a sense of freeness increases. The ability to open a window in one's environment is one of the simplest actions that signify control or freeness.

PROSPECT AND REFUGE

Appleton refers to the confluence of prospect (visual access) and refuge (enclosure) as "the ability to see without being seen." He describes this attribute as a fundamental response to the environment associated with protection and hazard surveillance (Appleton 1975). Although Appleton argues that the most appealing places provide prospect and refuge simultaneously, there are times when either just high refuge or high prospect may be very desirable. For instance, teenagers often prefer to gather in prospect-dominant open spaces where they can both see and be seen. Similar motivations have guided the design of large open plazas in Europe as gathering places for interacting with and

Figure 13-14: From hilltop views to cozy, enclosed spaces, all animals benefit from places of prospect and refuge.

watching others. Refuge, by contrast, is sought when people are ill, tired, or just want privacy to withdraw and replenish their psychological and physical resources. (See Figure 13-15 in color insert.)

In creating prospect and refuge in the built environment, the specific context matters a great deal. Context determines where on the continuum, from very open to very closed, the environment should lie, as well as the design features required to create both prospect and refuge. Appleton spends a great deal of time in *The Experience of Landscape* describing the multiplicity of ways to achieve prospect and refuge. He offers a rich palette of materials, light, openings, screenings, gaps, peepholes, changes in height, overhangs, implied horizons, and borrowed elements from external prospects, such as views of a tower or hilltop.

In Appleton's view, prospect and refuge are less powerful when hazard is absent. He writes, "To abolish the hazard all together is to deprive the prospect and refuge of their meaningful roles" (Appleton 1975, 96). This presents a problem for building design where the presence of hazards in any form is undesirable. Are there ways to create the illusion of hazards to provide a modest psychological tension that enhances the emotional appeal of the refuge? Hildebrand (1991) describes such an example in his analysis of Frank Lloyd Wright's Fallingwater. The house is cantilevered over a rushing brook that can be seen from numerous interior and exterior vistas, thereby reinforcing the value of the house as refuge (also see Chapter 16).

The experience of prospect and refuge is not static. In Appleton's view, the experience is enhanced by moving through the landscape where vistas suddenly open or close, where prospect is afforded by tree canopies or cliff overhangs, and where the horizon appears and disappears from view. The horizon, in particular, plays "a very special role in the imagery of prospect" (Appleton 1975, 90). Contemplation of the horizon stimulates the imagination with speculation of what lies beyond. From a biological perspective, the horizon is also the point at which important information appears, like gathering storm clouds or the setting sun, which motivate action to avoid hazards and seek refuge (Heerwagen and Orians 1993).

THE DESIGNED ENVIRONMENT

Biophilic qualities exist in the human built environment at various scales, as they do in nature: the small flower tossed gently into a stream, the stream as it flows through a sunlit valley, the sunlit valley as it moves to the ocean. Although different in size, they are interconnected though the entirety of nature. In the same way, a building, courtyard, neighborhood, and city each can have biophilic attributes in materiality, form, space, and connectivity to nature.

Clear parallels exist between the ways in which we

humans interact with the natural environment and with the built environment that we design, construct, and live in. We experience the natural environment by moving through solid and void, material and space, or as Kahn would say, "silence and light." As one begins to understand the concept of biophilia, it is important to explain how our responses to nature characterize the biophilic attributes of buildings. Rudolph Arnheim (1997) helps to illustrate the notion of solid and void:

> A building or complex of buildings seen from the outside has the all around completeness of solids. Dynamically it displaces space, as an object displaced water in the bathtub of Archimedes. It expands radially from its center. . . . In an interior the hollow matters more than the material walls.

As we discuss biophilic design and the built environment, we must look at form, material, and space.

Figure 13-16: The forest's principals provide paramount guidance in building design, where we learn lessons taught by canopies, branches, and even the sky.

THE BUILDING

> The pilgrim, perhaps drawn to Thorncrown by a sense of discovery, is struck by the silence of its setting. On approach, leaves obscure the distance. Around a bend in the path, the chapel appears, caught by the sunlight like the largest tree. It seems both man made and natural. (Robert Ivy 2001, 32)

Rightfully identified as a twentieth-century masterpiece by the American Institute of Architects, the inspiration for Fay Jones's Thorncrown Chapel in Eureka Springs, Arkansas, was the upper chapel of the Sainte Chapelle in Paris, a small light-filled Gothic space near Notre Dame on the Île de la Cité. In the case of the Sainte Chapelle, stone "trees" form a structural canopy protecting the congregation, with an early evening sky seemingly peeking through the branches of the ceiling. To the side, massive amounts of sunlight fall into the space through large stained-glass windows in a proportion unusual for a thirteenth-century church. Is it a forest, or is it a building?

Jones worked in Frank Lloyd Wright's studio as a young architect, and the influence of Wright and his fractal geometries on Thorncrown's design are apparent. Although a student of organic architecture, Jones did not believe in copying the forms of nature, but rather in using nature's principles. He viewed the elevation of the human condition as paramount in architecture.

> I am not trying to imitate in any way. Architecture should announce the presence of art. It is human made.—Fay Jones (quoted in Ivy 2001, 25)

Wright, a disciple of nature, maximized the use of natural materials, and often used thirty- or sixty-degree angles in patterns that became rich in emotive quality. Much like a leaf, with its pinnate venation moving out from the fold, these geometries built up into an organic whole. Jones's structural geometries likewise build upon each other to amplify a variation of the geometric theme. For Thorncrown, rhythm is enhanced using materials of a size that could be carried by hand to the site so that larger pieces could be built up from smaller ones.

Much like sitting under the forest canopy, Thorncrown blurs the boundary between inside and outside, and offers seemingly endless options for escape through the transparent glass spaces in its structure. As one is able to move in many directions in a forest, so a sense of freedom at Thorncrown enriches the soul. The feeling of being outside is amplified by natural materials, natural light, and large living ferns at the altar that bring nature indoors.

The canopy of the wooden structure, the immediacy of the forest outside, and the light undisturbed by screen or shade, creates a rich blend of sensory experience. Motion is constant, as trees drift in the wind, and clouds and sun interplay with shadows inside the space. Walking through Thorncrown adds human movement to the equation, compounding the sense of being under a forest canopy that sways and creates shifting patterns of light. Sitting in the pews of Thorncrown is like sitting on a rock in the sun-drenched forest, connected to nature. Here, the serendipity of a passing deer is expected, yet one feels safe.

Figure 13-17: Natural materials, sunlight, and woodland plants bring the immediacy of the surrounding forest inside.

THE SPACE

As one begins the architectural journey to Kahn's Salk Institute in La Jolla, California (built in 1965), the anticipation begins. It is not anticipation of the science that is housed within the walls, nor of the people whose minds invent knowledge. Instead, it is anticipation of the space, vista, material composition, attributes of nature, and knowledge that experiencing this icon imparts, making it meaningful in its own right.

According to Kahn, "All material in nature, the mountains and the streams and the air and we, are made of Light, which has been spent, and this crumpled mass called material casts a shadow, and the shadow belongs to Light."

Perched on a cliff overlooking the Pacific Ocean, the Salk Institute courtyard looks west to an endless, peaceful vista of deep blue sky. Named "roofless cathedral" by writer John Lobell in "Silence and Light," it offers views of enticing sky rather than ocean. At its center, a water-filled channel travels to the western edge of the space, enticing one to a sheer drop-off. However, the courtyard also provides refuge from the precipice, as do the canyon walls formed by the researchers' offices along either side. The structure feels solid and safe, yet is interspersed with voids as if in a forest—a sense enhanced by the warmth of natural wood upon the high walls.

Figure 13-18: The "roofless cathedral" reminds visitors of its surrounding natural environment, inviting us to its edge.

The power and beauty of this courtyard is immeasurable. Simple, complex, and sensory-rich, the quality of light within the space lifts the spirit. The color of the honed travertine stone, which one walks upon, gently reflects a spectrum of light that creates a soft radiance. Its subtle, natural patterns are not repetitive, but rather have a fractal relationship, as does the deliberately elegant rhythm of the slabs throughout the court. The slabs' patterns relate to the rhythm of the structure of the office blocks, as well as to the form marks and teak screens on the walls themselves.

Like the scale of a forest meadow, the articulating walls of the office blocks form a comfortably proportioned space. The walls evolve with the movement of the teak sun screens that shade the small offices from the sun. Concrete, wood, stone, water, light, and a few living things create a simple material palette in harmony with the ground and sky. Looking to the sea, this simplicity enhances the senses, and makes one much more aware of the movement of the sun, clouds, the sound of the water, and fresh breezes.

Figure 13-19: Simple, natural materials keep the structure in balance with the enveloping sun, clouds, water, and wind.

> By and large, what Kahn thought to be of primary importance [was] the past and the innate characteristics of materials, color, water, light, and nature itself. (Twombly, 2003, 10)

The space is quiet, but movement is assured through access to the ocean and the serendipitous coming and going of people, all enhanced by water flowing in a channel to the endless sky. Although simple, it is rich in sensory experience. In Wright's buildings, the ceiling—simple, light, and supported by mass—forms the sky, whereas here, the room's ceiling does not stop. One feels in control in this space, with options for how to move through it. This sense of freedom connotes a biophilic space. Not only does one feel the power of a connection to nature, but also to the human soul.

THE CITY

> When you've walked up the Rue la Paix at Paris,
> Been to the Louvre and the Tuileries,
> And to Versailles, although to go so far is
> A thing not quite consistent with your ease,
> And—but the mass of objects quite a bar is
>
> To my describing what the traveller sees.
> You who have ever been to Paris, know;
> And you who have not been to Paris—go!
>
> —John Ruskin, *A Tour Through France*

During a delightful Seattle lunch with a few alumni of Yale's *Bringing Buildings to Life Symposium*, the question arose, "What is a biophilic city?" Grant Hildebrand quickly answered, "Paris." With that single word, one could almost smell a fresh baguette, see the texture of the buildings, and feel the spray of the fountains.

The romance of this great, sensual city starts with

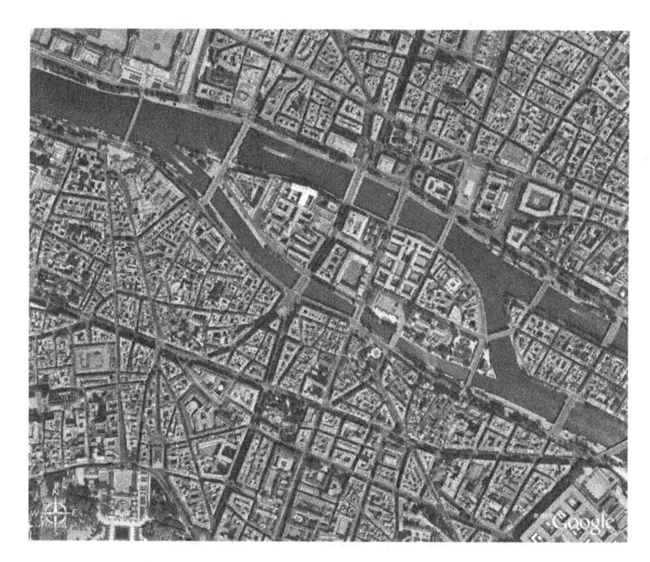

Figure 13-20: Paris is a city of movement vitally connected to, and complemented by, the Seine.

movement. It is the movement of the Seine, where refuge could be found centuries ago on an island in the river, the Île de la Cité. Here the meandering, constantly moving river creates the biophilic backbone of Paris. The urban plan responds to this natural element, forming around it, not over it.

Nothing is more wonderful than a walk in the Jardin des Tuileries on a misty morning in this magnificent city. Within its urban context, the garden is almost equivalent in scale to an African savanna. Open vistas, huge trees, gardens, and the crunching sound of the natural crushed stone, as one walks to the Louvre, offer direct access to nature. In contrast to the surrounding tall buildings, the experience actually suggests the savanna.

Paris originally formed according to the scale and speed of people walking or on horseback—according to the needs of living things, in other words. A morning stroll in the neighborhood of Montmartre, along the tight path, curving lane, and sudden jog of rue Ravignan, gives one the feel of this city's organic genesis. Although impacted by the automobile over the years, the creation of Paris's city center took place more than a thousand years before cars. This is evident in its dense arrangement of buildings with historic pedestrian tributaries flowing to meadowlike parks. Growing unchecked like a natural forest, Paris acquired its modern layout when Napoleon one day cut broad swaths of boulevards across it.

Figure 13-21: Buildings hover over streets like trees soaring over forest paths.

However, walking on lanes in the tightest portions of the city is much like walking on a trail in a dense forest. Spatially, the exterior street becomes the dominant element, much like Kahn's court at the Salk Institute. Here in the street, movement occurs constantly, and is sensed constantly. As Rudoph Arnheim notes,

> The hollow of the street canyon also accomplishes something that I shall soon describe as a prime quality of interiors, namely it acts as an exhilarat-

ing extension of man into surrounding space. Although man is only a small creature in comparison to the openness around him, he generates perceptual forces that permeate the environment. This enables him to experience the street channel as filled by a blown-up self-image, which invades space in all directions and also anticipates forward movement. (Arnheim 1997, 78)

The anonymous walk through a spatially undefined American suburb has none of the romance or serendipity experienced in Paris at Montmartre, the Left Bank, or the Île de la Cité. The bumper of an SUV suddenly appearing in the cul-de-sac has a different type of spontaneity than pedestrians brushing past in a Parisian sidewalk café.

Looking out from high atop Paris gives one an unusual organic sensibility. Its wholeness comes from historic street patterns, delicate modulation of rooftops, and subtleties in coloration. In the city's historic parts, satisfying rhythms come from the same textural variations found in nature. Even its defense walls evolved like a mollusk that sheds its shells as it grows. The serendipity of the Eiffel Tower, the Pompidou Centre, and Pei's pyramid at the Louvre enrich this sense of wholeness and inspire great affection for the city.

Figure 13-22: Paris's wholeness is realized through its patterns, colors, textures, sounds, and sights.

Modulations that subtly vary the city's consistent height, texture, and scale give Paris an overall fractal quality that points to long-term survival. With this resilience, it has been able to endure the introduction of new species, such as La Défense, the business district west of the city itself.

While strolling through the tighter parts of central Paris, one could imagine walking in river canyons etched from a vast plateau of stone. The city's mid-nineteenth-century reconstruction created new meadows and wider canyons within this dense "forest." Like the tributaries of a river that turn left or right depending on natural obstacles, Paris's original lanes flowed to destinations adjusted by centuries of ownership. The older rues curve to form slow opening vistas, with mystery and enticement around every corner.

Ah, the romance of Paris!—a biophilic city.

A WALK THROUGH A BIOPHILIC BUILDING

We end this chapter as we started—with a walk. This time, however, we travel through an imaginary building that conveys the attributes of nature.

As we approach our destination, we walk on a delightful urban street full of flowering native plants, shaded with trees. We hear a songbird and rustling leaves. The geometric sidewalk planters are part of a natural drainage system, cleaning the rainwater from the public realm and the building before it heads to a stream.

The building's exterior walls are carved in materials that remind one of the mountains nearby, casting fine shadows in the textural detail. Its form seems to be reaching for the sun, while delicate screens guard against the hot rays. We turn, step on local stone and get a glimpse of the lobby. We sense safety, place our hand on a warm wooden door handle, and enter.

The light is differential within the comfortable two-story space, as sunlight from above and behind casts a pattern on the wall and floor. The air inside is fresh. We feel a passing breeze and look to see a moving ceiling fan and open windows. The structure of the space is clear, much like a tree, as the forces of nature are expressed within the concrete columns and raw steel beams.

The lobby's walls have a subtle random pattern formed from wood with a bronze patina, and we see a small note about its origin from an old warehouse on this site. We smell coffee, see bright red tulips, and hear a fountain nearby. We look forward across the native stone floor, and seeing our friend through the glass-walled elevator, we smile.

Another breeze, a warm ray of sun, and movement. Attributes of nature, inside.

REFERENCES

Appleton, J. 1975. *The Experience of Landscape*. New York and London: Wiley.

Arnheim, R. 1997. *Dynamics of Architectural Form*. Berkeley: University of California Press.

Gehl, J. 2001. *Life Between Buildings: Using Public Space*. 5th ed. Copenhagen: Danish Architectural Press.

Gilden, D.L., M A, Schmuckler, and K. Clayton. 1993. The Perception of Natural Contour. Psychological Review. 100:460–478.

Hagerhall, C. M., T. Purcell, and R. Taylor. 2004. "Fractal Dimension of Landscape Silhouette Outlines as a Predictor of Landscape Preference." *Journal of Environmental Psychology* 24:247–255.

Heerwagen, J. H. 1990. "Affective Functioning, Light Hunger and Room Brightness Preferences." *Environment and Behavior* 22(5): 608–635.

———. 2006. "Investing in People: The Social Benefits of Sustainable Design." Paper presented at *Rethinking Sustainable Construction '06*, Sarasota, Florida, September 28–30.

Heerwagen, J. H., and G. H. Orians. 1993. "Humans, Habitats and Aesthetics." In *The Biophilia Hypothesis*, edited by S. R. Kellert and E. O. Wilson. Washington, DC: Island Press, Shearwater Books.

Hildebrand, G. H. 1991. *The Wright Space*. Seattle: University of Washington Press.

———. 1999. *The Origin of Architectural Pleasure*. Berkeley: University of California Press.

7Iltis, H. H. 1968. "The Optimum Human Environment and Its Relation to Modern Agricultural Preoccupations." *The Biologist* 50:114–125.

Ivy, R. A., Jr. 2001. *Fay Jones*. New York: McGraw-Hill.

Katcher, A., and G. Wilkins. 1993. "Dialogue with Animals: Its Nature and Culture." In *The Biophilia Hypothesis*, edited by S. R. Kellert and E. O. Wilson. Washington, DC: Island Press, Shearwater Books.

Krebs, J. C. 1985. *Ecology: The Experimental Analysis of Distribution and Abundance*, 3rd ed. New York: Harper & Row.

Lobell, J., 1979. *Between Silence and Light: Spirit in the Architecture of Louis I. Kahn*. Boston: Shambhala.

Mikiten, T. M., N. A. Salingaros, and H.-S. Yu. 2006. "Pavements as Embodiments of Meaning for a Fractal Mind." In *A Theory of Architecture*, edited by N. A. Salingaros. Solingen, Germany: Umbau-Verlag.

Orians, G. H., and J. H. Heerwagen, 1992. "Evolved Responses to Landscapes." In *The Adapted Mind*, edited by J. Barkow, J. Toobey, and L. Cosmides. New York: Oxford University Press.

Pallasmaa, J. 2005. *The Eyes of the Skin: Architecture and the Senses*. New York: Halsted Press.

Pinker, S. 1997. *How the Mind Works*. New York and London: W. W. Norton.

Sarhangi, R. 1999. "The Sky Within: Mathematical Aesthetics of Persian Dome Interiors." In *Nexus Network Journal* 1:87–98.

Smith, P. F. 1980. "Urban Aesthetics." In *Architecture for People*, edited by B. Mikellides. London: Studio Vista.

Spehar, B., C. W. G. Clifford, B. R. Newell, and R. P. Taylor. 2003. "Universal Aesthetic of Fractals." *Computers and Graphics* 27:813–820.

Twombley, R. C., ed. 2003. "Introduction: Kahn's Search." In *Louis Kahn: Essential Texts*. New York: W. W. Norton.

Ulrich, R. S. 1993. "Biophilia and Biophobia." In *The Biophilia Hypothesis*, edited by S. R. Kellert and E. O. Wilson. Washington, DC: Island Press, Shearwater Books.

Wijdeveld, H. Th., ed. 1965. *The Work of Frank Lloyd Wright: The Wendingen Edition*. New York: Bramhall House.

Wilson, E. O. 1984. *Biophilia*. Cambridge, MA: Harvard University Press.

Figures 3-6a and 3-6b: The ventilation system of Swiss Re's London Headquarters took inspiration from the water flow structures of marine sponges.

Figures 3-8a and 3-8b: Façade paint manufactured by STO is self-cleaning with rainfall thanks to the Lotus-effect®.

Figure 3-8c: Lotus leaves are self-cleaning thanks to a nano-rough structure which causes water to ball up and "pearl" dirt away.

Figures 3-15a, 3-15b, and 3-15c: Eastgate, an office building in Harare, Zimbabwe, uses ventilation principles from a self-regulating termite mound to cool without air conditioning. Now a major study called TERMES is underway to see if the geometry of the air channels (shown in plaster) could guide the next generation of naturally cool buildings.

Figure 3-20a: Frank Lloyd Wright's Samara House in West Lafayette, Indianna, paid homage to the winged seed.

Figure 3-20b: In full spring bloom, Wright's Samara House is cradled by biophilic plantings, which include maples and their beautifully designed samaras.

Figure 4-3: Still water as an interior reflecting pool creates a calming ambiance. Haworth showroom, Chicago, IL

Figure 4-5: Effective use of water and an indoor ecosystem. The Atrium at the Clubhouse, Huntington Lakes, Delray Beach, FL.

Figure 4-6: Exterior water garden exemplified by Pavilion and Reflecting Pool, Toronto, Ontario, Canada.

Figure 4-14: Highly effective design integrating architecture and adjacent natural water element. Volme River and Hagen Town Hall, Germany.

Figure 4-16: Storm water is led to overhanging fixtures at the roof edge, which guide the water to open downspouts. The water is collected in an open channel paralleling a walkway, then guided to a constructed on-site wetland. Sidwell Friends Middle School, Washington, DC.

Figure 4-19: Stormwater routing in a residential complex directs water to an exposed stream, designed to mimic a natural scene. Arkadien Asperg Housing Estate, Stuttgart, Germany.

Figure 8-6: The compelling nature of Frank Gehry's Disney Hall in Los Angeles is not just that it is a "flowering" form and that it catches the subtlety of changing light throughout the day, but that it creates sheltered places for the most beautiful gardens, with vistas and outdoor meeting and eating spaces for the offices.

Figure 8-8: Through its evidence-based design efforts, such as the Pebble project, the hospital design community has discovered that access to nature from hospital beds, staff desks and even emergency waiting rooms has measurable benefits in reducing length of stay, medicine levels, stress, and even anger, in patients, staff, and visitors.

Figure 10-5: Selection of appropriate plants such as soft miniature conifers offers two-year-olds a play setting of tactile, fragrant stimulation.

Figure 10-10: In this naturalized toddler play garden, soft-surfaced, curving paths allow children to move through and under the plants (overhead arbors) using wheeled toys, without fear of falling onto a hard surface.

Figure 10-14: The Forest Kindergarten Isarauen (literally translated as "the water meadows of the Isar," the nearby river flowing through Munich). The kindergarten, located in a Munich forest preserve, is organized as a family cooperative and attended by children 3 to 6 years old—the younger children for half a day, the older ones for the whole day. The modest two-room building serves as an administrative office and resting place. When children arrive in the morning, they make plans with the teachers, load needed artifacts on a small cart, and take off into the woods—returning to base for lunch. (See www.naturkiga-isarauen.de).

Figure 10-15: The nearby stream is an especially attractive, peaceful place for spontaneous cooperative play. From a distance, teachers keep an eye on the children, whose striking body language expresses a sense of belonging and agency over their environment.

Figure 10-19: With help from the nearby military, the Coombes imported geological samples from many regions of the country. As it was impossible to take the children to see rocks so far from home, they were brought to the school grounds for study and play.

Figure 10-20: The site of Blanchie Carter Discovery Park before restoration—a hot, unhealthy, boring desert.

Figure 10-21: Restoration of the Blanchie Carter Discovery Park included installation of manufactured play equipment surrounded by a grove of shade trees and vegetation chosen for seasonal variety, here seen in late fall. During the six hot months of the year, children feel comfortable and are protected from the harmful rays of the sun.

Figure 10-35: Shared greenspace at St. Francis Square, San Francisco, includes a play area, lawns, mature trees, areas of shrubbery, wide paths for wheeled toys, and night lighting.

Figure 10-37: Interior block greenspace, Southside Park cohousing community, Sacramento, California.

Figure 10-39: Part of the shared interior of the block at The Meadows, Berkeley, California. Backyard fences were removed in the 1970s.

Figure 10-48: Village Homes interior pathway following a drainage swale, lined with fruit trees, provides backyard access to the main greenway.

Figure 10-55: Urban alleys closed to traffic and parking can create near-home spaces for play and socializing. Neighbors celebrating the closure of the first alley in the Patterson Park neighborhood of Baltimore, 2006.

Figure 10-56: Children playing in a woonerf, a shared street, which permits access to slow-moving vehicles (Tel Aviv, Israel).

Figure 13-3: Colors, light, and patterns mingle to create rejuvenating spaces in all types of built environments.

Figure 13-5: The measured movement of metal and wood interplay in the main reception area at IslandWood, an environmnental learning center on Bainbridge Island, Washington.

Figure 13-7: Unexpected details inspire discovery about the larger significance of water at IslandWood, an environmental learning center on Bainbridge Island, Washington.

Figure 13-9: Repeated, rhythmic features organize the senses and enhance visitors' experiences at REI Seattle.

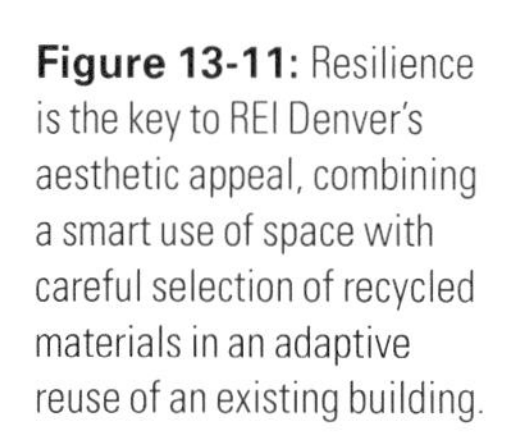

Figure 13-11: Resilience is the key to REI Denver's aesthetic appeal, combining a smart use of space with careful selection of recycled materials in an adaptive reuse of an existing building.

Figure 13-13: With large windows opening to Elliott Bay, Puget Sound, and the Olympic Mountains, boundaries blur at Mithun's office in a renovated pier on Seattle's waterfront.

Figure 13-15: At Mithun's office in historic Pier 56 in Seattle, transparency complements the privacy required for meetings.

Figure 14-5: Unified multilayer skin at Sidwell made from reclaimed fermentation barrels

Figure 14-4: Vertical and horizontal sunscreens, Sidwell Friends Middle School

Figure 14-6: Sidwell Friends Middle School, view from courtyard

Figure 14-7: Sidwell Friends Middle School, view of constructed wetland. Photograph © Peter Aaron/Esto.

Figure 14-10: View of Loblolly House from the west

Figure 14-11: View of orange glass bridge evokes the setting sun, Loblolly House

Figure 14-17: Middlebury College dining hall, exterior view

Figure 14-18: Middlebury College dining hall, interior view

Figure 15-6: The great window at Lyndhurst harbors small picture frames of the view within the larger expanse of the opening.

Figure 15-8: Frank Lloyd Wright populated the threshold of viewing with shapes implicating forms found both inside and outside of the house.

Figure 15-9: Wright's ornamented windows did not interrupt clear viewing through the lower portions.

Figure 15-10: Wright placed mementos of nature away from the window and into the center of the rooms.

Figure 15-11: In early modern architecture some interiors were brilliantly decorated in homage to nature.

Figure 15-14: The stringers of an interior staircase can dissolve into shapes innate to living forms.

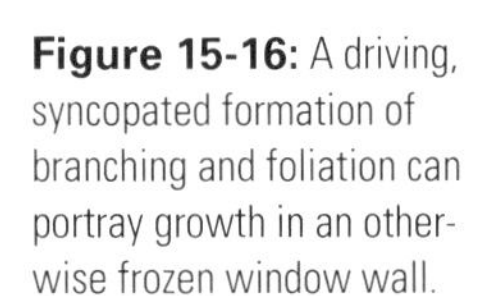

Figure 15-16: A driving, syncopated formation of branching and foliation can portray growth in an otherwise frozen window wall.

Figure 15-17: Geometry layered upon a ceiling transforms upward into a virtual flock of birds.

Figure 18-3: Highlands' Garden Village in Denver, Colorado, a mixed-use, mixed-income, higher density neighborhood of residential, commercial, and retail uses, set amongst a network of gardens and public green spaces.

Figure 18-6: New Housing in the South Bronx, New York City. The building rises from south to north to maximize southern light. The ascending roofs are gardened. The project sits near a local park.

Figure 19-1: Outdoor kitchen/breezeway. This area can be used during six months of the year as a multiuse space for dining, cooking, and other activities. The kitchen is on wheels and totally mobile, with "plug in" locations for hot water, cold water, and gas for cooking. Above on the balcony area, there is a solar cooker that cooks up to 25 pounds of food in 1½ hours, using only the sun and a little help by a human to keep the cooker on solar track. During party time, there are moments when guests are found dancing above the kitchen and on the flat roofs, using the breeze of the evening and the playfulness of a building that invites participation.

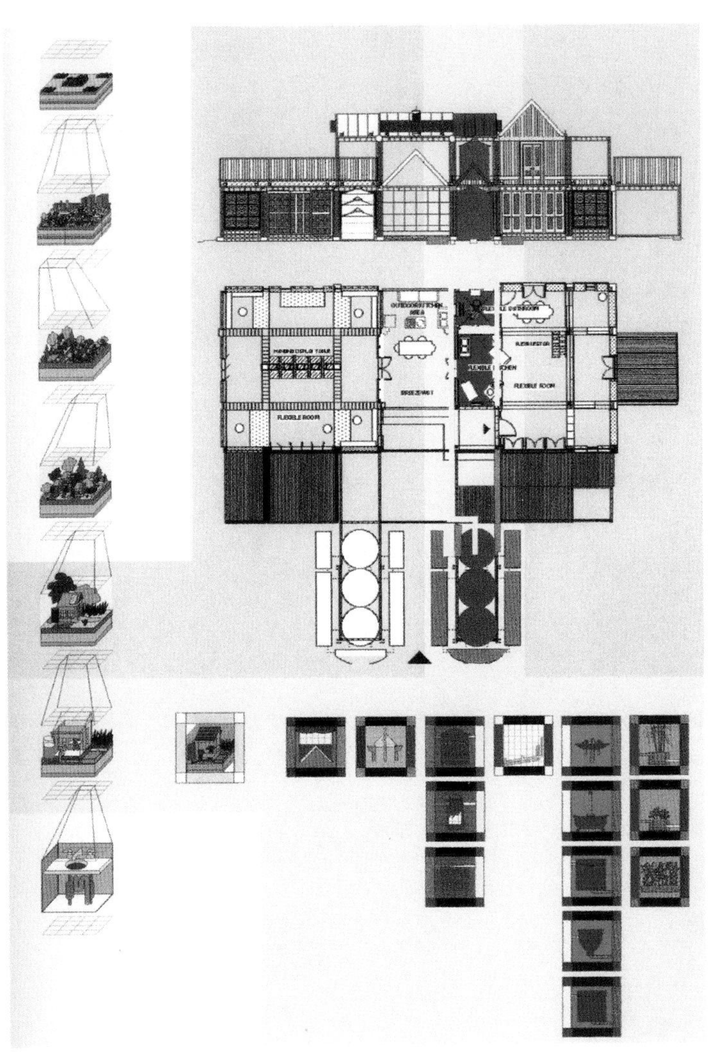

Figure 19-2: A plan showing how life-cycle events, in this case the water/wastewater life cycle, are designed as events. The color coding relates to the icons used as the elements in an object-oriented programming of water/wastewater performance in our computers—an attempt to partially bridge performance and art.

Figure 19-3: A view of the AGBD (Advanced Green Builder Demonstration) from the southwest, showing the roof-over-roof procedure, using the fact that precipitation always falls at an angle, thus catching water on both roof surfaces. The large tanks are cisterns flanked up the middle with wastewater treatment wetlands using flowering plants as the treatment system method. Photovoltaic panels are used as a shade method at the entry colonnade. All systems (energy, water, wastewater, and materials) are made visible so that the user or guest can feel a part of the experience, whether through disassembly for readjustment of structure, cutting flowers that need thinning for the wastewater treatment to function, following the water path from source to resource.

Figure 20-6: In an energy-based culture, cocreation with other consciousness becomes a powerful process of design, bringing energy and rightness into a place beyond what we can consciously conceive. In this church, the design was shifted by spirit guides into an inner focus rather than looking out into a garden—more powerful for meditation.

Figure 20-7: Even institutional structures such as banks can be transformed when we touch into the original heart of the institution—in the case of a community bank, as a means for a community to prioritize the manifesting of dreams. It also honors the great forests that cover the region, and the skills of the community woodworking industry. The design of this project started with the unique Pacific Northwest nurse log, which became the theme of the visible structural elements.

Figure 20-8: Honoring the lives of materials put into the making of our buildings shares the beauty and struggles of their lives, and makes them part of our own. Beach-combing, we often pick up things that jump out and connect with us. Making them part of our places can be good. We don't have to intellectualize why we had an urge to pick them up and drag them home. If they still attract us, use them!

Figure 21-1: A 10-foot'-diameter (3-m) Luminous 360 SkyCeiling at the CyberKnife Radiosurgery Center of Iowa in Des Moines helps reduce stress among patients undergoing treatment.

Figure 21-2: The Bank of Astoria in Manzanita, Oregon, designed by architect Tom Bender with SERA Architects, Inc., makes extensive use of natural materials.

Figure 22-3: Frank Lloyd Wright's grove of shade tree columns in the central space of the SC Johnson Administration Center

Figure 22-6: Diners in the lunchroom enjoy the connection to the lush indoor garden at the Sanitas headquarters in Madrid, Spain.

Figure 22-7: The central gathering area of the Sanitas headquarters in Madrid is filled with natural light, warm natural materials, and planted gardens.

Figure 22-10: Biophilic design sketch of vegetated facades for Council House 2

Figure 23-2: The meditation room at the Center for Well-Being is a quiet refuge, acoustically isolated from the basketball court above, finished in richly- textured materials, and lit by a subterranean light well.

Figure 23-5: At the Bank of America Tower in New York City, clear vision glass is tempered with a ceramic frit and waist-level hand rail to create a feeling of safety.

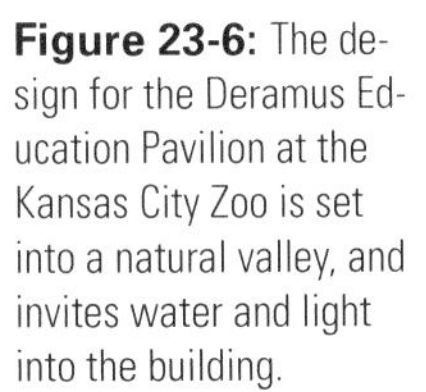

Figure 23-6: The design for the Deramus Education Pavilion at the Kansas City Zoo is set into a natural valley, and invites water and light into the building.

Figure 23-7: Interior View of the Deramus Education Pavilion illustrates the transition between natural and human environments.

Figure 23-8: A terrace-level green roof at Cook+Fox Architects orients the office toward the outdoors and draws green space into the sightlines of the workplace.

Figure 23-10: Exterior vView of the Isamu Noguchi Sculpture Court at the Bloch Building, Nelson-Atkins Museum of Art.

Figure 23-11: Interior View of the Isamu Noguchi Sculpture Court at the Bloch Building, Nelson Nelson-Atkins Museum of Art.

Figure 23-12: Exploring a modern concept of stewardship, the redevelopment of New York City's Front Street weaves together contemporary design, sensitive restoration, and subtle references to the neighborhood's whaling and maritime past.

chapter 14

Evolving an Environmental Aesthetic

Stephen Kieran

The subject here is not so much the substance of a sustainable architecture as its form, its potential for beauty.

In Nathanael Johnson's "Letter from Iowa" in Harper's May 2006 issue, the functional and aesthetic dilemma of the present environmental movement is drawn into focus through the agency of the hog:

> Bit by bit, scientific breakthroughs have emancipated the hog industry from the demands of nature, but each freedom comes at a price. Each new liberty for pork producers depends on further control, further domination of the pig. No one at the conference suggested what seemed the obvious answer: doing away with the causes of stress and lameness. But then, swine geneticists are innovators, not policy makers.
>
> In just a little more than a decade, the modern hog industry has produced a tower of efficiency-maximizing products, one stacked atop the next, each innovation fixing the problem the last fix created. It is a monumental if somewhat haphazard structure, composed of slatted floors and aluminum crates, automatic sorting scales and mechanized wet-dry feeders. It is constructed of Genepacker sows, Tylan antibiotic feed, Agro-clean liquid detergent, Argus salmonella vaccine, Goldenpig foam-tipped disposable A1 Catheters, CL Sow Replacer milk substitute, and Matrix estrus synchronizer. The scientists who add their discoveries to this edifice do not see themselves as its architects. As they see it, their job is not to shift the foundations of the hog industry but to build atop its tower of technology, masking what structural flaws they can with new construction, reaching ever upward.

What does our pig, here literally drawn in preparation for quartering, have to do with the evolution of a contemporary environmental aesthetic? In the eyes and minds of many, pigs are of course a proxy, a stand-in, for the opposite of beauty, and our pig is no exception. The parallel suggested here, however, is that the pig in his present state of evolution is a symbol for all that prevents the formation of a holistic environmental aesthetic today.

It seems appropriate to spend some time going back to origins, seeking in architectural terms the metaphorical "original hog," not in the interest of any thorough argument about a technological return to origins, as that seems neither likely nor desirable in any extended way. Rather, we do so in the interest of an environmental aesthetic, in the reestablishment of an aesthetic derived from man's connection to the natural world. This connection, termed "biophilia" by Edward O. Wilson, has important aesthetic dimensions that drive us toward affinity with not just nature itself, but with its representation and evocation. The work of Stephen Kellert and Grant Hildebrand articulated the importance that such abstractions from nature play in establishing meaningful connections to architecture.

To provide the groundwork to develop an environmental aesthetic derived from biophilic affinity, let us begin with what is widely reputed to be the first air-conditioned building: Le Corbusier's Cité de Refuge, built in 1933 in Paris. Le Corbusier is picked on only because of his iconic status and enormous influence within the modern movement. His 1933 building did, of course, prove prophetic, with few large new structures today, even in very temperate climates, daring to function without air-conditioning. Once sealed up and fully enclosed against the weather, the hog farmer's dilemma of an additive culture of innovation begins to climb the architectural mount. The early innovations paid scant attention to energy and horsepower. Once closed to the environment, all manner of new equipment was invented to replicate, with controls, the outside world within. Through the environmental movement, this is changing, with considerable attention invested in lower energy use. What has not changed, however, is the culture of solution by addition, the tendency cited in the "Letter from Iowa," to mask structural flaws with new construction. In lieu of energy-hogging machinery, we have substituted a living machine to process sewage, solar shading devices, and photovoltaic panels and solar thermal collectors to harness energy—all constructive, functional, and ethical, but often additive and therefore not yet aesthetic.

We have also invented a system, Leadership in Energy and Environmental Design (LEED), that even with all its enormous influence and benefits, sanctions architecture by addition, giving us points for good behavior, but no points for beauty. LEED has aided and abetted the design of a lot of really ugly buildings by a generation of point-counting A-students. LEED "bling" is here among us and threatens the very soul of what environmental design is attempting to accomplish, because it is not yet aesthetic. LEED, at this point in its evolution, is focused on the visible badges of environmentalism. LEED will only be truly successful when it is no longer needed.

Why is beauty important? James Wines has said, "If it isn't beautiful, then it isn't sustainable." The aim here is to speculate on the notion of an aesthetic derived from an integral, not an additive, relationship with the natural world. Nothing of beauty has ever been made by addition or by counting points. That said, there is a further aesthetic dilemma that derives from our ever-increasing separation from the natural world. Have we become so disconnected from the natural world that we have to develop an environmental culture before we can appreciate the beauty of an integral "natural" solution? In short, has evolution taken us so far afield that the resonance with nature that is at the very heart of biophilia has to be relearned before we can have an aesthetic response to the natural world as represented through the artifice of our shelters, our architecture?

The separation of man from nature is the motivation for the existence of architecture in the first place. The Renaissance theorist Filarete's depiction of Adam (and mankind in general) as a wimp seeking shelter from the elements is the proverbial motivation for the first house, the primitive hut, as traced in Joseph Rykwert's brilliant book *On Adam's House in Paradise*. The elements from which man has always sought both sustenance and shelter are the sun, air, and water. The simple sheltering of umbrellas and igloos has given way over thousands of years to ever more elaborate homes that have evolved into the antithesis of the primitive hut, sealing out rather than filtering in the elements. The consequences of this separation from the elements are by now well known. Adam's primitive hut, in its present single-family suburban form, has become a 15,000-gigajoule consumer of energy over a 50-year life cycle. The difference between the primitive hut and suburban home can perhaps be best illustrated by the coffee filter and the envelope. The filter is a smart membrane. It is designed to keep out what we do not want (coffee grinds) and to let in what we desire (liquid coffee extracted from the grinds). The envelope, a term widely used today to describe building skins, is the antithesis of the filter. It completely segregates inside from outside—keeping inside in and outside out.

To a large extent, the aesthetic being advocated here is about filters, about the development of an aesthetic language that selectively integrates rather than systematically segregates. How can this language evolve in contemporary terms? Returning for a moment to the primitive hut, as so many architectural theorists have done, the historic languages of architecture have almost always been justified by their origins in natural form. The French theorist Laugier's seventeenth-century justification of the classical language returns to his version of the original post and beam gabled hut derived from the tree, in his vision still alive and rooted to the earth (Rykwert 1981, 45). Sir James Hall's explanation of Gothic form similarly returns to the rooted tree but with a different vaulted, rather than gabled, interpretation, drawing the branches together into the vaults that characterize Gothic shelter (Rykwert 1981, 85). Two theorists with two languages derived from one natural form, the tree.

Further into the nineteenth century, however, we can already sense a sea change in this type of direct connection between architecture and nature. The German theorist Gottfried Semper's Carib hut is no longer literally rooted to the earth. Its columns have been cut and are fastened to a platform that elevates the floor of his hut above the ground (Rykwert 1981, 23). While the material palette of Semper's hut would certainly earn a lot of LEED points, its form and aesthetic are closer to prefiguring Mies's Farnsworth House than they are to

Figure 14-1: Farnsworth House

the more fully biophilic propositions of Laugier and Hall. Both Semper's and Mies's huts are lifted above the ground (Figure 14-1). Both introduce walls: textile mats for Semper and glass for Mies. Both center their huts around environmental systems similar to hearths: a fire pit for Semper and heating and air-conditioning equipment in the central core for Mies. In short, both represent an attitude toward nature that separates architecture, the artifice of man, from the living forms of the natural world.

For most of us in the developed world, this separation has become who we are. It is not that we have no biophilic affinities to the natural world, but we no longer see ourselves as literally within that natural world. One might intellectually appreciate the Aztec sculptor's representation of man inside the raptor, but confess to having a difficult time actually seeing oneself so thoroughly hybridized with nature. Given this sense of the separateness of man from nature, represented by our present preference for the envelope rather than the filter, how can we evolve toward not just the functional fact of integration but toward an aesthetic of integration?

Examples of KieranTimberlake Associates' struggles with this question can be observed in three projects: Sidwell Friends Middle School in Washington, DC; Loblolly House at Taylor's Island, Maryland; and Atwater Commons at Middlebury College in Vermont. At the Sidwell Friends School, we have undertaken a campus master plan and five projects. One of these is an addition and renovation to the Middle School that doubles the size of the existing 1950 structure. It is important to note that this project begins with an existing structure. Taking care of what you have already brought into the world is the first act of sustainability, one largely ignored by most architects practicing at the leading edge of the sustainable design movement. The existing landscape in this area of the campus was a biological and aesthetic scar in need of restoration.

The metaphorical vision at the outset for the regenerated landscape and architecture was that of the Quaker artist Edward Hicks, who painted dozens of versions of *The Peaceable Kingdom*. This is a Quaker vision of man at peace with the natural world, at one with its flora, fauna, water, air, and other inhabitants, all gathered around Penn's famous treaty tree. A central objective for the Middle School was the restoration and rejuvenation of the landscape and architecture, both backward and forward, into a modern Peaceable Kingdom, realizing the challenge of the Hicks painting. The commitment of the client to sustainability was never a question. It is in this instance even a religious belief that we are all obliged to sustain the world and all that inhabits it. But the aesthetic dimension of that commitment has been a topic of significant debate. It is the aesthetic dimension, however, as much as the facts of unsustainable initiatives, that is likely to determine the long-term viability of environmental design. Aesthetic form can attract, and attraction is always more potent than force.

For the moment, let us focus only on the formation of a new academic quadrangle at the new entry to the Middle School (Figures 14-2, 14-3). This quadrangle is a metaphor for all that the new program seeks to accomplish. Mathematically speaking, on the supply side of the water equation all water falling on the Middle School site was shuttled into poorly repaired underground piping and directed into the district storm sewer system. Our aesthetic and technical proposal is the retention of rainwater on the green roof, with the overflow directed down partially open leaders into a sloped spillway where it makes its way into a pond and rain garden at the building entry. With regard to *wastewater*, the opposite side of the water cycle, a living machine to process sewage was part of the program almost from the outset. What was unfamiliar, however, was the aesthetic proposition of sewage processing through the agency of the constructed wetland. The machine as a solution is always additive and therefore not aesthetic because it can be easily removed. At Sidwell, the machine is the constructed wetland, an integral landscape aesthetic that possesses the potential for beauty. This potential depends upon the readiness of the viewer to cross back into the inverted view of the natural world from within, as conveyed in the Aztec sculpture described earlier. At the same time, our aesthetic reaction is further enhanced by our innate biophilic aspiration toward ele-

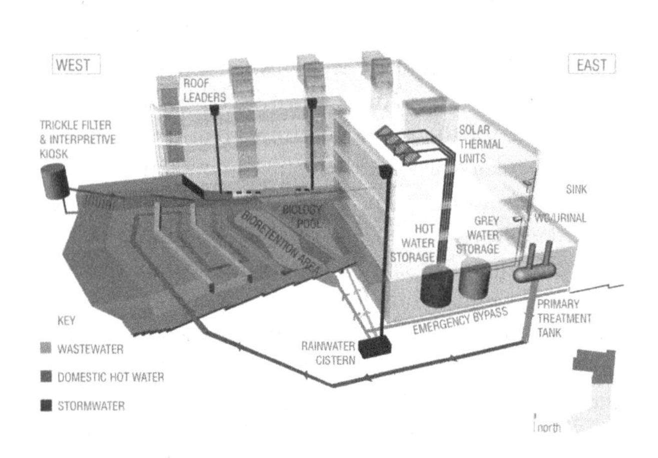

Figure 14-2: Rainwater and wastewater processing at Sidwell Friends Middle School

Figure 14-3: Rainwater overflow down sloped spillway into pond and rain garden, Sidwell Friends Middle School

vated views of this new world from the entry bridge and internal passages that line the courtyard.

Beyond the landscape of the courtyard, the material substance and aesthetic presentation of the building skin is another central biophilic aspect of the building. Just a decade ago, the word *façade* would have been used to describe the building elevation. *Skin* is appropriate here for its biological reference. Skin acts as a filter, not an envelope, which selectively admits and rejects the environment based upon the needs of the body across time. It sweats to provide evaporative cooling from the heat and forms goose bumps to close its pores to the cold. Skin is aesthetic because it mirrors, shapes, and contains the remainder of the body's organs—it is integral, not a façade that is completely disconnected from those organs.

The integral skin at Sidwell parallels biological skin in its functional layers. The two outer layers are wood, with the outermost layer organized to shield solar gain. For example, the wood on the west elevation is arrayed vertically and angled at 51 degrees north of west for minimal solar penetration and maximum penetration of daylight. Behind the wood solar shading is a wood rain screen wall designed to shed most water but remain open to the movement of air. The central aesthetic point here is that the solar shading is integral with the entirety of the wall assembly. It is part of a unified multi-layer skin made from very high grade reclaimed material, western red cedar from salvaged wine fermentation barrels. The source history of this material is referenced with a bench made from the barrels' base. Going one step further, the red cedar fins to the right of the entry extend all the way to the walkway, abstractly referencing the origin of the tree in the forest prior to its use in the wine barrel (See Figures 14-4 and 14-5 in color insert.)

The aesthetic inversion exists at several levels at Sidwell. The placeless site bordered by the existing building has been transformed into an academic quadrangle of sorts, but one very different from the traditional Oxford or Cambridge university quadrangle. While both the Sidwell Middle School and Peterhouse College at Cambridge are academic quadrangles formed as outdoor rooms to learn in, Peterhouse derives from the Christian monastic tradition and is intended to exclude all that is worldly. It is a vision of learning that focuses inward on the life of the soul and is sheltered from the outside world within the arcades surrounding the carefully cropped greensward. At Sidwell, the quadrangle is three sides rather than four, and welcomes students and the campus as a whole into its space and the building entry beyond. As opposed to the carefully controlled greensward at Peterhouse College, the constructed wetland at Sidwell yields an integral working aesthetic. The tiered wetland is constructed with low terraced walls to invite entry and participation. A reclaimed wood skin is developed to invite deep source knowledge at the secondary level of the wine barrel and at the primary level of the tree rooted in the forest. An existing great oak tree in the quadrangle comes to be experienced as the great Quaker Penn's treaty tree. It completes the restoration and rejuvenation of the prior environmental scar and transforms the landscape into a contemporary Peaceable Kingdom. (See Figures 14-6 and 14-7 in color insert.)

As an aesthetic type, the new Middle School court is at once familiar as an academic quadrangle, a place of learning and contemplation, yet unfamiliar as a wetland and wood. The intent in this portion of the Sidwell Friends project is to confront the tradition in Western education of the academic quadrangle, to simultaneously accept it but turn it on its head, inverting its focus outward to the natural world and its systems. Whether this inversion of the academic quadrangle comes to be viewed and experienced by the students and faculty with the same fondness we presently possess for the Peterhouse quadrangle remains to be seen, but it is our belief that only that which comes from the core out, that which is integral and fundamental, will have the potential to be perceived as beautiful.

Loblolly House, so named after the loblolly pines that dominate its site on the Chesapeake Bay, fuses the natural elements of site to architectural form. This house seeks to ground the artifice of architecture in the elements of nature. The five broad elements of nature that define the place of this house include the loblolly pine trees and the tall grasses, coupled with water, sky, and the west-facing sun. Loblolly House transforms these elements of nature into biophilic form. The evocative work of Ellsworth Kelly as he seeks to abstract

Figure 14-8: Loblolly House, section

water and land have been an inspiration to the representation of natural form in contemporary terms throughout this house. It is our innate kinship with abstract light, color, and form drawn from nature that motivates the substance of this house.

By way of distinguishing intent, if the Farnsworth House seeks to place man within nature but separated from it by a glass envelope, then Loblolly is a sieve at once within (Figure 14-8), above, and of the pines that ground the site. The foundations are themselves trees, timber piles, at once pragmatic and poetic—pragmatic because they minimize the disruption to the ground on this waterfront site, and poetic because the dwelling is literally founded on the tree. It is a house among and above the trees, a tree house. At the same time, the dwelling as seen from the east or land side is of the forest and in the forest. Earlier, more literal representations of this idea give way in the materiality of the wood board siding to the biophilic representation of the loblolly forest and the passage of light through that forest.

The pattern of the east wall was literally composed over a site photograph (Figure 14-9), with the abstraction of solid and void rendered in the staggered vertical board rain screen siding, sometimes positioned over solid wall and sometimes lapping over glazing to evoke the solids and voids of the forest, part open to the sky and part obscured by trees. The land side elevation of Loblolly House has no windows as holes in the wall, nor any windows as the wall. Rather it has the windows of the forest, only partially open to view and light from beyond, seen through the densely packed verticality of loblolly pines. This representation of natural form evokes a biophilic affinity between architecture and its context in nature.

Tall grasses form the floor of the site. These grasses become the literal and representational floors of the house through the use of deep green stained bamboo floors that carry the striations and color of the tall site grasses from the outside to within. A deep blue stained ceiling at the top floor evokes Ellsworth Kelly's representation of sky at the bridge between the house's two pavilions. Lastly, the dominant role that the setting sun

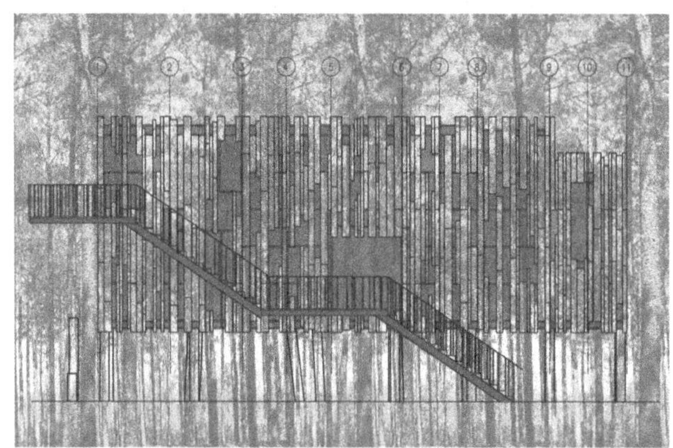

Figures 14-9: Loblolly House, pattern of the east wall composed over site photograph

plays in the daily life of this house is represented by the orange glass that clads this same bridge. This glass burns a bright orange from within as the sun sets, deepening the biophilic connection of this architecture to the natural world from which it evolves. It is here, at the bridge, that the sea green floors come together with the blue of the sky at the ceiling and the orange fire of the setting sun, all *experienced* within the forest and encapsulating the fusion of the house with its environment. (See Figures 14-10 and 14-11 in color insert.)

A final linkage between Loblolly House and natural form derives in part from a biophilic exploration of the cycle of life and death. As reluctant as we may be to discuss death in our own lives, that reluctance is dwarfed by our outright unwillingness to discuss and design for the death of our architecture. Loblolly House confronts not only the question of how we assemble our architecture but how we disassemble it, how we manage not only the beginning of life but its end. This is a largely ignored ethical obligation of the architect, and also a source of daily reference in our architecture to that most natural cycle of all events: inception, growth, maturity, and decline. We believe there is great aesthetic potential for deep biophilic connections to natural form by accepting and representing, in short, rendering aesthetic not only the manner of assembly but the disassembly, reuse or recycling of all the components of our buildings (Figure 14-12).

Loblolly House does this through a componentized assembly process that begins with an exposed, recycled aluminum frame that becomes the organizing fabric, the ground, of the house interior. It is bolted together for ease of assembly and disassembly. Floor and ceiling cartridges, which are fabricated off-site with all the building systems integrated, are bolted to the frame. Off-site-fabricated bath and mechanical room modules and exterior wall panels containing structure, insulation, windows, interior finishes, and the exterior "forest" wood cladding minimize site impact by optimizing off-site assembly. More important yet, the reading and celebration of assembly and disassembly in Loblolly House gives rise to a compelling aesthetic about origin and demise, perhaps our most deeply rooted connection to the natural world.

Lastly, Atwater Commons at Middlebury College, Vermont, offers a completed vision of a fully biophilic fusion of architecture and nature (Figure 14-13). The project is composed of three structures: two residence halls and a dining hall. Each reflects the College's environmental goals with careful attention to site strategy, water runoff, and material selection.

The residence halls (Figure 14-14) are naturally ventilated, incorporating through-floor suite plans, transom windows, and ceiling fans in all rooms. Ventilation is supplemented by attic fans, which exhaust through rooftop "chimneys." The dining hall incorpo-

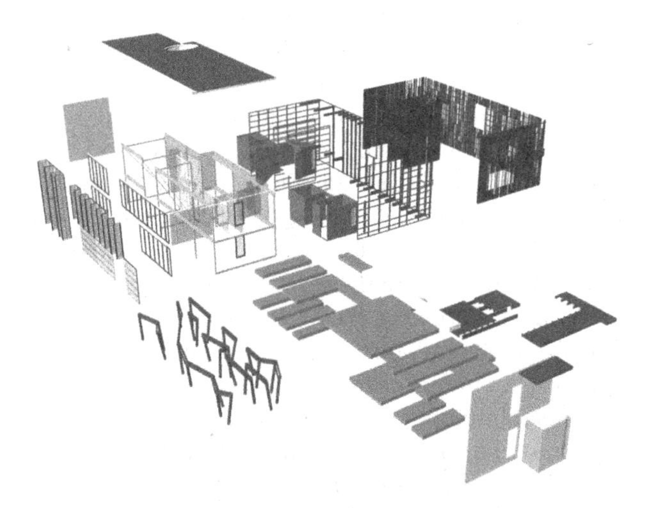

Figure 14-12: Axonometric diagram of Loblolly House components

Figure 14-13: Aerial view of Atwater Commons, Middlebury College

Figure 14-14: Atwater Commons, Middlebury College

rates a planted roof, providing excellent insulation, protection of the roofing membrane, and, most significantly, reduction of impervious surfaces on the campus.

The dining hall will be the focus here. Its sloping elliptical roof (Figure 14-15) allows the college landscape to extend literally through and across the dining hall structure. Planted with local flora, this roof rises gently from the entry, sloping upward toward Vermont's Green Mountains beyond. A passage opens beneath this roof—almost cavelike, branching in two directions: down beneath the hall to the faculty residence and village beyond and into the hall itself. This entire experience is almost geologic, beginning with the sense that the dining hall is itself an opening into and through the earth. Conical skylights, roof ventilation stacks, and fireplace chimneys enhance this sense of the hall as rooted to and emerging from the earth.

Figure 14-15: View of planted roof on dining hall

Figure 14-16: Dining hall interior view

Upon entering, however, geology gives way to an experience of the surrounding forest canopy that dominates experience in the hall. The dining hall is literally in the trees. Treelike biophilic form occupies and circumscribes the hall itself (Figure 14-16). Columns are grouped and angled in the center of the space like clusters of trees. Perimeter glazing is enhanced by the treelike form of randomly angled columns around the elliptical perimeter, mediating between the representational biophilia of the building and the actual surrounding forest. The open grid ceiling of the hall, positioned at random angles to the interior columns, is a visual extension of the tree canopies, screening the interior ceiling in the same way that the tree canopies screen the sky. The experience in total is one in which the views to the forest beyond actually merge with the treelike forms of the hall's perimeter and interior. Seen from the outside along the stone wall, the dining hall emerges from the ground and fuses itself with the forest. (See Figures 14-17 and 14-18 in color insert.)

The Sidwell Friends Middle School, Loblolly House, and Atwater Commons at Middlebury College are three deeply rooted explorations into how, in contemporary terms, we can evolve both backward and forward, not just toward a substantive environmental ethic in our architecture but, more importantly, toward an ethical aesthetic derived from nature. In different ways, the affinity evoked by each of these works derives from our innate biophilia. While the site of each structure provides a literal grounding in nature, it is the aesthetic abstraction and representation of natural form in each that extends and deepens our biophilic affinity. This affinity is an aesthetic that we must evolve beyond additive form toward an integral union between nature and architecture, the artifice of man.

REFERENCES

Rykwert, J. 1981. *On Adam's House in Paradise: The Idea of the Primitive Hut in Architectural History.* Cambridge, MA: MIT Press.

chapter 15

The Picture Window: The Problem of Viewing Nature Through Glass

Kent Bloomer

Let's take a look at a picture of a classic mid-twentieth-century modernist work of architecture taken from the outside in the picturesque setting of trees, rocks, and gardens. Such an image can produce the compelling spectacle of a connection or a healthy interaction between the world of man and the world of nature. It even suggests an architecture that displays a love of nature. But what about the other way around, looking outward, from a sheltered vantage point inside such a building, through a large flat plane of glass that provides a panoramic view of the outside? (See Figures 15-1 and 15-2.)

The desirability of viewing objects such as trees, gardens, and birds from within a residence, hotel, or workplace is beyond dispute.[1] Even the therapeutic power of viewing the natural environment is now acknowledged (Ulrich 2006). But can we therefore assume that viewing through a large modern glazed opening (let's call

Figure 15-1: Looking from outside, there may appear to be a lively equilibrium between house and nature.

Figure 15-2: From the inside, viewed through a large pane of glass, nature can seem to be subordinate and disconnected.

that a picture window) also provides us with a vital sense of connection, or an active understanding of our responsibility toward nature?

It is interesting to note that the plain, crystalline form of the picture window, as it has evolved today, coincided with both the ascent of the modernist project in architecture and the descent of architectural ornament. In the early- to mid-twentieth century, as a new, more mechanistic style of design emerged, industrialized settlement was in the process of increasingly occupying and gaining greater control over the natural environment. By the second half of that century, a certain "ideal" transparency was being developed between architectural interiors and the world outside, leading to a new type of relationship with nature. In America, a motorized suburbanization also promised to provide a more intimate connection to nature, trees, and the garden. But let us analyze that new relationship, particularly in regard to the contemporary popularity of the big "viewing" window itself and indeed the phenomenon of viewing in general.

While window glass is transparent, it is also hard, and for most practical purposes, impenetrable. We view through glass knowing that glass provides a powerful barrier and protection from heat, cold, wind, rain, insects, and animals. Indeed, glass, whether employed in sky, land, or undersea, is a marvelous triumph of man's protection from the immediate ravages of nature. But beyond the provision of shelter, what is so satisfying about viewing nature through large expanses of glass within the sealed fixed edges of mammoth openings? Does this attraction, this seeming instance of "biophilia," indicate that we are enjoying our control over, i.e., our dominion over and thus our secure distantiation from the "prickle" of nature; or do we imagine that we are truly bonding with or engaging the world outside (Kellert 1993, chap. 2)?[2] Glass is of considerable utilitarian value, but has its ubiquitous and commanding presence in the walls of today's architecture really brought us closer to cherishing the complexity, unpredictability, dangers, and grandeur of the natural world? Regarding the materiality of glass, that is, its sensuality as a medium, why do we go to such great pains to get clear glass, to sanitize it and make it so transparent that its visual substance disappears and thus virtually dematerializes? Curiously, with such means of viewing we might be looking at nature in a manner similar to the way we looked at animals in early twentieth-century zoos, their dangers held at bay by the slender bars of cages. Through glass, we observe the world outside comfortably and safely and without the challenges of actual engagement.

But what can we do to heighten our contact with the natural environment from within buildings, given that we must have our glass windows for any number of obvious practical reasons, as well as the fact that we are attracted to and enjoy viewing nature through glass?

Perhaps the crucial question in the light of the biophilia hypothesis is "Can we enhance the positive phenomenon of viewing nature through glass in a way that might heighten our connection to and possibly increase our love of nature?" And can we reduce the drawbacks of visual distantiation, physical separation, and even a sense of supremacy over nature by architecturally altering the design of today's typical picture window as well as the design of the immediate setting or framework of that window?

Consider the basic act of visual viewing, particularly staring, even without the intervention of glass. While informing, the mere act of looking is usually passive and only quasi-sensual. Viewing may provide a vicarious experience of the object being viewed without the trials of actual encounter. We might say that merely looking at something is somewhat "virtual" by lacking the component of action-reaction; for example, we can look at a mountain without climbing it. This, of course, does not mean that looking cannot evoke the excitement of a remembered or potentially more direct experience. But remembering and imagining are steps removed from actually engaging the object under consideration.

Consider also that our original and deepest sensual contact with the world around us was primarily developed in childhood through touch coordinated with sight, sound, taste, and smell. We discovered danger and delight by bumping into something. Over time, our visual perception of objects in the environment became largely a follow-up to our earliest encounters. Still, it is only through touching that we can again experience the simultaneity between action and reaction. I developed this argument 30 years ago in my book with Charles Moore titled *Body, Memory, and Architecture* by emphasizing that the entire system of touch that pervades both the inside and outside of our bodies, which J. J. Gibson (1966) called the haptic system, is a critical property in our experience of architecture's or nature's three-dimensional space. We were indebted to the seminal work of environmental psychologists. In the same study, we explored the "nature" of our own interior space, or the sense of a personal protected interior that we carry with us as we aggressively seek information about the world outside and beyond our personal space. We particularly focused on body imaging, that is, how we develop an image of our own bodies, including how we imagine our bodies relative to other bodies in space (Fisher 1970). An important finding was the notion that we possess a psychological boundary around our bodies (and by extension around our houses) that divides, or separates, our sense of a personal, possessed interior space from an exterior extra-personal space. This boundary is extremely sensitive and conditions our perceptions of the environment. It is also an elastic boundary that is subject to changes of shape, size, and hardness under different times and circumstances of encounter with the social and natural surround.

Such a psychological boundary arguably exists around the perimeter of vehicles, houses, and institutional buildings, or any vessel acting as a surrogate body. It is an intuitive condition that has traditionally informed the architectural design of the envelope, that is, the thick edge or section between the interior and exterior of a building. Certainly, at places of entry, visual statements about issues of social rank, safety, cultural belief, and the occupants' relationship to nature are played out by the shape, dimensions of setback, orientation, overhang, materials, decorations, mats, et cetera. Indeed, the passages through the psychological and actual boundaries of buildings, particularly important buildings, have forever been the most ritualistic moments of architecture. Principal windows and places of viewing have also been intimately dimensioned, shaped, and detailed to proclaim, sanctify, express, and allow a particular attitude toward our connection from within to the world outside. (See Figure 15-3.)

By combining those studies on haptic sensing, aggressive seeking, and body imaging, we concluded that

Figure 15-3: Great places of viewing can proclaim a particular attitude toward the natural environment.

our profound knowledge of the environment is corporeal and fundamentally developed from tangible experiences. From the standpoint of biophilia, let us assume that touching and the near-possibility of touching (haptic seeking) are fundamentally critical in establishing a firm connection, a "contact" with the natural environment. Yet, *touching is precisely what is negated by the pure picture window!*

THE ORNAMENTED PICTURE WINDOW

Consider that we can stimulate our sense of touch in the course of viewing through glass if we begin by thickening and populating the hard, glazed boundary between the inside and outside of a building. That is, we can invest the liminal transitional space of the window with material elements, including thicker or tinted glass elements, which might invite touch or simply imply something that is touchable in the course of viewing through the window. This was automatically the case in traditional window design, in which small panes were embedded within a grid of many mullions. (See Figure 15-4.) You still got the view, but the intimate threshold between being inside and outside a building was materialized with a wooden or stone grillwork. By touching or being able to imagine touching elements within the space of the threshold, you may heighten your sensual association with the world outside.

Put another way, by importing properties of the material environment into the glazed threshold, you deposit elements of matter implicated with the world around the window into the moment of divide between inside and outside. Arguably, the moment of divide is the most charged, ambivalent, and negotiable for belonging to both sides of the psychological boundary that informs our reaction to the environment. A further step, then, would be to design the shape of the mullions and incorporate additional material elements within the space of the window that begin to mimic, indeed to portend, some formations, complexities, and actions that are essential features of the world at large. (See Figure 15-5.) As the incorporated divide becomes more evocative and complex, it becomes more ambivalent; that is, it simultaneously implicates formations belonging to both interior and exterior places.

This is the classic function of ornament, to distribute material formations and rhythmic motifs into the spaces between things in order to heighten our sense of the world on both sides of a psychological threshold (Bloomer 2001, 61). Ornament thus performs as a sentinel or a bidirectional indicator of activity on each side of the threshold. It is a type of information. The educator-architect Charles Moore often spoke of the heightened perception given to viewing the distant ocean by inserting a bowl of water or a small pool between the viewer and the view, as compared to just staring into the distance over dry terrain toward the ocean.

If we consider the period of "modern" architecture in Western culture (the period to which we still belong)[3] as beginning around 1800 and developing throughout the nineteenth and early twentieth centuries, we can find any number of decorated windows that invested the glazed boundary with formations, particularly figures of ornament, capable of simultaneously evoking the inherent geometry of architecture and the "adherent" organic formations derived from the natural world out-

Figure 15-4: We may stimulate our sense of touching the world outside by looking through mullions and small panes of glass.

Figure 15-5: The shapes of the mullions may begin to mimic the formations found on trees.

The American tendency to view the environment through larger expanses of glass inspired another strategy of incorporating natural rhythms into the thresholds of windows as the tall building came into being. Louis Sullivan, considered by some to be one of the seminal composers of the modern skyscraper, inscribed ornament in the reveals of the window wall, that is, the inward face of the window frame perpendicular to the plane of viewing. (See Figure 15-7.) By looking through Sullivan's organic patterns of repetition, the viewer's peripheral vision was rhythmized in the act of looking outward into the land and cityscape. Indeed, one of the functions of ornament, beyond its capacity to portray complex formations innate to nature, is to impress and suffuse its objects with an amount of temporal rhythm. The term *temporal* here refers to types of visual organization that suggest time and changefulside, particularly those found in trees and leafage (Bloomer 2006).[4] Sometimes this was simply achieved with patterns on curtains, and other times with the shaping of mullions and the incorporation of more complex geometric details.

Indeed, ornamenting windows for viewing was seminal to modern architecture (only to have been professionally condemned and forgotten in the last 60 or 70 years of the rapid growth and colossal mechanization of design in the later modernist movement). For example, Alexander Jackson Davis, one of America's most gifted nineteenth-century architects, was a great inventor of practical window mechanisms who, early in the century, "anticipated such developments of the modern age as strip windows and window walls" (Peck 1992, 9). However, his innovative talent in mechanics did not stifle his inclination to express rhythms and figures found in the natural environment within the glazed boundaries of windows. Indeed, one of the most brilliant compositions of viewing through "pictures" resulted from a collaboration with Tiffany at Lyndhurst in Tarrytown, New York, in which the picture itself is a literal detail (a picture frame within a picture frame) enshrined by the splendor of geometric pattern and the polychrome foliated tracery embellishing the great arched window. (See Figure 15-6 in color insert.)

Figure 15-7: By distributing ornament along the reveals of the window frame, the viewer's peripheral vision is "rhythmized."

ness in contrast to stasis. In systems of ornament, such rhythmic patterning is generally quite minimal in area compared to the frozen units of geometry typically found in the overall shape and proportion of buildings. Ornament's intimate rhythmized detail thus incorporates an amount of sensation that originates from living patterns in nature into the basic inorganic forms of architecture.

Sullivan's student Frank Lloyd Wright obviously understood the vitality of viewing through ornamented windows. (See Figures 15-8 and 15-9 in color insert.) It is important in examining Wright's window designs to observe that he does not foreclose the option of clear viewing, especially in the lower eye-level portions of his windows. Even in Wright's more complex designs, you can still view the outside as directly as you can through an unornamented larger plane of clear glass. Like A. J. Davis in his great window at Lyndhurst, Wright understood that attentive viewing does not have to be panoramic in scale.

CONCLUSION

The placement of ornament within the critical thresholds of viewing from within buildings establishes a visible and touchable moment of mediation between inside and outside. By exploring the psychological boundary between the interior of buildings and the natural world, the claim that an "ideal" connection can be achieved by merely looking through clear, simply framed, and expansive units of glass can be critiqued and refined. Paradoxically, by inserting an amount of "picture" in the "picture window," we might articulate and effect a greater bond between the places in which we live and work with the surrounding nature.

The danger in omitting the ornamented window from the study of the biophilic merits of viewing from buildings is the possibility that the popular "naked" picture window may be applauded and declared sufficient, despite the fact that it provides a sanitized vision and might even promote a false feeling of fulfillment predicated upon an *illusion* of experiencing and being connected to the natural environment.[5]

ADDENDUM: THE PROBLEM OF VIEWING NATURE THROUGH "STONE"

The complete absence of any kind of visual sighting of the natural environment establishes an extreme instance of sensual disconnection from nature in deep interior space. Clearly, with that in mind, Wright also situated figures of ornament away from the exterior walls and in the sequestered interiors of his buildings, more as mementos, rather than as direct mediations with "pictures" of the natural world outside. (See Figure 15-10 in color insert.) Indeed, the practice of colonizing the blind spaces deep inside buildings with "cosmic" ornament preceded the classical architecture of antiquity with its colonnaded and decorated center places. That ancient tradition was still brilliantly recalled in many examples of early modern architecture prior to the extremes of reductivism governing the design of interiors in the period of late–twentieth-century canonical modernism. (See Figure 15-11 in color insert.)

In fact, many of us spend at least part of our days back away from the outer walls, all too often in nasty white boxes further debased with a plethora of written messages, computer screens, and way-finding signals (like a digital clock or exit sign). Others spend all day sequestered in such quarters. Occasionally there are potted plants presenting bits of the outdoors. Moreover, unlike the quasi-immateriality of large glass windows that can at least provide an illusion of connecting to nature (which I have contended could be more positively biophilic by incorporating rhythmized physical elements within the window and its frame), the perceivable edges of deep interior space are most often governed by impervious building components creating an essentially deflated world-picture. That fact proposes that there is no alternative for the designer dedicated to biophilic values, other than to contest the material reality of the box.

Logically, the spatial confinement of the box proposes an amount of *going against* its inherent materiality and by extension against the materiality of the building qua building from which the box issues. Building the walls with richer, allegedly natural, and seemingly less commercial or manufactured materials could, by itself, merely edify the fact of confinement. This suggests moving in a direction that is the reverse of artificially materializing the picture window. It suggests dematerializing critical moments within the surface of the blind box with formations (see Figure 15-12) that (from the standpoint of construction) are intrinsically nonexistent, in order to subvert the hard structure. For biophilic purposes, such imaginary formations, such *in-formation*, would be spirited, originating elsewhere and embedded within the intrinsic structural elements to proclaim the vitality and rhythm of nature. (See Figure 15-13.) Of course "going against" the pure primary structure of building is anathema to the core ideology of late-twentieth-century architecture, which idolized the physical elements of construction. By declaring that the expression of tectonic form and its authority over subordinate space is the defining essence (medium) of the art of building, the modernist canon strained to identify architecture as a limited phenomenon (as a highly specialized profession). Extreme dedication to such "limitations" can promote a type of idolatry that inclines toward the worship of an inorganic geometry rather than the more organic rhythms of life rooted in biophilia. (See Figure 15-14 in color insert.) Consider that a well-hewn wooden beam in a blind box, while charming and rustic, is twice removed from the living exuberance of a tree. At least this writer believes that a deep interior within a work of architecture that *only* aggrandizes the material elements signifying the economics of construction, whether plastic or rustic (its rugged supports, blocks of stone, and geometric paving), may produce marvelous and even elegant spaces—but they are works begotten from the finality of life rather than the emergence of life per se. Visually disconnected from the vitality of nature, they become the stuff of tombs.

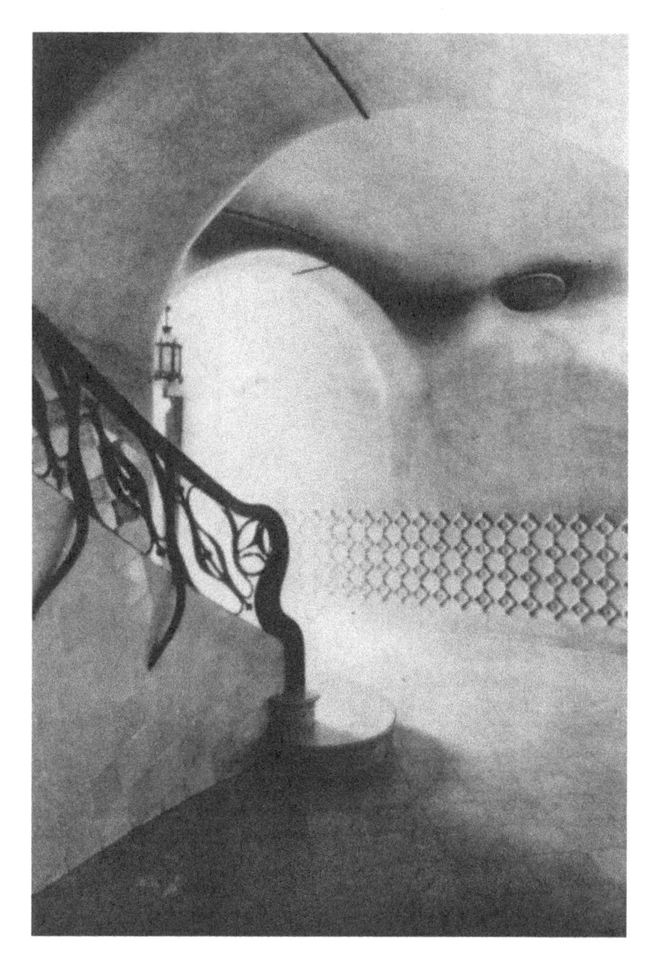

Figure 15-13: Elements of repetition and organic railings can be spirited through dark interiors.

Figure 15-12: Imaginary elements mimicking the formation of foliage can challenge hard concrete walls.

Yet, paradoxically, tombs such as the Theban tombs of ancient Egypt, have served as the birthplaces of seminal and powerful, perhaps biophilic ornament! The large blocks of stone and massive vaults of antiquity were well suited for the body of an eternal dwelling for the afterlife of the dead, that is, for their immortality. The necrophilia implicit in the material gloom of those

Figure 15-15: The necrophilia of an ancient Egyptian tomb was opposed by images of renewal in a band of bud-blossom-bud ornament.

tombs was mitigated by ornament upon the stone and polychrome friezes portraying rhythmic sequences of bud and blossom. (See Figure 15-15.) As a consequence, the powerful stone walls of those dark sanctuaries were ultimately challenged by figures of ornament that virtually dematerialized the stone in order to magnify the importance of *renewal* implicit in the spirit of foliation. Thus, the strong materiality and massive structure constituting the seminal architecture of death allowed vital formations evoking life to subvert its essential power.

For the purpose of this brief paper, let me isolate a few vital actions capable of being manifested in the dynamic line work of ornament, springing from formations more frequently found in nature than in the statics (the frozen geometry) of construction that usually define the interiors of late-twentieth-century architecture. First is rhythmization (especially a driving, syncopated formation of rhythm) (see Figure 15-16 in color insert); second is a spectacle of changefulness, sometimes portrayed as cycles of growth and decay, or which at any one moment may appear as a visible pattern of metamorphosis (see Figure 15-17 in color insert) (a metamorphose); and third is a composition of dynamic entanglement or competition between different species of things. All three of these actions, taken together or individually, convey measures of indefiniteness, temporality, impermanence, mystery, ambivalence, and growth. None of them necessarily assert the order and harmony that is generally assumed to be a positive emblematic property of basic architecture, but all of them proclaim life, which is the subject of biophilia.

These biophilic qualities are typically found in great ornament, a visual tradition that speaks in a manner more akin to calligraphic writing than to the architectonic shaping of space typically and necessarily found in buildings. Ornament presents visual percepts gained and imagined from without that for centuries have been brilliantly suffused, indeed have been "essential" properties in the richer understanding of architecture that flourished prior to the radical sanitization of design that has dominated the built environment for the last 50 years. Indeed, the primal function of ornament has been forever to mediate between the contrived spatial province of the man-made world, and the living immensity of the natural world-at-large.

Endnotes

1. Window viewing from a house or a hotel room is universally valorized in real estate and resort marketing.
2. Kellert observes that a biophilic response per se does not necessarily promote a protective or restorative attitude towards nature.
3. The term *modern architecture* is used here to identify ide-

ological developments in architecture after the French Revolution and the subsequent period of industrialization.

4. Figures of ornament, especially in Western culture, tend to evoke nature via foliation.
5. The term *sufficient* is used here to mean "mission accomplished"; that is, the naked picture window is all that is needed in building design to provide an affective vision of the world outside, a vision capable of promoting restorative action and a will to further nature's well-being.

REFERENCES

Bloomer, Kent. 2001. *The Nature of Ornament*. New York: W. W. Norton.

———. 2006. "A Critical Distinction Between Decoration and Ornament." In *Decoration*, edited by Emily Abruzzo and Jonathan D. Solomon. New York: 306090 Books.

Fisher, Seymour. 1970. *Body Image in Fantasy and Behavior*. New York: Appleton-Century Crofts.

Gibson, James J. 1966. *The Senses Considered as Perceptual Systems*. Boston: Houghton Mifflin.

Kellert, Stephen. 1993. "A Typology of Biophilia Values." *The Biophelia Hypothesis*, Washington, DC: Island Press.

Peck, Amelia, ed. 1992. *Alexander Jackson Davis, American Architect, 1803–1892. Exhibition catalogue*. New York: Rizzoli.

Ulrich, Roger S. 2006. "Human Well-Being in Patient Care Settings." Lecture at Symposium on Building Design, Whispering Pines, Rhode Island, May 11.

chapter 16

Biophilic Architectural Space

Grant Hildebrand

Unlike many of the other contributors to this book, I cannot claim to have authored significant research; my work has focused on the application of the research of others to the field of architecture. It follows that I am enormously indebted to those others, and especially to Jay Appleton, Nicholas Humphrey, Judi Heerwagen, Gordon Orians, Stephen and Rachel Kaplan, and Roger Ulrich.

Those who are reading this book will be familiar with the principle that the survival of our species depended, tens of thousands of years ago, on our ancestors' intuitions; those who were driven, by pleasure or relief of discomfort, actively to seek food and to procreate and to care for their young, will have yielded more abundant descendants. Such too is the case with our physical habitat: those of our ancient ancestors who were drawn to settings that offered survival advantages would, over time, prevail. We, of course, are their progeny, and given the slow rate of genetic modification, it is likely that we are still innately drawn to settings whose characteristics hold some survival advantage, even though that survival advantage may no longer have any practical value for us.

The architect, however, might usefully turn the postulate around. Suppose we could identify, at some useful level of abstraction, some survival-advantageous architectural characteristics that commonly recur in settings we find attractive. If we could identify such characteristics and could design those characteristics into our buildings, there is at least the solid possibility that we could, thereby, really make those buildings more widely satisfying. And such a design tool would have useful advantage over other theoretical bases for our architectural decisions, because it would have a real grounding in a body of empirical and theoretical work in cognate fields.

For an overly long time, I have been trying to identify such survival-advantageous characteristics that commonly recur in settings we find attractive. I am not yet happy with the terminologies of all the characteristics I describe, and I would not argue that my work in its present state is complete and definitive. But abundant support from related disciplines and professions, and a growing body of empirical evidence, lead me to think that the work can now claim to be, at the least, a defensible hypothesis worthy of further attention.

I do not present this approach as either exclusive or comprehensive. Many buildings I admire bear little relationship to this approach, and some others illustrate one or two of the characteristics I describe but no others. I believe this approach can be a useful design tool, but I do not suggest that it is a unique path to architectural quality.

Provisionally, then, I identify five survival-advantageous characteristics that seem applicable to architecture. I call them prospect and refuge (Jay Appleton's terms, necessarily paired; Appleton 1996); enticement; peril; and complex order. Let me describe the possible bases of their appeal and then illustrate what they look like when we see them in buildings.

COMPLEX ORDER

All "higher" animals, including ourselves, must continuously process a vast quantity of sensory information. Our retinal cells alone receive, at every waking moment, an array roughly equal to a digital camera's pixels—and that accounts for just one of our senses. We must sort this plethora of information into some kind of order, instantaneously—and we do: we enter a room (having already necessarily classified "doorway" and "solid wall"), and in microseconds we assess the presence or absence of chairs, tables, food, drink, sources of light, areas of darkness . . . and, especially, others of our species.

We also benefit, however, from attending to distinctive features in the informational array. In that room we entered, for example, we will obtain some advantages if, among others of our species, we can distinguish man from woman, our child from another's, our child ill or injured as distinct from our child in health. Some of us will benefit by distinguishing salesperson from client, CEO from second vice-president.

If we return to the premise that survival-advantageous characteristics should be appealing, then we should find some innate satisfaction in these processes of ordering and distinguishing. It should follow, then, that we would like sensory material that is rich in order, and rich too in variation or complexity. If we accept that natural selection, in a sense, "designs" species, we have been designed to like order and complexity.

But we have been designed, apparently, to like them only as a pair. Order alone is monotony, "not enough to keep mind alive"; complexity alone is chaos, "a mess." So, for convenience, we might join the two characteristics as ordered complexity, or *complex order*, to save ourselves two syllables.

Figure 16-1: Beverley Minster, Yorkshire; the western crossing

Figure 16-2: The Musée d'Orsay, Paris

Our enjoyment of complex order, like all such behaviors, is independent of pragmatic purpose; any practical advantage is, in a sense, a happy accident. Thus, while that urge guides our useful responses to the world around us, it also leads us to create tangible manifestations of itself. All cultures about which we know anything at all have created and enjoyed complexly ordered human movement and complexly ordered sound. And there is substantial empirical evidence that we are genetically programmed to respond positively to complexly ordered sound (music) but not to chaotically complex sound (noise). One might argue, similarly, that consciously or unconsciously, we distinguish architecture from "just building" by the evident order and complexity of its materials and spaces (see Figures 16-1 and 16-2). Nor is complex order exclusive to the realm of professional design; preferred vernacular buildings, even townscapes, can be understood in the same way (see Figure 16-3).

Figure 16-3: Arlington row, Bibury, Gloucestershire

As architects, clients, and users of buildings, we can think about designing order into our everyday rooms, corridors, and streets. And we can think about variations on ordering themes that reward our urge to attend and examine. In these ways, our architectural settings may be, or may become, analogous to dance and music, and such analogies may be not be trivial; rather, they may be fundamental to our emotional well-being.

PROSPECT AND REFUGE

For 99 percent of our species' life, we lived in settings of entirely natural elements, and we have an affinity for scenes in which natural features predominate. Roger Ulrich has shown that views of such scenes reduce stress among university students facing an exam; that even pictures of such views significantly shorten recovery times in hospital recovery rooms; that prison inmates with views to nature report for sick call less often. But not all natural settings are reassuring. We respond with anxiety to many literary portrayals of natural settings, and the thought of being left in the open, today, on the African savanna is terrifying. There and elsewhere, long ago, our instincts must have led us to find or build a more specifically supportive dwelling place. What must it have provided?

Lacking so many of coping devices of other animals, we must have a haven to conceal and protect us against climate and predators. Jay Appleton calls this the "refuge"; we seek it as urgently as we seek food and water. But we must get food and water too, and in safety. We need a place to hunt and forage, offering open views over long distances, ideally brightly lit to illuminate resources and dangers, and directionally lit to cast information-laden shadows—our fondness for sunlight may derive from its usefulness for this purpose. Such a place lets us hunt animals, gather plants, and find water, while revealing threats that demand flight to the refuge. Appleton calls this the "prospect."

Refuge is small and dark; prospect is expansive and bright; they cannot coexist in the same space. They can occur contiguously, however, and they must, because from the refuge we must be able to survey the prospect, and from the prospect we must be able to retreat to the refuge. Appleton has shown that the parks we design as places of beauty and rejuvenation inevitably typically-include many refuge-and-prospect juxtapositions. The Alhambra's Court of Lions (Figure 16-4) is a remarkably clear example of the presence of such characteristics in a much-loved architectural setting—it is an elegant distillation of the haven and the foraging ground, replete with surrogate animals at the water source.

Refuge and prospect need not be exterior characteristics only, and I believe they are most important, and most effective, in a building's interior. I have argued elsewhere that their consistent presence in the interiors of Frank Lloyd Wright's houses may explain, in part, the extraordinary affection those houses claim. The 1904 Edwin Cheney house in Oak Park, Illinois, is typical. The fireplace occurs at the very center of the house (Figure 16-5), in a small subspace, under a low ceiling, with opaque walls on three sides, and low levels of light. This is an *interior* refuge, small, low, dark, warm, cozy. Forward of it, at the left in the illustration, the ceiling planes of the larger and more brightly lit part of the living room are much higher than the flat ceiling of the fireplace zone. These ceiling planes continue over the dining room (in the distance in the illustration) and the library (outside the photo to the left). This area is the *interior* prospect, larger in all dimensions and more brightly lit, with views to the outside in three directions. Each of these two spatial conditions can be seen, surveyed, and accessed from the other. These characteristics are not unique to Wright's houses. I have cited the same characteristics in buildings by Adolph Loos, Jorn Utzon, and Mario Botta, and in the work of Seattle colleague Wendell Lovett, who long ago independently arrived at a belief in the value of refuge and prospect under a different terminology: he calls the refuge "the cave," the prospect "the meadow."

Men and women apparently seek a different balance between the two extremes: women are more oriented toward refuge and men, toward prospect. These preferences appeared with remarkable consistency in several architectural design studios with 50–50 gender

Figure 16-4: The Alhambra, the Court of Lions

Figure 16-5: Frank Lloyd Wright, architect, the Edwin Cheney house, Oak Park, Illinois; the living room

distribution at the University of Washington, and Judi Heerwagen and Gordon Orians find them as well in landscape paintings by contemporaneous women and men. Since few buildings are used exclusively by one sex or the other, this suggests that there should be a range of choice among refuge and prospect conditions—and common sense would suggest that varying needs relate not only to gender and, obviously, to individual personality, but also to time of day, time of year, and time of life. Thus, a range of choice among many refuges and many prospects in any setting is likely to be of real value, yielding a malleable surrounding that can accommodate changing emotional needs.

Such a range of choice is not provided by a collection of rooms of similar plan dimensions, uniform ceiling height, and relatively uniform distribution of light; it requires a more complex spatial composition. Interior refuge is established by largely opaque boundaries that define spaces of relatively small plan dimensions, with, often, a ceiling palpably near the top of one's head. But a view to an adjacent interior prospect is essential. The interior prospect is opposite in every way: opaque surfaces, if any, must be at greater distances, the ceiling may be significantly higher, a broad arc of view must be available. And the refuge must seem darker, the prospect brighter; contrast of light quantity is essential to the prospect/refuge model. This merits emphasis. Although vast areas of glass in buildings of recent decades have admitted large quantities of relatively uniformly distributed light, we need to remind ourselves that the dark place of concealment is important too. That is where we procreate, sleep, meditate, and recover from illness and injury; it is our haven in times of vulnerability.

ENTICEMENT

Stephen Kaplan has described a preference for certain natural scenes:

> The most preferred scenes tended to be of two kinds. They either contained a trail that disappeared around a bend or they depicted a brightly lit clearing partially obscured from view by intervening foliage. (Kaplan 1987, 8)

"A brightly lit clearing partially obscured . . ." is close to being a definition of prospect as seen from refuge, but it differs in that significant aspects of the scene are hidden; this is true too, of course, of the trail that disappears around a bend. In each case, there is the "promise that more information could be gained by moving deeper into setting." Kaplan proposes the term *mystery* for this characteristic.

> A scene high in mystery is one in which one could learn more if one were to proceed farther into the scene. . . . What it evokes is not a blank state of mind but a mind focused on a variety of possibilities, of hypotheses of what might be coming next. It may be the very opportunity to anticipate several possible alternatives that makes mystery so fascinating and profound. (Kaplan 1979, 50)

There is, again, a survival value: we better our chances if, finding evidence of a promising setting, we have an urge to discover, in relative safety, whether it offers advantages or dangers. I would add one qualification to Kaplan's definition, however. Moving toward the light is important to either scene: if progress is from dark to light, we will see before we are seen, and so will ensure relatively safe exploration. But if the path takes us from light to dark, other creatures could see us before we see them, and that is less pleasant—is, in fact, a staple of horror movies. I propose the term *enticement* rather than mystery, for view and access to a setting brighter than the one we occupy, whose features are only partly revealed.

Figure 16-6: The Ryoanji shrine, Kyoto

Figure 16-7: The Cathedral of St. Louis, St. Louis, Missouri; the sanctuary

Figure 16-8: Orvieto; a street leading to the Duomo

Figure 16-9: Wendell Lovett, the Cutler-Girdler house, Medina, Washington; the corridor

The architectural equivalent of the "brightly lit clearing partially obscured from view by intervening foliage" is represented by the view to the Ryoaji shrine near Kyoto (Figure 16-6); an interior example is the sanctuary of the Cathedral of St. Louis in St Louis, Missouri (Figure 16-7). The approach to the Piazza del Duomo in Orvieto (Figure 16-8) and the central hallway of Wendell Lovett's Cutler-Girdler house in Medina, Washington (Figure 16-9), are examples of the architectural trail "that disappears around a bend."

Enticement, then, partly reveals an information-laden scene; we must explore to discover more information. Whether the information is hidden by solid material, intermittent screening elements, or both, there must be clues that the concealed material is interesting enough to make exploring worthwhile. This, of course, is a judgment call; clues adequate for some may be inadequate for others. In any case, however, movement must be from relative darkness to relative brightness. If enticement demands movement from dark to light, how can we reverse the path, as most architectural configurations require at some point? We can ensure a brightly lit zone of enticement at another terminus, or several, so the sequence includes multiple enticements; this strategy can yield wonderfully rich spatial sequences. We can also design the path that links such a sequence so that it has no places of concealment—no alcoves, no corners—or, if places of potential concealment are unavoidable, they must be well lit so they too are prospect spaces; so that as we move toward them we see without being seen.

PERIL

Why do we build structures such as that shown in Figure 16-10? The prospects they offer are extraordinary, but what we really enjoy in each case is the thrill of the audacious experience. And the word *thrill* is the key. It is paradoxical: it involves two emotions, fear and pleasure, which are normally mutually exclusive. In these and all such settings, thrill is what we seek and enjoy.

Why? Appleton argues that survival requires sensitivity to danger, and this again invokes the pleasure-response rationale:

> If we were to be interested only in those features of our environment which are suggestive of safety, coziness and comfort, and not at all concerned with those which suggest danger, what sort of recipe for survival would that be? Seeking the assurance that we can handle danger by actually experiencing it is therefore itself a source of pleasure . . . (Appleton 1996, 85–90)[1]

I suggest for this characteristic the term *peril. Peril,* as defined here, differs from *anxiety*. In situations of anxiety, there may or may not be unseen dangers whose avoidance is not entirely within our control; hence, our fear is unalloyed by pleasure. In settings of peril, real dangers are fully evident, but they are dangers we can control, even if only by the exercise of care and skill—thus the appeal of such purely natural settings as Niagara Falls, the Grand Canyon, and the Matterhorn. Such settings present apparent and dramatic peril, but in all cases, we control the degree of risk, and in that controlled confrontation, we find a thrilling elation.

Figure 16-10: Tour Eiffel, Paris

The tall office building offers a similar opportunity. The profusion of balconies of Seattle's Northern Life Tower (Figure 16-11) dramatizes the thrill of elevation, and of a view over the void. Many cities now have incentives for what is often called "sculptured massing," a progressive recession of wall plane as the building ascends, and the condition carries with it opportunities for balconies or terraces. It may be that the time for the office or condominium balcony has returned.

Figure 16-11: The Northern Life (now Seattle) Tower, Seattle, Washington

Interior opportunities for the exploitation of peril can be found in elevated passageways across or adjacent to large interior spaces, though these obviously mandate a building with a generous interior volume. There are medieval examples in the galleries and the walkable triforia of any Romanesque or Gothic cathedral. Modern materials have enormously increased such opportunities; Mario Botta, at the San Francisco Museum of Modern Art, employs a transparent bridge floor to induce the thrill of peril into a museum experience.

The very expensive and high-style Reid Dennis house in Sun Valley, Idaho, by architect Arne Bystrom, illustrates a sophisticated use of the concepts. To the south, the entire length of the Dennis house is a single high and bright space (Figure 16-12) that opens to a series of terraces, with the mountains in view beyond. The north edge of this interior prospect, however, is established by what one might call a building within a building, at left in the illustration, that is opposite in every way. It includes, on the lower level, a dining room, a kitchen, and two bedrooms; above are three more bedrooms. These spaces are small, much less brightly lit, with low ceilings and dark surfaces; these, clearly, are the internal refuges. Each upper floor bedroom, furthermore, is a self-contained refuge and prospect, for the northern two-thirds of each bedroom is all opacity, concrete solidity, dark tones, wood and warmth (Figure

Figure 16-12: Arne Bystrom, architect, the Reid Dennis house, Sun Valley, Idaho; the great southern interior prospect space at right, the bedrooms at left

Figure 16-13: The Dennis house, a bedroom; the interior refuge

Figure 16-14: The Dennis house, a bedroom; the interior prospect, from which one looks through intervening foliage, and along a trail that disappears around a bend, to the southern interior prospect space, and the exterior beyond

16-13), while the southern third of each (Figure 16-14) is a white and light peninsula with a view to the interior prospect that in turn looks to an exterior prospect of the terraces and the valley beyond. And the path from the bedroom-refuge to the bedroom-prospect is a trail that disappears around a bend, while from the bedroom-prospect the great southern space is seen, as it were, through intervening foliage.

A CASE STUDY: A RETIREMENT HOME

The characteristics described here are especially important to retirement homes, hospices, hospitals—settings that are under unusual pressure to provide comfort and contentment, since their occupants are relatively confined. I want to show how some quite modest changes in an actual retirement home design can provide some of the characteristics described.

The plan (Figure 16-15) is that of a two-bedroom unit in an existing Seattle retirement home of the mid-1990s. Its floor area is about 950 square feet; the ceiling height is eight feet throughout. (Above the ceiling a continuous 12-inch-high space houses ducts, electrical conduits, and the sprinkler system.)

Apart from any survival-advantage characteristics, some plan flaws are obvious. From the entry, one looks directly into living space and also, through the kitchen, into the main bedroom if its door is open; if the bathroom door at left is open, the toilet is visible. A few steps ahead, the entire living space and the small bedroom can be seen, so when callers appear at the door, the whole unit is on view. The kitchen is cramped, the corner cabinet is less than ideal, and where does one put a dishwasher? In the main bedroom, the toilet is seen from the bed if the bath door is open. The balcony is a real amenity, but it is too shallow to hold a table-and-chairs arrangement.

Furthermore, the three major rooms are of about the same size and exactly the same ceiling height; each bedroom is lit by a window centered in the exterior wall, and in the living room a sliding door is similarly located. Therefore, there is no meaningful variety or contrast among the spaces.

The proposed revision (Figure 16-16) accepts the unit's existing dimensions and volume, the basic wall lo-

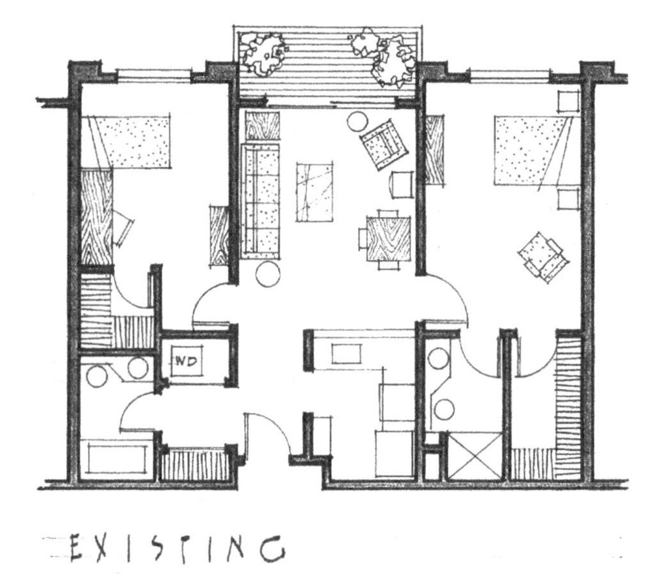

Figure 16-15: A two-bedroom unit in a Seattle retirement home of the 1990s, as built

cations and plumbing, the corridor, and the structure. Similar furniture items are shown in each plan, and at generous actual sizes—each sofa as shown, for example, will sleep an NBA guard. A queen-size bed is shown in the main bedroom, a twin-size bed in the extra bedroom. The revision is entirely straightforward: Materials are commonplace and inexpensive, surfaces meet at right angles—there are no elaborate plan complications like those of the Dennis house, no such demands on materials and craftsmanship. A bolder architectural vocabulary probably could yield a richer result, but comparing two straightforward configurations can demonstrate that the characteristics herein described can be developed in ordinary materials and a pedestrian geometry.

Simple things first. The revision provides a wider and deeper corridor recess; the inside entry hall is broader by three feet. From this more spacious entry, none of the main spaces is seen. At left, when the bath door is open the toilet is hidden, as is the other toilet from the main bedroom. The deck is deeper, front to back, by two feet, and can now accept a dining table for three—four in a pinch. Revised columns at its sides widen the deck and the living space by a foot, at the cost of six inches of breadth in each bedroom. The living space depth is increased by a foot by revising the kitchen, which now has more counter space, no maddening corner cabinet, and room for a dishwasher. From both bedrooms, small windows give views to the deck, and in the main bedroom, a little interior window opens to the living space, with a hinged closing panel. Closet space is unchanged in the bedrooms. The entry closet is smaller, unfortunately—we live in an imperfect world.

A sliding door to the main bedroom is shown; it is likely to be open most of the time, and when it is it will be invisible. When it and the little hinged panel between bedroom and living, are closed, the main bedroom becomes a private micro-apartment for one occupant, while the other works late in the living space, or meets with a business associate or an old high-school friend.

Now, what about prospect and refuge, enticement and peril, and complex order?

Figure 16-17 shows, in each plan, the interior refuge areas as shaded, the interior prospect not. In the existing unit, no interior refuge is shown, and no interior prospect, because the ceiling height is a constant 8 feet,

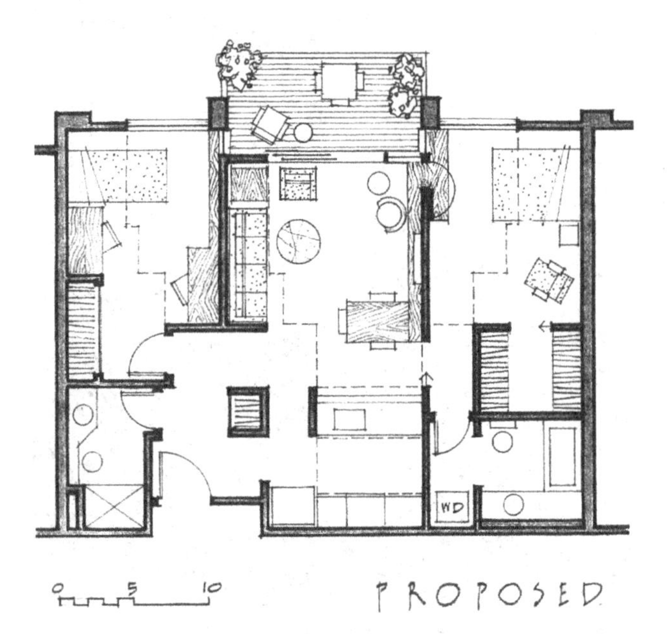

Figure 16-16: A proposed revision of the unit plan. Dashed lines indicate the edges of dropped ceilings.

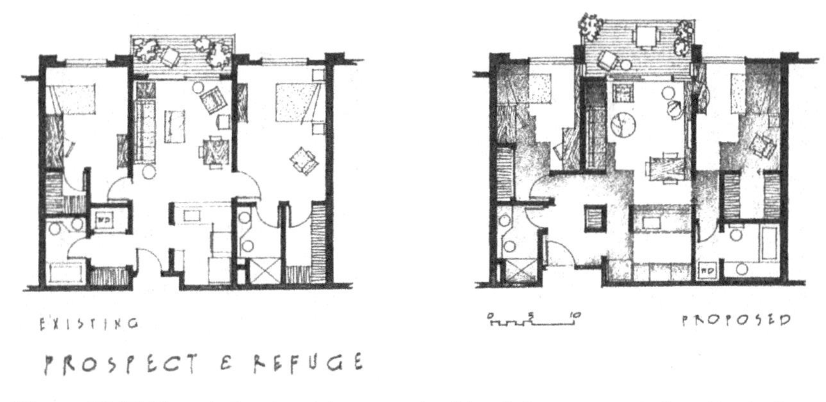

Figure 16-17: The as-built unit and the proposed revision. Refuge areas are indicated by shading.

and the side walls in the three major rooms are equally lit by the centered windows. In the proposed design, the ceilings in the shaded areas are at 7 feet 6 inches, those in the unshaded areas at 8 feet 10 inches. (The higher ceilings are obtained by locating ducts, electrical runs, and sprinkler pipes in lower ceiling areas; floor-to-floor height is unchanged. The average ceiling height of the revision, interestingly, is 8 feet 3 inches, 3 inches more than existing.) In each bedroom, a lower ceiling is above the likely bed location. The main window in each bedroom is moved toward the corner to create darker and lighter zones in the room; and the lighter zone includes a small window to the deck that brings in a bit more light. Thus walls of refuge areas, distant from window edges, are softly lit, while walls of prospect areas are washed with light from contiguous windows. And, obviously, all interior refuge areas open to one or more interior prospects. The left third of the living space is similarly dark and cozy, while at right is a higher and brighter prospect-space whose wall is washed with light. At the far end of this wall, the interior window opens the interior prospect to that of the adjacent main bedroom, and vice versa.

Figure 16-18 indicates arcs of view to the exterior from three significant vantage points. All three major rooms now enjoy a view of the deck, the exterior prospect-claiming platform. Both bedrooms have windows on two walls rather than one—which may seem a

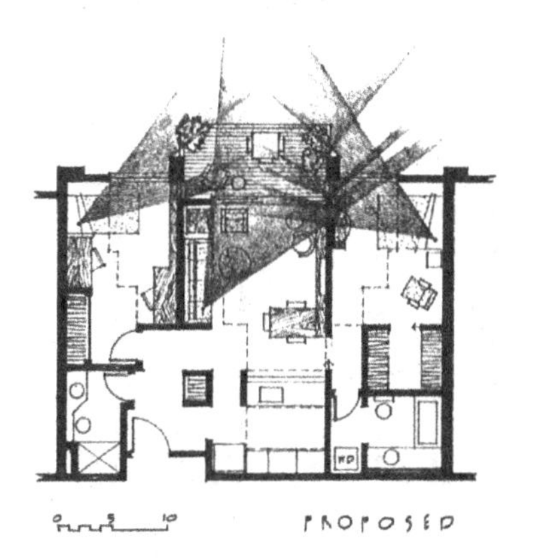

Figure 16-18: The as-built unit and the proposed revision; a comparison of views available from key locations

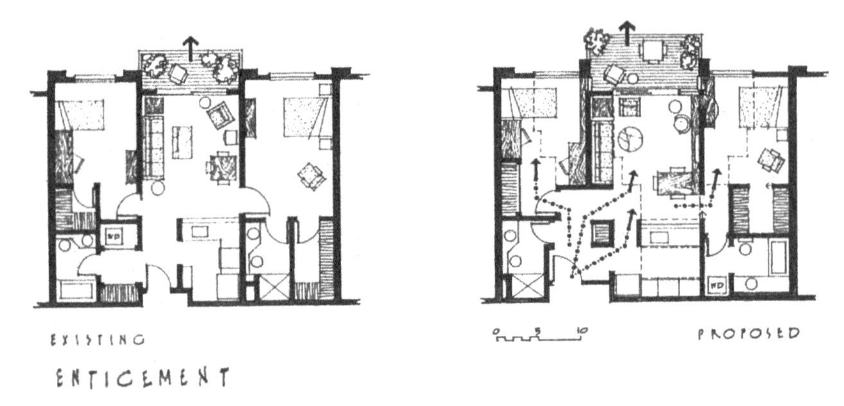

Figure 16-19: The as-built unit and the proposed revision; trails that disappear around bends, leading toward the light

minor point, but it doubles the axes of outlook. The little opening between the living space and the main bedroom opens a view from each room to the other, but also *through* each to the exterior.

Enticement is indicated in Figure 16-19. The existing unit offers none. The proposed revision has four enticing trails that disappear around luminous bends, and the deck's perilous thrill can enliven breakfast, lunch, martini time, dinner, and the small hours of the morning.

What of complex order? In the existing design (Figure 16-20), the outside wall of each major room is a symmetrical composition—a window is centered in each wall. No other ordering relationships are evident, nor can I find any complexities. In the proposed design, the outside wall of the living space is symmetrical—or is it? No, not quite. . . . The bedroom windows are clearly not symmetrical within their respective walls, but there is a richer symmetry, in that the bedrooms, taken together, are mirror images of one another. Or are they? Their more complex relationships—their similarities, their differences—will be discovered and understood only through movement and memory, as the mind holds a series of musical notes to create melody. The left wall of the entry has a stepped alignment that is like—and not like—the stepped changes of the ceiling planes in the major rooms. The steppings of those ceiling planes align from room to room, as shown by the dotted lines—but not entirely; the steppings of the

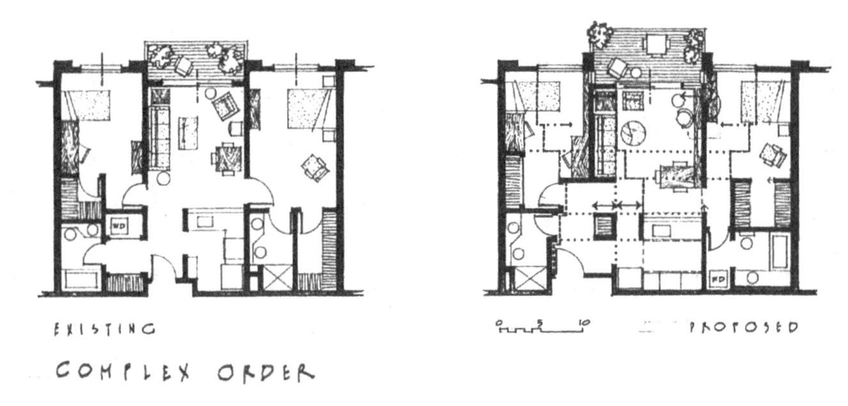

Figure 16-20: The as-built unit and the proposed revision; repetitive themes and variations thereon

living space ceiling relate to, yet differ from, those of the bedrooms. Other dotted lines and arrows identify other alike-yet-different relationships. No such observations of complex order can be made of the existing unit. And if we were to draw the walls as vertical planes, as we would see them in actuality, we could show how similar complexities and orders can be worked out in the vertical dimension. The proposed design is, perhaps, a long way from being a Rachmaninoff concerto, but it might begin to claim vague analogies to a simple melody.

The proposed design would be more costly than the existing. More wall and ceiling surfaces must be built and finished, with many more corners and edges. The sliding door and the hinged panel in the main bedroom are extras, as are the little exterior windows in both bedrooms. But all appliances, fixtures, flooring, windows, other doors, hardware, electrical provisions, and 75 percent of wall and ceiling surfaces, are unchanged, and the cost of the larger deck is probably nil. Furthermore, the major costs of the total building—land, fees and permits, excavation, structure, elevators and stairs, electricity and plumbing, heating and air-conditioning, lobbies and corridors, the group dining space and its kitchen, administration spaces, exterior walls, and the roof—are unaffected by unit plans. So the premium for including many units like the one proposed will be, at most, an additional 1 percent of the building's total cost. Whether such spaces are worth the cost could probably be discovered by making a computer graphics movie of such a living unit, and a similar movie of an existing one, and asking a test audience to compare the appeal of each. Nevertheless, some fairly solid theoretical and empirical material, including that described in this chapter, suggests that such units are likely to justify themselves in the market.

In ways such as these, then, perhaps, buildings might offer meaningful surrogates of those appealing characteristics of nature that once, long ago, gave us better odds for survival, and, although no longer useful for that purpose, may still bring us pleasure. If buildings can do that, it may be of real importance that they do so. Because, dear as the natural environment is to our emotional well-being, it is now hardly the environment in which we live. We are today an urban-dwelling species. Buildings are where we eat, and sleep, and entertain, they are where we live, and work; they are where we hold conferences; they are where we write books, and chapters for books. It seems to me, therefore, that our study of the value of the natural cannot stop with the natural, but must somehow affect the everyday, the not-nature rooms and corridors and meeting halls where we spend the vast majority of our lives. I am interested in the degree to which buildings might, at some serious level, be effective surrogates, microcosms, of what Wordsworth has called that imperial palace whence we came.

NOTES

1. Martin Mador suggests as well, although Appleton does not, that the adrenaline rush generated by such settings may also be evolutionarily useful in preparing us to deal with potential adversities.

REFERENCES

Appleton, J. 1996. *The Experience of Landscape*. London: Wiley.

Kaplan, S. 1987. "Aesthetics, Affect, and Cognition in Environmental Preference from an Evolutionary Perspective." *Environment and Behavior* 19(1): 3–32.

———. 1988. "Perception and Landscape." In *Environmental Aesthetics*, edited by Jack L. Nasar. Cambridge: Cambridge University Press.

chapter 17

Toward Biophilic Cities: Strategies for Integrating Nature into Urban Design

Timothy Beatley

CITIES OF NATURE

Not long ago an advertisement appeared in the real estate section of the *Washington Post*. It was promoting a new development, and it provided some telling insight into our popular view of cities. The advertisement read, "The *NICE THING* about the city is that it eventually ENDS" (emphasis in original.) The image juxtaposed sidewalks, a fire hydrant, and other essentially grey surfaces in the foreground (bad), with the bucolic images of forest and farm field in the distance (good). The implications were clear—if you want any meaningful exposure to nature, quickly exit the city. Nature was out *there*, not in the city, not close to where most people live. Such advertisements convey much of what our popular attitude toward cities is, and our impressions of them.

There are elements of truth, to be sure, to the sentiments expressed in this advertisement—there is in fact too much pavement, too many bleak grey neighborhoods, too many cars—but these sentiments are wrong in some profoundly significant ways. Cities are inherently embedded in complex ecosystems, nature is all around us in cities if we look, and the extent of urban biodiversity is often quite considerable. In Chicago, some 7 million birds pass through the city during peak migration times. Some of the oldest and most impressive trees can be found in places like New York City, where there is (in Queens) an oak tulip tree more than 400 years old. Underfoot, the diversity is even more impressive: 51 species of ants were recently categorized in the city of Philadelphia, for instance, enough to make any urban myrmecologist proud.

Cities, moreover, can be designed and planned to be profoundly more "natureful" and organic, providing opportunities for extensive and deep contact between urban residents and nature, if we choose this direction. Thankfully, there are now many compelling models and examples, many greening ideas and techniques, that can be employed in building and rebuilding, or should I say *growing*, biophilic cities, and what follows is a brief sampling of some of them. I draw extensively from urban design and planning practice in North American, European, and Australian cities.

An initial and important observation is that *green urbanism*, as I often refer to it, must be seen to occur at multiple geographic levels. It's an inherently multiscale project, where progress and good work can happen at metropolitan and regional levels, down to the level of a home or building (indeed, even the interior spaces of those structures). Table 17-1 summarizes some of the more specific greening ideas and techniques that might be applied at these different primary scales (meant to be illustrative more than exhaustive).

TABLE 17-1 Biophilic Urban Design Elements Across Scales

Scale	Biophilic design elements
Building	Green rooftops Sky gardens and green atria Rooftop garden Green walls Daylit interior spaces
Block	Green courtyards Clustered housing around green areas Native species yards and spaces
Street	Green streets Urban trees Low impact development (LID); vegetated swales and skinny streets Edible landscaping High degree of permeability
Neighborhood	Stream daylighting, stream restoration Urban forests Ecology parks Community gardens Neighborhood parks/pocket parks Greening grayfields and brownfields
Community	Urban creeks and riparian areas Urban ecological networks Green schools City tree canopy Community forest/community orchards Greening utility corridors
Region	River systems/floodplains Riparian systems Regional greenspace systems Greening major transport corridors

Source: Modified from Girling and Kellett (2005)

Green features of individual urban buildings and projects can cumulatively contribute much to the green fabric of cities and neighborhoods. Green or ecological rooftops provide an example. In some of the green urban cities I have studied in Europe, a long and extensive history of mandating and subsidizing these green features has resulted in their number and presence such that they affect in meaningful (and positive) ways urban climate, biodiversity, beauty, and aesthetic experiences. In the city of Linz, Austria, there are now more than 300 green rooftops, in addition to other features. Seen as they should be, as pieces in a larger urban mosaic, these patches help to form important elements of a neighborhood's "green grid."

Green rooftops and other urban ecological features can also help to reconnect us to our native landscapes. One recent creative green rooftop design can be seen in the new Ballard branch of the Seattle library (Figure 17-1). The sloping rooftop has been planted with 14 species of grasses native to the Northwest. Such species as woolly yarrow, red-creeping fescue, long-stoloned sedge, and Oregon stonecrop connect it to the indigenous landscape in a way that many green rooftops don't. And the relatively high bushy grasses can be seen from the sidewalks and public spaces around the building, injecting a wild vegetated grassy view in an otherwise heavily built-up area. From within the structure there is a creative periscope for viewing the roof, as well as a stairwell leading up to a 360-degree observation deck that provides a spectacular view of the sea of grass on the top of this different sort of library. In all, 18,000 plants have been planted on the roof, all low-water species.

As these creative architectural and building practices demonstrate, there are many things that can be done at the level of the building. But it is important to move beyond the sense that simply providing a view of nature out of windows, or incorporating plants and greenery within the interior of living spaces, is enough. These are positive steps, to be sure, but they don't allow for the deeper connections to, and the personal experience

Figure 17-1: Ballard Branch of Seattle Library, Seattle Washington

of, the natural world that we need. However biophilic individual buildings are, they will not by themselves lead to biophilic cities and a broader biophilic culture, unless they are situated and configured in a way that permits extensive and deep outside experiencing of nature—in one's backyard, or in a nearby forest, along a river or creek.

Indeed, those urban environments that I believe have the highest biophilic qualities, are places where it is possible in fact to move from one scale—from a street or backyard—to progressively larger and more diverse ecosystems at larger scales. Ecological connections are critical, but pedestrian connections are essential as well. For children (and families and adults) to enjoy these spaces and environments, it must be possible to walk or bicycle to them and in them. Overcoming the tremendous obstacles presented by cars—dangerous multilane arterials with fast-moving vehicles being the norm in many places—is a huge challenge. Some green neighborhoods have been designed to overcome these through pedestrian overpasses and underpasses. Hammarby Sjöstad, for instance, a new ecological neighborhood in Stockholm, features much nature very nearby for most residents, and a highly connected, walkable environment. Even more admirably, two green ecoducts connect the neighborhood with a larger forested park, providing in this case safe passage to a large mysterious natureful world beyond the boundaries of one's specific block or neighborhood.

At regional, bioregional, or metropolitan levels, importance must be given to preserving and restoring large interconnected green systems—forests, rivers and riparian networks, farmlands—that set the larger template in which fit green systems at smaller scales. There is a long tradition, particularly in Europe, of planning such regional ecological networks, and guiding regional growth patterns to ensure the existence of and access to these larger networks. In German, Dutch, and Scandinavian cities, for instance, importance has been given to bringing about compact urban form, often along transit lines, but within a large regional network of green spaces that in many cases come into the very center of cities (see Beatley 2000, 2004). In Copenhagen, its famous regional "fingers" plan, with large green wedges that extend to the center, dates to 1947. In Helsinki, large green wedges have been designed similarly. Keskuspuisto, Helsinki's central park, is perhaps the best example—it extends 11 km from old-growth forest at the city's edge to the center of this compact, fairly dense city. Large blocks of greenspace and natural landscape, in close proximity to dense populations and easily reachable by public transit, is a hallmark of many of these green urban cities. In Hannover, Germany, an 80-km long "green ring" has been recently completed, connecting very large blocks of greenspace and a diverse set

Figure 17-2: The Eilenriede forest, a part of the Hannover Green Ring, Hannover, Germany

of ecosystem units that surround the city. At its center, the network includes the large (650 hectare) and beautiful Eilenriede forest (Figure 17-2). Regional and urban-scale green networks serve many functions, of course, including climate modification and urban heat island mitigation (German cities protect forested riparian areas because of the positive movement of fresh air urban areas), habitat conservation, water quality protection, carbon sequestration, and sustainable wood production, but providing recreational benefits and access to nature for urban population is a major goal.

GREEN NEIGHBORHOODS

Regional green systems and urban ecological networks in turn provide a framework in which greening strategies and biophilic design can occur at the neighborhood level, the scale at which everyday life and living occurs. The configuration of buildings at the block and neighborhood levels becomes critical in shaping the kind of experiences and access to nature that residents enjoy. There are many contemporary good examples and compelling models of what biophilic neighborhoods might look and feel like, and what their main biophilic design elements might consist of. My own studies of green projects in Europe suggest not only the importance of layout and design at this level, but also the many different and creative ways this can be done. Some of the best projects involve clustered housing, where (relatively dense) dwellings are configured in ways that allow a network of green spaces, including courtyards, and semi-formal common spaces, then connect with and lead to other green areas nearby. Recent good examples include Eva-Lanxmeer in Culemborg, Netherlands; Vauban in Freiburg, Germany; Vikki in Helsinki; Hammarby Sjöstad in Stockholm; and Kronsberg in Hannover; among others. Common design elements include internal green courtyards; green areas at the block level that restrict or prohibit car access, and connect to larger natural areas; regional systems of greenspace; extensive mobility options designed-in, especially walking and bicycling but also safe and fast public transit (fast trams as a key design element in Kronsberg, for instance); green features serving as major functional elements (e.g., stormwater collection in Kronsberg) as well as addressing recreational and aesthetic needs; and a diversity of green areas and green features (tree planting and preservation of trees, green rooftops, community gardens, water features, and natural habitats). Overall, these projects demonstrate convincingly both that such robust and extensive greening can be comprehensively designed and integrated into new developments and neighborhoods, and that they can enhance tremendously the quality of life at the same time advancing goals of sustainability.

I recently took a group of students to see one of the oldest and most written about Danish cohousing projects, Trudesland, which demonstrates the impressive level of green access that can result from these types of clustered housing design. Not only is there a delightful and car-free pedestrian pathway or ped-way in the center (essentially the public spaces shared between the attached houses), but there are extensive green spaces behind the homes as well, with a relatively wild green area, actually only a few hundred meters away, for kids to play in. The wild space had high grass and flowers, tire swings, and large climbing ropes strung from tree to tree. Our host on that day, a Trudesland resident, described the remarkable degree of mobility and freedom her four-year-old enjoyed growing up in this neighborhood. She admitted the only problem was sometimes finding out at any given time where—in which house—the four-year-old could be found. Beyond these green areas around and near the houses, residents are able to walk or bicycle to larger green parks, as well as to the center of town where there are shops, restaurants, a grocery store, and a train station for traveling to the center of Copenhagen if one wished. (For an overview and discussion of cohousing see McCamant, Durrett, and Hertzman 1993.)

A similar story can be told in other green European neighborhoods studied by this author. In Understenshöjden, an ecovillage in Stockholm, clustering has allowed for a remarkably small development footprint (the homes have been designed and built to look as though they have been dropped in by helicopter!) and the preservation of a marvelous forested environment, again near transit and near the center of the city. This is on top of a host of green building features, including solar hot-water heating and use of nontoxic paints and other sustainable materials. Interspersed between and among the buildings are small tables and outdoor eating areas where families informally gather. Unpaved paths meander through the woods connecting the homes to each other and to a peripheral common parking area. The dominant feeling of this place is of living (compactly) in a native forest, though in a very urban setting.

New urbanist–style projects and neighborhoods in the United States (and elsewhere) offer some biophilic potential as well, and have been especially promising in their attempt to create a (relatively) more compact urban form, their interest in mixed uses, and their pedestrian investments and comparatively high degree of pedestrian connectivity. Many of these neighborhoods, however, are situated in car-dependent suburban settings and don't incorporate much nature (though there are usually neighborhood parks and public squares). Some promising exceptions exist, including Civano in Tucson, which does a splendid job, I believe, helping to tie residents to the incredible flora and fauna of the Sonoran desert. Though still rather car-dependent, considerable desert habitat has been set aside, residents are able to hike on desert trails, and priority has been given to the use of native desert vegetation in landscaping (including an on-site nursery that has successfully relocated cacti, mesquite trees, etc.).

Moreover, nature can be a prominent design priority in revitalizing and redeveloping urban neighborhoods. The Greenwich Millennium Village is one excellent example. Here, a dense new ecological village has been built in the heart of London, but with an extensive and impressive ecology park (Figure 17-3). A network of wooden walkways, with benches and bird blinds, gives residents unusual access to this restored ecosystem and its biodiversity. Part of a connected pedestrian and bicycle network (including the Thymes Trail), adults and especially children have an easily reached and safe natural area to visit and explore. For those living in the neighborhood, this natural element is central and ever-present.

Figure 17-3: Greenwich Millennium Village, London, UK

I have documented other impressive Scandinavian examples as part of my ongoing work on European green urbanism. The Western Harbor redevelopment project (Västra Hamnen) in Malmö, incorporates a number of impressive natural elements. Compact, dense, and highly walkable, developers here were required to incorporate green features and to satisfy a minimum number of green design points. Most innovatively, this new urban district contains a marvelous network of water features, with water flowing and trickling and extensive aquatic plants ever present. Downspouts in Western Harbor lead to a marvelous open-channel water system that snakes its way through dense housing, including terrific vegetation and bubbling water sounds throughout. Green roofs, gardens, and courtyards are also features, and an impressive 100 percent of the energy needs for the district are provided from local, renewable energy sources. Another way to envision reconnection to environment, the Western Harbor shows the value and ability of harnessing nature's free services and ecological bounty. Here, a wind turbine, roof- and façade-mounted solar hot-water heating panels, and utilization of the heating and cooling benefits of seawater and underground aquifers (through a heat-pump system), are indicative not only of a sustainable form of development, but one in tune with the climatic, geologic, and other intrinsic natural characteristics of that particular *place on earth*. There are, of course, many elements of existing urban neighborhoods that represent special opportunities for inserting or restoring nature—vacant lots, building sideyards and setback areas, parking lots, and parking areas, among many others.

RETHINKING URBAN INFRASTRUCTURE: GREEN STREETS AND BEYOND

Biophilic urban design at the neighborhood (and city) level also requires profound rethinking about infrastructure and infrastructural needs. The Western Harbor project, for instance, turns the power grid on its head—energy infrastructure here is in the form of resilient on-site production, restorative and renewable, not the usual kind of energy infrastructure. Roads, bridges, tunnels, ports, to name a few, could all be profoundly reconceived and re-imagined through a biophilic lens. The so-called green bridge in London, for instance, connects two pieces of an otherwise fractured ecology park, almost like a magic green carpet of mature trees and greenery (and no cars) floating over and above several lanes of congested urban traffic. A sewage treatment plant in Seattle has become (partly) a park and hiking trail, while a recycling facility in Phoenix has been redefined as an opportunity to teach about waste—there are increasingly many good examples to be found. (See Beatley 2005 for a discussion of these projects.)

Streets must be reconceived as not only (or primarily) infrastructure for the conveyance of cars and traffic, but as places that harbor native plants and biodiversity, that collect and treat stormwater, and where pedestrians can experience intimate contact with nature as part of their daily routine. The rise in the use of low-impact development (LID) techniques to address stormwater has provided new opportunities to profoundly rethink yards, streets, and alleys. In Seattle, for example, under the leadership of the Seattle Public Utility (SPU), an effort has been made to show the natural alternatives to conventional street-sidewalk-and-yard designs, demonstrating LID methods through retrofitting existing streets. Beginning with its Street Edge Alternatives (SEA) program, wide auto-dominated (suburban) streets have been converted into narrow, wavy, vegetation-filled green streets with sidewalks, where before there were none, and a seemingly endless diversity of wild flowers and greenery. The street has become a series of rain gardens collecting and treating stormwater and nourishing this verdant scene, where sterile conventional turf grass lawns existed for the most part before (Figure 17-4). Seattle has now gone beyond converting single streets to creating entire "green grids" of connecting and intersecting roadways that together set the baseline condition for these green neighborhoods.

Figure 17-4: Seattle Street Edge Alternative, Seattle, Washington

Projects like SEA in Seattle are certainly a viable alternative way of thinking about stormwater infrastructure—a biophilic approach going beyond the conventional pipe in-ground, engineering philosophy prevalent (still) in most cities. And the evidence suggests that such LID systems are highly effective at containing stormwater and controlling urban pollutants, and of course enhancing amenity (and economic) value of urban neighborhoods. The notion of a "green street" is now commonly incorporated into city plans, but its meaning is typically restricted to the planting of street trees or median-strip vegetation, for instance. These are often positive contributions, of course, but we should stretch our imaginations and visions further about what a street *could* be. In Seattle, an initiative in the Belltown neighborhood, Growing Vine Street, is demonstrating some remarkable new ideas about reenvisioning streets. Here a portion of Vine Street, sloping down to Elliot Bay, has been reconfigured (Figure 17-5). Space for cars has been taken back and reallocated to the pedestrian—water is collected from rooftops, sent to visible (and artistically dramatic) cisterns, in turn feeding a marvelous, multileveled "runnel" or open water channel. Here is one redesign where an auto-oriented street becomes a community gathering place, a place to stroll, a place to

Figure 17-5: Growing Vine Street, Seattle, Washington

enjoy trees, flowers, greenery, a garden for growing food (it integrates a Seattle P-Patch garden), as well as a novel system for collecting, treating (and celebrating) stormwater. And in the end, the redesign helps build a reconnection to place, and a strengthening of this increasingly vital neighborhood.

Examples like the Mole Hill community in Vancouver, B.C., show, furthermore, that food production can and should be designed into new neighborhoods as well, and this is another way in which biopic urban design can restore natural connections (and provide another response to rethinking conventional urban infrastructure). Here, an urban alley has been largely converted into raised-bed gardens and a lush edible landscape (Figure 17-6). The new ecological district called Viikki, in Helsinki, is a marvelous example in that region. There, linear community gardens have been incorporated into the spaces between major blocks of housing. Residents are entitled to garden plots adjacent to their homes, and in a number of cases the growing vegetables or flowers can be seen from balconies and windows. Common areas, areas in and around buildings, become redefined, again, as opportunities for experiences and physical earthly delights that in most typical urban neighborhoods would be absent or infrequent.

Restoring the natural hydrology of urban areas is a major social and engineering challenge, to be sure, but has tremendous potential for reconnecting us to the natural rhythms of life. The sounds of water, the sights, smells, and tactile sensations of water, especially for children, are indeed life-enhancing and ought to be viewed as an essential element of any urban neighborhood. Cities like Zurich have been active in promoting stream daylighting (bringing streams and creeks back to the surface) and urban stream restoration, and examples exist as well in a number of American cities (for instance, the recently completed daylighting of a portion of Ravenna Creek in Seattle). One of the boldest and most ambitious proposals has been to daylight a portion of Strawberry Creek, which would run directly through a portion of downtown Berkeley. In a consulting report by Mason-Wolf associates, a range of engineering alternatives were examined, including a "full flow restoration" option. A series of photo simulations of what the green results might look like are impressive and persuasive: Center Street in Berkeley is transformed from an environment largely of parking lots and pavement to one of verdant riparian wildness in the heart of the city, with trees, meandering water, vegetated stream banks, and more pedestrian spaces (and fewer car) as well.

Figure 17-6: Mole Hill, Vancouver, British Columbia, Canada

Herbert Dreiseitl has been one of the most enthusiastic and inspired designers when it comes to reintegrating water back into the urban fabric, and his projects are groundbreaking indeed. I have found myself on more than on occasion sitting long hours on a train to visit and document his work. One of my favorite projects can be found in the German town of Hattersheim, outside of Frankfurt. Here, Dreiseitl has converted the steps of the town hall into a gurgling flow form, transforming a built object into something that appears intrinsically organic and irresistible to passing school kids who play, seemingly for hours, in the flowing water (see Figure 17-7). The water then flows through the town's main plaza, moving underground in several places and popping up occasionally, becoming a natural stream at the edge of the town center. Few urban modifications have as much transformative potential as projects like this, reinserting natural rhythms, sounds, and sights, and creating delightful magical spaces in the urban realm. Americans might more frequently profess to love cities if they looked and felt more like this.

Figure 17-7: Hattersheim Town Hall, Hattersheim, Germany

Part of the planning and design challenge will be to pursue a variety of creative means to propel people outside. This will have to include investments in bicycle mobility and pedestrian amenities and infrastructure of all sorts (traffic-calmed streets, pedestrian over- and underpasses, sidewalks), as well as creative public and neighborhood art projects, making street festivals and block parties easier, and generally making it more interesting, intriguing, fascinating, and safer to be out and about. Breaking the sedentary hold on our interior lives, and through neighborhood and community design getting people outside, will do much by itself to advance a biophilic agenda. Incorporating *edible* opportunities into cities and neighborhoods is also key—both the conventional (cafés, neighborhood restaurants) and less-than-conventional (edible landscaping where kids and adults alike can forage, pick, and savor apples, blackberries, and whatever else will grow). The latter, I've discovered from firsthand experience, provides an element of fun to being outside that can be extremely useful in implementing our charge of "no child left inside."

My own research suggests that there are a variety of design and planning ideas for moving us in the direction of spending more time outside. Partly this is a function of designing-in fine-grained spaces at the neighborhood level—small pocket parks, plazas, green courtyards, but being even more creative about these spaces. Informal gathering spaces can and have been provided in sometimes unconventional ways and places (e.g., consider the Intersection Repair initiative, spearheaded by the Portland, Oregon group, City Repair). Much of the planning agenda is about retrofitting existing urban and suburban neighborhoods to permit more vibrant outdoor life. In many residential neighborhoods there are opportunities to pull down fences and connect backyards to create habitat-rich collective spaces, and to create urban and suburban hiking and nature trails. This has been done to a limited extent (e.g., the N Street Cohousing project in Davis, California), but finding ways within our planning systems to facilitate this, indeed encourage it, would be worthwhile indeed. Ensuring that every new (and existing) neighborhood is connected to a regional bikeway, greenway, or trail system, would do much to facilitate nonautomobile, outdoor living, and the return of "free-range children."

There have also been considerable new efforts at rethinking schools, in the urban planning and design literature and community. The notion of smaller neighborhood-centered schools, and schools that allow students to walk or ride their bikes to them, are regaining favor. And while there are many impediments to this (including state facility guidelines that have encouraged ever larger and more peripherally located schools), rethinking schools also offers the opportunity to advance biophilic design. With the compelling evidence of the pedagogic effectiveness of outdoor learning, schools could be reimagined as neighborhood and community natural systems, places where educational facilities are sensitively integrated into and fit within (protect and help to restore) these natural systems. Almost every school, even those without much nature on-site, will have access to the larger ecological networks mentioned earlier. In my own home city of Charlottesville, Virginia, our 20-mile-long Rivanna Greenway encircles the city, with most city schools within just a short walk of trailheads and greenway openings (Figure 17-8).

Figure 17-8: John Holden, a member of the Rivanna Trails Foundation, holds up a map of the Rivanna River Greenway, Charlottesville, Virginia

Some of the most compelling school examples can be found in Australian cities. Few schools are doing as much as the Noranda Primary School, which I had the great pleasure of visiting. Located in the Bayswater Council in northern Perth, this school has placed a priority on preserving a significant natural area, a beautiful bushland, on the school grounds and to incorporating natural heritage and bushland conservation values into the curriculum of the school. Specifically, behind the main school buildings lies an impressive remnant of forested bush. Remarkably intact, though degraded in parts, the land is home to an abundant and diverse flora and fauna—grass trees, red gums, even orchids are there, including at least one species of rare orchid. There are many school activities that utilize the bush, and it has essentially taken the place of some of the more conventional forms of school equipment typically seen on school grounds. There is a Bush Wardens program, where participating students are involved in a variety of activities aimed at learning about and caring for this natural area. The bushland is the site of daily walks by the students and is utilized by many of the classes in teaching particular subjects. Students in all grades, whether or not they are participating in the Bush Wardens program, are taught about the bush, and the school has commissioned a special curricular manual "Our Bushland Classroom" to help in this pedagogical mission. And so this is a different notion of a school—a biophilic school, if you will—where much of the teaching is about nature and where nature becomes the classroom. The surrounding residential neighborhood joins in the enjoyment and appreciation of this impressive site of local nature as well.

THINKING BEYOND URBAN PARKS

We perhaps also need to rethink the greenspaces around us and to greatly expand our notion of what these spaces are, and might be, and how they function.

For many, urban greenspaces fall into a single mental category: a park, with its cut grass, benches, play equipment, and perhaps a water feature. Such places in neighborhoods and cities are useful and enjoyed by many, but I think we need to expand our concept of what parks might or could be. Ensuring that there are wild spaces and wild elements in every urban park we create is one step. But perhaps every neighborhood needs not just a small park, but a community forest, and perhaps as well this forest should be viewed as part of the working landscape—managed for its wild and natural conditions, but also a source of sustainable wood and lumber, partially supplying the material needs of the residents. Such community forests exist, and although acquiring and protecting natural areas such as these within cities can be very expensive, they serve many ecological, social, and even economic functions that justify and compensate for these high costs. The community forest in Arcata, California, is one example—a magnificent stand of redwoods providing an impressive degree of nature within a close walk for many. At the same time, timber is harvested, at a level below its annual growth and therefore sustainable, and the income helps pay for the forest's management and for the purchasing of additional land to add to the forest.

A second example can be found in the London ecological project BedZED—Beddington Zero-Energy Development. While this new green neighborhood incorporates many ecological design elements, timber used in its construction is harvested from community wood lots in the nearby borough of Croydon, and wood waste derived from local urban tree trimmings and cuttings is burned in its combined heat and power plant. The borough itself has even become certified under the Forest Stewardship Council as a sustainably managed forest, quite an unusual step. The *city* as a *forest* (or an orchard, or a wild habitat)—what a marvelous recasting of the spaces and places in which most Americans (and increasingly much of the rest of the world) live!

Green urban design and biophilic cities must also be about appreciating the nature that persists or has emerged in the many leftover spaces that can be found there. These include abandoned rail lines, vacant lots, and former industrial sites. Considerable biodiversity emerges in such places, and many of these spaces, moreover, are famously beloved by the children and others who visit them and spend much time in them and with them. Bob Pyle's eloquent recollections of his time along the High Line Canal are testament to the immense value of these places (Pyle 1993).

To be sure, there are many potentially important elements that comprise the green urban fabric of cities, including street trees, pocket parks, green courtyards, green and permeable paving (indeed, taking out the pavement wherever possible, or what the Europeans call "desealing"), canals and median strips and utility corridors, hedges and greenwalls, among many others. These are the patches and corridors that fill in the green matrix, between the large ecological features found in a metropolitan area (rivers, mountains, shoreline) (see Low et al. 2005). These spaces, in concert with the larger green grid, provide a host of benefits and also provide opportunities for directly and personally enjoying, visiting, strolling, exploring, uncovering, and pondering nature. Not the "nature" far away, as in the form of a national park or wilderness, but the common nature that is everywhere around us. We must design and plan for everyday nature, design it in from the beginning, and work to retrofit existing urban neighborhoods to afford this exposure.

Partly, the charge is to do what we can to let nature reemerge, as it is wont to do, in many of the green spaces around us. This can happen through both action (active planting and habitat restoration) and inaction (no-mow zones have been designated in many neighborhoods). The City of Brisbane, Australia, has implemented a "Greening the Gaps" initiative, an effort at actively bringing nature back in many of these spaces. One of the first outcomes has been a collaboration between the city and Powerlink, a corporation responsible for electricity transmission, whereby funds are provided for revegetating transmission easements. This could be done everywhere. Utility easements and corridors of all sorts, and a natural gas easement behind my own home, serve as important habitats and linear connections, though perhaps not commonly thought of as part of our green urban fabric.

There are also now marvelous examples of former industrial lands that have gone through a process of extensive ecological restoration. Landscape Park in

Duisberg-Nord, Germany, is one of the best examples. Here a former steel mill has been converted to a popular, highly frequented regional park—with trees planted in coal storage areas, a windmill moving canal water through a cleaning system, and gardens shaped from the foundations of the structures. One can scuba-dive in a cooling tank, climb to the top of the aging steel mill structure, rappel from huge foundation pillars, or walk or bicycle along an extensive network of trails. A combination of planted and volunteer nature, mixed with these remnants of an industrial past, make for a feeling that one is visiting a kind of industrial Mayan ruin. There is a wondrous, magical feel to this place, demonstrating the value of preserving, adaptively reusing, and celebrating these leftover lands.

And there is a new sense that these leftover lands do in fact harbor patterns of nature, combinations of flora and fauna that are special and unique and worth appreciating and saving. At the green rooftops center in Malmo, Sweden, for instance, they have begun demonstrating the notion of "brownfield roofs"—ecological rooftops that accommodate and highlight these industrialized natural systems and ecosystems (Setterblad and Kruuse 2006).

ORGANIZING URBAN LIFE AROUND NATURE

Urban planners and designers need to think more about how to design in and facilitate a sense of natural wonder in our neighborhoods and living areas. Gentle forms of nudging have taken some creative directions. Movable metal display stands with placards that provide information on native species of birds and other flora and fauna can be found in some places (at the Hammarby Sjöstad in Stockholm, for instance).

Just as essential as the urban form and physical design features are the organizational and programmatic strategies for growing a green neighborhood, and for actively engaging the people and families who live there. Gardens, community orchards, spaces for gathering and exploring are all important, but so are the many things that might be done to challenge residents to understand more deeply their place in the world and to connect more intimately with the other people and creatures that inhabit that space. Every new homeowner should be given, I believe, an ecological owner's manual that describes the ecology and unique environmental conditions and history of their neighborhood and watershed. It should help new residents identify nature flora and fauna, and when and where they are likely to encounter them in their neighborhoods.

Accompanying this manual might be a (formatted) nature journal, in which residents are encouraged to join in the ancient practice of phenology—watching and recording the seasonal changes around them, and in this way connecting to place and picking up on the native nuances of weather, ecology, and the nature around them. This past spring, I started such a journal myself, observing and recording natural phenomena and changes in my neighborhood. The experience has been a highly beneficial one, causing me to notice many things that otherwise blend into the background of everyday life. On many occasions, the simple drawing of something (a wood nymph butterfly or a purple thistle) triggers the noticing of beauty and detail and the scurrying off to my guidebooks to identify and learn more about what I am seeing. Looking, noticing, paying attention, living with a more mindful outlook serves, I believe, to connect us to our place-home, and in the end to care more about these places.

Becoming native to place is a key aspect of green urbanism, and one that requires new relationships and new ways of living. Cooking (and eating), landscaping, and recreating, can be seen as acts of place-strengthening and expressions of place-commitments, as opportunities to re-earth and reconnect. A membership in a CSA (community-supported agriculture) should be a standard feature, as well as a strong invitation to join the local bird club or native plant society, and to attend the next star-gazing party or fungi foray. Indeed, the organizing of such place-strengthening events ought to be viewed as an important neighborhood function. We should rethink the basic equipment that a new (or existing) home can be given, as well as what neighborhoods need. Along with the ecological owner's manual, perhaps basic equipment should include a bicycle and a pair of walking shoes for getting around the neighborhood, city, and region at a slower pace; a butterfly-catching net (Bob Pyle would like this);

and perhaps a telescope, among others. Similarly, there are new "equipment" needs at the neighborhood level, providing opportunities to both connect to nature and strengthen through sharing and shared experiences the bonds between neighbors. My suggested list for what equipment might be needed at this neighborhood level includes some of the following: a bat detector, an outside collective baking oven (one or more); a tree house; a native seed–collecting and storage facility; and a community apiary.

The challenges extend, moreover, to not simply educating but activating neighborhoods to appreciate and celebrate the nature around them. How do you instill in neighborhoods a sense of awe and wonder at what is all around them—above them, beside them, underfoot? Organizing neighborhoods in ways that facilitate wonder and awe and experiential hands-on learning is a major task. Perhaps designated *neighborhood docents* personally take on some of these tasks, helping to organize fungi forays, neighborhood bird-watching and butterfly-catching excursions, and so on. Perhaps each house on a street or block becomes the center of expertise for a particular natural element or sphere. (Does every neighborhood need and deserve a resident ornithologist, entomologist, or mycologist, perhaps?)

Perhaps parents (and others) might take on the task equivalent to volunteer lifeguard at the neighborhood pool—a kind of nature exploration lifeguard, watching over neighborhood kids, allowing free nature play in the neighborhood under a somewhat more actively watchful eye of one or more parents. This may allay the common concern that parents have about what Rich Louv calls the "bogeyman syndrome," which helps keep kids indoors on the couch in front of the television or computer screen. Occasional help and active mentoring might be part of this job (and the self-learning that is required for this to happen). There needs to be a recognition about and discussion of the explicit reasons why such things are important—that such things are not just "extras" but part and parcel of what it takes to grow a decent, caring human being; as important as, perhaps more than, the standard things taught in schools today. Without these experiences, we are unlikely to have the care and stewardship we need, but perhaps more importantly, many soon-to-be-adults will have missed out on things that will make their lives profoundly more enjoyable, more meaningful, richer in so many ways. This needs to be said, as often as we discuss only the cost savings from natural drainage techniques, the energy savings from planting trees in urban environments, the property value increases that clearly accrue from green urban amenities, and so on.

For several years, I have been administering a "what is this" slide show/quiz, asking my students to identify common species of local flora and fauna. I have been astounded and discouraged that few of them are able to recognize or identify even very common species of birds (e.g., mockingbirds), trees (sycamores, poplars) and invertebrates (even our state insect, the eastern tiger swallowtail, goes unrecognized by most students). Mike Archer, dean of sciences at the University of New South Wales in Australia, similarly bemoans this lack of knowledge about native species. He and coauthor Bob Beale have written the provocative book *Going Native*, suggesting a host of creative ways to rebuild this knowledge and reconnect Australians to their incredible native biodiversity. They speak of designs to "reintegrate" people with nature, for instance, neighbors joining lots together to create natural habitats, similar to some of what has already been mentioned. Perhaps the most intriguing idea, and a charge to architects, is to design homes that convert urban wildlife from being a nuisance to being an opportunity to learn and reconnect (Archer and Beale 2004, 334–335):

> Another option we suggest is to share homes themselves with wildlife. People often complain about possums in the roof doing unseen 'things', yet at the same time they complain about their square-eyed children spending hours in front of the television watching junk. Why not construct houses so that they actively accommodate native animals such as possums, bats and native bees? Imagine a house—as suggested by biologist Nick Mooney—constructed with a central well from ceiling to floor that had large one-way glass windows enclosing a space with artistically distributed vegetation (nourished by skylights in the roof and soft lights at night) as well as logs. In this in-house refuge, possums could make nests, mate, raise babies, feed, feud and provide hours of fascinating evening viewing for the human family. Even watching parrots

feed in native trees on the outside of a large picture window is a visual and aural treat to start off the working day.

These are stimulating ideas to think about from a design perspective, and a real challenge to my architecture and building colleagues. We have considerable examples of relatively superficial building add-ons, like bird boxes, but can we design a new home that itself serves as a habitat and viewing and education device, as well as an amazingly fun and interesting place to live?

REFORMING URBAN PLANNING SYSTEMS

While there are now a number of cities where investments have been made at neighborhood, community, and regional levels in biophic urban design, there is little consensus about this within the planning profession, and city planning design standards codes rarely take biophilia explicitly into account. In 2006, the American Planning Association published a definitive planning treatise *Planning and Urban Design Standards* (APA 2006). Tellingly, although it is 720 pages in length, "biophilia" is not to be found in the index and is not explicitly considered in this otherwise immense and comprehensive document. And not only is it true that conventional planning and regulatory standards and requirements rarely stipulate biophilic urban design, they often represent obstacles or impediments to many of the creative green ideas discussed earlier. Engineering standards that specify overly wide streets and roads, giving deference to (fast-moving) motorized traffic, or conventional stormwater system design standards that stipulate highly concretized approaches, actually impede the possibility of injecting more nature into cities.

Green urban design is also often resisted based on the additional costs that may be incurred (say, for a green rooftop). Experience increasingly shows, however, that even judged on strictly narrow economic grounds these measures pay for themselves when a (slightly) longer time frame is taken into account (a green rooftop serves to extend the economic of a building roof).

Biophilic-oriented reforms in our planning and standards and regulatory systems seem in order then, and some cities have indeed made positive steps. Local subdivision ordinances commonly mandate the setting-aside of a certain portion (e.g., 15 percent) of a developable parcel for parks and open space. It is also now fairly common for cities to apply landscaping requirements to new development projects, and the results of these can be helpful from a biophilic perspective. The City of Chicago has been implementing a fairly impressive Landscape Ordinance, for instance, that mandates the planting of shade trees and screening hedges, which has had a visible impact in greening urban neighborhoods there.

Urban park standards must, moreover, be modified to better take into account the sorts of natural areas and natural experiences we aim to give urbanites—a shift from more formal turfgrass parks (largely ecologically sterile) to wild spaces, even those that might be quite small. No cities to my knowledge have sought to establish any kind of per capita or proximity standard for these forms of urban nature, but this would be a worthwhile exercise. Should not nature be found within a half-mile walk of every home or work location, and one or more significant nature features (a small forest, a median-strip prairie, a green rooftop) within eyesight and within a few hundred meters' walking distance of every building and flat?

While there has been considerable interest, especially by new urbanists, in the notion of ped-zones or some delineation of the likely distance and area people will walk, this idea should perhaps be extended to better and more directly address Bob Pyle's notion of "home range." Our neighborhoods (both new and existing) might be judged according to whether every resident child's "home range" interacts with or includes at least one wild and natural place of "experience," in the manner that Pyle means it (Pyle 1993). We frequently draw pedestrian zones on planning maps, and so it would not be a major stretch to extend the idea to include access to a neighborhood zone of natural wonder, to areas where "radical amazement" (in the words of one of Rich Louv's interviewees) can occur (see Louv 2005).

Planning and land-use regulatory systems will need to be reformed and modified to facilitate and make possible many of these green-urban ideas. Features such as green rooftops are commonly mandated in European cities, and financial subsidies and technical assistance provided for their installation. In some American cities, density bonuses are now provided to encourage such green elements. Portland, Oregon, for instance, has adopted a density bonus for installation of ecoroofs: the greater the portion of the rooftop covered by green, the higher is the allowable density. This is a promising trend.

TABLE 17-2 Malmö's Green Points system: Developers building in the Western Harbor chose a minimum of 10 of the following 35 green measures:

No.	Measure	No.	Measure
1.	A nesting box for every dwelling unit.	19.	In the courtyard or adjoining apartment buildings, at least 5 m^2 of orangery and greenhouse space per dwelling unit.
2.	One biotope for specified insects (plant biotopes excluded) per 100 m^2 courtyard area.	20.	Bird food in the courtyard all the year round.
3.	Bat boxes inside the plot boundary.	21.	At least 2 different traditional cultivated fruit and soft fruit varieties per 100 m^2 courtyard space.
4.	No hard standing in courtyards—all surfaces permeable to water.	22.	House fronts to have swallow shelves.
5.	All non-hard surfaces in the courtyard, to have soil deep enough and good enough for vegetable growing.	23.	The whole courtyard to be used for growing vegetables, fruit and soft fruit.
6.	The courtyard includes a traditional cottage garden, complete with all its constituent parts.	24.	The developer/landscape architect to cooperate with ecological expertise and to shape the overall idea and the detailed solutions together with the associate. First the choice of associate has to be approved by BoO1/the City of Malmö before it can be counted as a green point.
7.	Walls covered with climbing plants wherever possible/suitable.	25.	Grey water to be purified in the courtyard and reused.
8.	1 m^2 pond for every 5 m^2 hard standing in the courtyard.	26.	All biodegradable domestic and garden waste to be composted and the entire compost output to be used within the property, in the courtyard or in balcony boxes and suchlike.
9.	Courtyard vegetation specially selected to be nectar-yielding and to serve as a butterfly take-away.	27.	All building material used in constructing the courtyard—surfacing, timber, masonry, furniture, equipment, etc—must have been used before.
10.	Not more than 5 plants of one and the same species among the courtyard trees and bushes.	28.	At least 2 m^2 permanent growing space on a balcony or in a flower box for every dwelling unit with no patio.
11.	All courtyard biotopes designed to be fresh and moist.	29.	At least half the courtyard to be water.
12.	All garden biotopes designed to be dry and lean.	30.	The courtyard to have a certain color as the theme for its plants, equipment and material.
13.	The whole courtyard made up of biotopes modeled on biotopes occurring naturally.	31.	All trees in the courtyard to be fruit trees, and all bushes fruit bushes.
14.	All storm water captured to run aboveground for at least 10 m before being led off.	32.	The courtyard to have topiary plants as its theme.
15.	Green courtyard, but no lawns.	33.	Part of the courtyard to be allowed to run wild.
16.	All rainwater from buildings and courtyard paving to be collected and used for watering vegetation or for laundry, rinsing, etc., inside the buildings.	34.	At least 50 wild Swedish flowering plants in the courtyard.
17.	All plants suitable for domestic use, one way or another.	35.	All roofs on the property to be green, i.e. vegetation-clad.
18.	Batrachian biotopes in the courtyard, with hibernation possibilities.		

Source: Persson, 2005, p.51

Figure 17-9: Western Harbor Water System, Malmö, Sweden

The Western Harbor district in Malmö, Sweden, mentioned earlier, provides an example of how green elements can be mandated, while at the same time allowing flexibility for developers and designers. A green-space factor required that a minimum of half of each lot be left in a natural and undeveloped condition. Also, an innovative green points system was applied. Developers there agreed to utilize a minimum of 10 green items from a list of 35 options (and were also free to come up with and propose their own "points"). Table 17-2 lists these 35 green points, ranging from installation of bird and bat boxes to courtyard vegetation schemes that provide nectar for native butterflies, to incorporating water features, to green roofs. I especially like point 34: "At least 50 wild Swedish flowering plants in the courtyard"(!) (Persson 2005, 51). Making this new urban area a "habitat-rich city district" was a key goal, and the green points while not perfect have certainly helped to bring this about. The overall green result is impressive and considerable. There are green roofs (a high proportion of the buildings have them) and vegetated courtyards, and new and restored biotopes throughout (see Figure 17-9).

Protecting leftover areas of nature in already developed parts of the city sometimes conflicts with other equally important sustainability goals, of course—notably the need to reuse brownfield sites, to promote urban infill and densification in existing urbanized locations as a way of combating low-density sprawl at the urban periphery, and to achieve density and intensity levels sufficient to support public transit and pedestrian life—and this is another obstacle. This requires some degree of sensitive and sensible balancing, although our current planning systems don't include many mechanisms for such. When New York Mayor Giuliani proposed selling off some 600 of the city's small community gardens for new affordable housing, there was a major public outcry. These were (and are) important and beloved spaces, small green pockets in dense places like lower Manhattan. In the end, most of the gardens were protected (although some were sold).

The case of the New York City community gardens demonstrates vividly the precarious tenure of these important leftover spaces, and the need to provide avenues and structure in the planning/regulatory system to accommodate and protect them. New York considered (but never adopted) a special land-use classification that would strengthen their protection. Perhaps some similar measure is generally needed as part of our community planning and regulatory systems. Perhaps at least a zoning designation (an "Area of Potential Environmental or natural Significance," or "APES") that would at least trigger an on-site assessment of its potential to the neighborhood and surrounding area as important and worthy of cautious deliberation before clearance and development would be allowed.

Perhaps new methods should be devised whereby preserving these remnant natural spaces—at the same

Figure 17-10: Community Gardens, New York City

time working to creatively accommodate density and infill in other non-nature-diminishing ways—is made easier. In some places and in some circumstances, governments are provided the right of first refusal; this might be one way to acquire and set aside these important areas when they are on the verge of being developed. Often neighbors would like to have had the opportunity to buy (collectively or individually) leftover lots but are not even aware of their impending sale or loss until they hear the sounds of chain saws or see the construction equipment appear. Providing some institutional means (developing an arrangement with a local bank) for securing funding to buy such lots, perhaps through the tool of a community land trust or revolving fund, could be a priority.

An initial step, in the spirit of biophilic cities, is to gauge and evaluate the nature present in such places. If it is considerable, and if the spaces are being visited, enjoyed, and actively used, erring on the side of protection makes some sense. In some cities (Toronto, for instance), comprehensive reurbanization studies have sought, in a thorough and systematic way, to indicate where additional urbanization and reurbanization is appropriate, and where, in part because of the loss or diminution of informal green spaces, it is not. A second step is to ensure that whatever is built in a reuse or reurbanization project positively contributes to a net increase in urban nature.

THE VISION OF BIOPHILIC CITIES

Stepping back a bit, it is an interesting question to consider what marks or distinguishes a city as *biophilic*. Different cities, in different parts of the country and world, for instance, are said to have different qualities, different paces of life. Can the spirit or essence of a city be more biophilic or less biophilic? And how do we know this is the case? Heerwagen and Gregory (see Chapter 13) make a strong case for the biophilic qualities of Paris, its sensory qualities, natural elements, historic street patterns, and dynamic wholeness. There are many cities, of course, that exude similar natural and organic qualities, and understanding them through a biophilic lens is extremely useful indeed. It will make for interesting future research and perhaps a ranking list for biophilic cities.

There is certainly evidence of more (or less) interest in nature in some cities compared with others: active efforts throughout a city to restore and repair ecosystems, and to manage the built environment so as to support native flora and fauna (e.g., Chicago's "lights out" program for turning high-rise building lights off during key periods of bird migration), investments in education about native flora and fauna (support for natural science museums, in-school and out-of-school nature program funding), support for programs to involve citizens in active, hands-on habitat repair and restoration (as in Australian cities like Brisbane), and some measure of the numbers of people involved in such programs, et cetera. There are other specific measures, I'm sure, but you get the idea.

While it's partly a mystery how to nurture a biophilic sensibility in cities, I know it is possible. Contrasting examples can be seen in how two Texas cities—Austin and Houston—have responded to the presence of Mexican free-tailed bats. The story in Austin has become a bit famous by now. Reaction to the discovery of the bats roosting in the underside crevices of the Congress Avenue Bridge, with the undisputable help of Bat Conservation International (based in Austin), has moved relatively quickly from fear to celebration, and nightly watching of the bats.

Austin might be said to have gone a bit bat-crazy. It has named its local hockey team after them (the Austin Ice Bats). The local newspaper, the *Austin American-Statesman*, has constructed a bat observation deck and viewing area (they actually call it the "Statesman Bat Observation Center"). Several companies now offer bat-watching dinner cruises. The bat fest, a major public event, is held every year on Labor Day weekend. Bat Conservation International estimates that 100,000 people come to the bridge to see the bats each year, with $10 million in ecotourism revenue generated. Purported to be the "world's largest urban bat colony," there is clear and palpable pride in their presence, and hundreds of residents converge on summer nights to watch the spectacle of an astounding 1.5 million bats circling, diving, feeding.

While this level of interest and fascination probably does not carry over to many other things in the natural world in and around Austin, it is nevertheless an impressive display of the biophilic impulse, and surely a good sign of having found at least a partially biophilic-city. The reaction to the bats in Houston, where they have been similarly found in smaller numbers, has been more muted, reserved, and from some residents downright hostile. As a Texas friend recently related, Houstonites are worried about those pesky bats doing such unholy things as swooping down and drinking from backyard pools. A concern about rabies (and a recent case of such) has dominated much of the public discussion there. Residents living near the Waugh Drive bridge (crossing Houston's Allen Parkway) where the bats actually live year round, have expressed anxiety and fear. As nearby resident Janet Jenkins states: "We've seen them fly out by the thousands. It sounds kind of scary. . . . We all have fireplaces and they can get in. I'm concerned now. Maybe those bats aren't so great. I thought they lived in their spot and we lived in our spot." (Bryant 2006). Fear seems to eclipse any awe or wonder, and there is a desire for safe separation from nature rather than closeness to it. The Buffalo Bayou Partnership and passionate staff of the Texas Parks and Wildlife Department have been working hard to build understanding and interest, including organizing local bat-viewing tours, but it will likely take considerable time and may in the end be a lot harder there to awaken these latent biophilic impulses.

A city's biophilic credentials are strengthened, moreover, when it makes real and significant efforts to save and protect and steward over the special ecological places and qualities present. Containing growth, steering roads and development away from sensitive lands, imposing ecological controls on new growth, and buying and otherwise safeguarding important natural areas, should all be seen as signs of the biophilic commitment of a city. In this regard, Pima County's (Tucson's) Sonoran Desert Conservation Plan, which aspires to protect and manage an unprecedented 2 million acres of its precious biological patrimony in a biologically coherent regional preserve system, says much. To its credit, the community has embraced in a new and impressive way the need to steward over these lands. William Shaw, who has chaired the science committee, notes a remarkable shift in the local culture, a sort of mainstreaming of a desert conversation ethic there. A commitment to public education is part of the Desert Conservation Plan, and it has a small staff of environmental educators—quite unusual—who have been working with schools and the general public (there's even a "kids' desert plan") to build an awareness of this amazing natural setting.

Australian cities seem far ahead of their American counterparts when it comes to this somewhat more amorphous notion of biophilic spirit. Perhaps because these cities are so much younger, they have an almost wild quality. In cities like Perth, in Western Australia, even the most established parks are homes to tremendous native flora and fauna. Kings Park in Perth is a great example. Here, just 1.5 km from the central business district is a relatively large park (a bit over 400 hectares) with two-thirds left in native bush. Perth's "Bush Forever" program has as its goal "keeping the bush in the city" and has designated a network of close to 300 urban bush sites, together comprising more than 50,000 hectares of land. Few neighborhoods will be very far away from areas of remnant bushland and native biodiversity (Figure 17-11).

How we talk about cities and places, and the words and language we use, also provide important cues. Our

Figure 17-11: Bold Park, Perth, Western Australia

language systems are important, and here again the Australians may be ahead of us. There the "bush" conjures up a real and tangible and visceral meaning. While it may have a different look depending on whether it is Sydney or Brisbane, these are meaningful words and descriptively accurate ways of talking about what it is in the city that is important to protect. Our American city vernacular is more problematic—we speak of open space (do we imply empty space?) and green spaces, for instance, but this language doesn't evoke very much for us, and doesn't indicate much about our emotional attachment to these urban lands. It's easy to accept the loss of "open space," difficult to get worked up about development slated to take place on it.

While Australian cities suffer from most of the same planning ills as American cities—too much low-density development, too much dependence on cars—proximity to and accommodation for these wild areas is impressive. And it is refreshing to see friendly competition between cities about which harbors the most biodiversity. Perhaps (again) because they have so much nature, so much biodiversity, in and around these cities, they tend to devote much time and staff to managing, protecting, and educating people about it. Even these cities could be doing much more, but these are certainly hallmarks of a biophilic city.

Finding ways to actively involve citizens and residents in the task of caring for and repairing nature in cities is another key part of the mission of growing biophilic cities. Andrew Light thinks of this as part and parcel of an urban environmental ethic, part of what it should mean to be an ecological citizen of a particular city or place. And much progress in activating people to be involved can be made. Again, in Australia, much importance is given to these kinds of programs, commonly called urban bushcare groups. In Brisbane, there are now 120 bushcare groups, involving more than 2,000 active volunteers (as of December 2005), and these groups are supported as a matter of official city policy through commitment of revenue and staff. There are many benefits beyond the ecological restoration results themselves, of course. Ku-ring-gai Council, a locality in the Sydney metro area, notes this in their description of their Bushcare Volunteer program: "In addition to the environmental and educational benefits, joining Bushcare is an opportunity to make friends, become a part of a social group that shares the same concerns and to change community perception about bushland."

Australian cities support these nature initiatives to a remarkable degree. There are typically one or more bushcare officers who help to organize these volunteer efforts, equipping and coordinating and training volunteers. Local councils sponsor bush walks and conduct seminars and workshops and training on a host of bush issues; much of this activity is funded through a specific environmental levy. In Ku-ring-gai, for instance, recent bushcare seminars have included topics such as possums and gliders, native orchids, grasses and ground covers, and bush tucker cooking. Many (perhaps most) local governments in Australia, moreover, operate their own community nursery, propagating native plants from locally collected seeds; these plants are distributed free to the public or at a very small cost, with general encouragement to plant and appreciate the special local flora of their community.

Local council bushcare staff are also typically active in helping citizens in bush restoration efforts in and around their own homes. Under Ku-ring-gai's backyard buddies initiative, staff will visit and provide planting and regeneration advice. An even more radical notion is being implemented in this jurisdiction. Council staff have been attempting to interest homeowners in restoring native fauna, in particular in breeding and releasing blue-tongue lizards. Most interesting, (à la Archer and Beale 2004), homeowners' swimming pools are being re-envisioned as new homes for native aquatic species. Peter Clarke of Ku-ring-gai Council recently reported on this experience in the Ku-ring-gai "Bushcare News" (Clarke 2006, 4):

> Elvis Claus has successfully turned his 'boring' suburban pool into a magnificent native pond teeming with rainbows and gudgeons. The really exciting part of this is that the fish have bred themselves and we have used the population in Elvis's pool to populate other ponds in Ku-ring-gai. We now have another pool being converted into a pond in Lindfield and if this trend continues I hope that Ku-ring-gai one day will be known as the Kasmir of the south.

The council is also breeding and distributing Pacific blue-eyes, a native fish, which is especially effective at controlling mosquito larvae (and so in turn is perhaps part of the solution to getting kids and families outside).

CONCLUDING THOUGHTS

As these many city and neighborhood examples demonstrate, it is indeed possible to combine urban living and a life close to nature. Cities already harbor much more nature than we commonly acknowledge, and there are now a host of creative planning and design tools, techniques and concepts that can be applied to make urban neighborhoods profoundly greener and more biophilic. A sustainable future will of necessity require an urban future, I believe, but this does not (indeed cannot) mean that this future is one disconnected and detached from nature and natural systems. The choice between urban and natural, as depicted in that *Washington Post* real estate advertisement, is a false choice and an unnecessary and outdated dichotomy. Biophilic cities and biophilic urbanism transcend this dichotomy and present a compelling new vision for a rapidly urbanizing world.

REFERENCES

American Planning Association. 2006. *Planning and Urban Design Standards*. Hoboken, NJ: Wiley.

Archer, Mike, and Bob Beale. 2004. *Going Native*. Sydney: University of New South Wales Press.

Beatley, Timothy. 2000. *Green Urbanism*. Washington, DC: Island Press.

______. 2005. *Native to Nowhere*. Washington, DC: Island Press.

Bryant, Salatheia. 2006. "Bats Make Some Uneasy in Wake of Rabies Case: Despite News of Boy's Infection, Creatures Mostly Avoid People, Experts Say." *Houston Chronicle*, May 11, sec. A.

Clarke, Peter, 2006. "Backyard Buddies Update," *Bushcare News*, Ku-ring-gai Council, Pymble, New South Wales, Australia, Autumn.

Dreiseitl, Herbert., and Dieter Grau, eds. 2005. *New Waterscapes: Planning, Building and Designing with Water*. Basel: Birkhäuser.

Girling, Cynthia, and Ronald Kellett. 2005. *Skinny Streets and Green Neighborhoods: Design for Environment and Community*. Washington, DC: Island Press.

Karbabi, Barbara. 2005. "Bats over Bayou: The Park People Take Sightseers to the Waugh Drive Bridge, Where the Furry Creatures Hang Out Until Dusk; The Stars Come Out at Night and Take Flight." *Houston Chronicle*, November 15, sec. A.

Light, Andrew, 2006. "Ecological Citizenship: The Democratic Promise of Restoration," in Rutherford H. Platt, ed. *The Humane Metropolis: People and Nature in the 21st Century* City, Amherst, MA: University of Massachusetts Press.

Louv, Richard. 2005. *The Last Child in the Woods: Saving Our Children from Nature-Deficit Disorder*. Chapel Hill, NC: Algonquin Books.

Low, Nicholas, Brenan Gleeson, Ray Green, and Darko Rodovic. 2005. *The Green City: Sustainable Homes, Sustainable Suburbs*. Sydney: UNSW Press.

McCamant, Kathryn, Charles R. Durrett, and Ellen Hertzman. 1993. *Cohousing: A Contemporary Approach to Housing Ourselves*. Berkeley, CA: Ten Speed Press.

Persson, Bengt, ed. 2005. *Sustainable City of Tomorrow*. Stockholm, Sweden: Formas.

Pyle, Robert M. 1993. *The Thunder Tree: Lessons from an Urban Wildland*. New York: Lyons Press.

Setterblad, Martin, and Annika Kruuse. 2006. "Design and Biodiversity. A Brown Field Roof in Malmö, Sweden." Unpublished paper.

chapter

18

Green Urbanism: Developing Restorative Urban Biophilia

Jonathan F. P. Rose

Throughout the history of urbanization, there has been a tension between the economic, environmental, social, and security benefits of dense cities and the human desire to be connected to nature. Early cities were usually enclosed in walls to keep them secure. The effort and expense of building and maintaining the walls led to maximizing the use of space inside the walls, and thus to greater building density. And yet a walled city could only survive if connected to the fruits of nature, water and food.

Most early religions and cultures were deeply rooted in nature, and provided a system for humans to understand their place in the universe. As urbanism grew, this sense of place moved from the natural realm to mediating between the human-made realm and nature. And yet the purpose of a sense of place remained the same: to give the resident a sense of his or her place locally, regionally, and in the universe. And because we are of nature, our sense of place has always included elements that speak of nature's proportions and cycles.

The Enlightenment brought with it many things, but new forms of governance and economic growth led to cities having both a sense of security (arising from more stable governance) and increased prosperity. These gave rise to cities' growth beyond their walls and to an intermediate zone between city and nature—the suburbs. For example, in the mid-nineteenth century, this economic growth, security, and the desire to be connected to nature was expressed by the expansion of the dense urban core of the City of Vienna. In 1857, Emperor Franz Joseph I tore down the walls surrounding the city, and built the Ringstrasse, giving rise to new leafy neighborhoods and broad boulevards, developed along with the rise of a new middle class and an influx of peoples from all over Europe.

At the turn of the twentieth century, the desire to

live in communities near nature was framed by Ebenezer Howard in his seminal book *Garden Cities of Tomorrow*. Howard, an Englishman who had spent time in America and was deeply influenced by its late-nineteenth-century Utopian movements, proposed a network of dense garden towns and villages, none larger than 32,000 people, surrounded by greenbelts of preserved agricultural and natural lands.

A generation later, in 1925, Le Corbusier in *Urbanisme* proposed the replacement of older cities with "towers in a park," an ultimately heartless, placeless strategy that, after World War II, gave rise to urban renewal. This strategy wiped out whole neighborhoods, eliminating their historic fabric, and replaced them with a failed monoculture of isolated low-income public housing projects, or boring middle-income apartment buildings.

And yet the essential human desire to live in community, while connected to nature, persists. And the wanton environmental destruction and placeless character of post–World War II suburbanism calls for us to find an urban solution—one that is dense, so as to reduce the amount of land that it takes up, that is economically and environmentally efficient, and yet satisfies the deep human need for a sense of place that includes contact with nature. A new urbanism is emerging that addresses these concerns. Called *green urbanism* in a book of that title by Timothy Beatley, this urbanism is green because the buildings themselves are green, filled with daylight and fresh air, and because the context is green—gardened, tended by the hands of the residents.

Green urbanism follows a basic principle of ecology that organic systems are integrated at multiple scales. Thus, it proposes to garden individual buildings, to garden communities and neighborhoods, and to link neighborhoods with pathways of nature. Its ultimate goal is to be restorative of both people and place.

Restorative design has two intrinsic implications: to restore the human and natural environment and to restore the self. The goal of green urban development should be to achieve both. As we shall see, the restorative aspects of greener communities also provide quite positive marketing features. And why not? If restorative design really does appeal to our biophilic impulses, then it should be more appealing to residents.

Let us begin with density. Just as older cities needed density to support the cost of building and maintaining walls, so modern cities need density to make their rich social, cultural, and physical infrastructure economically viable. For example, mass transit can only be economically viable if there is a sufficient density of riders to make it cost effective. And density supports both diversity and clusters. We know that biological systems thrive with diversity; so do economic and social systems. But human cultures also need clusters of similarity within pools of diversity, places in which knowledge can accumulate and grow. And thus, the economies of our cities are typically driven by both extraordinary social and economic diversity, and also by clusters of economic services. Originally, these clusters might have been trade- and place-based, such as seaport industries clustered around a harbor or manufacturing industries clustered around cheap energy sources, such as mill rivers or coal. Now clusters tend to be knowledge-based, such as with biotech, finance, or software industries. Thus, for example, biotech industries tend to cluster around research universities and medical centers. Even in a time of electronic communication, we find that people do best living and working in communities, and that density is essential for the benefits of increased proximity.

Dense urbanism is also the greenest and most energy-efficient way of dealing with twenty-first-century population growth, and suburbanism is the least. For example, if one combines the energy used by a home and the energy used in transportation getting to and from the home, an urban multifamily home consumes one quarter of the energy used by a suburban home.[1] So energy consumption and climate change are deeply causally related.

Many people have chosen to live in the suburbs rather than the city to be closer to nature. And yet contemporary suburban life does not necessarily provide an intimate relationship with nature. In fact, urban residents typically spend more time walking outdoors than suburban residents, who must spend so much of the day in their cars. "For trips less than one mile, mixed use communities generate up to four times as many walk trips" (Ewing 2005). For example, a study found that a 5 percent increase in the walkability of a neighborhood provided a 32.1 percent increase in time spent in physical activity, with 6.5 fewer vehicle-miles traveled (Handy 2005)). If we green our streets, create urban parks, and create green pathways along rivers, waterfronts, and other parts of the natural connective tissue of our cities, then urban residents are likely to have more daily pedestrian contact with green space than suburban residents.

The biophilia hypothesis suggests that there is a deep and natural desire for humans to be connected to nature, because we evolved from nature. In fact, there should be no doubt that we are part of nature, as we are totally dependent upon air, water, food, and sunlight for survival. We could not live without breathing air or drinking water. (The only aspect of us that seems to believe that it is separate from nature is our minds.) Much work, described in previous chapters of this book, has been done to elucidate how important a connection with nature is to our health and productivity. We know that contact with nature deeply enhances children's ability to pay attention, a critical component of social and emotional learning. Patients heal quicker in biophilic environments. Professor Roger Ulrich, Ph.D., in a 1984 study, compared the healing rates of two sets of patients recovering from surgery. Half of the patients' rooms had views of a brick wall; the other half had views of trees. The study, which has been duplicated many times, showed that the patients with the view of trees from their hospital beds spent fewer days in the hospital and required less pain medication—the biophilic power of the view itself was a healing force. As this desire to connect with nature becomes increasingly understood, it is also being expressed in real estate markets.

Private developers are increasingly recognizing the human desire for contact with nature, and incorporating biophilic elements in individual urban buildings. In urban apartment buildings, terraces and balconies have traditionally provided outdoor space. These have always been very popular with renters and buyers. But contemporary urban developers, recognizing the desire of residents to connect more deeply with nature, are now paying greater attention to communal contemplative and green spaces. Developers are greening roofs; filling lobbies with trees, fountains, and gardens; and creating quiet meditative "Zen gardens" as building amenities.

> To escape the frenetic pace of city life, Troy and L. Camille Thornton need only slip out through the glass door of their bedroom in SoHo. There, tucked away on an expansive deck, sits a verdant garden of rose of Sharon, lavender and andromeda, along with Hinoki cypress, Japanese maple and cherry trees, artfully displayed in jardinieres and wooden planters.
>
> "Every morning when I wake up, I look out at the garden, and it gives me peace," said Ms. Thornton. . . .
>
> . . . Developers–many of whom are catering to an increasing number of families choosing to stay in the city but still wanting suburban-style amenities like yards–are incorporating generous and often innovative outdoor designs into their buildings. (Marino 2006)

In August of 2006, an article in the *Denver Business Journal* reported:

> As evidenced by the cranes across downtown Denver's skyline, more and more people are making the decision to live downtown.
>
> They want to be part of the urban environment and are willing to give up their large homes and lawns to settle into a simpler lifestyle, at least to some degree. Even though these buyers are drawn to the energy and excitement of living in the urban core, they still want to be able to appreciate the serenity of nature.
>
> So, many downtown developers are willing to sacrifice valuable pieces of real estate—which could be used to build another townhouse, penthouse or condo—and use it to add a touch of nature to their projects. The city of Denver requires that at least 3 percent to 5 percent of new residential projects include some type of open space, whether it's a hardscape, a lawn or a garden. In response to that mandate and the requests of their buyers, some de-

velopers are dedicating as much as 15 percent to 25 percent of their projects to landscaping.

> "The outdoors is becoming a much more significant part of the overall architecture of new residential projects in Denver, and it is adding to the overall ambiance and appeal of where people are choosing to live downtown," noted Curtis W. Fentress of Fentress Bradburn Architects. "People desire high-end features with their new downtown homes, and one of the most important features is beautifully designed landscaping. With a tranquil place that is part of the overall design, people can enjoy a peaceful setting even in the hustle and bustle of the city."
>
> Featuring landscape architecture in an urban residential project has only recently gained a strong foothold in Downtown Denver developments. Until recently, most downtown living spaces offered outdoor balconies that were little more than 50 square feet in size, barely enough room for a smattering of flower pots. That trend has changed dramatically, as landscape architects and developers are finding creative ways to bring nature to the city. (Kessler and Rubba 2006)

These include large terraces, Zen gardens, green roofs, brownstones and town homes with private backyards, borrowed landscapes, four-season design, and water features. Park and water views have always added value to urban real estate, reflecting the consumer's desire to have a prospect of nature. The market is expressing a preference for both natural and contemplatively designed spaces.

An article in the *AIA Journal of Architecture* also noted the trend:

> Finding meditative space in a dense urban environment can be more difficult than its rural equivalent: finding a needle in a haystack. The defining qualities of urban life—throngs of people, endless traffic, poor air quality, deafening noise, and continuous commotion—are antithetical to a meditative state, which is characterized by quiet and calm. As urban stressors increase, space and time for quiet reflection seem to diminish. . . .
>
> With thoughtful planning and the inclusion of natural elements, such as waterfalls, richer vegetation, and man-made sound absorbing materials in urban plazas and parks, small meditation zones may bloom throughout our cities. (Harris 2006)

Green urbanism shares much in common with emerging complexity theory. One of the signatures of complex adaptive systems is self-similarity across scales. For example, golden mean proportions that were often used in the design of sacred spaces occur at the molecular level of crystal organization, and at the solar system level of the spacing between planetary orbits. These patterns remind planners, designers, and developers that we too must repeat and integrate healthy proportions across scale. Viewing biophilia as a human instinct and perhaps even a human need, we should look at how we might integrate it the level of the room, the home, the street, the neighborhood, the community, and the larger bioregion. As biophilia has emerged from our coevolution with nature, how might we develop communities that are more resonant with our biophilic needs?

Perhaps answers can be seen in three examples of our work, which will be further described: David and Joyce Dinkins Gardens, a low-income multifamily project in Harlem, which is greened with green roofs and gardens; Highlands' Garden Village in Denver, which provides a wide range of parks and gardens as the fabric of a denser mixed-income community; and the Via Verde, winner of the New Housing New York competition, which takes this work further by connecting the gardens on the ground with the gardens on the roofs, literally weaving the buildings and gardens together.

These green projects are supported in an urban framework of green infrastructure. At the neighborhood scale, our communities need neighborhood parks, playfields, and community gardens; at the borough scale, networks of green spaces interconnected by walking and bike paths, such as the proposed Bronx River Greenway; and at the regional scale, connective pathways such as the Hudson River Greenway, which one day might reach from the tip of Manhattan to Albany, New York. And thus, we see these projects as cells in an organism of a greener neighborhood, borough, and city, both dense and gardened.

Public examples of biophilia are familiar to us—large urban parks, smaller neighborhood parks and squares, and reclaimed waterfronts. The Trust for Public Land (www.tpl.org) has determined that many of our best cities are comprised of 20 percent parks and open space. Financing structures are being established to support a biophilic form of community development. Enterprise (www.enterprise.org), a national low-income-housing nonprofit, has created the Green Communities Program, a superb affordable-housing green-building incentive system of design guidelines and financing.

DAVID AND JOYCE DINKINS GARDENS

Following these guidelines, the Jonathan Rose Companies, along with Harlem Congregations for Community Improvement (HCCI), a Harlem, New York–based community organization, is developing David and Joyce Dinkins Gardens (see Figure 18-1), named after New York City's first African American mayor and his wife. This 86-unit low-income housing project will serve youth aging out of foster care and low-income families. The project, designed by Richard Dattner Architects, is being constructed (2006–2007) on an abandoned toxic lot in the heart of Harlem's Bradhurst neighborhood. The building is shaped by an environmentally responsible design focused on not only restoring the block with a new building, but also restoring the lives of residents through thoughtful supportive services, communal gardens and contemplative spaces. Many of the project's residents will have been in and out of hopelessness for much of their lives. The goal of the project is to give them not only a housing unit, but also a sense of "home" rooted in a community.

David and Joyce Dinkins Gardens has many green features, including innovative energy-conserving systems, the use of low-toxic materials, and horizontal sunshades to shield windows from direct southern sunlight. The biophilic design elements include sunlit elevator corridors, a private green roof garden for the residents, a teaching garden adjacent to a youth construction

Figure 18-1: David and Joyce Dinkins Gardens. This model green affordable housing project in Harlem, New York City, demonstrates the feasibility of biophilic design on a tight budget, bringing nature into neighborhoods that most need it. Common areas open onto green gardens on and in the rear yard.

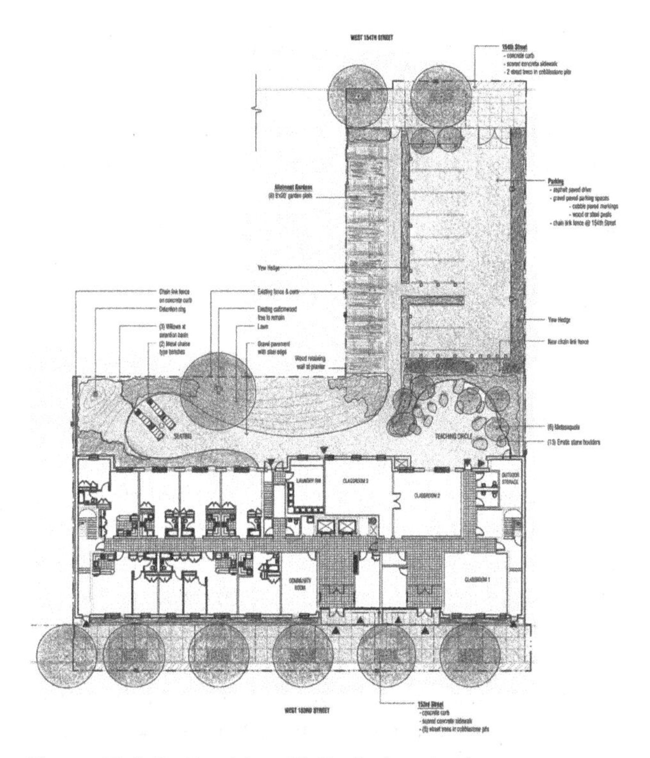

Figure 18-2: David and Joyce Dinkins Gardens site plan

trades academy on the ground floor, and a public community garden for the surrounding community. Nature is made palpable by a rainwater harvesting system that funnels water from the roof into storage tanks to be used for garden irrigation. An outdoor classroom features native vegetation and places to gather (see Figure 18-2).

HIGHLANDS' GARDEN VILLAGE

The best private opportunity to create biophilic development occurs in the planning of a larger community, organized around a range of biophilic experiences. Many singular biophilic elements can be combined in a holistic, integrated community. The Jonathan Rose Companies' Denver affiliate, Perry Rose, created Highlands' Garden Village, near Denver, a mixed-use neighborhood of residential, commercial, and retail uses interspersed among gardens and public green spaces. The 27-acre site is an urban-infill project, located in the middle of two historic neighborhoods (see Figure 18-3 in color insert).

Highlands' Garden Village is a new community built on the site of the old Elitch Gardens, a privately owned amusement park, theater, and display garden that delighted Denver residents for over a century. When it closed in the early 1990s, the owners first tried to redevelop it as a big box retail center. The neighborhood protested, and a community-based preservation concept was developed with city support. The owners then sought development proposals, and Perry Rose was selected to develop the site.

The site plan, designed by Peter Calthorpe and evolved through an extensive community input process, created the framework for a mixed-use development consisting of environmentally responsible single-family houses, townhouses, market-rate and affordable senior and multifamily apartments, cohousing condominiums, live/work loft units, office, and retail, set within a pedestrian-friendly fabric of parks and walkways. The proximity of different uses encourages people to walk. The site's location near downtown and on bus lines makes travel convenient and reduces transportation costs. The environmental and social qualities of the plan, including its urban infill location, its mix of uses and extensive public realm, and the green buildings, provide a model for the reuse of vacated urban sites including declining malls, brownfields, and other abandoned urban areas. Civic uses at Highlands' Garden Village include pocket parks, community gardens, public plazas, the renovation of the 1926 carousel pavilion that once sheltered Colorado's first merry-go-round, the restoration of the historic Elitch Theater (built in 1891), and construction of a charter school (Denver Academy of Arts and Technology). The civic spaces host an outdoor movie series, concerts, and private events.

The development plan (see Figure 18-4) was designed to weave nature throughout the community. The design process began by working with existing green features such as the Rocky Mountain Ditch, a 150-year-old irrigation ditch running from the mountains, which carries water through much of Denver. The ditch had run under the old roller coaster, which had been fenced off to keep children from playing under it. Protected from mowing, a natural environment grew up around the ditch, including a resident fox. The site plan was organized to maintain a finger of nature running through the project on either side of the ditch.

Unfortunately, the Rocky Mountain Ditch Company's owner, the Adolph Coors Company, claimed that maintaining the ditch aboveground would be a liability, even though the ditch runs aboveground for miles in either direction of the project. Thus, the ditch had to be buried, although the vestigial boundaries of the wild land around it are preserved, in what the neighbors named Mary Elitch Park. The park now has a rich landscape and extensive community gardens, and forms a spine of parkland through the development.

The second large green area in Highlands' Garden Village was a formal park, which during the Elitch Gardens days had been cultivated as a flower display garden. As part of the development, the park was preserved and replanted, and the historic shell of the old carousel building was restored (see Figure 18-5) to become the contemplative heart of the project, with a walking labyrinth inscribed into its floor. The building serves as a communal sacred and performing space.

The Highlands' Garden Village is now complete—a $100-million community consisting of 52 single-family

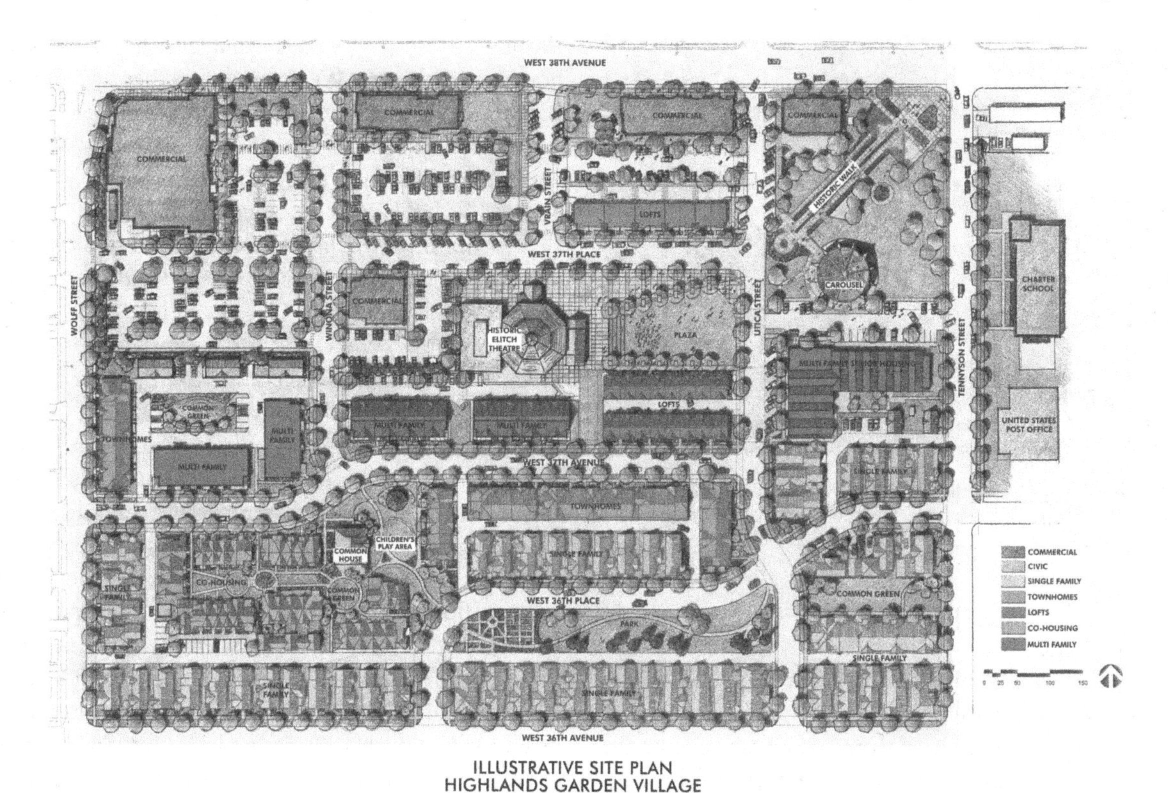

Figure 18-4: Highlands' Garden Village site plan

homes, 38 townhouses, 26 live/work condominiums, 33 cohousing units, 63 senior apartments, 74 multifamily apartments, 20,000 sq. ft. of civic spaces, and 70,000 sq.ft. of office and retail, including a 28,000 sq.ft. organic grocery store anchor (Sunflower Market). The economic impact of the project can already be seen in the rise of property values in the area, the increase in retail sales, and the continued development of new and rehabilitated housing through private investment in the area surrounding the Highlands' Garden Village development.

Just as there is a coevolving relationship between computers (hardware) and programs (software), so is there a coevolutionary relationship between the places that we build and the culture that we use to live in those places. Highlands' Garden Village was conceived of as a garden village, but it took a village of resident gardeners to make the gardens come alive. A garden club was formed so that the gardens would be the work of the residents, rather than outside contractors. In addition to community garden plots, the project also includes formal display garden, butterfly gardens, xeriscapes containing drought-resistant plantings, contemplative gardens, children's play gardens, and gardens that line streets and walking paths. Staff from the nonprofit Denver Botanic Gardens were engaged to organize and train the volunteers in garden design and maintenance. Additional programming includes farmers' markets, intergenerational theater, Christmas caroling, concerts, and many other activities to support the residents living in Highlands' Garden Village.

The Denver Urban Renewal Authority (DURA) researched the economic value created by these biophilic strategies. The study indicated that single-family home values in and adjacent to the project grew at an average annual rate that was 5 percentage points higher than the city as a whole during the eight-year period following the project's inception. As a result of the increase in home values, the assessable value of surrounding prop-

Figure 18-5: Highlands' Garden Village: A preexisting carousel building, set in a park, was renovated and a labyrinth added to make it the spiritual heart of the community.

erty appreciated by an additional $30.8 million over the same period—generating an additional $158,000 annually for the city treasury.

The same study also analyzed sales tax receipts in the study area and found that in the year following build-out of the residential portion of Highlands' Garden Village, area retailers experienced a 19 percent increase in overall sales, compared with only 1 percent growth citywide. The increased sales generated an additional $260,000 in additional sales tax for the year. Taken together, the increased property values and retail sales have resulted in more than $400,000 per year in additional tax revenue to the city. This $400,000 is over and above the incremental taxes generated by the project itself.

VIA VERDE

In the summer of 2006, the New York City department of Housing Preservation and Development (HPD) and the New York Chapter of the AIA ran a competition for a re- conception of green affordable housing. The winner would win a site in the South Bronx upon which to develop the project, and the subsidies needed.

Our firm entered the competition, named New Housing New York, with a codeveloper, the venerable Phipps Houses, one of the nation's oldest affordable housing developers. We brought together two architects, Richard Dattner Architects, architect of our 153rd Street project, and the New York office of Grimshaw, a brilliant, very green British firm. And because we believed that gardens would be a key element of the design, we added Lee Weintraub, who has been the landscape architect for many of our projects, including David and Joyce Dinkins Gardens and Highlands' Garden Village. The resulting project, named Via Verde for its unifying pathway of green, was selected as the competition winner.

The plan is organized around a ribbon of buildings, starting with low-rise townhouses on the southern side, rising to mid-rise buildings on the west, and a high-rise tower on the north (see Figure 18-6 in color insert). The shape of the ribbon provides for a rising plane of south-facing rooftops, all of which are gardened. The ground-floor courts are connected to the first level of community gardens by an amphitheater, the most public use. As the gardens rise, they become increasingly private, ending with a meditation garden at the highest level. Other green features include slim floor plates to enhance cross ventilation and natural daylight, energy-efficient systems, and nontoxic materials. The project also includes a 20,000 sq. ft. community health center and a coop organic health food store.

CONCLUSION

When we restore our cities and make them greener, we help restore our citizens and their capacity to be members of a community. And it seems that their connecting with nature enriches our lives, making us a bit more whole. We see the yearning of our residents to place their hands in the dirt everywhere we work, from the long-term Harlem Community gardeners on 153rd Street, to the new immigrants living in the South Bronx, to the young suburbanites moving into inner-city Denver for a richer and more integrated life. Urban gardens bring a sense of peaceful productivity, of connection to the larger whole. Whether walking, sitting, thinking, reading, or listening to a concert, gardens are a gateway to feeling the interdependence of humans and nature. There is an increased interest in contemplative spaces, whether in higher-end developments aimed at highly stressed financial and marketing executives, or for families just emerging from homelessness. The market demand, resident satisfaction, and increased real estate value that arises from biophilic design is likely to fuel continual interest.

If there is a purpose for our innate connection with nature, perhaps it is to help humans find our place on earth and in the universe. When experiencing elements of nature, many express a greater sense of wholeness and an innate recognition of the interdependence of all things. Perhaps if we more deeply experience this connection, we will be less abusive of nature and treat the earth more responsibly. Developing biophilic buildings and communities is a step along the restorative pathway.

ENDNOTES

1. Jonathan Rose Companies study, New York City, 2003.

REFERENCES

Beatley, Timothy. 2000. *Green Urbanism: Learning from European Cities*. Washington, DC: Island Press.

Ewing, Reid.2005. "Can the Physical Environment Determine Physical Activity Levels?" *Exercise and Sport Sciences Reviews* 33(2), April, 69–75.

Handy, Susan, Xinyu Cao, Patricia Moktarian. 2005. "Correlation of Casualty Between the Build Environment and Travel Behavior: Evidence from Northern California Transportation Research." *Art. D.* 10(6), November: 427–444.

Harris, Caroline G. 2006. "The Zen of Cities: Meditative Spaces in Urban Environments." *AIA Journal of Architecture*, January.

Howard, Ebenezer. 1945. *Garden Cities of Tomorrow*. London: Faber and Faber.

Kessler, Agatha, and Dennis Rubba. 2006. "Bringing Nature to the Urban Jungle." *Denver Business Journal*, August 11.

Le Corbusier. 1946. *Urbanisme*. Paris: Éditions Gonthier.

Louv, Richard. 2005. *The Last Child in the Woods: Saving Our Children from Nature-Deficit Disorder.* Chapel Hill, NC: Algonquin Books.

Marino, Vivian. 2006. "Rethinking the Balcony." *New York Times*, July 2.

chapter

19

The Greening of the Brain

Pliny Fisk III

INTRODUCTION: BIOPHILIA AND AN ECOLOGY OF THE MIND

In our cell phone era, the ever-evolving neocortex of the human brain craves instant gratification. Yet those very impulses perpetuate a distancing from nature's eternal cycles, putting our support systems at risk. How do we respond to this disjunction between the human brain and "nature's brain"? Do we go back to the woods, or try to cope in innovative ways with this clashing of planetary forces? This chapter draws on neuropsychological research to suggest that sustainable building design has the potential for even deeper global ramifications than we might think. While sustainable buildings continue to respond directly to the environment, they may also satisfy the human brain's natural need for stimulation by involving nature's cycles on a miniaturized scale proven to elicit brain response. This approach is quite different from that currently proposed by biophilic design and aims to draw on how this life-cycle scaling is key, and needs to be brought into the biophilia conversation.

The model outlined here is informed by the theory that our relationship with life-cycle events—moments or behaviors in which we directly encounter the life cycles of water, energy, food, air, and materials—is, in effect, what we are designing when we design buildings, especially what we are calling sustainable buildings. If we shrink life cycles within and around the building to a scale easily recognized by the brain, what we are designing is a significant trend in our evolution as a species. The argument is based on a neuropsychological understanding of how humans engage with critical elements of our environment that, as biophilia scholars point out, are often remote from our everyday experience. Biophilic design promotes, in part, the creation of buildings that aid in reconnecting humans with the life around us. Our concept extends biophilia's principles to suggest that buildings might be designed to *mimic* and *illuminate* these life-cycle events—even seem to speed them, electronically and otherwise—causing humans to experience resource flows and cycles, understand resource dependencies at an evolutionary level, and adapt behavior accordingly. The goal becomes a combined effect as a society, which Ray Kurswell would call "singularity" or Gregory Bateson might today refer to as an "ecology of being." The attempts discussed below are the beginning of a much deeper exploration.

A NEUROLOGICAL BASIS FOR DESIGN

Early humans, like other animals, organized around what might be referred to as resource events existing in relation to what was directly visible in time and space around us: We saw food, sourced and transported it, then disposed of the remains. Our ability to predict conditions of change from the patterns around us was limited. As our interpretation of resource events eventually became more representative of time past and time future—evidenced in our prehistoric paintings—our brains evolved to perceive sequence and seasons, and to respond to mistakes with a more sophisticated trial-and-error adaptive strategy. These perceptions evolved into the unique human trait of critical thinking, located in the neocortex, which makes up the majority of the human brain.

The neocortex is responsible for our senses, parts of our motor functioning, spatial reasoning, conscious thought, and language. According to neuroscience, the neocortex is also responsible for interval pattern recognition—the understanding of the durations between repeating events—responding to activity sequences and controlling our ability to adapt when confronted with new ones. This is in contrast to the part of the brain associated with the circadian clock, those daily and seasonal rhythms focused on in biophilic design. The neocortex can quickly develop feedback loops that reinforce or discard past conditions and also propose entirely new ones.

There is evidence that the neocortex part of the brain tends to seek new stimuli to feed itself: Its food for evolutionary growth is the new, the different, the challenge of solving, of patternizing in rapid response sequences (Biederman 2006). Recent discoveries have shown that when properly and sufficiently stimulated, this part of the brain actually grows new neurons (Gould et al. 1999). In the twenty-first century, our neocortex's stimulation hunger is satiated at least partially by participation in the world of electronic information technology, which takes us into make-believe realms unconnected to much of the actual physical world around us.

Today we face not the simplistic resource events of prehistory, but life-cycle events of mammoth proportions, such as climate change and the long-term toxic effects of industrial and technological processes, and we have begun to understand their significant effects on humans and planetary life in general. Thanks to our advanced neocortex's ability to record and propose alternative action strategies, science is able to project potential environmental catastrophes, but in many cases what our brains are willing to see and predict, even if we immediately act, outstrips nature's ability to respond. Success or failure on nature's scale will not be evident sometimes until decades or centuries later, as exemplified in the world response to ozone depletion due to CFCs. This discrepancy between nature time and human neocortex time may indeed be at the core of our increased disconnect from nature and its processes. Further, the more humans satisfy the neocortex with technology removed from natural processes, the more the brain could be said to evolve away from synchronization with nature, perpetuating negative life-cycle events.

Yet the same nascent evidence of the neocortex's response to the resource-unconscious phenomenon of information technology, in light of the work of biophilia scholars who have delineated the connections between humans and the key life-support capabilities of the natural processes around us, suggests that we have the potential to virtually redirect our own brain evolution to incorporate awareness of the resource events around us. Designers in particular—architects, landscape architects, urban planners, et cetera—may be able to practically instruct the neocortex to conceptualize the resource problems of our everyday world before there is planetary devastation.

To accomplish this, within an individual lifetime the

neocortex should be stimulated enough by engagement with life-cycle events in the built environment, at a time pattern closer to that which the neocortex craves, that the gap between natural processes and human consciousness begins to close. Environmental psychologists such as DeLong and Lubar have identified new conditions (not yet attributed to a specific physical area of the brain) suggesting that humans perceive a strong relationship between space size and time (DeLong and Lubar 1979). Larger space has been shown to slow perceived time, while smaller spaces speed perceived time up.[1] In this case, in addition to circadian rhythms influencing our synchronization with natural processes connected to the larger world around us (a primitive brain function), encapsulating macro-level natural processes on a small scale may increase our ability to synchronize the brain with nature. A space—and the events within that space—might then be designed so that occupants witness more thoroughly their interaction with a resource and the life cycle that creates it; turning on a faucet, for example, triggers an understanding of a rainwater cistern the water comes from, the rain that filled it, and the life cycle of water that we rely on. This space-time correlation may form a critical link to the time-interval element of the neocortex, speeding up or slowing down how we perceive events sequenced in time (DeLong et al. 1994) and potentially satisfying our evolutionary need to stimulate brain growth. Sustainable design, in this eventuality, can bridge the widening gap between human brain capacity and the key life-support capabilities of the natural processes around us. In other words, the neocortex fulfills its evolutionary potential as an advanced, internal consequence-mapping tool, its drive for information sated by engagement with the life-cycle events made explicit in the design that surrounds us, its dominance directed to resource-related reasoning that contributes to continuing life on earth.

DESIGNING FOR THE NEOCORTEX

It is useful to refer to a concrete example of a building construed to function within the hypothesis proposed here as well as within the overlapping realm of biophilia. The Advanced Green Builder Demonstration (AGBD), completed in 1996 in Austin, Texas, was designed with the support of the influential Austin Energy Green Builder Program to function as the state's demonstration of green building. The building, conceived as a flexible dwelling for a family of four, integrates passive ventilation, daylighting, and earth-based cementatious materials that act as thermal mass. In addition, it incorporates several event cycles that become a series of visible and tactile points of contact between humans and the life-cycle events that they depend on—in particular, energy, water, and material cycles. The building attempts to "miniaturize" these life cycles, or locate the processes that support life within the site boundary so that the processes are no longer removed and abstract. Thus, it has the potential to trigger brain functions that might better connect us to these significant environmental sequences. The building, then, could be said to extend our perceptions and connect us to the resources we use on a deeper level than previously imagined.

The AGBD functions as an armature for life-cycle events that both support people and reinforce our dependence on the cycles that are nature driven. The photovoltaic systems that generate electricity from the sun use the roofs at the entrance to the building, becoming both a very necessary summer shade and the primary electrical energy generator, with a functional, visible presence. The manner through which energy is transported and processed in the building is similarly transparent, as the carrier beams show the electrical conduits, protected as an open carrier of energy with vines surrounding, becoming part of the transport process. The indoor-outdoor kitchen—with herb garden close by—includes stove, refrigerator, back-up water-heating element, and solar cooker. It allows the cook to engage inside or outside with the natural surroundings while emphasizing the use of the sun. The location of the large solar cooker, perched above on a balcony, is specifically meant to yield a solar-cooking experience. Cooking food, food storage, and preservation—often the most energy-intensive elements of a home—become an adjustable, rollable, and therefore convenient part of the everyday environment. Here the life cycle of energy is miniaturized, from sun-source to end use, creating a consciousness of use and waste at a scale satisfying to the neocortex.

Similarly, the water cycle is not hidden but displayed. Additional roofs, built to provide supplementary surface area for rainwater harvesting, literally reach for the sky in an effort to collect each drop of dew or precipitation. The source, then, is celebrated, and the story continues to be told through the gutters; these transport conduits from the harvesting surfaces, sharing some of the same distribution channels as the energy system, illustrate the resources' interconnectedness and the common resource event of transport. At the building's entry, water is collected in cisterns and wastewater is treated in a celebratory and aesthetically pleasing fashion, using a flower and reed bed that reminds one of the nutrient effects of wastewater on plants, making them vibrant and available during most seasons. The occupant becomes part of the life-cycle experience by trimming the flowers so root depth does not interrupt the wastewater's passage into the reed bed. Likewise, water availability is measured by hand: How is the water storage performing? Use the simple test of temperature on the skin, touching the cool metal tank to judge water level. Even the bathroom is designed to provoke reflection on and immediate comprehension of natural processes; the playful water pole upon which the sinks and shower and commode can be all coordinated together or separately technically enables the user to use the shower, sink, and toilet simultaneously. Again, the neocortex is engaged due to convenience, aesthetics, and time coordination brought about by a scalar condition.

The material event cycle is also highlighted, becoming as much a part of the biophilic experience as the energy or water cycles within the building. The building visibly supports vines and other plantings until the observer is not quite sure whether the building was built for humans or for nature. All materials mimic the region's metabolic conditions, whether virgin (mesquite, caliche, straw) or recycled (fly ash cement, rubber tires, bottom ash, recycled steel rebar). Through many visible, accessible, reconfigurable joints, sourcing new components to reconfigure the space at will becomes part of the design motif. Stairs are detachable; columns and beams invite users to unbolt and reattach in a new way; the kitchen, as mentioned, is mobile, while the bathroom revolves according the best fit and combination. This flexibility underlines the building's expression of a deep respect for nature in its acknowledgment of changing habits and times; the decisions it reflects are recognized to be temporary, and nature is recognized to supersede it. The building is a microcosm of the life-supporting functions usually associated only at the planetary scale—whether water, energy, food, or materials—distilled on a 5,000 sq.ft. site. It supports not only the physical but also the psychological and neurophysiological needs of humans, feeding the neocortex.

While limited space precludes a deeper discussion, we put forward that today's technical capacity offers the means to both speed up and miniaturize for the brain many key life cycles, no matter their scale, and to account for our individual and combined impact both past and future. This has vast planetary implications. New technologies offer promise in this area. RFID tagging, in which information on each object and each process is "taggable," may have implications for the life cycles noted above. In addition, the modern Swiss army knife—aka the cell phone—may now read embedded patterns and barcodes that have the potential to make our iconography sing with levels of information only dreamt of before. These approaches must not simply lead *us;* we can lead *them* into a patterned information world that is connected to our evolution. And in the process, we may include the planetary partners that ultimately determine our survival. (See Figures 19-1, 19-2, and 19-3 in color insert.)

The following are presented as a preliminary distillation of the above hypothesis as design principles, the launch of a "greening of the brain":

1. Consider life-cycle events in a building—direct interactions with the natural life cycles of water, air, et cetera—as mimicking the life-cycle events around us, and treat them with the same awe and respect as the natural life-cycle events so that people can engage in and respond to these cycles through design.
2. Identify the full range of life cycles and life-cycle events in and around our buildings and consciously cover all life-cycle phases (or in behavioral terms, "events") from source to resource.
3. Conceive of the life cycle as successions of resource events able to be balanced, with the user as part of the balancing act, so that people understand both the parts (the individual events) and the whole.
4. Differentiate between the two significant scales of

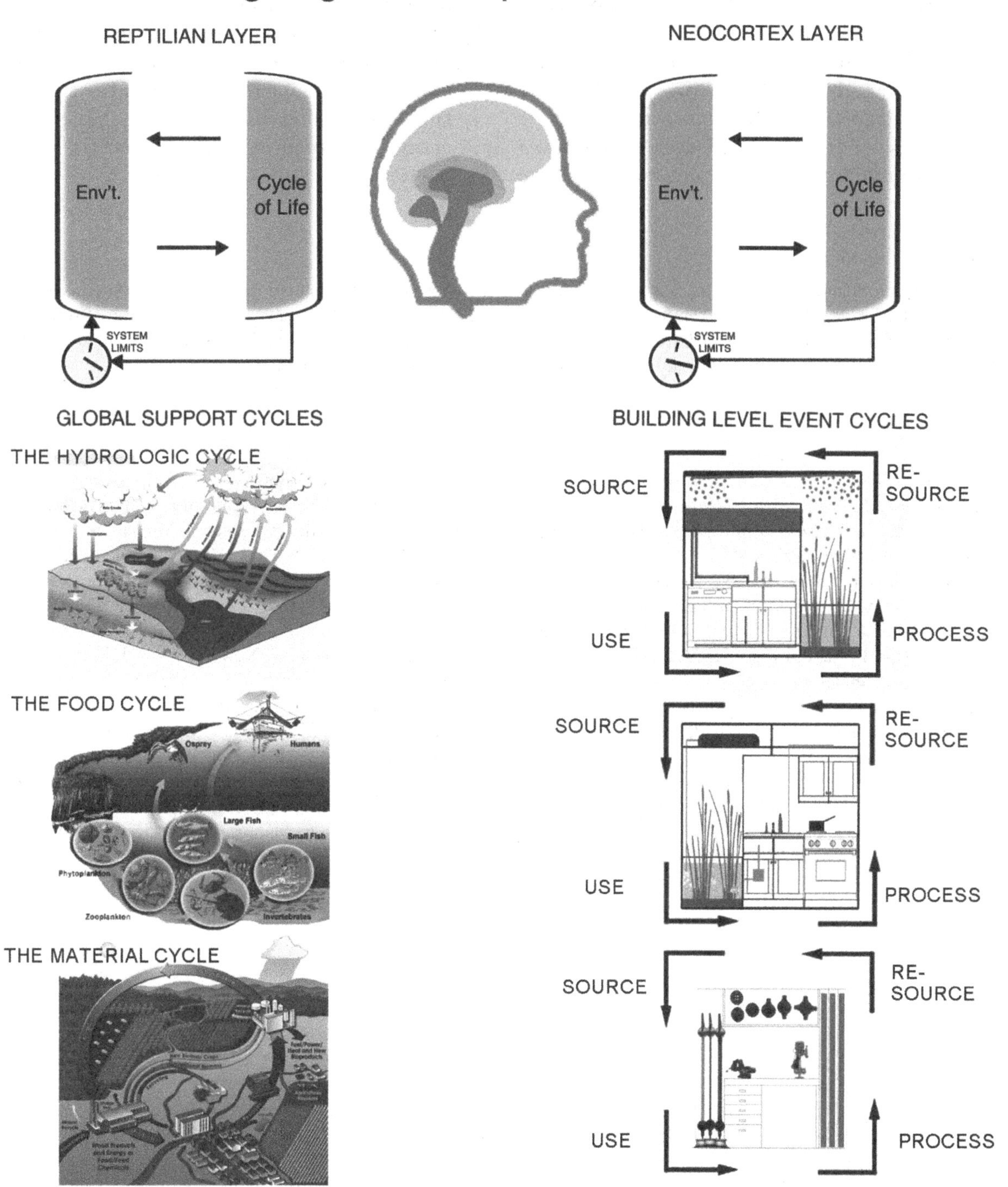

Figure 19-4: Demonstrating how scalar activities match brain functional areas

human brain activity, the circadian and the interval, so that life-cycle involvement can occur at both levels.

5. Go beyond these seasonal and diurnal (day-night) circadian brain rhythms by entering into the interval time function of the brain's neocortex.
6. Bring the scale of everyday life-cycle events into a time synchronization with the time intervals of the neocortex through two- and three-dimensional means and miniaturization.
7. Project from past to future and from locus to region the effect of our actions, not just at our own scale but also at the scale of community. Consider simulation and gaming environments so the neocortex is enticed to participate with the life cycles that support us.

Endnotes

1. Another brain function that seems equally relevant in the work discussed here is the human tendency to speed up time at smaller and smaller two-dimensional images. Where three-dimensional time at smaller scale models showed direct proportion to scale (i.e., ⅙ scale is equivalent to ⅙ time), screen images when in use can also be demonstrated to retain information but do so faster and with more accuracy than larger screens.

References

Ashby, William Ross. 1952. *Design for a Brain.* Boca Raton, FL: Chapman & Hall.

Bateson, Gregory. 1972. *Steps to an Ecology of Mind.* Chicago: University of Chicago Press.

Biederman, Irving, and Edward A. Vessel. 2006. "Perceptual Pleasure and the Brain." *American Scientist* 94(3): 247–253.

Brickey, J. 1994. The Effects of Pattern Scale in the Near Environment on Preschool Play Behavior. Master's thesis, University of Tennessee.

DeLong, A. J., and J. F. Lubar. 1979. "Effect of Environmental Scale of Subjects on Spectral EEG Output." *Society for Neuroscience Abstracts* 5:203.

DeLong, A. J. et al. 1994. "Effects of Spatial Scale on Cognitive Play in Preschool Children." *Early Education and Development* 5(3): 237–246.

Durham, D. F. 1992. "Cultural Carrying Capacity." *Focus* 2:5–8.

Durning, A. B. 1989. *Poverty and the Environment: Reversing the Downward Spiral.* Washington, DC: Worldwatch.

Gould, Elizabeth, Alison J. Reeves, Michael S. A. Graziano, and Charles G. Gross. 1999. "Neurogenesis in the Neocortex of Adult Primates." *Science* 286(5439): 548–552.

Kursweil, Ray. 2005. *The Singularity Is Near: When Humans Transcend Biology.* New York: Viking Penguin. Rheingold, Howard. 2003. *Smart Mobs: The Next Social Revolution.* New York: Basic Books.

Teilhard de Chardin, Pierre. 1959. *The Phenomenon of Man.* New York: Harper & Row.

Von Foerster, Heinz. 2003. *Understanding Understanding: Essays on Cybernetics and Cognition.* New York: Springer.

Wright, Karen. 2006. "Times of Our Lives." *Scientific American* 16(1): 26–30.

chapter

20

Bringing Buildings to Life

Tom Bender

LIVING ARCHITECTURE[1]

Over the last 30 years, we have demonstrated that we *can* create places with souls—places that are alive, that deeply reconnect us with the rest of Creation. We can again create places that move our hearts, that give us peace and nurture, and that continue to unfold and enrich our lives the more we are part of them. When the pieces are right, everyone entering such places breathes a sigh of relief. Their legs get rubbery and they want to just sit and soak in the energy (Bender 2000b).

These places are filled with a powerful stillness. They act as energized portals to connect us to the rest of Creation. They nurture us–and our institutions–with the breath of life, with the strength of truth. They inform our sciences, and help us transform knowledge to wisdom (Bender 2004a). (See Figure 20-1.)

Figure 20-1: Connection, connection, connection. With the rhythms of the earth, the planets, the sun, the stars, the moon, the winged and four-legged. No longer can we stand to be apart. This house started with a love of camping, of being open to life. The living room is half a square. The other half is ocean. Why have a separate bedroom? It's nice to have a soft place to curl up together any time of day.

This architecture comes from a profoundly different place and culture than our currently dominant one. It is based on life-enhancing values, not greed. It is based on a physics that extends into the energetic realms, not just the material ones. It is based in the awe arising from connection with the sacred. It's just natural to build this way when we're not divorced from the rest of Creation.

As opposed to the engineering model central to our present culture, this architecture is an integrative rather than analytic process. Dealing with wholes, rather than fragmenting distinctions, its process is a good model for operation in our new culture.

There has been a wonderful group of architects developing this integrative architecture with little recognition from the profession or academia. Jim Swan's "Spirit of Place" conferences go back to the late '80s (Swan 1991; Swan and Swan 1996). Malcolm Wells' underground buildings and tree-covered bridges (Wells 1981, 1991), Athena Steen's wonderful "Houses That Sing" women-built straw bale house projects (Steen, Steen, and Bainbridge 1994; Steen and Steen 2001), and Kelly Lerner's straw bale projects in China and Mongolia have all developed elements of this "new architecture." Chris Day's Steiner Schools (Day 1990), Sun Ray Kelley's organic buildings, Ianto Evans' cob building and Natural Building Colloquia are other valuable examples (Evans, Smith, and Smiley 2002)—as are Pliny Fisk's early work with indigenous building materials, the whole Rudolph Steiner architecture movement (Pearson 1994), Baubiologie, and Carol Venolia's "Building with Nature" newsletter, going back to 1989, to name but a few.

This "living architecture" dissolves the demarcation long existing in our culture between architecture and landscaping, between building and "nature," between what we contribute to the design of a place and what is contributed by other life. The sacred, the earth, the sun, the spirit sung in making places are all present and alive.

Living architecture, in a culture aware of the role of life-force energy, fulfills a different role and holds a different focus than one in a materialistic culture. It focuses on place, not space—as our existence extends far beyond a space-time realm. It focuses on relationships rather than structure, as dynamic interconnectedness, not unchangeable rigidity, is paramount. It focuses on meaning instead of aesthetics, as inner rather than surface characteristics are of central value. The "design principles" of materialistic architecture are subsumed by higher priority needs in an energetically based society.

Take aesthetics. Aesthetics is a visual, surface consideration, comparing to existing standards. It becomes central in a materialistic culture specifically because it is easy to fake. Architecture becomes a "visual art," concerned only with the surfaces of things. How things look, not what they are. Surfaces are what people are conditioned to pay attention to by advertising. People will buy an apple because it looks beautiful. Pesticide-free, "slightly blemished" apples don't get bought. Aes-

Figure 20-2: Mirrors distort. They focus our attention on outer rather than inner qualities—often in our groggiest states. A window into a garden can connect us to our surroundings instead of reminding us of a hangover. Hide mirrors until needed—here on the inside of a medicine cabinet door.

thetics focuses production of food on looks, not nutritive value. As a result, wheat has dropped from 90 percent protein in 1900 to 9 percent now, and apples now contain 80 percent fewer vitamins and minerals (McTaggart 1992). The nutritional value of architecture has likewise diminished. Beauty and aesthetics are superficial, dangerous, and distracting from far more important things. Even the I Ching warns of that. (See Figure 20-2.)

As we regain understanding of the interconnection of life-force energy and the sacred, and the role of arts in those realms, we are no longer dependent on artistic concepts of materialistic cultures. In sacred art, many of our familiar concepts are transformed. Beauty, for example, becomes an *offering* to spirit, a *vehicle* for opening our hearts, and *a way of knowing* truth—a special harmony and synchronicity within the oneness of spirit. Within beauty, recognition of truth on a deep level is possible.

But this beauty is not the predictable harmony tying to past conditions. It cannot and does not relate to artistic conventions, principles, yearly fads, or historical styles. Those are all related to echoing the familiar. The *deeper* essence of beauty is perception of living truth as an ongoing process of unfathomable creation. This is the beauty of inner purpose, coherence, and joyful existence that emerges to our amazement, as we explore and discover new and unfamiliar things. It is a beauty that deepens through unceasing rediscovery, and opens an intimacy among all that it touches.

This beauty is a measure of love. And love is the root force in the manifestation of life. This beauty is not something intentionally creatable. It is a gift of spirit to an open heart.

History gives many examples of these differences. The Temple of Hathor in Dendera, Egypt, is in a 900 ft. sq. precinct surrounded by a mud-brick wall 30 to 50 feet thick and up to 80 feet high, containing more than *five million cubic feet* of mud brick. Historians never mention this wall. It isn't stone, it isn't carved, it doesn't call attention to itself. The energy and intention this simple structure embodies, however, gives far greater power to the temple in the enclosure than any of the design elements in the temple itself.

Egyptians often buried mud bricks from older temples under the footings of new ones they were building. It was a device for linking with the energy of the older temple. Other temples, such as Horus at Edfu, carved their genealogy on the temple walls—again, not to prove their parentage, but to connect with the accumulated energy from previous temples that remains accessible in the energy realms. Sanctuary layout, such as at Hathor at Dendera, surrounds the central sanctuary with chapels to other deities in connection with the main one, to connect to their energy. And the yearly ritual calendar up and down the Nile linked the energy of different temples, locations, and deities to each other and to the people (Bender forthcoming 2007).

At the First Cataract of the Nile, archaeologists have revealed superimposed remnants of dozens of mud-brick temples to Khnum, rebuilt again and again and again after destruction by periodic floods over the millennia. With Egypt's skill in masonry construction, all this seems puzzling.

Other cultures have beautiful Origin stories. The Egyptians say simply, "We come from the mud." Mud from yearly Nile floods, and the life-giving fertility it brings, is *truly* the basis of life in that region. Durability, or physical magnificence, or aesthetics of a temple honoring what sustained their life was more than irrelevant. It would have prevented the primary sacred role the temple represented. These temples were *offerings*—in gratitude and celebration—renewed as their fields were, *made from* and *giving thanks to* the mud and its giving of life. This has vital power not possible in cosmetically "pretty" buildings. *Meaning* is important here, not aesthetics.

Inca "architecture" has a similar lesson. Architectural historians may debate the reasons for the geometry of the Intiwatana, the Sacred Rock, or the Slide elements of Machu Picchu. Such aesthetics, however, were never a consideration. It didn't matter what they looked like. The energetic dimension of Machu Picchu and of all its shrines and meditation seats was paramount, not the physical.

What was built was merely retaining walls to give access, and low walls to keep people from falling off of what were the tip-tops of 200-foot-high needles of crystalline white granite. These high points focused the life-force energy of a site already empowered by tectonic pressure against a huge crystalline granite batholith sheared by two fault lines. Those conditions *meant* powerful support for connection with the world of spirit. Aesthetics were irrelevant. The Sacred Rocks were empowered as energetic linkages to the *apu*, or spirits, in the great peaks surrounding the site (Bender 2006b, 12).

Similarly, in a Head Start Center I designed some years ago, the kitchen ended up right in the center and open to all the other spaces. This likewise had nothing to do with architectural theory or aesthetics. What it had to do with was the importance of welcoming kids with the smell and promise of good food. It meant giving parents an opportunity to catch their breath with a cup of coffee and peek around the corner to see that their kids are OK without them. It made the cook as central a staff person as teachers—giving hugs, telling teachers what was going on behind their backs. The *right placement* of the kitchen had meaning far more powerful than any aesthetics of how it looked (Bender 2000b, 44).

The truly rampant diseases in our materialistic culture were not of the body, but *diseases of the spirit.* They arose from lack of self-esteem and mutual respect, being of value to our community, or finding meaning in our lives. These diseases manifested in rape, substance abuse, addictions, violence, crime, obesity, isolation, depression, and despair—things possible in any culture, but overpowering in ours. They arose from the root violence of our deepest cultural values—our separation from the love of others caused by denying existence of the spirit world.

Healing those diseases of the spirit requires that we give primacy to the emotional, energetic, and spiritual well-being of all. In our surroundings this requires the honoring of the materials, the elements and forces of nature, the rhythms and cycles of life. It requires limiting our wants to ensure the fulfillment of other forms of life. These are all possibilities inherent in building done with reverence, which comes from love, which comes from intimacy. It involves, in architecture, acknowledgment of the energetic dimensions of people, place, and of all Creation and their internurture. It involves the creation of truly "living architecture" as surroundings that enable, enrich, celebrate, connect, and become part of our truly living in fullness. (See Figure 20-3.)

Figure 20-3: Living architecture invites, energizes, relaxes, and connects. It uses and honors native materials and human skills. It draws nurture from, rather than fighting, the climate, the site, the surroundings, the spirit of the culture and place. It starts with the question, "What are the most important opportunities to nurture and enrich our lives?"

People often ask, "Well, what can I do to design this way?" Usually they're asking, "Where's a checklist of what to put in a building to make it 'right'?" There isn't one, fortunately. Because all that rational stuff can be copied and faked. Architecture, like everything we do, is merely a mirror, reflecting what is inside us. Every time I've tried to design something that isn't part of me, I've failed. That is why I talk more about changing how we *are*, not how we design. That's the root that grows new trees. I've written several books giving visual examples of how those "differentnesses" have been manifested in the past, and how design interweaves with life-force energy and the sacred (e.g., Bender 2000b).

But *you* have to do it differently than I did. And *I* have to do it differently every time, because it is the living creation of things that is involved, not copying what worked before.

* * *

Biology used to be the realm focused on "living matter." But now, even our physics has to acknowledge that "living" includes buildings, rocks, invisible dimensions of the universe, and realms of pure consciousness shared by all that exists—physical or not. *Wholeness* is perhaps a better way now to think about how to approach design.

Without embracing these realms, "reconnecting architecture with nature" deals only with symptoms, not causes. We, and our architecture, are already part of nature. That's the first and most important thing to shift in our viewpoint. We just don't like what that reveals of *our* nature! Rationalistic design and sciences, and their limited concept of "natural processes" *are the problem.* It is our consciousness of *mind*—of rational processes, of logic, of separation, not of unity—that is reflected in "dead" architecture. It is the primal consciousness operating through the realms of qi energy that brings life to architecture.

Institutions based on greed cannot create either a living architecture or a sustainable society (Bender 1993c). An economics that discounts the future cannot create a livable future.[2] The cities and urban culture we know are creations of a culture whose basic premise is unworkable in the conditions of our future. The possibility, or desirability, of their continuance is uncertain. It seems wise to at least be aware of and explore alternatives, and not assume that the city is the best, the only, or even a possible, sustainable solution.

Our education system trains us to perpetuate and serve this falseness. *It* is fundamentally flawed (Bender 2001d). Our professions also serve this falseness. As we move beyond that world, a living architecture, a living culture, and a truly joyful and creative existence reveal themselves as a natural part of life. It is time, perhaps, to paraphrase the Balinese, and say, "We have no architecture. We only do the best we can."

It doesn't take an advanced degree in shamanism to move our architecture into this realm. It begins simply with opening our hearts, letting go of the intellectual, verbal and physical noise of our culture, and listening to what is really needed in every thing we design. It takes form as we honor and celebrate the particular lives of materials, of ecological and human communities, and cycles of the seasons and the stars. It comes to life as we empower the intention to connect deeply and humbly with all that surrounds us and join with it to create an ever newer and ever more wonderful universe (see Bender 2000b, 2000a, 2001a, 2003b, 1986b).

It manifests almost unconsciously, in my own rainy climate, in wanting to celebrate water running off a roof, in collecting and enjoying it in a pond, and then using the pond overflow to recharge the aquifer, avoiding imposing our storm water on others. It manifests in covered walkways where people frequent, and windows positioned to welcome and celebrate the sun on its occasional visits. It is *wanting* to be surrounded by trees and birds, not making a landscape plan, that gets the right things to happen.

* * *

The "reconnecting architecture with nature" issues central to biophilic architecture are architectural expressions of a deeper out-of-trueness of a culture. That culture has today entered into a phase of intense change. Indeed, the time of that culture may well be at its end. It is important to biophilic architecture to examine and learn from the forces actively transforming its base culture. It is likely more productive to engage and assist the changes inherent in those forces than to overlay cosmetic changes. Regardless, these forces and the world they lead to can provide a depth of understanding of basic biophilia issues and how to effectively address them which can substantively improve its effectiveness. Some of those forces include the following:

1. The end of oil
2. Vulnerability of current cultural patterns revealed by 9–11
3. Fundamental changes in the physical sciences
4. Qi energy
5. Sustainable economics
6. Rationality versus wholeness
7. Culture change, deep connection, and the sacred

THE END OF OIL

Depletion of oil, and consequent change from growth to enduring patterns of culture, power, and architecture is forcing transition to a surprisingly wonderful world of sustainability.[3] The falseness that was the economic heart of our culture has exposed itself, and we are beginning to restructure our lives based on deeper and more rewarding goals. Those "new" patterns are profoundly and desirably different from our current ones (Bender 1986b, 2000a, 2000b, 2001a, 2001b, 2002b, 2002c, 2002d).

VULNERABILITY OF CURRENT CULTURAL PATTERNS REVEALED BY 9–11

9–11 was not an aberration, but an epoch-changing event. The changes engendered from its revealing the extreme vulnerability of a complex, centralized economy, and the subsequent self-revelation of the abusive, inhumane, fossil-fuel-dependent, and exploitive nature implicit in a global culture based on corporate profiting are only beginning.[4] Consequent events have also shown fundamental "out-of-trueness," or "diseases of the spirit" of our culture which have generated the social sickness underlying such events.

FUNDAMENTAL CHANGES IN THE PHYSICAL SCIENCES

Quantum nonlocality is forcing us to acknowledge instantaneous faster-than-light communication throughout *all levels* of Creation. This demands acknowledgment of faster-than-light realms integrally tied to our everyday world; energies totally outside of our conventional physics; an integral consciousness shared by all that exists; and our individual cores existing eternally outside of space and time. Needless to say, this represents a enormous shift for our culture.

QI ENERGY

Of particular interest to architecture is the validation this is bringing to *qi*, or life-force energy. It has been the central basis of sciences, world view, and healing arts in virtually every culture on earth except our own. Working with qi has been the central basis of the architecture of most cultures.[5] Qi is now viewed as a faster-than-light magnetic standing-wave diffraction energy existing congruent with our space-time physical realm. (See, for example, Tiller 1997; Tiller, Dibble, and Kohane 2001; for an introductory overview, see Bender [forthcoming 2007].) It appears to be a central mechanism for the observed nonlocality phenomena.

Research on biochemistry and consciousness has shown that our brains are only one element in consciousness. Schempp, Marcer, Gariaev, and Tertishny's recent work pushes this to the level of DNA and diffraction patterns associated with qi energy and the totally different integral consciousness it manifests (see McTaggart 2007).

Qi energy is a central part of the soul of place and art. (See "Sacred Art, Sacred Space" and other articles at www.tombender.org.) The Maya say that there is life-force in everything, but that *we* have to imbue the things *we* create with that energy. This becomes important in building partly because nobody wants a "dead body" sitting around in an otherwise living world. It is important also because of the energetic role of buildings in connecting with the sacred, in nurturing our health, and in enabling our connectedness with other life.

Combined with intention,[6] life-force energy forms the template upon which our material world takes shape in its wonderful complexity. It is vital to supporting our physical as well as our emotional and spiritual health. It is blocked by artificial building materials, intensive use of electromagnetic devices, and cultural practices based on taking from others (Bender 1998b).

We can locate our buildings on good natural concentrations of qi, as the Japanese did with powerful temples such as Kiyomizu. Qi energy can also be called directly into a place, enhanced, and worked with by individual intention and group ritual as well as by design. It forms the glue that keeps a community healthy. We are discovering the connections with the spirit world inherent in a qi-energy-based world, and how places can be made specifically to work with individual and community qi and to act as access points to the spirit realm.

Qi energy does, and doesn't, show in physical form. Two identical-looking objects may have embodied within them vastly different energy. Qi becomes "visible" more through the choices of what is attempted, the values embodied, the placement of things, the underlying intention immanent in a design. A building that is windowless because the designer didn't care about connection with the rest of nature, versus a kiva or a windowless meditation space focusing on deep connection with spirit, have powerfully different effects on people within them. (See Figure 20-4.)

Figure 20-4: A font formed from a single 4,000-pound crystal of columnar basalt was used here to energetically create a gateway into sacred space leading both into the sanctuary and to a sacred garden. It also honors the native rock of the area and the geological processes by which the area was formed.

SUSTAINABLE ECONOMICS

A new economics has emerged in recent years. By taking a more wholistic and comprehensive perspective, it has been able to achieve *order of magnitude* (10 times) the effectiveness of conventional economics. Implicit in it is a reordering of our priorities to better achieve our individual and social goals, including the energy, material, social, and psychological dimensions of our architecture. (For overviews, see Lovins, Lovins, and Hawkins 1999; and Bender 2002d. Also Bender 2002a, 2003a, 1996e, 1998b, 1996c, 1996a.)

RATIONALITY VERSUS WHOLENESS

We have been brought up in rational consciousness, and to value rational process as a tool. Many of us have also learned, with some difficulty, its limitations and the value of nonrational processes. Both generally, and specific to biophilia, there are limits to what can be achieved by a collection of largely *rational* beings *rationally* discussing a process based on *love* of life. Love is of the *heart*, not the mind. A rational approach to "life-loving architecture" can't get there. It can show you the gate, but it can't let you inside.

Talking about love is as far from experiencing it as talking about sex is from experiencing it. Love is *becoming One*. Rationality is an analytic making-of-distinctions, or *becoming separate*. There is a vital and core dimension of biophilic architecture that cannot be attained by rational means.

Our space-time-focused rational consciousness blocks the *deep knowing* of the primal *unitary* consciousness that connects every atom of our being to all the dimensions of existence (see Somé 1994; Bender 2006b, 15). Reopening this, obviously, opens new vistas.

Learning to set aside our rational compulsions and tap into our shared unitary consciousness gives us important new tools. They help discriminate which material possessions help to attain our goals and which distract from and inhibit our highest goals. They help us design with respect for the needs and aspirations of *all* Creation. And they help us deal with the silences and absences in the places we create—death, illness, age, equity, fairness, the sacred, sustainability—the things our culture hasn't wanted to acknowledge or deal with, but which contribute vital elements to the (w)holiness of the places we make (Bender 1998a).

We created a memorial garden in a church project recently, whose theme is that death is the compost out of which richer life emerges. Its image is a rotting nurse-log—a stump out of which a new sapling has taken root. Honoring death, in its deeper meanings.

CULTURE CHANGE, DEEP CONNECTION, AND THE SACRED

In the last generation, direct experience of the spirit realms has broken through religious traditions that have outlawed it since the days of the Roman Empire. Reconnecting with the oneness of Creation is bringing new purpose, new meaning, and new richness in our lives. From it, we are beginning to transform our culture to sustain and support the well-being of all life. This is enabling us to reopen our hearts—long closed off from the pain an exploitive society caused us, others, and all life.[7]

As we begin to connect in humility rather than arrogance with the deeper traditions of other cultures, we are finding with joy that art and architecture can fill a profoundly different role than in a culture of material growth and greed. What occurs as we shape our surroundings is vastly different in a sacred culture (see Bender 2006a, 2006b, 2001b, 1991, 1996d).

For example, in discussing pottery-making in the Dagara tribe, Malidoma Somé (1994) talks about more rewarding alternatives than working to get money to satisfy our wants. Most work in the Dagara village is done collectively. The purpose is not so much the desire to get the job done but *to raise enough energy* for people to *feel nourished by what they do*. The nourishment does not come *after* the job. It comes *before* the job and *during* the job. "We are nourished first," he says, "and then the work flows out of our fullness."

The indigenous notion of abundance that underlies such work practices is profoundly different from that in the West. Villagers are interested not in accumulation but in a sense of *fullness*. Abundance, achieved through that sense of fullness, has a power that takes us away from the worry characteristic of our culture. So even the process associated with the making of art is profoundly different.

SACRED SPACE, SACRED PLACE [8]

All art changes dramatically as we move into this deeper realm. The sacred, we are rediscovering, is a vital part of everyday life. It underlies, but is distinct from, religious expressions of the sacred—which often tend to separate us from others with different traditions and from personal experience of the sacred.

The sacred emerges simply in our lives. Whenever we allow ourselves to know someplace, someone, or something intimately, we come to love them. We see among their inevitable warts and wrinkles the special and wonderful things that they are, and their existence becomes as precious to us as our own. Loving them, we come to hold their existence inviolate—or sacred—and any action that would harm them becomes inconceivable. Loving them, we open our hearts. And in that open heart we discover the oneness of all creation. Openness, intimacy, knowledge, and love are the essential foundations upon which any healthy existence and any true sustainability must be built.

A church can be sacred space. But so can a bank, a bathroom, a place of eating or of community. It depends on our intentions, and how we design the place to reveal the sacred interconnectedness of our lives.

The purpose of sacred art is to bridge between our finite world and the infinite—to activate and carry us into that realm, and to help us understand the dance of creation in the realms of spirit. It is an avenue through which to connect with wondrous unknown things, and to transform our own souls. Its goal is revelation.

The outer product, as in Navajo sand painting, Tibetan sand mandalas, or Inuit scrimshaw, is often only by-product—forgotten, discarded, or carefully erased after the process of creation in communion with the sacred. The true goal is the *inner* product—the transformative re-experiencing of the oneness of all creation, and the specific accessing of healing, understanding, and personal growth that occurs in the process. The outer product is a record of, and sometimes a vehicle for, that experience, and at times a means that can assist others with the same process.

This is true architecturally as well. Temples in certain Indian traditions were left untouched once built, to return to dust in their own time, allowing new opportunity for the creative process. The Ise Shrines in Japan are built with relatively impermanent techniques. For more than 1,000 years, they have been built, consecrated, left untouched for 20 years, then rebuilt and reconsecrated on an alternate site. The importance, in this case, lies in the cyclic ritual process of continual reinvigoration of the skills, patronage, and rituals of honoring the forests, trees, tools, and expertise through which the building arts continue to give to the sacred (Itoh 1965).

Sacred art (or architecture) cannot be meaningfully evaluated from outside. It requires participatory experience to even be aware of the realms within which it is operating. Brad Keeney, an anthropologist who has trance-danced with the Bushmen for many years, is emphatic that an inexperienced outside observer has *no* means of comprehending what is occurring in the process. Not knowing that American Indian "shuffle-dancing" or Sufi dancing can be trance-dancing with the ancestors, or that the Maya and Bushman community trance rituals actually *do* bring the ancestors from the spirit world into their village ceremonies leaves us tragically blind to the functions of their art, dance, architecture, and music (Keeney 2003). We need to experience such ways of connecting with the spirit realms in order to have any concept what it is like to live *in spirit*. This does not preclude objective evaluation of what is experienced, but gives an essential experiential basis from which to understand.[9]

Sacred space is an energetic, not material, phenomenon. It is a place of linkage between our material world and the world of spirit. It may have material attributes that we see as sacred places, or it may exist almost entirely on an energetic level. In cultures such as the Huna in Hawaii, or in the Amazonian rain forests, climate does not require fancy enclosures, and direct connection with spirits in plants, or rocks, or natural places is the norm (see Amaringo and Luna 1990; Gebhart-Sayer 1985; Wesselman 1995, 1998, 2001). Not much "architecture," but the energetics of sacred space are scrupulously adhered to.

Conditions of physical space, and energetic dimensions attached to physical space or symbols, may influence and call into being sacred space, but the material is not the primary operational level. The *energetic* or spirit realm is the operational level of our universe, where things manifest into material existence, where our eternal energy selves exist, and where we connect directly with other life and other existence.[10] (See Figure 20-5.)

Figure 20-5: Sculptures in the Cave Temples of Ellora, India, give powerful connection to the sacred. Special geometries, coherence of the rock, tantric trance-work, ritual empowerment with life-force energy and other techniques work together to give this capability.

Indeed, the primary operational dimension of our visible universe is turning out to be magnetic/electrical plasma, directly linked to the qi energy standing-wave diffraction energy realm.[11]

In culture after culture, rocks, places, statues, and buildings are *empowered*, or energetically linked to deities, spirits, ancestors, healing powers, or specific energetic realms. That role is the primary function of most of the "great architecture" of the world, and its design occurs in the energetic realm rather than the physical (Bender 2000a, 2006b). In denying the existence of life-force energy (*qi*, *prana*, *mana*, *baraka*, etc.), we have kept ourselves totally blind to this dimension of architecture. (Also see Figures 20-6, 20-7, 20-8, and 21-2 in color insert.)

BIOPHILIA

Interestingly, my dictionary defines *bio-* as "relation to or connection with life, vital phenomena, or living organisms." Biology is a distant third there. *Life* has far more involved than planting trees around a building. *Vital phenomena*, amazingly, is the qi-energy realm rearing its wild head. And then, of course, *philo*is defined as "loving." Loving the livingness of reality. And that doesn't operate in the "rational" world. This means it's important to focus on the vital *qualitative* changes involved in biophilia, rather than merely quantitative micro-tuning.

More research and documentation in regards to the importance of deeper connection with the rest of nature in our buildings and communities is both needed and welcome. Its true value is low, however, if not also based on experience in nonrational process. It is also important that *real*, *meaningful*, and *deep* connectivity be included in the research, rather than just "pictures of nature" pinned to the end of a patient's bed, or fake skylights giving "illusions of nature." A real skylight costs no more than a fake one, can reduce rather than increase energy use, and connects a patient with reality, not illusion. It is mind-boggling to see illusions promoted as solutions. There is a profound difference between the two, in spite of what today's culture would wish us to believe. It is important to test and compare environments where "nature" is truly doing the healing, not just pin-up illusions.

* * *

Living architecture is part of a living culture—a transformation to a nourishing and integrally connected universe with profound new potentials. This is indeed a new realm we are entering, and a wonderful time to be alive!

NOTES

1. This material diverges significantly from that presented in presymposium papers, in my symposium presentation, and in the chapter originally prepared for this book. Post-symposium discussions indicated issues and perceptual gaps that were more urgent to present. PDFs of the presymposium papers ("Sacred Art, Sacred Space"; "Living Architecture"; and "Places Touching Spirit"), with images, are available at <www.tombender.org>, along with many other papers referenced below.
2. "Foreclosing Our Future" (Bender 2002c) shows the falsifying role of PNV accounting in economics that discounts the future. "Fixing Failed Forests," (Bender 2002b) shows the implications in forestry of using holistic economic analysis. "Learning to Count What Really Counts" (Bender 2002d) gives the comprehensive picture of that economics.
3. See, for example, Bender 1973, 1975; Bender et al., 1974; and O.E.R.P., 1974. Current "peak oil" discussions largely repeat what was assembled in this seminal period. Global warming is merely the flip side of the same issues.
4. See "True Security (Bender 1982). "The End of Nuclear War" (Bender 1986a) anticipated events such as 9–11 as well as the profound need for and implications of positive change implicit in such events and their causes. "Ten Easy Pieces - of a Better World" (Bender 2001c) lays out ways of dealing with some of the root issues. See also Bender 2004b, "We Have Found the Terrorist—The Terrorist Is U.S."

   For an overview of the implications of current corporate guidance of our culture, see the following:

   Race for the World (Bryan et al. 1999). A gleeful layout of the strategies of consolidating the world's wealth in the fewest possible hands.

   Confessions of an Economic Hitman (Perkins 2004). Most people, including myself until recently, have been unwilling to believe the intentionality of exploitation of people and planet because the system is so beautifully invisible. False economic projections (by Perkins and others), then unrepayable "development" loans (which also pay for a military in each country to "protect investments" and control the people), then the IMF and World Bank's murderously exploitive "Structural Adjustment Programs"—all of which lead to:

Planet of Slums (Davis 2006). Slums now constitute almost 80% of urban populations in developing countries—equal to a third of the global urban population. They represent unimaginable poverty, and result from conscious exploitation. That exploitation pays for the urban amenities you enjoy. The details here will probably make you sick.

Learning to Count What Really Counts (Bender 2002). I'll add this in here, as it's hopeful, short, easy to read, and lays out the amazing 10-fold economic benefits of economics based on ecology, systems, qi energy, and the sacred. Unlikely? There is plenty of data supporting it. Can a forestry economics of 40-year tree harvesting rotations to be honest when it takes 20 years to even fully capture the sunlight falling on the ground?

5. For comprehensive references to the underlying aspects of qi energy, see *Building with the Breath of Life* (Bender 2000a), which also provides details of working with chi energy in design. For historical examples, see Bender (2000a), Chapter 3. *Silence, Song and Shadows* (Bender 2000b) provides a right-brain introduction to working with qi energy in design, and our need for the sacred in our places.

   The DVD/video *Cave Temples of India* (Bender 2004a) provides a historical case study of sacred sciences based on qi or prana, and their use architecturally in creating powerfully interactive places. *Building Architecture of Sthapata Veda* (Sthapati 2001) gives an outline of the South Indian sacred sciences. *Principles of Composition in Hindu Sculpture* (Boner 1962) covers the mandalic geometries used; *Tantra: The Cult of the Feminine* (Van Lysebeth 1995) and *Yantra* (Khanna 2003) cover the practices involved.
6. For a sensitive discussion of the role of intention, see Chopra (2003). For its role in working with qi, and in architectural applications, see Bender (2000a).
7. For some of the unexpected dimensions of these changes, see Bender 1999, 1993b, 1993a, 1992, 1987. For a personal account of such changes, see "Shedding a Skin That No Longer Fits" (Bender 1996b).
8. See "Sacred Art, Sacred Space" at www.tombender.org for images and more detail.
9. And with quantum nonlocality, there is no such thing as an objective, noninvolved observer.
10. See Tiller, Dibble, and Kohane 2001; Wesselman 1995, 1998, 2001. "Silence, Song and Shadows" (Bender 2000b) gives succinct examples of how to address these spiritual connections with the rest of nature in the design of our surroundings.
11. Space research is bringing back emphatic evidence that the primary order of our universe is electric/magnetic-based plasma, not neutral-charged material. See Talbott and Thornhill, 2007; and Thornhill and Talbott, 2007.

REFERENCES

Amaringo, Pablo, and Luis Eduardo Luna. 1990. *Ayahuasca Visions.* Berkley, CA: North Atlantic Books.

Bender, Tom. 1973. "*Living Lightly.*" Monograph.

———. 1975. "*Sharing Smaller Pies.*" Monograph.

———. 1982. "True Security." *Rain,* October–November.

———. 1986a. "The End of Nuclear War." December.

———. 1986b. "Putting Heart into Our Homes." *Yoga Journal,* September.

———. 1987. "Sacred Building." Spirit of Place Conference.

———. 1991. "Making Places Sacred." In *The Power of Place,* edited by James Swan. Wheaton, IL: Quest Books.

———. 1992. "In Beauty We Walk."

———. 1993a. "Cities of Passion, Cities of Life."

———. 1993b. "Towards a Sacred Society." *Urban Ecologist,* Spring.

———. 1993c. "Transforming Tourism." *Earth Ethics,* Summer.

———. 1996a. "Big Changes Are Easier Than Small Ones." *North Coast Citizen,* September.

———. 1996b. "Shedding a Skin That No Longer Fits." *In Context,* no. 44, July.

———. 1996c. "Some Questions We Haven't Asked." *In Context,* no. 44, July.

———. 1996d. "The Spiritual Heart of Sustainable Communities," *In Context,* no. 44, July.

———. 1996e. "Unexpected Gifts of Sustainable Community." Solar Energy Association of Oregon.

———. 1998a. "Ending the Silences: Changing Community Chi." September.

———. 1998b. "It Gets Even Stranger from Here On." *Building with Nature,* no. 19.

———. 1999. "Our Need for the Sacred in Our Surroundings."

———. 2000a. *Building with the Breath of Life.* Manzanita, OR: Fire River Press.

———. 2000b. *Silence, Song and Shadows.* Manzanita, OR: Fire River Press.

———. 2001a. "Hospice: Gateways of Life and Death." *Healing Ministry*, November–December.

———. 2001b. "Portals to the Spirit World." *Shaman's Drum*, September.

———. 2001c. "Ten Easy Pieces—of a Better World." *Daily Astorian*, October 31.

———. 2001d. "Terminal Ed: Our Schools Are Dead."

———. 2002a. "The Economics of Wholeness." *Magical Blend*, November.

———. 2002b. "Fixing Failed Forests."

———. 2002c. "Foreclosing Our Future."

———. 2002d. *Learning to Count What Really Counts*. Manzanita, OR: Fire River Press.

———. 2003a. "Economics, Architecture, and Banking." *Green Money Journal*, April; *Sustainable Business Insider*, June; *Green Money Journal* "Greatest Hits" issue, Winter 2004.

———. 2003b. "Putting Heart Back into Our Homes." In *Another Kind of Space*, edited by Alan Dearling with Graham Meltzer. Enabler Press.

———. 2004a. *Cave Temples of India*. DVD/VHS. Manzanita, OR: Fire River Press.

———. 2004b. "We Have Found the Terrorist—The Terrorist Is U.S." October.

———. 2006a. "Sacred Art, Sacred Space."

———. 2006b. "Places Touching Spirit."

———. Forthcoming 2007. *The Physics of Qi*. DVD/VHS. Manzanita, OR: Fire River Press.

Bender, Tom, and Joel Schatz, with Office of Energy Research and Planning, Governor's Office, State of Oregon. 1974. "Cosmic Economics."

Boner, Alice. 1962. *Principles of Composition in Hindu Sculpture*. Leiden: Brill.

Bryan, Lowell, et al. 1999. *Race for the World*. Boston, MA: Harvard Business School Press.

Chopra, Deepak. 2003. *The Spontaneous Fulfillment of Desire*. New York: Three Rivers Press.

Davis, Mike. 2006. *Planet of Slums*. London/NYC: Verso.

Day, Chris. 1990. *Places of the Soul*. London/S.F.: Aquarian Press.

Evans, Ianto, Michael Smith, and Linda Smiley. *The Hand-Sculpted House*. White River Junction, VT: Chelsea Green.

Gebhart-Sayer, Angelika. 1985. "The Geometric Designs of the Shipibo-Conibo in Ritual Context." *Journal of Latin American Lore*.

Itoh, Teiji. 1965. Japanese Environmental Design. Ms.

Keeney, Brad. 2003. *Ropes to God*. _Philadelphia PA_: Ringing Rocks Foundation.

Khanna, Madhu. 2003. *Yantra*. Rochester VT: Inner Traditions.

Lovins, Amory, L. Hunter Lovins, and Paul Hawkins. 1999. *Natural Capitalism*. Boston: Little, Brown.

McTaggart, Lynne. 1992. *What Doctors Don't Tell You*. New York: Avon.

———. 2007. "DNA Double Helix: Our Body's Recording Studio and Radio Station." *Global Intelligencer*, February.

Office of Energy Research & Planning, Governor's Office, State of Oregon. 1974. *Transition*.

Pearson, David. 1994. *Earth to Spirit*. San Francisco CA: Chronicle Books.

Perkins, John. 2004. *Confessions of an Economic Hitman*. San Francisco CA: Berrett-Koehler.

Somé, Malidoma. 1994. *Of Water and Spirit*.New York NY: Tarcher/Putnam.

Steen, Athena, and Bill Steen. 2001. *The Beauty of Straw Bale Homes*. White River Junction, VT: Chelsea Green.

Steen, Athena, Bill Steen, and David Bainbridge. 1994. *The Straw Bale House*. White River Junction, VT: Chelsea Green.

Sthapati, V. Ganapati. 2001. *Building Architecture of Sthapata Veda*. Chennai, India: Dakshinaa.

Swan, James, ed. 1991. *The Power of Place*. Wheaton, IL: Quest Books.

Swan, James, and Roberta Swan. 1996. *Dialogs with the Living Earth*. Wheaton, IL: Quest Books.

Talbott, David, and Thornhill, Wallace, 2007. *Thunderbolts of the Gods*. DVD. Portland, OR: Mikamar Publishing.

Thornhill, Wallace, and Talbott, David. 2007. *The Electric Universe*. Portland, OR: Mikamar Publishing.

Tiller, William. 1997. *Science and Human Transformation*. Walnut Creek, CA: Pavior.

Tiller, William A., Walter E. Dibble Jr., and Michael J. Kohane. 2001. *Conscious Acts of Creation*. Walnut Creek, CA: Pavior.

Van Lysebeth, Andre. 1995. *Tantra: The Cult of the Feminine*. Newburyport, MA: Weiser.

Wells, Malcolm. 1981. *Underground Designs*. Amherst, NH: Brick House Pub.

———. 1991. *Gentle Architecture*. New York: McGraw-Hill.

Wesselman, Hank. 1995. *Spiritwalker*. New York: Bantam.

———. 1998. *Medicinemaker*. New York: Bantam.

———. 2001. *Visionseeker*. New York: Bantam.

chapter

21

Biophilia in Practice: Buildings That Connect People with Nature

Alex Wilson

This chapter[1] examines how biophilia can inform building design. The diverse chapters of this book have defined biophilia, exposed readers to the fundamentals of this concept, and articulated the many benefits of biophilic design. Here we affirm the important link between sustainable design (or green design) and biophilia, and we address some of the many ways in which biophilic design can be incorporated into our buildings.

Applicable to all buildings where people live, work, learn, or heal, biophilia is referred to by Stephen Kellert as "the missing link in sustainable design." While many of the leading examples of green design incorporate aspects of biophilic design, many, unfortunately, do not—something that should be remedied as we move forward in the green building movement.

A REVIEW OF BIOPHILIA

Harvard biologist Edward O. Wilson, Ph.D., coined the term *biophilia* in his book by the same name (Wilson 1984), arguing that human beings have an innate and evolutionarily based affinity for nature. He defined the term as "the connections that human beings subconsciously seek with the rest of life."

Kellert defines the concept of biophilia in *Building for Life* (Kellert 2005, 50) as "a complex of weak genetic tendencies to value nature that are instrumental in human physical, material, emotional, intellectual, and moral well-being. Because biophilia is rooted in human biology and evolution, it represents an argument for conserving nature based on long-term self-interest."

Judith Heerwagen, a psychologist whose research

has focused on the relationship between buildings and psychological well-being, argues that "biophilia evolved to guide functional behaviors associated with finding, using, and enjoying natural resources that aided survival and reproductive fitness—and avoiding those that are harmful." Biophilia, she suggests, evolved as an adaptive mechanism to protect people from hazards and to help them access such resources as food, water, and shelter. This translates in present conditions into the strong preference people exhibit for features that suggest those evolutionary roots. "People will fight to keep biophilic features," Heerwagen says, describing competition in commercial buildings for offices with views to the outdoors. In workstations without views, people adapt by surrounding themselves with potted plants, images of nature, and nature-focused screen savers on their computers.

WHY BIOPHILIA MATTERS

We care about biophilia in building design—or we should care—for two primary reasons. First, it is becoming increasingly clear that biophilic elements have real, measurable benefits relative to such human performance metrics as productivity, emotional well-being, stress reduction, learning, and healing. And second, from an environmental standpoint, biophilic features foster an appreciation of nature, which, in turn, should lead to greater protection of natural areas as well as efforts to eliminate pollution and maintain a clean environment. Both the measurable benefits of biophilia and the less tangible arguments are discussed in much greater detail throughout this book and are briefly summarized below.

Healing

The most clearly demonstrated benefits of biophilia are related to health and healing. If the biophilia hypothesis is correct, all human beings have carried its stamp on their genes for millennia. Indeed, the historical record reflects that the potential for biophilic features to produce positive, measurable outcomes on human health and healing has been understood for centuries. As long as 2,000 years ago, according to Richard Louv, Chinese Taoists recognized that gardens and greenhouses were beneficial to health. Leonard Maeger, writing in the *English Gardener* in 1699, recommended spending time in a garden: "There is no better way to preserve your health." (quoted in Louv 2005, 45). In 1860, the pioneering British nurse Florence Nightingale wrote in *Notes on Nursing* that "variety of form and brilliancy of colour in the objects presented to patients are an actual means of recovery"(Nightingale 1860, 59).

More recently, Roger Ulrich quantified the medical benefits of views of nature. In a landmark study Ulrich showed that patients recovering from gallbladder surgery recovered more quickly and required less pain medication if they had a view of trees outside their windows than if they looked out on a brick wall (Ulrich 1984). Such benefits have clear economic advantages.

According to Ulrich, there are a number of ways in which biophilic design may alleviate pain: "Exposure to nature appears to reduce pain through different types of mechanisms, including distraction and stress reduction," he says. "Distraction theory holds that pain absorbs attention; the more attention devoted to pain, the greater the experienced intensity. If patients are diverted by or become engrossed in a pleasant nature view, they allocate less attention to pain, and accordingly the intensity is reduced" (pers. comm.).

"A second mechanism," says Ulrich, "is suggested by the well-documented finding that viewing nature effectively lowers stress. When stress is lessened, levels of stress hormones, such as norepinephrine, often are lowered as well, and this may alleviate the experienced intensity of pain" (pers. comm.).

When contact with nature involves exposure to natural light or sunlight, yet another pain-reduction mechanism may come into play. A recent study of hospital patients in Pittsburgh showed that those in bright, sunny rooms took fewer strong pain relief medicines and had lower levels of stress than patients undergoing the same type of surgery, but located in rooms that received less daylight due to the presence of another building 75 feet away. The authors, including Ulrich, speculate that the differences were due to elevated serotonin levels in the patients housed in the bright rooms. Sunlight exposure appears to increase concentrations of serotonin, a neurotransmitter that inhibits pain pathways in the central nervous system (pers. comm.).

Despite the limited available data, many hospital planners have taken the message of nature contact seriously. At the CHRISTUS St. Michael Health Care Center in Texarkana, Texas, for example, every patient room looks out on a natural outdoor scene. The Bronson Methodist Hospital in Kalamazoo, Michigan, includes a garden atrium that the hospital's website describes as incorporating light, water, and greenery "to connect patients and visitors with the healing powers of nature."

When patients cannot be provided with an actual view of nature or direct contact with nature, representing nature in photographic images and other artwork has also been shown to be beneficial—though the results are not quite as dramatic. Nature photographs and artwork of natural scenes are common in the more progressive hospitals today. Expanding on this concept is

Simulating Nature with Luminous SkyCeilings

When it's not possible to put people in actual contact with nature, the next best thing may be to provide an illusion of nature that achieves similar calming benefits. For such applications, The Sky Factory, based in Fairfield, Iowa, offers the luminous SkyCeiling™.

The SkyCeiling is a ceiling-mounted, backlit grid of translucent acrylic panels with high-resolution photographic transparencies mounted on a modular grid of aluminum extrusions that simulate skylight framing. Full-spectrum (6,000 Kelvin) fluorescent lamps above the SkyCeiling turn the system into a realistic view of the sky, often with some tree branches showing at the edges (see Figure 21-1 in color insert). "We convince the mind that there's a real skylight up there," according to company founder Bill Witherspoon. "Once the mind is convinced, it triggers a psychophysiological response . . . a powerful sense of ease and well-being." He notes that this input can be received even through our peripheral vision; we do not have to be looking up at the ceiling to benefit from it (pers. comm.).

Introduced in 2002, close to 2,000 SkyCeilings had been installed by mid-2006, with roughly 70 percent going into healthcare facilities, according to Witherspoon. "It's kind of a no-brainer," says Witherspoon. "We have people who are captive observers of ceilings, and they're under tremendous stress."

The benefits of such a view do come with an energy penalty. While the fluorescent lamps are high-efficacy T-5s, the translucent panels block a significant portion of the light. Just how much light is blocked depends on the photo; Witherspoon guesses 30–35 percent. The system uses one lamp (56 watts) for every 8 sq. ft. of luminous SkyCeiling, or 7 watts per sq. ft. (75 W/m^2). "This is not considered sole-source lighting," says Witherspoon.

The cost of a SkyCeiling system is fairly high—about $95 per sq. ft. ($1,000/m^2), not including installation, according to Witherspoon. For a typical 6 ft. × 8 ft. (1.8 × 2.4 m) system for a hospital laboratory room, the cost will be over $5,000. Installation is straightforward and compatible with standard ceiling grids and standard wiring.

A relatively new feature is the integration of dimmable and programmable controls. Luminous SkyCeilings can be programmed to brighten and dim on a daily cycle and vary seasonally. This feature can be important in patient rooms, where the benefits of circadian rhythms are beginning to be understood.

The company also produces luminous Virtual Windows™ for walls. These are 1¼ inch-deep (44 mm), edge-lit, wall-mounted panels that look like clear windows looking out on attractive natural scenes. The Virtual Windows are commonly installed in pairs with some separation between, which helps to simulate binocular, three-dimensional vision.

For more on The Sky Factory, visit www.theskyfactory.com.

the SkyCeiling™, an illuminated ceiling system that provides an illusion of an attractive sky scene that helps people relax (see sidebar).

Visual images can affect health either positively or negatively. A 1992 study Ulrich was involved with examined rates of recovery from heart surgery with different wall treatments in the recovery rooms. Rooms had either bare white walls or various types of artwork, including photographs of deep, dark forests, photographs of open landscape vistas, and rectilinear abstract art. Ulrich and his fellow researchers found that the closed forest images resulted in little difference to patients compared with the blank wall, while the open landscape scenes dramatically reduced pain and anxiety. Significantly, the abstract art *hindered* patient recovery; in fact, according to Ulrich, the negative effect of the abstract art was so significant that the researchers discontinued that aspect of the experiment in the interest of patient health (pers. comm.).

Attention and Learning

In Chapter 11, journalist Richard Louv suggests that nature may be useful as a therapy for attention deficit hyperactivity disorder (ADHD) and that lack of contact with nature may be one of the contributors to the dramatic rise in ADHD among children in recent years. He refers to this idea as "nature-deficit disorder." While much of the evidence Louv cites is anecdotal, it is compelling—and leads him to conclude that "yes, more research is needed, but we do not have to wait for it." He argues that we should be providing much greater contact with nature in learning environments.

Various studies, including several by the Heschong Mahone Group, have shown a correlation between daylighting or views to the outdoors and performance in schools (Heschong 1999, 2003). If borne out by future investigations, such findings could provide powerful incentive to incorporate biophilic design features into schools.

Productivity, Creativity, and Satisfaction

In almost any building type, there are benefits to improving the performance and satisfaction of the people working or living there (see Figure 21-2 in color insert). We often lump the wide-ranging benefits of human performance under the rubric of productivity. While measuring productivity is difficult, there is growing interest in doing so. Researchers from the Rocky Mountain Institute and Carnegie Mellon University have compiled reports of significant improvements in productivity as a result of green building features, including daylighting and views to the outdoors. A field study of the Philip Merrill Environmental Center in Annapolis, Maryland, a highly biophilic building located on the shores of the Chesapeake Bay, showed very high satisfaction scores for daylight, views, and connection to nature. The scores were among the highest in a large-scale building evaluation database managed by the Center for the Built Environment at the University of California–Berkeley. The study also showed that the occupants were very proud of the building and the environmental values it conveyed (Romm and Browning 1994; Carnegie Mellon 2005)

A number of researchers have examined whether there is a connection between creativity and childhood contact with nature. Louv cites various studies that show connections between time spent with nature during childhood and creativity as adults. In her 1977 book, *The Ecology of Imagination in Childhood* (Spring Publications, reprinted in 1993), Edith Cobb reported on her studies of childhood experiences of some 300 autobiographical descriptions of childhood written by people who gained recognition in adulthood as creative thinkers. "She concluded," writes Louv, "that inventiveness and imagination of nearly all of the creative people she studied was rooted in their early experiences in nature" (Louv 2005, 92–93).

Appreciation for Nature

From an environmental standpoint, one of the most compelling reasons to incorporate biophilic design features in buildings is to inspire interest in—and appreciation of—nature. This appreciation, in turn, can motivate people to protect the environment and preserve natural areas.

Richard Forman, a professor of landscape ecology at Harvard University and a widely published author in the landscape design and planning fields, argues that, in addition to the anthropocentric benefits of buildings, biophilic design offers significant benefits to nature it-

TABLE 21-1 Biophilic Design Strategies and Priorities

General	
Address biophilia early in the design and planning process.	By considering biophilic design strategies very early in the design process, opportunities relating to building siting, architectural form, internal layout, interior design, and landscaping can more easily be achieved.
Address biophilic design with all buildings, but especially those for children, the elderly, and the infirm.	Views of natural scenes are particularly important for calming children and instilling in them an appreciation of nature; for the elderly and infirm, natural scenes can ease discomfort and promote healing.
Integrate teaching of ecology into buildings.	Interpretive signage and displays about natural features can help people understand and appreciate what they see.
Seek ways to integrate biophilic design into existing as well as new buildings.	Many of the biophilic strategies from this list can easily be incorporated into existing buildings, though not always to the extent possible in new buildings.
Help get the message out.	Conveying the importance of biophilic design to the design community and specific market segments, such as education and healthcare, will take concerted effort by the green building community.
Design landscapes and buildings for a sense of mystery.	This strategy encourages building occupants to explore, discover, and learn from the complexities of nature. This is especially important for spaces designed for children.
Foster attachment to place.	Visually, ecologically, historically, and culturally connecting a building to the locale helps connect occupants to a place and, in doing so, inspires them to protect that area.
Landscape and site design	
Provide open space around buildings.	Enough cannot be said of the importance of open, naturalized or planted space around buildings—spaces that put building occupants in closer touch with nature. Native plantings are preferred to support diverse ecosystems.
Maintain existing trees and native landscapes.	Protecting trees and native landscapes during land development and construction is often the most cost-effective way to achieve natural landscaping. Preserving natural ecosystems is almost always preferable to creating new landscapes.
Provide plantings and pleasing natural settings around buildings.	Well-designed landscaping should be visible from occupied spaces in buildings. As many windows as possible should look out over plantings, water elements, and other natural features.
Build pathways through naturalized and landscaped areas.	Walking and biking pathways can be provided along restored native landscapes within both residential and commercial developments; pathways can connect these developments with the larger community. Push beyond swaths of Kentucky bluegrass to provide ecologically rich landscapes.
Replace impervious landscape surfaces with diverse native plantings.	Vegetated, naturalized areas that allow rainwater and snowmelt to infiltrate the ground are both more environmentally responsible and more pleasant to view and explore.
Provide living walls on building exteriors.	Bringing nature closer to building occupants is one of the features of living walls—typically vines that climb on screening held away from building walls or on the walls themselves. Such vegetation can save energy by providing shade but may also block beneficial daylight.
Building design	
Provide views to nature.	Windows should be designed and placed to afford easy viewing of natural, outdoor scenes.

(Continued)

TABLE 21-1 *Continued*

Blur the transition between interior and exterior spaces.	Where feasible, extend living and working spaces into the surrounding landscapes through terraces, courtyards, balconies, covered porches, gazebos, and benches situated along pathways. Create transitions to these spaces that invite their use.
Avoid interference with key sightlines.	In designing glazing systems, deck railings, and other features that could interfere with views of nature, carefully plan the sightlines and avoid interference whenever possible.
Provide high levels of daylighting.	Where practical, glazing should be vision glass, offering views to the outdoors and creating rhythmic patterns of living light, shadows, and sparkle that vary throughout the day. Even skylights should be vision glass so that clouds and weather patterns can be seen; to avoid glare, consider tintable glass for skylights—as provided by the glazing product SageGlass® (www.sage-ec.com).
Provide operable windows.	Providing building occupants with control over their own immediate environments can expose them to the smells, temperature fluctuations, and feel of nature, including the smells of flowers in the spring and summer.
Provide green roofs.	Incorporate green (vegetated) roofs onto low-slope roofs and provide both visual and physical access to those roofs.
Incorporate vegetated atria and interior planting beds.	Bringing nature inside buildings is the idea behind atria and planting beds. Open, vegetated areas within buildings, sometimes extending several stories in large commercial buildings, provide building occupants with a respite from the typical indoor environment. In hospitals, such atria have been shown to promote healing and reduce stress. Provide pathways through planted areas to allow building occupants to experience close contact with nature.
Consider incorporating *living walls* and other living systems for air and water purification in buildings.	Living, vegetated wall systems are being promoted as a way to remove air pollutants. Living systems for wastewater purification have been successfully incorporated into some buildings. Both can provide biophilic benefits.
Consider incorporating water features in buildings.	Water features can provide both visual and acoustic benefits, reminding occupants of a waterfall or spring rain.
Create a sense of complexity—yet order—in building design.	The relationship of variety and intricacy within an underlying natural pattern of order is an important element of biophilic design.
Address both spaciousness and refuge in building design.	As is demonstrated in many of Frank Lloyd Wright's buildings, varying ceiling height can create spaces that mimic the outdoors (open, daylit spaces) and areas of refuge to provide a sense of security of containment (more constrained spaces with lower ceilings).
Incorporate organic forms into buildings.	A wide range of shapes and forms that mimic nature can be used to add depth and variety to spaces.
Interior design	
Decorate with potted plants.	Using potted plants and small gardens as part of the interior design strategy will put building occupants in closer contact with natural features. With creativity, nature can be woven in throughout an interior space, even with furniture.
Provide natural materials and nature art in buildings.	Especially where actual views to nature are not feasible, natural building materials (wood grain, patterned stone, etc.) and artwork of nature scenes can be used to elicit biophilic response.
Configure office spaces to enhance views of nature.	Workstations should be positioned so that workers can see out windows and benefit the most from natural lighting, interior gardens, and other biophilic features.
Provide interpretation as part of the interior design.	Use signage and other interpretive features to explain biophilic features and functions so that they will be better appreciated, managed, and understood.

self. "Structures can be designed to provide habitat for targeted rare species, to enhance surrounding natural systems, to attract the richness of fine-scale nature on the texture of building surfaces, and even to educate people—leading to nature protection elsewhere" (pers. comm.).

The potential of buildings to inspire and motivate people about the importance of natural systems is particularly important with children. The National Wildlife Federation (NWF) Schoolyard Habitats program (www.nwf.org/schoolyard/) provides educators and school administrators with a framework for using the school grounds as an interdisciplinary teaching resource that also enhances natural habitats on the school property. To date, NWF has certified some 2,000 schoolyard habitat sites in 49 states in the United States. According to the organization, studies have found dramatic improvements in student behavior, attendance, attitudes, and performance in schools with environment-based curriculum such as NWF Schoolyard Habitats.

BIOPHILIA AND BUILDING DESIGN

Efforts to put people in closer contact with nature can focus on building design, landscape design, interior design, or any combination of the three. Many of the strategies are simply common sense. Once the benefits of biophilic design are understood, the strategies for achieving it are fairly intuitive. A sampling of biophilic design strategies is presented in Table 21-1.

BALANCING BIOPHILIA WITH OTHER GREEN DESIGN PRIORITIES

The SkyCeiling system is a popular strategy for easing stress, particularly in healthcare facilities—but it comes with a penalty of increased energy consumption. Incorporating this biophilic feature may make it more difficult to achieve energy conservation goals. Other strategies, such as large glazing areas of high-visible-transmittance glass, operable windows, and indoor-outdoor spaces that connect people with nature, may carry even more significant energy penalties.

On a different level, providing large open areas around buildings—to serve the evolutionarily based desire to look out on savannalike vistas that many biophilia proponents suggest we have—may conflict with the strategy of high-density development, or may encourage development of the most beautiful greenfield sites.

These conflicts are real, but they are surmountable. By understanding these potential conflicts and working with integrated design teams to address them, all of these goals can be achieved. Designers may need to work a little harder to maximize energy efficiency elsewhere in the building to compensate for some energy penalties with biophilic designs, and building owners or developers may have to invest more in ecological restoration and landscaping to turn urban brownfield sites into beautiful biophilic assets, but these are doable. Biophilic design involves understanding potential conflicts and achieving the right balance.

At the same time, significant synergies can be achieved with biophilic design. Green (vegetated) roofs, for example, can afford contact with natural features in an urban environment while also reducing the volume and impacts of stormwater runoff and helping to mitigate the urban heat-island effect. Restoring damaged ecosystems around a building benefits the ecological health of the area, and walking or jogging trails around a corporate office may benefit worker health. Increased glazing areas (key to biophilic design), when implemented effectively, can reduce energy use for electric lighting and cooling, and natural ventilation (in some climates) can reduce energy consumption for heating, ventilating, and air-conditioning.

Integrated, whole-systems, green design is a process of balancing all of these issues—and biophilia should be one of the issues considered in that process.

JUSTIFYING COSTS

Convincing clients to spend the money necessary to incorporate biophilic features is a challenge. Robin Guenther, FAIA, principal of New York City–based Guenther 5 Architects, which specializes in healthcare design, says that biophilic features often seem like decoration or ornamentation. "People haven't connected them to some core human need," she says. She often has trouble convincing her healthcare clients to invest in such strategies (pers. comm.).

While many of the benefits of biophilic design may be hard to attach specific value to, the benefits are real and ultimately quantifiable, according to various experts. Vivian Loftness, FAIA, of Carnegie Mellon University, argues that both the benefits of biophilic design and problems experienced with conventional design can be measured. There are real costs associated with headaches, asthma, and depression, according to Loftness. "You can actually translate those problems into dollars" (pers. comm.).

An interesting question comes up in the implementation of biophilic design: to what extent is it necessary for the biophilic elements to be real? Are artificial representations of nature—such as SkyCeiling, artificial plants in a building, and wall-hung images of nature—as good as the real thing?

Some suggest that it is not views of nature, per se, that elicit the positive responses to biophilic design, but something about those views, objects, or images. James Wise, an associate professor of psychology and adjunct professor of environmental sciences at Washington State University–Tri-Cities, suggests that it is mathematically defined fractal patterns that produce these results (pers. comm.). Fractals are complex geometric shapes that appear to repeat at finer scales; such shapes are often found in nature and can be defined mathematically. Wise believes that the beneficial psychological effects of fractals have the same evolutionary basis as other aspects of biophilia but that these benefits can be achieved by fractals alone, obviating the need for actual images of nature. The implication is that we should incorporate fractal-patterned fabrics, wall coverings, and artwork—as well as fractal patterns in nature (such as clouds, ocean waves, tree branches, or ferns)—into our buildings. Nikos Salingaros and Ken Masden (see Chapter 5) also discuss the connection between fractal geometry and biophilia as it relates to biophilic design.

The relative merits of real versus simulated nature is a hot topic of debate. Guenther is of two minds about this. On the one hand, she has a negative reaction to the representations of artificial nature. "It's a little too kitsch, a little too contrived," she suggests. On the other hand, she has healthcare clients who swear by the benefits of products like SkyCeiling, and her research into biophilia and simulating natural features has lessened her concerns. "It doesn't have to be believable to have an impact on people," she says (pers. comm.). The general feeling of biophilic design experts is that the artificial representations of nature aren't as good as the real thing, but they are beneficial.

NEXT STEPS FOR INTEGRATING BIOPHILIC DESIGNS INTO BUILDINGS

Moving forward with the important concept of biophilic design could be significantly boosted through three efforts: research into biophilia and human health and performance, education about biophilic design, and incentives to spur the implementation of these concepts.

There is clearly a need for more research into the human performance benefits of biophilic design. Given the magnitude of the benefits that can be realized through biophilic design—especially the healing benefits—it is remarkable that there hasn't been more interest in carrying out research to prove such associations. With healthcare design, Guenther puts a high priority on "continued research into the benefits of light and nature on healing." Research to date has been hampered by the lack of buildings to study that incorporate biophilic features, but that is changing, she says.

The evidence collected to date is compelling, though integrating biophilic design strategies into buildings on a more widespread basis will require significantly more scientific data showing tangible benefits of these features. Federal and state agencies should take the lead in funding this research, but health maintenance organizations (HMOs) and insurance companies should get involved as well. Loftness has been working to convince the National Institutes of Health (NIH) and the National Science Foundation (NSF) to fund such efforts. "The NIH should be jumping in with two feet to study the long-term effects of buildings on health," she says (pers. comm.).

Even as research is carried out, efforts should be directed toward education about biophilic design. Architecture schools can play a big role in this, as can continuing education programs for the design community and healthcare community. Workshops, conferences, and webinars on biophilic design should be offered on a wide level for both the design community and specialized building segments, such as healthcare and education.

Finally, there are opportunities for spurring the integration of biophilic design into buildings. The LEED® Rating System currently rewards certain features that relate to biophilia, including daylighting and green roofs, but there may be opportunities for more directly recognizing biophilic designs. LEED version 3, which is currently under development, could offer points for biophilic features. Version 2 of the *Green Guide for Health Care* rating system has expanded its "places of respite" credit based on the growing body of knowledge about health benefits of both direct and simulated contact with nature.

ENDNOTES

1. This chapter was adapted from an article in *Environmental Building News*, July 2006, Volume 15, No. 7; www.BuildingGreen.com.
2. Special thanks to Jenifer Seal Cramer and Benjamin Shepherd for input on Table 21-1.

REFERENCES

Heerwagen, J. H., and Leah Zagreus. 2005. "The Human Factors of Sustainable Building Design: Post Occupancy Evaluation of the Philip Merrill Environmental Center." Center for the Built Environment, University of California, Berkeley.

Heschong, Lisa, principal author. 1999. "Daylighting in Schools." Pacific Gas & Electric Company.

Heschong, Lisa, principal author. 2003. "Windows and Classrooms: A Study of Student Performance and the Indoor Environment." California Energy Commission.

Louv, Richard. 2005. *Last Child in the Woods.* Algonquin Books.

Romm, Joseph, and Bill Browning. 1994. "Greening the Building and the Bottom Line," Rocky Mountain Institute.

Ulrich, Roger S. 1984. "View Through a Window May Influence Recovery From Surgery." *Science* (224):420–421.

chapter 22

Transforming Building Practices Through Biophilic Design

Jenifer Seal Cramer and William Dee Browning

Between now and 2025, the population of the United States will increase by 70 million—the equivalent of the populations of New York, Florida, and California combined. To accommodate this growth, 100 billion square feet of new residential space will have to be constructed. According to the Brookings Institution, half of the buildings in which Americans will live in the year 2030 do not yet exist. This represents a $25 trillion building boom that is changing the face of this country (Brooks 2006).

How different would our built environment look if the building industry embraced biophilic design and green development for all future construction? This is a not a simple rhetorical question. We believe that fully embracing biophilic design will change the way we configure our home, work, and other spaces. If we get it right, we will have far more life-enriching places in which to live.

> Architecture is desperately in need of a conceptual, theoretical, and philosophical reunion with nature. This does not simply mean more urban greening or conservation efforts; instead it refers to what Le Corbusier once hailed as a "new spirit." Rather than continue to design buildings as hermetic compositions of abstract geometry, architecture should see their structures as narrative fusion of ideas and elements that connect shelter to the natural environment.
>
> —James Wines, president of Sculpture in the Environment (1994)

Many of our most cherished buildings and landscapes include prominent biophilic features only vaguely recognized by occupants and users, although they nonetheless exert powerful effects. This chapter explores how we might take this emerging knowledge

Figure 22-1: Elk Rock Gardens at Bishop's Close in Portland, Oregon.

(quantitative and qualitative) of biophilic design and work toward transforming design practice to better integrate this theory and its elements into our man-made environments.

PRODUCTIVE ENVIRONMENTS

> We will know that we have it right when we walk into the lobby and feel it through our skin.
>
> —Deborah Butterfield, sculptor

So much of a lifetime is spent in buildings that people tend not to be conscious of how a space affects them. It is the rare, exceptional space that is remembered as warm, nurturing, or inspiring. Given that most Americans spend more than 90 percent of their lives *within* buildings, it is important to determine the effects the indoor environment has on us.

One of the worst accusations that can be hurled at architects, however, is to say that through their design they have undertaken social engineering—that the places that they create dictate behavior. Architects have explicitly moved away from social messages, in effect trying to shed responsibility for the psychological and social implications of spaces. This was largely in response to the rather spectacular social disasters of places like Cabrini Green in Chicago and Pruitt-Igoe in St. Louis. But these horrific low-income housing projects failed because the designs fundamentally did not reflect human nature and psychosocial needs.

In 1977, Christopher Alexander and others published *A Pattern Language* (Alexander et al. 1977). This groundbreaking book contains more than 200 spatial

Figure 22-2: Central interior greenhouse space adjacent to the conference table of Rocky Mountain Institute Headquarters, Snowmass, Colorado.

patterns and design elements intended to lead to buildings and communities that enhance human well-being. This book continues to be a seminal text in design education. The patterns are largely based on observations of how people use certain spaces and definitions of their spatial qualities.

In 1994, Rocky Mountain Institute (RMI) published *Greening the Building and the Bottom Line* (Romm and Browning 1994). This study documents eight cases in which efficient lighting, heating, and cooling measurably increased worker productivity, produced better sales per square foot, decreased absenteeism, and/or improved the quality of work performed. Productivity gains from energy-efficient design can be as high as 6–16 percent, providing savings far in excess of the energy savings. Efficient lighting, in particular, can measurably increase work quality by reducing errors and manufacturing defects. Although the companies in the case studies undertook programs to increase the energy efficiency of buildings, they also inadvertently increased worker productivity.

The companies profiled and many others undertook the energy efficiency retrofits for good economic reasons. For example, a three-year payback, typical of lighting retrofits, is equal to an internal rate of return in excess of 30 percent. Such a return is well above the "hurdle rate" of most financial managers. By cutting energy use by $.50 or more per square foot, a retrofit will also significantly increase the net operating income of a building. These gains, however, are tiny compared to the cost of employees. In a typical building, salaries are greater than energy and operating costs combined.

In 1990, a survey showing a breakdown of costs per square foot of the stock of U.S. offices[1] was released (BOMA 1991); in the last decade and a half, these numbers have all increased. As updated in the April 2005 issue of *Environmental Building News*,[2] current average costs per year for a typical U.S. office are as follows:

Salaries and benefits	$318.00/sq. ft.
Technology	50.00/sq. ft.
Mortgage/lease	16.00/sq. ft.
Energy	2.35/sq. ft.
Churn	1.00/sq. ft.
Total	**$387.35/sq. ft.**

With salaries and benefits at $318.00 per square foot, a 1 percent increase in productivity equals $3.18 per square foot, a 5 percent increase in productivity equals $15.90 per square foot, and a 10 percent increase in productivity equals $31.80 per square foot. These numbers add up quickly. For example, in a 44,000-square-foot building, a 5 percent productivity increase equals $699,600 per year.

Productivity increases can be measured in several ways: production rate, quality of production, changes in sales per square foot or per shopping cart, and changes in absenteeism. (Absenteeism is a major concern for companies in Europe, where it is very difficult to fire employees. Unhappy workers will go absent.) Some of the research to date has focused on countable units of output. Other work, such as many of the examples collected in the Carnegie Mellon University Building Investment Decision Support (BIDS) cases,[3] focuses on cognitive performance or task performance, which, while not directly output related, are helpful in understanding potential areas for productivity gains. All of these can be improved if people suffer fewer distractions from poor visual acuity, poor thermal comfort, and similar factors.

It has been generally believed that *any* change in a worker's environment will increase productivity. Research done at Western Electric's Hawthorne plant in Chicago from 1929 to 1932 has been interpreted to show that experiments to monitor the effect of a workplace change on productivity can be complicated by interaction between workers and researchers. This led to the widespread belief that changes in working conditions affect productivity only because they signal management's concern, the so-called Hawthorne effect. Any gains were believed to be only temporary.

What was less well known was that the experimental methods and results from this work were extremely questionable. The research pool included only five subjects, who, along with their supervisors, were being rewarded for gains in productivity and could monitor their own production rate on an hourly basis. Despite these flaws, the pervasive mythology of the Hawthorne effect has led researchers for over 70 years to ignore the effects of building design on productivity—even though a major 1984 study found direct correlation between specific changes in the physical environment and worker productivity.[4] This work by Buffalo Organization for Social and Technological Innovation, along with materials collected by the National Lighting Bureau, pointed in the direction of clear connection; however, they did not receive wide attention.

The 1994 study by RMI is thus of particular importance. It has been cited in more than 500 articles in print media as well as in national broadcasts. It also led to other documented cases and a major biophila study funded by the U.S. Department of Energy. It should be noted that the measures described in the RMI study were not energy-conservation changes but, rather, measures to increase energy *efficiency*. Both activities lower energy consumption, but conservation implies a decrease in service. Energy efficiency must meet or exceed the quality of service that it replaces. It should also be noted that the decisions to undertake energy-efficiency improvements were based solely on projected energy and maintenance savings, not on any desire to increase productivity, as it was not believed to be possible to do this by altering the building. In all of the examples, productivity had always been monitored by the companies. Some companies were aware that the measures implemented improved the quality of spaces; however, none of the cases involved a change in management style. The gains in productivity observed by the companies were an unanticipated effect. (See Figure 22-3 in color insert.)

Subsequent studies by Lisa Heschong of Heschong Mahone Group have investigated connections between daylighting and productivity. These studies move beyond anecdotal case studies and involve large data sets that allow good statistical analysis. One study documented a 40 percent gain in retail sales in daylit grocery stores (Heschong 1999). Another study found increases in academic performance among schoolchildren in daylit schools (Heschong 1999), while a third found increases in office worker productivity in daylit spaces. The results of these studies were controversial, and the sponsor asked the team to do follow-up research. This research strengthened the conclusions in the retail and schools studies. In the second school study, it was found that increased daylighting could decrease performance if it caused overheating. When revisiting the office study, Heschong concluded that the daylighting may not have been as important in increasing cognitive performance as the view to nature out of the window (Heschong 2002, 2003).

Possible gains in productivity have become one of the key drivers for the green building movement, and many are studying it, including Carnegie Mellon University's Center for Building Performance and Diagnostics. Energy savings and other measurable environmental performance improvements are typically the main economic arguments for undertaking green buildings. Increased productivity, while considered very important, in many cases is not a deciding factor. While the research to date has compellingly recorded gains in productivity, the often asked follow-on question is, how can we predict the gains?

Beginning to craft such a hypothesis was the intent of the aforementioned U.S. Department of Energy funded study of a new Herman Miller plant in Zeeland, Michigan. Elements from the early work on biophilia were used to define the research agenda. Judith Heerwagen, James Wise, and others studied the conditions in the facility, conducted surveys of the occupants, held focus groups, and analyzed the organization's Total Quality Metrics data. The data indicate that the workers in the new facility achieved a gain in production. The researchers also found that the workers in the daytime and swing shifts were more satisfied with the building than the night shift. During daylight hours many occupants have good access to daylight and views to the restored prairie landscape. At night these qualities are missing.[5]

Much of the biophilia research to date has focused on human response to different landscape conditions. Work by J. Appleton, Gordon Orians, and Judith Heerwagen has led to a list of spatial patterns and physical elements that occur in preferred landscapes (Appleton 1975; Heerwagen and Orians 1993). These include patterns called enticement, peril, prospect and refuge, mystery, and complexity and order. Steven and Rachel Kaplan have subsequently published a book on the incorporation of these patterns into the design of parks (Kaplan, Kaplan, and Ryan 1998).

In 2002, Ole von Uexküll, Benjamin Shepherd, and Corey Griffin at the Rocky Mountain Institute compiled a database of studies related to biophilia and design. This list included 246 references and contacts with 19 scientists from six countries. In 2003, Marissa Yao, of the Yale School of Forestry and Environmental Studies, expanded and further analyzed the biophilia database (Yao 2003). From this analysis, a preliminary list of 13 biophilic conditions emerged:

1. Peril
2. Enticement
3. Access to water
4. Natural ventilation
5. Prospect and refuge
6. Complexity and order
7. Local, natural materials
8. Dynamic and diffuse daylight
9. Educational about biophilic aspects
10. Visual connection between interior and nature
11. Physical connection between interior and nature
12. Material connection between interior and exterior
13. Frequent, repeated spontaneous contact with nature

PLACES THAT ENHANCE THE HUMAN AND NATURAL ENVIRONMENT

The new challenge for designers is to create places that *enhance* the human and natural environment. The green building movement has successfully brought daylighting, low-impact and natural materials, and other features into more developments. The emphasis has been largely to lower energy costs and environmental impact, and much of the discussion has included arguments about the benefits of capturing gains in productivity. The next step in the green development movement is to design life-enriching, restorative buildings and landscapes that elicit a positive sense of nurture and well-being (or one could say, a biophilic response).

Given what we know to date, we predict that there are three categories that would help define biophilic buildings:

1. Nature in the space
2. Natural analogs
3. The nature of the space

Nature in the Space

Incorporating plants, water, and animals into the design of a space is one way to create a biophilic environment. This is nothing new in homes, as cultures around the world have almost always had domestic gardens, houseplants, cut flowers, fish bowls, and pets. There are many historical precedents for fountains, garden courtyards, and other measures in large buildings. There are even plenty of prototypes in modern commercial settings: the suburban office park with low-rise buildings set among lawns, trees, and shrubs; the landscaped atrium found in many hotels; and the aquarium in doctors' offices. While in some ways this is the most easily understood

Figure 22-4: Elements of biophilic design grace the atrium of the HealthPark Medical Center, Fort Myers, Florida.

of the biophilic design elements, it requires some space and a maintenance budget.

Bringing nature into the space can involve a series of different strategies. Large features include planted terraces, courtyards, and atriums; green roofs that are visible from occupied spaces; fountains; and water features. Smaller features include potted trees, cut flowers, and aquariums.

Landscape paintings and photographs are another way of bringing nature into a space, through representation. Research has documented lowered blood pressure and stress rates among test subjects shown paintings and photographs of natural areas that have a number of features identified from the preferred landscapes research. For example, in a Swedish study, cardiac patients had posters placed at the foot of their beds—either abstract paintings, two nature scenes, or a blank poster board. The patients with the natural scenes had better recovery response than the others (Ulrich 1992). In other research, people working in windowless spaces used significantly more nature décor than those in comparable spaces with windows, to compensate for the lack of connection to outdoor nature views (Heerwagen and Orians 1986).

In fact, some intriguing research study has emerged in the health industry when patients are exposed to actual natural scenes. One is Roger Ulrich's study of comparative recovery times of cholecystectomy surgery patients, in which some of the patients had a view of trees and shrubs, while others had a view of a brick wall. The patients with the view to nature had a shorter average recovery period, took fewer pain-killers, and had fewer nursing calls (Ulrich 1984). This research, and other subsequent studies on healing times and stress recovery, led to the use of "healing gardens" as elements in many new hospitals (Cooper Marcus and Barnes 1999).

Natural Analogs

Natural analogs are design features that evoke some aspect of nature. This includes ornamentation, use of natural materials, and biomorphic forms.

The use of leaves, flowers, fruits, nuts, seashells, and animals as inspiration for architectural ornamental is almost as old as human architecture. In many cases, this natural imagery also has symbolic value—for example, symbols of institutional strength (lions, oak trees, etc.) or religious significance (lotus blossoms, olive branches, etc.). Images of flowers, leaves, and birds are very common in textile patterns. Prior to the modernist movement, ornamentation drawn from nature was extremely common.

While much architectural ornamentation was stripped during the modernist movement, there are still ways that natural analogs are used. The modernist movement celebrated the "honest" use of materials: wood stained and finished to show the grain; stone cut and polished to enhance color and pattern; and fabrics woven to show the inherent texture and color of natural fibers—linen, wool, silk, and cotton. To this palette, the green building movement has added cork, bamboo, ag-fiber-board, and other natural materials. (See Figure 22-6 in color insert.)

We also see natural analogs in structural elements based on living objects. Historic examples would be the papyrus reed columns in Egyptian temples or the forest grove formed by the columns in Gothic cathedrals. Modern examples include Frank Lloyd Wright's grove of shade tree columns in the central space of the SC Johnson Administration Building; Eero Saarinen's use of seashells and bird wings as inspiration for famous airport terminal buildings; and Santiago Calatrava's use of torsos, limbs, and bones as inspiration for large public buildings. All of these buildings have clear references to natural form, hence the term *biomorphic design*.

Figure 22-5: Leaves and floral patterns adorn this column capital.

The Nature of the Space

Exploring human response to spatial patterns as a way of evoking a biophilic response is an area most in need of research. One of the first efforts to translate the spatial patterns found in *preferred landscapes* into buildings was undertaken by Grant Hildebrand. In *The Wright Space*, Hildebrand explores the use of these spatial patterns in 36 houses across the span of Frank Lloyd Wright's career. It is apparent that Wright's use of these patterns was intuitive, as there is little reference to these spatial patterns in his writing or in the work of many of his interns. Hildebrand further explored the use of these patterns in buildings in *The Origins of Architectural Pleasure* (Hildebrand 1991, 1999). In *Patterns of Home*, several of the coauthors of *A Pattern Language* took the lessons learned from working with the larger set of patterns and compiled a smaller set for residential design (Jacobson, Silverstein, and Winslow 2002).

Codifying of spatial patterns can be found in many traditional geomancy systems. Buried among the layers of mysticism in feng shui and Ayurvedic design, there are insights about local climatic responses and occupant psychology. Most of these patterns are based on long-term observation within a specific cultural context. The research into the spatial patterns in preferred landscapes attempts to reach a deeper, universal or non–culturally specific understanding of spatial patterns.

There are plenty of historic and contemporary examples of the use of these spatial patterns in architecture. *Refuge* is a pattern in which the occupant's back is protected and a lowered ceiling height over the refuge space enables the occupant to *safely* look out from this sheltered space. The inglenook next to the fireplace is a classic refuge space. *Prospect* is the ability to see out across the landscape from a raised place. A balcony, for example, is a prospect space.

Prospect and refuge can be found together in many places. The raised sheltering front porch of a craftsman bungalow is a good example of prospect and refuge. Translating these into large commercial or institutional buildings is also possible. For example, in an oncology center patients may spend several hours receiving infusions while sitting or lying in chairs in a large room. With patients who are already feeling compromised and vulnerable, it is very important that perceptions of comfort are addressed. In the infusion rooms, the prospect and refuge spatial pattern, for example, can be created fairly simply by having a partial height wall behind the patient chair and slight soffit or lowering of the ceiling just over the patient chair area. Then, from this protected area, the patient has a view into a bigger space. These spaces are even more powerful when they are designed with a view out to nature (Browning and Bannon 2006, 5).

Determining the simplest spatial form of the patterns from preferred landscapes, identified by Hildebrand and others, will result in a powerful set of biophilic design tools.

MODERN EXAMPLES OF BIOPHILIA IN DESIGN

> Design is a healing art that provides the opportunity to enhance people's lives using elements of nature as a gift.
>
> —Clodagh

While biophilic design and its spatial patterns are being studied, some designers and developers are using their intuition and early biophilic knowledge to bring connections to nature into their projects. These progressive designers from around the world are building on the foundation of green development's focus on resource efficiency, environmental sensitivity, and community and cultural responsiveness, while being attentive to the bottom line and real estate market indicators.

Sanitas Corporation's Headquarters, Madrid, Spain (See Figure 22-7 in color insert.)

Sanitas is Spain's leading private health insurance company and a part of BUPA, the foremost insurance company in the United Kingdom and Europe. Sanitas set out to design a headquarters building that represented the company's health-centered goals—satisfying physical, social, and environmental requirements. To achieve their vision, Sanitas held a design competition for the project and selected Ortiz Leon Architects of Madrid. The result is a glowing gem among a flood of conventional sprawling office parks on the outskirts of the city.

The twin oval-shaped buildings have plants and trees throughout the atria gardens and are surrounded outside by landscape gardens featuring native plants. A roof garden with nestled nooks for seating offers a quiet refuge for lunch breaks or meetings. The welcoming slate rock fountain trickles water leading to a pathway to the building entrance.

With a commitment supported by the corporate leadership, from the outset architect Iñigo Ortiz designed the buildings to have natural elements and biophilic attributes as well as green building features such as passive solar design, natural ventilation, appropriate materials, daylighting, and good indoor air quality. Interestingly, Metrovesca, Spain's leading real estate developer, was so impressed by the project it hired Ortiz Leon Architects to integrate the same features into the plans for its nearby speculative office building, Alvento. Due to its superior and thoughtful design, the project leased up before construction was even complete, in a real estate market flooded with other more conventional product offerings.

Figure 22-8: Prisma's inner courtyard features a flowing stream and rich layering of space.

Gewerbehof Prisma, Nuremberg, Germany

Prisma is a richly unique example of biophilic design woven throughout its "walls." Built on a restored brownfield site, the urban infill project consists of three buildings around a green courtyard. Two of the buildings are connected by a beautiful greenhouse atrium.

Designers Joachim Eble and Herbert Dreiseitl created this long, public atrium sanctuary filled with plants and water features, natural daylighting, and fresh air, making occupants feel as if they are outdoors. At the same time, these features passively moderate the indoor climate. This development also has exhibited superior energy performance and has leased up quickly in a difficult real estate market.

The International Netherlands Group, Amsterdam, Netherlands

ING Bank in Amsterdam is a very unusual place. The half-million-square-foot headquarters of the country's second largest bank, previously known as Nederlandsche Middenstandsbank, is one of the most remarkable buildings in the world. Featuring 10 interconnected towers, it is largely daylit, highly energy efficient, and architecturally innovative. Its angular forms and many cleverly integrated building amenities include local materials, plants and gardens, artwork, and flowing water. Architect Anton Alberts' anthropomorphic building geometries were drawn from the teachings of Austrian philosopher Rudolph Steiner.

The bank's board laid out a vision for the building: It would be "organic" and would integrate "art, natural materials, sunlight, green plants, energy conservation, low noise, and water." While one of the board's requirements was that the building be energy efficient, it could not cost "one guilder more" than conventional construction. And it didn't.

The bold new image of the bank—resulting from the building—is credited with elevating International

Figure 22-9: People love the ING building so much that weddings are held on the gardens over the parking garage.

Netherlands Group from No. 4 to No. 2 among Dutch banks.

Council House 2, Melbourne, Australia

The Mayor of Melbourne is hoping that the recently completed Council House 2 (CH2) will change the way buildings are designed and constructed in his city and around the world. This US$38.7 million 10-story mixed-use building is hailed for its myriad of innovations and ecological design, led by renowned architect Mick Pearce:

- Reuse of an existing office building and urban land.
- Vegetated facades.
- Optimized solar orientation.
- Computer-controlled louver screens to neutralize western sun.
- Fabric "shower towers" that feed chilled air into the ground-level retail spaces.
- Rainwater harvesting to provide water for secondary uses such as toilet flushing and irrigation.
- An innovative, mixed-mode natural ventilation system that includes vegetation filtering, chilled ceilings, thermal mass, and night-flush.
- Phase Change Material thermal storage in large, battery-like cells—a world's-first installation at this scale of this leading-edge European technology. (See Figure 22-10 in color insert.)

Lindsay Johnston, chair of the Royal Australian Institute of Architects national environment committee, noted that the "lack of knowledge is not the obstacle to environmentally responsive or green buildings. It is the lack of commitment by society that allows the greater part of our cities and built environment to be procured by a method that is driven by short-term dollar gain rather than long-term quality."[6]

Embassy Suites and Hyatt Hotels, Marketing Concepts Emerging

In addition to the design and development community, hotel giants such as Embassy Suites and Hyatt are taking their own twist on biophilia to attract weary travelers. In the fall of 2004, these hotels launched high-end advertising campaigns that featured plant and garden images that evoke a desire to be in close contact with nature. One ad shows a businessman looking down from an upper balcony office to a lush open garden below. By placing these ads in leading newspapers such as the *Wall Street Journal*, these hotels are competing for guests by illustrating a peaceful, nature-based refuge. "It seemed natural to appeal to business travelers' senses through an ad campaign that shows we understand their plight and offer refuge through our core features," says John Lee, vice president of brand marketing for Embassy Suites. "The planted atrium is a core attribute to our brand." His competitor, director of advertising for Hyatt Hotels Corporation Johanna Vetter, comments, "Hyatt does all it can to incorporate the natural setting into its hotel design" (Pliska, 2005).

ROADMAP TO TRANSFORMATION

> Our emotional freedom, our spirit, is nurtured and supported by those environments which are themselves alive.
>
> —Christopher Alexander (2002, 372)

As we reach toward these more life-enriching patterns to design and enhance our man-made environments, diverse disciplines are coming together to help articulate *how* to transform practice. This transformation not only brings together the design community of architects but also brings in others: engineers, ecologists, botanists, biologists, community planners, educators, landscape architects, real estate developers, affordable housing specialists, physicians, epidemiologists, physicists, interior designers, psychologists, hydrologists, socially responsible investors, educators and students (from daycare through university), artists, marketing specialists, spiritualists, and more. We are possibly witnessing an emergence of a completely new transdisciplinary field. The scientific research shows that biophilia is not merely "a nicety." As Judith Heerwagen has stated, "it is a physical need" for us as humans to grow and perform at our best. This reasoning may be why so many are seeking a better way to build and create community.

As a comprehensive strategy for transformation is developed, more empirical research is needed, as well as collaboration among the disciplines and training to build on the foundations of green development. Clear biophilic design attributes or principles are beginning to emerge. This is a good first step, a clear and sensible definition of the terms. From here, illustrative patterns of biophilia can be described to create a kind of "kit of parts" for biophilic design. The U.S. Green Building Council has expressed interest in incorporating biophilic design patterns into future iterations of the Council's Leadership in Energy and Environmental Design (LEED) standard. These tools can be used to help educate not only architects and designers but also building owners, educators, healthcare providers, and others who want their architects to incorporate this kind of design into their spaces.

There is a danger, however, in turning this design approach into visual shorthand—for example, just simply adding a garden. As noted above, healing gardens in hospitals have recently become popular. But often they are token, isolated foreign inserts, not integrated into the fabric of the building design. Design integration is critically important for many reasons, but in large part so that the biophilic patterns are not deemed appendages that can be easily cut off if the budget shrinks. This approach also creates a more seamless, well-thought-out result for the space.

Coupled with the biophilic design patterns, this transformation requires:

- A shift from a philosophy of control of nature to working in concert with nature
- An understanding of ecosystems services and natural capital preservation and restoration, as well as
- A mindset that embraces the principles of green development, appropriate renewable technology, natural capitalism, biomimicry, and whole-systems thinking

Simultaneous with the transformation of the design community, a transition in the general marketplace needs to occur. From those who deliver real estate product (developers) to those who market it (brokers) to the consumer, all levels of the marketplace can be engaged. The myriad of benefits of biophilic design need to be quantified and shown to these market sectors. Conventional market research simply asks questions about historic market performance of comparable real estate products, or "comps." This practice can be one of the largest barriers, because comps may not give an accurate reading of the appeal of the new, more biophilic real estate development. Conventional market research can hinder innovation and cause risk-averse real estate developers and financiers to avoid considering this new way of building. Sometimes called the "rearview mirror" approach to market research, the industry evaluates new product by using just traditional methodology. To leapfrog ahead of this stalemate to more creative thinking, it is important to illustrate the approach specifically with both successful physical examples in the commercial, retail, residential, resort, and institutional sectors (even if they represent only aspects of these spaces) *and* robust research data.

It is ironic that the marketplace already captures value based on biophilic attributes of some locations. A study by the National Association of Homebuilders found that preserving trees on a home site, while adding $1,500 to the cost of construction, increased the value of the home by $5,000. In a study by the MIT Center for Real Estate, homes located within 100 feet of a small neighborhood park had increased property values (Miller 2001). And we even see this in the price of hotel rooms—the water view is typically more expensive than the parking lot view to the "rear."

Connecting with the investor and the financial community is a vital step toward successful transformation as well. There is increasing interest in the Socially Responsible Investment (SRI) community, and biophilic developments would be value-aligned assets. From 1995 to 2003, the amount of money invested in SRI increased more than threefold to $2.16 trillion—11.3 percent of all the money under professional management in the United States (Social Investment Forum 2006). The SRI sector is looking for real estate investment that has community, health, and environmental benefits. Investment vehicles like the Rose SmartGrowth Fund have emerged in the last year to help finance progressive development. Perhaps in the future there will be a Restorative Biophilic Building Fund of $500 million.

To be wholly transformative, the discussion of biophilic design needs to move beyond the four walls of buildings to restorative landscapes and communities. The connections in our communities now are largely sterile vehicle corridors and telecommunication nodes. As Macon Cowles, leading environmental attorney and former chair of the Boulder, Colorado, planning board, said, "Organic growth of communities formerly resulted in cities and towns that looked like a creature, or an organism, and worked with the same efficiency. Such growth has been replaced by the stamped coinage of sprawl: office parks, roadways, subdivisions, shopping malls, and the desultory placement of public buildings." Biophilic design is a broad undertaking and our thinking will be more robust if we embrace this larger challenge of connections as we move forward with development.

ENDNOTES

1. Data from the 1991 BOMA *Experience Exchange Report*, showing national means for downtown 100,000–300,000 square foot private-sector office buildings in 1990. Areas are net rentable space; income ($21) is for the office area only, versus $16.68 for the entire building including retail space, parking, and so on. The energy costs, other costs, and income, are probably somewhat higher for new offices than for the stock average described here, which is based on a sample of hundreds of buildings totaling more than 70 million square foot (BOMA 1991, 95). The authors are grateful to BOMA for graciously making these proprietary data available.
2. Thanks to Alex Wilson and the staff of *Environmental Building News*.
3. The Carnegie Mellon University, School of Architecture, Building Investment Decision Support tool can be accessed at http://cbpd.arc.cmu.edu/bids/.
4. For a survey of some of the literature on the flaws in the Hawthorne effect research—and a major study that came to a different conclusion—see Michael Brill et al., *Using Office Design to Increase Productivity*, vol. 1 (Buffalo, NY: Workplace Design and Productivity, Inc., 1984), 224–225. See also William J. Dickson and F. J. Roethlisberger, *Counseling an Organization: A Sequel to the Hawthorne Researches* (Boston: Harvard University Press, 1986). The best investigation of the original Hawthorne work can be found in H. M. Parsons, "What Happened at Hawthorne?" *Science*, March 8, 1974, 922–932, and H. McIlvaine Parsons, "What Caused the Hawthorne Effect? A Scientific Detective Story," *Administration & Society*, November 1978, 10(3): 259–283.
5. This study was initiated by Rocky Mountain Institute, the U.S. Green Building Council, the U.S. DOE, and Herman Miller. The material in this case study is based on James A. Wise, Judith Heerwagen, David B. Lantrip, and Michael Ivanovich, "Protocol Development for Assessing the Ancillary Benefits of Green Building: A Case Study Using the MSQA Building," in NIST Special Publication 908, *Proceedings of the Third International Green Building Conference and Exposition—1996*, edited by A. H. Fanney and P. R. Svineck (Gaithersburg, MD: National Institute of Standards and Technology, 1996, 63–80), a site visit to the Herman Miller SQA building, and personal communications with the Battelle researchers, architect William McDonough, and Keith Winn and Joseph Azzerello of Herman Miller.
6. Quoted in an article on CH2 in *Architectural Review Australia;* cited in Huston Eubank, "State of the World: High-Performance Building," *Urban Land: GreenTech*, October 2005, 54.

REFERENCES

Alexander, Christopher. 2002. *The Nature of Order.* Berkeley, CA: Center for Environmental Studies.

Alexander, Christopher, Sara Ishikawa, Murray Silverstein, Max Jacobson, Ingrid Fiksdahl-King, and Shlomo Angel. 1977. *A Pattern Language: Towns, Buildings, Construction.* Oxford: Oxford University Press.

Appleton, J. 1975. *The Experience of Landscape.* New York: Wiley.

Building Owners and Managers Association (BOMA). 1991. *Experience Exchange Report.* Washington DC: Building Owners and Managers Association.

Brooks, David. 2006. "US Fast Becoming Nation of New Suburban Villages." *New York Times*, January 19.

Browning, William D., and Jeffrey E. Bannon. 2006. Notes from the Environmental Opportunities Charrette for the Oncology Center at Hackensack University Medical Center, Hackensack, New Jersey. Browning + Bannon LLC, February.

Cooper Marcus, Clare, and Marni Barnes. 1999. *Healing Gardens: Therapeutic Benefits and Design Recommendations.* New York: Wiley.

Heerwagen, Judith, and Gordon Orians. 1986. "Adaptations to Windowlessness: A Study of the Use of Visual Décor in Windowed and Windowless Offices." *Environment and Behavior*, 18(5): 23–29.

———. 1993. "Humans, Habitats, and Aesthetics." In *The Biophilia Hypothesis*, edited by S. Kellert and E. O. Wilson. Washington, DC: Island Press.

Heschong, Lisa. 1999. An Investigation into the Relationship Between Daylighting and Human Performance: Detailed Report. Heschong Mahone Group for Pacific Gas & Electric Company. August 20.

———. 2002. Reanalysis Report, Daylighting in Schools: Additional Analysis. HMG Project #0008, NBI PIER Element 2, Final Reports, Task 2.2.1 through 2.2.5, File name: 2D2.2.5b_021402.doc. For the New Buildings Institute. February 14.

———. 2003. Windows and Offices: A Study of Office Worker Performance and the Indoor Environment. P500–03–082-A-9. Heschong Mahone Group for the California Energy Commission. October.

Hildebrand, Grant. 1991. *The Wright Space: Pattern and Meaning in Frank Lloyd Wright's Houses.* Seattle: University of Washington Press.

———. 1999. *Origins of Architectural Pleasure.* Berkeley: University of California Press.

Jacobson, Max, Murray Silverstein, and Barbara Winslow. 2002. *Patterns of Home: The Ten Essentials of Enduring Design.* Newtown, CT: Taunton Press.

Kaplan, Steven, Rachel Kaplan, and Robert L. Ryan. 1998. *With People in Mind: Design and Management for Everyday Nature.* Washington, DC: Island Press.

Miller, Andrew Ross. 2001. Valuing Open Space: Land Economics and Neighborhood Parks. Master's thesis, Massachusetts Institute of Technology.

Pliska, Shane. 2005. "Biophilia: Selling the Love of Nature." *Interiorscape Magazine*, January–February.

Romm, Joseph J., and William D. Browning. 1994. *Greening the Building and the Bottom Line: Increasing Productivity Through Energy-Efficient Design.* Snowmass, CO: Rocky Mountain Institute.

Social Investment Forum. 2006. *2005 Report on Socially Responsible Investment Trends in the United States: 10-Year Review.* Washington, DC: Social Investment Forum, Industry Research Program.

Ulrich, Roger. 1984. "View Through a Window May Influence Recovery from Surgery." *Science* 224:420–421.

———. 1992. "How Design Impacts Wellness."' *Healthcare Forum Journal*, September–October.

Wines, James. 1994. *Earthword Journal*, no. 5, February.

Yao, Marissa. 2003. The Natural Environment and Human Health: Epidemiological Evaluations of Case Studies on Biophilia and Green Design. Yale School of Forestry and Environmental Studies, May.

chapter 23

Reflections on Implementing Biophilic Design

Bob Berkebile and Bob Fox, with Alice Hartley

As the authors of this collection have made clear, humans, like all other living beings, are wired to respond to their environment. The natural world has imprinted on us, biologically and psychologically, certain affinities and aversions that we are only just beginning to understand with our conscious minds. The fact that those of us who shape the built environment—a circle that includes not only architects but developers, planners, and policy makers—have long been missing the terms to describe these principles makes them no less real and fundamental. For better and for worse, the buildings around us provide plenty of examples in which to see how people react to environmental cues in the places they live, work, and learn. Since we can little change these responses, it is in our best interest to understand them as an important and richly interesting layer of the human-nature relationship.

The scholars, scientists, and designers included in this volume have given us a new language for interpreting the built environment. The principles described here can inform—and thereby start to transform—contemporary building practices by suggesting strategies that designers can weave into their visions of what is innovative, bold, and beautiful. However, great architecture has always been born from the artist's soul as much as from the scientific mind. How can inspiration for a new quality of building and community lead to innovation in architecture? What changes will we see in both the creative planning and practical execution of design?

Biophilia is a set of ideas we can start to understand on many scales. From brain chemistry to building design to city and regional planning, the influence of this understanding can nourish and help restore the human-nature relationship. While this new approach will lead

Figure 23-1: The Center for Well-Being at the Ross Institute, an innovative institution in East Hampton, New York

us to greater environmental *responsibility*, its core dynamic is about environmental *response*—creating places that respond to and celebrate the natural world, while evoking a positive response from their inhabitants.

Tucked into a scrub oak forest on eastern Long Island, the Center for Well-Being at the Ross Institute is an example of a building that surfaces slowly from the landscape. The anchor building for a campus dedicated to educating global citizens, the Center aims to express the school's holistic philosophy of integrating body, mind, and spirit. On entering, students remove their shoes, and through stocking feet feel the changing textures of stone, wood, bamboo, and tatami. Like a complex fabric, textures of stone and patterns of light weave together in a way that sharpens and elevates the visitor's awareness of space. A core of local South Bay quartz defines the Center's circulation: at the building's top level, it is a stone oven in an organic cafeteria filled with the aroma of baking bread and chatter of birds; at the street level, it becomes an open hearth and gathering spot for the community. One level below ground, the shaft sinks into a tranquil pool rippled by a school of koi. The essential elements of fire, water, and earth permeate the interior and instill a sense of being present in the landscape. Housing both athletic and performance spaces and more informal social areas, the Center is a place for training, connection, and transformation. Beyond the dimensions of a photograph, to experience the building is to reawaken the senses and become keenly aware of one's surroundings (see Figure 23-2 in color insert).

WE CAN AIM HIGHER

The rapid growth of green buildings has been incredible to witness. In recent years, more and better-performing green buildings have opened their doors, challenging others to meet a higher standard. Progressive-minded towns and cities have set examples with their own municipal buildings, and have passed legislation encouraging or requiring privately owned buildings to follow. By all indicators, awareness is growing among both professionals and the general public: Membership in the U.S. Green Building Council surpassed 10,000 companies and organizations in 2007, and in just five years, attendance at its annual Greenbuild conference grew from a few hundred to more than 13,000. Entire industries have sprouted up to supply and advise clients committed to sustainable design, gathering momentum for "market transformation" with powerful economic repercussions.

With public concern over climate change rising sharply—along with the realization that we will feel its impacts in our lifetimes—green buildings seem to be moving past a critical inflection point. The most main-

Figure 23-3: In Kansas City, the Anita B. Gorman Conservation Discovery Center is dedicated to increasing knowledge, understanding, and compassion for Missouri's natural resources. Urban dwellers can visit a Living Machine™ that treats wastewater biologically and learn "life skills" to help them reconnect with the state's natural resources.

stream media outlets have picked up on the tide of interest in environmental design and dedicated features, columns, and even entire issues to the topic. Many people making greener choices are moved by a profound sense of responsibility and concern for the state of the earth we leave to future generations. For others, however, the call to less-harmful, low-impact living sounds either unappealing or unaffordable. How can sustainability become the new standard practice, while continuing to raise the bar of human health, environmental integrity, and occupant happiness?

Given the urgency of these issues, we argue that ecological literacy should assume a permanent place in the training and licensing of design professionals. This change is critical for public health and the ecosystems that support our lives, and it can also happen more quickly than one might assume. Twenty-five years ago, before the term *universal design* had gained currency, handicapped accessibility was a specialty field, not a basic element of building code and design education. Whether out of ignorance or arrogance, much of the built environment had been made inaccessible to those permanently or even temporarily disabled. It took a deliberate act of widening the common perspective—putting ourselves in another's shoes and realizing that we, too, may one day need accommodations—to change the standard. Likewise, the realization that a small fraction of the world's population inequitably consumes resources and creates waste, in a pattern unsustainable on a global scale, urges us to open our eyes to the bigger picture. Bringing the vast majority of humanity, and the next seven generations, into our present perspective will again radically change our standards. We hope to one day see "green building" fade away as a specialty field; then we will know that this movement has driven a fundamental shift.

Rather than just an obligation, however—and we do believe wholeheartedly that architects and planners have a responsibility to comprehend and consider the impact of their decisions—biophilia represents an abundantly creative moment in design. Along with literacy in biological and ecological principles, the next generation of designers might gain fluency in the emotional landscape of the built environment: the way spaces can reassure, uplift, calm, or refresh their occupants. This awareness will invite us to think about shaping the emotive experience of landscape and architecture—not in order to manipulate, but to answer design challenges in a deeper, more resonant way and reconnect with the natural world. What, for example, would the designers of a psychiatric hospital want to understand about the emotional frequency of their proposed materials and circulation plan? How do these considerations change when the same team designs a day care center?

For the Bank of America Tower in New York City, the vision was to create a daylight-infused workplace and the most transparent possible connection between indoor and outdoor environments. Seeking to dissolve

Figure 23-4: The LEED® Platinum Bank of America Tower at One Bryant Park is clad in exceptionally transparent, low-iron, low-e glass.

this boundary led to the choice of extremely clear, low-iron floor-to-ceiling glass, with a "low-e" coating for improved energy performance. Imagining themselves in this work environment, however, the architects realized that, while thrilling, absolute transparency to the natural elements could feel too dangerous, especially at heights of almost 800 feet. The solution was to add a "frit" pattern of small ceramic dots, silk-screened directly onto the glass curtain wall. Densely patterned near the floor and ceiling, the frit fades away to clear vision glass in the center 5 feet of each panel (see Figure 23-5 in color insert). While helping block heat gain to the interior, the frit lets the human eye make sense of the transparent plane and adds a feeling of security. The pattern also dapples light and shadow into the interior, recalling the experience of being outdoors. To reinforce the perception of safety, a railing at waist height was added. Giving the occupant's hand a natural place to rest, the rail helps mediate exposure to the elements with a layer of human-scale, tactile reassurance.

While few people may consciously understand or fully recognize architects' efforts to master biophilic design, they will respond to environments that employ these principles. What we are collectively proposing, based on the conclusions drawn in these pages, is that designers can and should invest the time to understand how their choices affect people on a mostly subconscious level. To spend time in these biophilic spaces and communities will be to understand them; the experience will speak for itself. Over time, we expect that more and more compelling statistics on human health and well-being will follow.

Returning to the question posed above, biophilia can be a key driver in making green building the new standard of practice. Biophilic design speaks persuasively to two audiences: those who aspire to ever-higher pinnacles of quality experience as well as those waiting for a "good" reason to change. As others here have shown, places that conquer the common dysfunctions of distracting, even toxic building environments—and, on the positive side, that support health, productivity, and creativity—will have very real value in the marketplace. And more architects may finally become genuinely interested in looking at design as environmental response—because beyond just challenging us to calculate a project's ecological footprint, biophilia challenges our imaginations.

SECOND NATURE: A RETURN TO BUILDINGS THAT SUPPORT LIFE

> A human being is part of a whole, called by us "the universe," a part limited in time and space. He expresses himself, his thoughts and feelings, as something separated from the rest—a kind of optical delusion of his consciousness. This delusion is a kind of prison for us, restricting us to our personal desires and to affection for a few persons nearest us. *Our task must be to free ourselves from this prison by widening our circles of compassion to embrace all living creatures and the whole of nature and its beauty.*
>
> —Albert Einstein

These collected visions of biophilic design, vividly illustrated and thoughtfully articulated, take us back to the simplest questions: How will people feel in this space? What belongs here? Can I make sense of this place? In answering these questions, designers' intuition can help guide the way. As some of the few remaining generalists in a highly specialized world, architects and other design professionals are perhaps uniquely prepared to start restoring the human-nature relationship. As Richard Louv has described, children's earliest education is—or usually was—in the laws of nature, an understanding acquired consciously and subconsciously through outdoor play. The design professions continue this study, a formal education in the laws of physics, material properties, and patterns of human organization. Architects, trained to think spatially and to synthesize multiple perspectives, can use these strengths to imagine how people will experience a building or a neighborhood. If we take time to immerse ourselves in a greater awareness of biological processes, natural history, and human nature, our efforts can make a real difference in deepening people's connections to the environment.

The Deramus Education Pavilion at the Kansas City Zoo was imagined as a portal or place of transition between natural and human environments (see Figures

23-6 and 23-7 in color insert). A center for public education and events, the building serves to orient visitors to the conservation mission of the zoo. The profile of the relatively large building is subdued by its setting in a natural valley, the lowest part of the site. The landscape is an integral part of the building; at the same time, the building is integral to the landscape. At every opportunity, gardens, water, and light are invited into—and sometimes through—interior spaces. Much like a geode, a fracture in the building's domed copper roofline opens to reveal a glass "lantern" that reaches up to gather light and announce the point of entry. The pavilion collects light in a variety of ways—direct and indirect, subtle and dramatic—in response to the different types of spaces within. By appealing to both subconscious, sensory experience and the agenda of visitor education, the building imparts a sense that humans are a part of nature, just as nature is an essential part of our habitat.

Advancing the field of green design, it turns out, may look a lot like returning to things humankind used to just know: how to take advantage of the sun's heat and light, how shelters can store or shed water in response to climate and geography. Throughout most of human history, forms of shelter were in balance with the natural environment—not voracious consumers of energy, water, and materials. This also made them finely tuned expressions of a place, reflecting a vernacular wisdom gained over many iterations and generations. In a previous era, architects were also master builders, and consequently design was guided by an intimate knowledge of construction techniques and materials. In the wake of the Industrial Revolution, we have lost our knack for regionally perfected problem-solving and the elegance, simplicity, and integrity that used to be second nature.

In the fall of 2006, the Cascadia Region chapter of the U.S. Green Building Council issued a challenge to all green building professionals. With the intention of raising the bar of "sustainable design," the Living Building Challenge sets out simple but ambitious targets for a new type of building. Responding to the local environment, these buildings will generate all their own energy from renewable resources; capture and treat all their water on-site; and use resources efficiently, for maximum beauty. The 16 criteria set forth in the Challenge are radically streamlined and strictly performance-based; all 16 are required. Beyond the admirable goal of lightening a building's impact on the environment, the Challenge calls for buildings that reconnect themselves to the rhythms and systems of the natural world. Informed by local ecology, such buildings will engage with nature in ways that bring the built environment to life—in form, function, and spirit. In inviting the industry to answer this challenge, the hope is to transform the meaning of true sustainability and to embrace a parallel transformation in architecture itself.

A principal theme of this book has been that our natural intelligence can be reclaimed. In this seemingly simple assignment lies a deeper challenge: learning to listen to and reconnect with other living things, freeing ourselves from Einstein's "optical delusion" or the prison of perceived separation. In the legacy of Descartes, Bacon, and other figures of the Western scientific revolution, generations of architects have continuously and aggressively sought to separate themselves from nature. This approach has designed us into the destructive relationship we see today. The renewed relationship we must now cultivate promises to restore our own health, while healing the wounded condition of ecosystems everywhere.

Retraining ourselves in the habit of integrated thinking, in which we again see ourselves as a part of nature, will give us a new perspective on our place in the environment as a whole. Displaying this kind of highly interconnected perspective, the great Buckminster Fuller was known to overhear a person commenting on a beautiful sunset and to point out that, in fact, the sun was not setting but, rather, the planet was rotating to eclipse their view. To comprehend the condition of the whole is to understand *integrity*—a quality of design that values the genuine, and by which each small decision serves to reinforce a larger vision.

Karan Grover, the architect who designed the CII-Godrej Green Building Center in Hyderabad, India, once explained that he never set out to design the first LEED Platinum building outside the United States. Rather, the creative process was more like a dialogue between physical laws and spiritual principles; in every decision, the designers took a holistic view of the Center's mission to "make the world a better place to live in." The building that resulted aspired first to resonate with architectural tradition and the natural world, and then secondly, found that it met enough credits to merit a LEED Platinum rating.

Like Grover's complex, our built environment can return to a balance that respects the earth for its own sake, rather than viewing it as a resource to be exploited. It is only in relatively recent times that people have lost a sense of reverence—once born, at least in part, from fear—for the material earth as a spirit-infused domain. A new perspective, in which we give up mastery over nature in favor of a respectful partnership, is the moral and spiritual foundation for constructing fundamentally different buildings, neighborhoods, and cities.

With a renewed awareness of place, designers will tend to go about their work with an innate sense of connectedness. The question of whether future generations can live happily with our choices will serve as an inner compass for all design decisions. In imagining this new ethic, we are heartened to remember that, long before the fathers of Western scientific philosophy, our predecessors always regarded nature as a fundamental partner in human endeavors.

REBALANCING THE MODERN ENVIRONMENT

However, as the authors here have suggested, reclaiming this wisdom will look quite different the second time around. Traditional buildings, without adaptation, would, in general, have trouble accommodating modern lifestyles. And few people would advocate that we should—or even can—spread out to the remaining undeveloped land in order for each building to meet its own needs for waste treatment, energy, et cetera, within the property line. This becomes an issue of equity: The developed nations cannot delude themselves into believing that those in the developing world want to lead ecologically balanced, "low-impact" lives forever. If people around the world want the same conveniences and all have equal rights to pursue them, the ecological footprint of this standard of living must perform an incredible shrinking act. As humanity's expectations and population have grown, our vision of environmental sustainability must radically (and quickly) evolve into a different picture than the snapshot from pre-industrial times. Those of us in the developed world have an obligation to make the first move, starting at home.

The great urbanization under way around the world can not be overstated. In China alone, 10–15 million people move from rural areas to cities every year,[1] an annual mass migration greater than New York City's entire population. This urbanization is occurring at a rate unprecedented in the history of human civilization. In the midst of this building boom, green open space might be considered an afterthought at best. In the United States and other developed countries, cities are also growing rapidly, putting pressure on ecological infrastructure, farmland, energy, and transportation systems. For most of us, whether in developing or industrialized nations, retreating to the countryside is not an option. Cities are a key part of the answer—*if* they succeed in providing mass transit, dense development, and good stewardship of their food-, waste-, and watersheds. Urban areas can take a lesson from outback living and strive for self-sufficiency: buildings and districts that make their own power, purify their own wastes, and adapt to bioregional conditions. We need to see a rapid transition back to buildings that practice good local citizenship, where there is no throwing or flushing things "away" and where energy is so clean that people will welcome it into their backyards and onto the skins of their buildings.

The green infrastructure strategies described here are one way of guaranteeing nature a permanent place even in dense and growing cities. As documented by Tim Beatley and others, many cities are starting to recognize the value of investing in "working landscapes." Even in city centers, natural features that treat storm water or filter pollutants from the air are perhaps more likely to receive protection than parklands that "only" provide quiet and recreation. Of course, natural areas that perform a municipal service can simultaneously offer us restorative places and create habitat for wildlife. Such places may become hallmarks of the next generation's sustainable cities.

In urban areas, we can also look for new ways of embracing scaled-down versions of nature, to adapt the experience of nature to the dense conditions of modern cities. A growing number of studies show that even small patches of greenery can have a restorative effect on indoor environments. A green roof or view to street trees can provide daily contact with living things and the benefits of lowering stress, relieving pain, and improving concentration and productivity. Knowing that some contact with the outdoors is essential, designers and scholars might seek to understand how people react to intensively cultivated experiences of nature, whether literal or figurative. They might also cultivate a mastery of techniques that incorporate as well as simulate natural experiences of light, shadow, breezes, thermal patterns, et cetera within the constructed environment. Finally, biophilic designers can explore a more abstract dimension, creating places that "work" in instinctively logical, meaningful ways, inspired by nature's own strategies. While wild, unchoreographed experiences of nature remain essential—whether through exposure to majestic mountains or to the enormity of the night sky—small doses of nature can greatly benefit the growing ranks of urban dwellers.

At Cook+Fox Architects' office in Manhattan, a 3,600-square-foot green roof greets visitors and gives employees daily exposure to a microecosystem of sedums, insects, and occasionally birds (see Figure 23-8 in color insert). Visible from most of the studio, the roof marks the changing seasons and reveals varying conditions of sun, wind, and water that might otherwise go unnoticed in an urban environment. Like a living, green horizon, the roof breaks up the city's vertical mass and the man-made landscape of cars and concrete buildings. Its presence invites observation and has a way of engaging the office with the outside world, from the budding of small *Talinum* flowers to the serendipity of an alighting butterfly. Twenty-five volunteers from the firm donated the labor for its installation–an investment that goes beyond the simple benefits of green space.

A NEW ETHIC FOR EXCELLENCE IN DESIGN

> The house is a machine for living in.
>
> —Le Corbusier

By now it has been shown that good design goes beyond outward form and beauty. Guided by a modern ethic, the regenerative "living buildings" now being created aspire to a different (and arguably more enlightened) aesthetic.

What are some of the sea changes driving these new currents of thought? Technology has attained a permanent place in our lives and most certainly shapes modern buildings and development patterns. We also can no longer afford to accept ignorance of our industry's environmental impacts. We know too much to ignore the decline in biodiversity and ecosystem integrity in progress in all regions of the world—and we do so at our own peril. Early indicators of a global warming are showing us just how dependent we are on "ecosystem services" that we have long taken for granted, from pollination to flood protection. Climate change will affect the next generation in profound ways, the consequences of which remain to be seen. If the Modernist spirit celebrates the age of machines and general triumph over nature, what new creative expression will emerge from the era in which we realized how much we actually stand to lose?

If Modernism has explored abstract form and space on a more or less blank canvas, the next architecture might turn these statements into a conversation with the essential elements of a place. The driving ethic may be seen as a new "minimalism" imperative: stripping away mechanical breathing apparatuses, weaning off energy created far away or long ago, using nothing that can't be infinitely recycled. When architects pare down to the least possible degree of intervention, and draw instead on natural, free endowments of sunlight, water, and other elements, what beauty will emerge from the landscape's healthy glow? We are starting to see the answers all around us. Some of its traits are a sense of place and orientation, a sensory language and mastery of light, a scale that resonates with human nature, a transcendent sense of lightness. This architecture may still strive to convey transparency and innovation, but tempered—to quote Cramer and Browning—with the honest warmth of natural elements.

The Class of 1945 Library at Phillips Exeter Academy is one of Louis Kahn's masterworks of "silence and light." Vaulted spans of concrete frame its light-filled central hall. Rather than imprisoning their charges, however, these walls reveal the library stacks through perfectly circular and perilously huge voids. Like giant

Figure 23-9: In Louis Kahn's library at Phillips Exeter Academy, study carrels are placed adjacent to the exterior, connecting students to the environment beyond.

portals, these openings bring to life the otherwise brutal ambition of the building's powerful frame. At the personal scale, Kahn's use of teak and white oak humanizes the library environment: along the perimeter, sliding wood panels in individual study carrels allow students to screen out distractions or, equally important, refresh their gaze on trees and figures in the distance. With an intuitively grasped organization, attention to the parts of the building you touch, and ways for users to control their environment, Kahn understood the subtle aesthetics of natural simplicity.

In Kansas City, Missouri, in the folds of a rolling hillside, an addition to the Nelson-Atkins Museum merges landscape and sculptural expression (see Figures 23-10 and 23-11 in color insert). In a solution that allows the museum to expand while respecting its original 1933 building, the addition carves out space underground while adding a graceful, contemporary form above. Instead of a building in the conventional sense, architect Steven Holl set out to create "an architecture fused with the landscape," which invites visitors to engage with both the art and the architecture. Transcending the separation between indoors and outdoors, surface and subterranean spaces blend as the galleries descend into the hillside. Five translucent glass pavilions, conceived as "lenses," emerge from the ground to capture and diffuse daylight throughout the building, including the underground galleries. At night, they illuminate the museum's sculpture park like serene paper lanterns.

Modern architecture, as practiced by some, gave us "high-tech" buildings that celebrated a new material aesthetic. Many have aged less than gracefully, now showing decades of wear and weather and overexposed visions of eternal youth. In the pursuit of perfection, many of these buildings reveal an inherent vulnerability to obsolescence, which now appears analogous to the industrial metaphor that defined the spirit of their times.

In contrast, when we study buildings that we love in a timeless, enduring way, we realize that these places convey a richness that runs deeper than the luster of "mint condition." This itself is a biophilic quality: we feel comfortable among objects and places that wear their age well. Like an exquisitely crafted Shaker chair, design of the built environment can result in both simple beauty and structures that gracefully weather a lifetime of use. Those places that show their age in a real and genuine way will possess, to echo Janine Benyus's lyrical phrase, an authenticity that we recognize as beauty.

In contrast to the sleek glass tower above, it was important for the lobby of the Bank of America Tower at One Bryant Park to touch the earth with solid, natural materials. Walls of pale Jerusalem stone, from which generations have constructed the dwelling-places of human civilization, are embedded with fossils from even more ancient life. Along the bank of elevators, the building's core of circulation activity, deep red leather paneling lines the walls. With thousands of people passing each day, many hands will touch the walls—a tactile instinct that, as we now know, design can choose to resist or embrace. Over time, human touch will patinate the leather, giving its color and texture a natural richness. Unlike materials that must be kept in pristine condition, leather actually becomes more beautiful with age and the weathering effect of repeated use.

A few miles from the Bank of America Tower, the South Street Seaport Historic District represents a rich chapter in New York's history—when the wealth of a former era circulated by cargo schooner and where Herman Melville set the opening chapter of his famous American odyssey, *Moby Dick*. In this neighborhood, the redevelopment of Front Street explored modern concepts of stewardship (see Figure 23-12 in color insert). Eleven historic but dilapidated brick warehouses, some on the verge of collapse, have been repurposed into shops and apartments that blend modern loft aesthetics with the complex patina of age. Rather than a sterile restoration, the redevelopment makes careful incisions for new windows and courtyards, bringing in light, air, and a pervasive sense of orientation. At the same time, imperfections and layers of urban residue have been left exposed, revealing the ghosts of the neighborhood's past lives. Woven into this historic fabric, three new buildings juxtapose contemporary materials with references to artifacts such as ships' rigging and skeletons of whales which, while no longer tangible, are integral to the Seaport's legacy. With a human scale that knits together a lively streetscape, the neighborhood is also tied to larger rhythms of the waterfront and the earth. Ten geothermal wells, sunk 1,500 feet into the ground, provide nearly all heating and cooling for the development's 14 buildings, utilizing modern technology that helps preserve natural and cultural resources for future generations.

A sense of place, purpose, and connection, along with the underappreciated senses of touch, sound, and smell, all deserve to be recognized for their contribution to design excellence. As we have seen, reawakening these senses adds depth and opens new creative territory in design of the built environment. As we come closer to mastering the experiential dimensions of biophilic design, more effortlessly and more often our designs will convey the kind of experiences found in nature—the way entering a cathedral can feel like walking into a solemn forest. Honoring the unseen Architect(s) of the diverse and magnificent natural world, we can learn to live not as tyrants over creation, but so that our existence restores and actually enriches the natural environment.

CONCLUSION

Long before the term *biophilic design* existed, many of the most notable examples of this concept were created by indigenous people and gifted architects. One such place is the much-acclaimed Thorncrown Chapel, designed by the late E. Fay Jones, FAIA, which has attracted millions of visitors to its isolated, woodland setting on a sloping hillside in the Ozark Mountains of Arkansas.

"Let the outside in" was a principle of Jones's mentor, Frank Lloyd Wright. This principle, while essential to the chapel, was greatly enhanced by a key design decision that Jones made in response to the beautiful but fragile site. He decided to construct the structure of local materials, none larger than could be carried into the woods by two workmen, to avoid site damage during construction. To blend with the bark of the surrounding trees, all of the structure's framing lumber was hand-rubbed with a gray stain. The cross braces were joined with hollow steel joints, creating airy trusses that relate to the surrounding forest. Seeing "the potential for light play on the structure," Jones enlarged the roof-ridge skylight to increase the sense of drama. As a result, the chapel is bathed in dappled forest light and ever-changing shadow patterns. Throughout the changing hours of the day and seasons of the year, it never looks quite the same. Looking up, the visitor sees the complex trusses as an extension of the forest canopy and many perceive a crown of thorns.

The flagstone floor is surrounded with a low rock wall, giving the feeling that the chapel is part of its Ozark mountainside. The walls are just clear glass, sheltering visitors yet creating an open-air sensation. In honoring Fay Jones and Thorncrown with its Twenty-

Figure 23-13: Fay Jones's Thorncrown Chapel conveys a sense of communion with the forest.

Five Year Award, the American Institute of Architects noted that millions have come to feel profoundly moved by the chapel: "At Thorncrown there is a great sense of peace. You feel calm. It is the special genius of this place, its humility, if you will, that Thorncrown Chapel captures and quietly celebrates."

If we aspire to restore as well as reenchant the experience of the built environment, our buildings and communities must engage with nature on a higher level, in the kind of spirit that came naturally to Wright and Jones. Whereas the main tide of sustainable design strives, however nobly, to minimize buildings' impact on the earth, regaining our equilibrium is a much greater task. Coming to our senses—all of them—promises to awaken a new awareness among those who influence design. While achieving true "living buildings" is no small dream today, we should be encouraged to remember that we are capable of rapid change, and in fact have already seen an accelerating pace of leadership. As Buckminster Fuller once said, "The only way to make significant change is to make the thing you're trying to change obsolete." Rising to this challenge, the authors of this collection have skillfully set the stage for further research and exciting new forms of collaboration.

Biophilic design, as a fundamentally different approach to the built environment, takes on special meaning for practitioners in the field. While our industry will learn a great deal as this concept continues to grow, let us also acknowledge that some of its secrets may be left unarticulated. This is not necessarily a shortcoming. The impulse to explore an idea through space, materials, and light is what motivates the creation of architecture, and inspiration can emerge from both natural and human experience. Just as poetry is needed to express certain shades of the soul, architecture is an essential dialogue between spiritual and material realms, a conversation that can be trusted to teach us new ways of living in balance.

ENDNOTES

1. Wu, Harry X. and Li Zhou. *Nov 1996.* "Rural-to-Urban Migration in China," *Asian-Pacific Economic Literature*, 10(2): 54–67.

REFERENCES

Fuller, Buckminster. 1969. *Operating Manual for Spaceship Earth*. New York: Simon & Schuster.

The Living Building Challenge. http://www.cascadiabc.org/lbc.

Louv, Richard. 2005. *Last Child in the Woods*. Chapel Hill, NC: Algonquin Books.

Wilson, E. O. 2006. *The Creation: An Appeal to Save Life on Earth*. New York: W. W. Norton.

Zacks, Stephen. 2007. "The Magic Lantern." *Metropolis*, March, 99–103, 147–148.

Contributors

Timothy Beatley

Timothy Beatley is Teresa Heinz Professor of Sustainable Communities, in the Department of Urban and Environmental Planning, School of Architecture at the University of Virginia, where he has taught for 20 years. Much of Beatley's work focuses on the subject of sustainable communities and creative strategies by which cities and towns can fundamentally reduce their ecological footprints, while at the same time becoming more livable and equitable places. His most recent books on this topic are *Green Urbanism: Learning from European Cities* (Island Press, 2000) and *Native to Nowhere: Sustaining Home and Community in a Global Age* (also published by Island Press, December 2004). Beatley holds a Ph.D. in City and Regional Planning from the University of North Carolina at Chapel Hill.

Tom Bender

Tom Bender is one of the founders of the "green architecture" and "sustainability" movements. His work has spanned the gamut from technical tools, such as solar design, to the spiritual roots of our cultural problems. His "Factor 10" economic principles have been endorsed by the European Union, the World Business Council for Sustainable Development, and the United Nations Environmental Program. Bender's depth of experience in application of subtle energies in architecture has been a vital element in the acknowledgment of qi energy now occurring. His community work shows how we can heal the diseases of the spirit that plague our culture and restore our connection with the rest of nature. <www.tombender.org>

Janine Benyus

Janine Benyus is a biologist, innovation consultant, and author of six books, including *Biomimicry: Innovation Inspired by Nature*. Her favorite role these days is "Biologist-at-the-Design Table," helping innovators consult life's genius in the creation of well-adapted products and processes. Her company, the Biomimicry Guild, offers biological consulting, research, workshops, and field excursions to such clients as Arup Engineers, General Electric, Gensler Architects, Herman-Miller, HOK Architects, IDEO, Interface, Kohler, Nike, Seventh Generation, and Procter and Gamble. Benyus is currently creating a Google of Nature's Solutions—a digital library of biological literature organized by design function and a "biology-taught-functionally" course for architects, engineers, and designers. To help naturalize Biomimicry in the culture, she founded the nonprofit Biomimicry Institute, whose programs include an open-research Biomimicry Challenge and an Innovation for Conservation program that uses proceeds from bio-inspired products to conserve the habitat of the mentor organisms.

Bob Berkebile

Any list of accomplished, influential environmentalists and preservationists includes Bob Berkebile. Highly regarded by fellow professionals and recipient of numerous awards, Berkebile focuses on improving the quality of life in our society with the integrity and spirit of his firm's work. He is a founding principal of BNIM Architects in Kansas City and brings more than 45 years of experience to the architectural profession. By combining his design and leadership skills, Berkebile has consistently created new approaches to holistic, integrated community building. He utilizes education, integrated design, historic preservation, quality housing, community dialogue, public safety, and neighborhood conservation as tools to restore social, economic, and environmental vitality. Berkebile is a former board member and current advisory board member of the U.S. Green Building Council. He currently serves on the boards of the Nature Conservancy and the Center for Global Community, the New Earth Organization and the Athena Institute. Berkebile was the founding chairman of the AIA's National Committee on the Environment and has conducted numerous sustainable design charrettes and workshops for the White House, the National Park Service, the U.S. Department of Energy, the Federal Emergency Management Administration, and the Canadian Provincial Architects. He has been a juror and/or guest lecturer at numerous universities including Harvard, Rice, Stanford, and Cambridge University, United Kingdom.

Kent Bloomer

Kent Bloomer is a sculptor, architectural designer, Adjunct Professor of Architecture at Yale University, and principal at Bloomer Studio, New Haven, Connecticut. He is the principal author of *Body, Memory, and Architecture* (Yale University Press, 1975) and author of *The Nature of Ornament* (W. W. Norton, 2001). Bloomer's major projects include the roof ornament at Harold Washington Library Center, Chicago, and the foliated trellis at Reagan National Airport, Washington, D.C.

Tina Bringslimark

Tina Bringslimark (M.A., Ph.D.) recently completed her doctoral dissertation within the Department of Plant and Environmental Sciences at the Norwegian University of Life Sciences. She has been interested in environmental issues since the early 1990s. After some years of studies on the environment, she began to study psychology. On her way to the Ph.D., she earned a master's degree in environmental psychology from the Norwegian University of Science and Technology.

Hillary Brown

Hillary Brown, AIA, is principal of New Civic Works, which assists public and institutional clients in sustainability planning and greening infrastructure and facility capital programs. Government clients have included the City of New York, the New York Power Authority, the City of Salt Lake, the State University of New York, City of New Haven public schools program, and Battery Park City Authority, among others. As founder of New York City's Office of Sustainable Design within its Department of Design and Construction, she oversaw the city's 1999 *High Performance Building Guidelines*, and more recently envisioned and coauthored its *High Performance Infrastructure Guidelines*, both projects of the Design Trust for Public Space. Brown has served on the national board and currently the New York Chapter board of the U.S. Green Building Council. She teaches sustainable design at Princeton and Columbia University Schools of Architecture. Brown was a 2000 Loeb Fellow at Harvard's Graduate School of Design, and a 2001 Bosch Public Policy Fellow at the American Academy in Berlin. She is a Senior Fellow with City University of New York Institute for Urban Systems, and a past Fellow with Second Nature, a national organization fostering sustainability in higher education. Brown is a graduate of Oberlin College and Yale University School of Architecture.

William D. Browning

William Browning is a partner in Terrapin Bright Green LLC, a strategic consulting and policy firm with offices in Washington, D.C., and New York. He is a cofounder of the Washington, D.C. based green development consulting firm, Browning+Bannon. In addition, he founded Rocky Mountain Institute's Green Development Services in 1991 and has led or supported innovative design and development efforts for scores of clients, including the Sydney 2000 Olympics, Wal-Mart, the White House, the Pentagon, Monsanto, Hines, and George Lucas. He coauthored *A Primer on Sustainable Building* (1995), an introduction to green building; "Greening the Building and the Bottom Line," a 1994 study of increased worker productivity in energy-efficient buildings; and *Green Development: Integrating Ecology and Real Estate* (1998), an acclaimed textbook. His papers have been published in *Urban Land*, *Architectural Record*, *Progressive Architecture*, and AIA's *Environmental Resource Guide*. He serves as an editorial advisor for *Environmental Building News*, *Environmental Design & Construction*, and *Green@work* magazines. He was a founding member of the U.S. Green Building Council's Board of Directors and an honorary member of the American Institute of Architects.

Clare Cooper Marcus

Clare Cooper Marcus is Professor Emerita in the departments of Architecture and Landscape Architecture, University of California–Berkeley, where she teaches courses in the social and psychological implications of design. Her areas of special interest include housing design, open space design, children's environments, restorative landscapes, and healing gardens in healthcare. Marcus has published numerous articles in academic and professional journals and is the author/coauthor/editor of five books, including *Housing As If People Mattered* (with Wendy Sarkissian), *People Places* (with Carolyn Francis), *House as a Mirror of Self*, and *Healing Gardens* (with Marni Barnes). Her work has been translated into Chinese, Japanese, and French. In the last few years, she has been invited to lecture on healing landscapes in healthcare facilities in the United States, UK, Sweden, Denmark, Iceland, the Netherlands, Australia, and New Zealand.

Pliny Fisk

Fisk cofounded the Center for Maximum Potential Building Systems in 1975 and currently serves as codirector. The Center is recognized as the oldest architecture and planning 501(c)3 nonprofit in the United States focused on sustainable design. In addition, Fisk serves as Fellow in Sustainable Urbanism and Fellow in Health Systems Design at Texas A&M University, where he holds a joint position as signature faculty in Architecture, Landscape Architecture and Planning. In 2002, Fisk was awarded the U.S. Green Building Council's first Sacred Tree Award in the public sector category. He is also recipient of the Passive Solar Pioneer Award from the American Solar Energy Society, the Herrin Distinguished Fellow award from Mississippi State University, the Presidential Team Award for the sustainable relocation of towns displaced by the Mississippi flood, and the National Center for Appropriate Technology's 15th-Year Distinguished Appropriate Technology Award, recognizing significant work in the field of environmental protection. Fisk's special contributions in the research field have been principally in materials and methods; from low-cost building systems development referred to as open building, to wide-ranging material development that includes low-carbon and carbon-balanced cements and many other low-impact materials. He was instrumental in developing the first input-output life cycle assessment model for material flow in the United States and connecting it to a geographic information system so that human activities can be placed into the context of natural systems on a national scale. The model represents greenhouse gases, criteria air pollutants, and toxic releases of over 12.5 million businesses. He has also developed an alternative land planning and design methodology referred to as ecobalance design and planning. Fisk received B.Arch., M.Arch., and M.L.Arch. degrees from the University of Pennsylvania. His graduate studies focused on ecological land planning under the guidance of Professor Ian McHarg. His work was also influenced substantially by Russell Ackoff in the various disciplines associated with the systems sciences.

Robert Fox

In 2003, Robert F. Fox, Jr. joined with Richard Cook to form Cook+Fox Architects, a firm devoted to creating beautiful, environmentally responsible, high-performance buildings. A founding partner of Fox & Fowle Architects, Fox guided that firm to a prominent position of national leadership in sustainable high-rise buildings and urban design. Fox led the team that created the original "Green Guidelines" for the Battery Park City Authority in Lower Manhattan; in addition, he is a founding member and former Chair of the U.S. Green Building Council/New York Chapter, a member of the President's Council at The Cooper Union, and a member of the "Green Team" for Interface Corporation. Fox serves on the advisory board of the Center for Health and the Global Environment at Harvard Medical School, is the CoChair of the Sustainable Design Committee of the Real Estate Board of New York, and serves on numerous other advisory committees. Fox received a Bachelor of Architecture degree from Cornell University and a Master of Architecture degree from Harvard University. He and his wife Gloria live in New York City and the Hudson River Valley.

Howard Frumkin

Howard Frumkin is Director of the National Center for Environmental Health/Agency for Toxic Substances and Disease Registry (NCEH/ATSDR) at the U.S. Centers for Disease Control and Prevention. He is an internist, environmental and occupational medicine specialist, and epidemiologist. Before joining the CDC in September 2005, Frumkin was Professor and Chair of the Department of Environmental and Occupational Health at Emory University's Rollins School of Public Health and Professor of Medicine at Emory Medical School. He founded and directed Emory's Environmental and Occupational Medicine Consultation Clinic and the Southeast Pediatric Environmental Health Specialty Unit. Currently serving on the Institute of Medicine Roundtable on Environmental Health Sciences, Research, and Medicine, Frumkin is interested in public health aspects of urban sprawl and the built environment; air pollution; metal and PCB toxicity; climate change; health benefits of contact with nature; and environmental and occupational health policy, especially regarding minority workers and communities, and those in developing nations. He is the author or coauthor of over 160 scientific journal articles and chapters, and his books include *Urban Sprawl and Public Health* (Island Press, 2004, coauthored with Larry Frank and Dick Jackson; named a Top Ten Book of 2005 by Planetizen, the Planning and Development Network), *Emerging Illness and Society* (Johns Hopkins Press, 2004, coedited with Randall Packard, Peter Brown, and Ruth Berkelman), *Environmental Health: From Global to Local* (Jossey-Bass, 2005; winner of the Association of American Publishers 2005 Award for Excellence in Professional and Scholarly Publishing in Allied/Health Sciences), and *Safe and Healthy School Environments* (Oxford University Press, 2006, coedited with Leslie Rubin and Robert Geller). Frumkin received his A.B. from Brown University, his M.D. from the University of Pennsylvania, his M.P.H. and Dr.P.H. from Harvard, his Internal Medicine training at the Hospital of the University of Pennsylvania and Cambridge Hospital, and his Occupational Medicine training at Harvard. He is board-certified in both Internal Medicine and Occupational Medicine, and is a Fellow of the American College of Physicians and the American College of Occupational and Environmental Medicine. Frumkin was born in Poughkeepsie, New York, and is married to Beryl Ann Cowan, an attorney and psychologist. They have two children, Gabriel (age 19) and Amara (age 15).

Bert Gregory

Bert Gregory, FAIA, president and CEO of Mithun, is a national leader, speaker, and advocate for sustainable building and urbanism. His perspective reaches beyond traditional architecture to merge science and design—an interdisciplinary approach for the future that creates lasting places for people. Under Gregory's leadership, Mithun has become renowned for setting new standards in the development of resource-efficient structures and communities across the country. Mithun has received four AIA COTE Top Ten U.S. Green Projects awards, an AIA National Honor Award for Regional and Urban Design, and two ASLA National Honor Awards for excellence in planning and analysis. A USGBC LEED® accredited designer, Gregory has served as AIA Seattle's President, and is currently serving on the national USGBC LEED for Neighborhood Development core committee, the Washington Clean Technology Alliance Steering Committee, the Cascade Land Conservancy Board of Directors, the Seattle Mayor's Urban Sustainability Advisory Panel, and the State of Washington Governor's Climate Change Challenge advisory team.

Terry Hartig

Terry Hartig (Ph.D., M.P.H.) has studied the health resource values of natural environments for more than 20 years. He completed graduate training in environmental psychology and social ecology at the University of California–Irvine, and postdoctoral training in social epidemiology at the University of California–Berkeley. He currently works as an Associate Professor of Applied Psychology with the Institute for Housing and Urban Research and the Department of Psychology of Uppsala University in Sweden. He also holds an Adjunct Professor position with the Norwegian University of Life Sciences, where he participates in a university-wide Nature and Health initiative through affiliations with the Departments of Plant and Environmental Sciences and Landscape Architecture and Spatial Planning. He also leads a working group on the health benefits of nature experiences within a 22-country networking project funded by the European Science Foundation.

Alice Hartley

Alice Hartley joined Cook+Fox Architects in 2005 and is responsible for writing, editing, and outreach related to the firm's work. She also coordinates green materials research, educational resources, and internal sustainability initiatives. She serves as Senior Editor for Terrapin Bright Green, a consulting and strategic planning firm affiliated with Cook+Fox. Prior to joining Cook+Fox, Alice worked for the nonprofits Green Map System, Sustainability Institute, and Rocky Mountain Institute. She is a Vice Chair of o2nyc, a local ecodesign network. She graduated with honors from Dartmouth College and is a LEED Accredited Professional and Master Composter.

Judith Heerwagen

Heerwagen is a psychologist whose research and writing have focused on sustainability, biophilia, and the evolutionary basis of environmental aesthetics. Prior to starting her own business, Heerwagen was a senior research scientist at the Pacific Northwest National Laboratory and a research faculty member at the University of Washington, College of Architecture and Urban Planning. Her work at both PNNL and the University of Washington focused on the human factors of sustainable design. Heerwagen has been an invited participant at conferences and at national meetings sponsored by the National Academy of Sciences, the National Institute of Medicine, the General Services Administration, and the American Institute of Architects. She has lectured widely on environmental human factors and is the author or coauthor of numerous articles and book chapters on workplace, creativity, biophilia, and habitability. She was recently selected as a 2005 environmental champion by Interiors and Sources Magazine. She has a B.S. in communications from the University of Illinois and a Ph.D. in psychology from the University of Washington.

Grant Hildebrand

Following a professional degree from the University of Michigan in 1957, Grant Hildebrand began several years of professional practice with such firms as Minoru Yamasaki and Albert Kahn. In 1964, he began a career in teaching at the University of Washington. In 1974, he saw published by MIT a pioneering study of industrial architecture, *Designing for Industry: The Architecture of Albert Kahn*. In 1978, he became interested in the work of the English geographer Jay Appleton, who argued that the appeal of certain landscape characteristics is based in part on the survival advantages they offer. In 1988, Hildebrand inaugurated a course in the architectural implications of such an approach, for which he was appointed Chettle Fellow at the University of Sydney the following year. This interest led to the publication in 1991 of *The Wright Space: Pattern and Meaning in Frank Lloyd Wright's Houses*. Hildebrand expanded this work toward a general critical theory in *Origins of Architectural Pleasure* (1999), now being republished in translation in China. He has now retired from teaching but continues to write; his monograph on *Frank Lloyd Wright's Palmer House* was released in March of 2004.

Stephen Kellert

Stephen R. Kellert is the Tweedy Ordway Professor of Social Ecology at the Yale University School of Forestry and Environmental Studies. His work focuses on understanding the connection between human and natural systems, with a particular interest in the value and conservation of nature and designing ways to harmonize the natural and human built environments. His awards include the Outstanding Research Award for contributions to theory and science (2005, North American Association for Environmental Education); National Conservation Achievement Award (1997, National Wildlife Federation); Distinguished Individual Achievement Award (1990, Society for Conservation Biology); Best Publication of Year Award (1985, International Foundation for Environmental Conservation); Special Achievement Award (NWF, 1983); and being listed in *American Environmental Leaders: From Colonial Times to the Present*. He has authored more than 150 publications, including the following books: *Building for Life: Designing and Understanding the Human–Nature Connection* (Island Press, 2005); *Kinship to Mastery: Biophilia in Human Evolution and Development* (Island Press, 1997); *The Value of Life: Biological Diversity and Human Society* (Island Press, 1996); *The Biophilia Hypothesis* (edited with E. O. Wilson, Island Press, 1993); *The Good in Nature and Humanity: Connecting Science, Religion, and Spirituality with the Natural World* (edited with T. Farnham, Island Press, 2002); *Children and Nature: Psychological, Sociocultural, and Evolutionary Foundations* (with P. Kahn Jr., MIT Press, 2002); and *Ecology, Economics, Ethics: The Broken Circle* (edited with F. H. Bormann, Yale University Press, 1991).

Stephen Kieran

Stephen Kieran is a partner at KieranTimberlake Associates LLP, an award-winning and internationally recognized architecture firm noted for its research, innovative design, and planning. Kieran received his bachelor's degree from Yale University magna cum laude, and his Master of Architecture from the University of Pennsylvania with honors. He is a recipient of the Rome Prize, American Academy in Rome, 1980–81. In addition to his activities at the firm, Kieran is currently an adjunct professor at the University of Pennsylvania's School of Design, where he and his partner, James Timberlake, lead a graduate research studio. He has served as Eero Saarinen Distinguished Professor of Design at Yale University and Max Fisher Chair at the University of Michigan, and has taught at Princeton University. He and Timberlake were the inaugural recipients of the prestigious Benjamin Latrobe Fellowship for architectural design research from the AIA College of Fellows. They have coauthored two books: *Manual: The Architecture of KieranTimberlake*, published by Princeton Architectural Press in 2002, and *Refabricating Architecture*, published by McGraw Hill in 2004, which examines how manufacturing methodologies are poised to transform building construction.

Vivian Loftness

Vivian Loftness is an internationally renowned researcher, author, and educator with more than thirty years of focus on environmental design and sustainability, advanced building systems and systems integration, climate and regionalism in architecture, as well as design for performance in the workplace of the future. From 1994 to 2004, she was head of the School of Architecture at Carnegie Mellon University. Supported by a university-building industry partnership, the Advanced Building Systems Integration Consortium, she is a key contributor to the development of the intelligent workplace, a living laboratory of commercial building innovations for performance, along with authoring a range of publications on international advances in the workplace. She has served on six National Academy of Science panels as well as being a member of the academy's Board on Infrastructure and the Constructed Environment, and she has given three Congressional testimonies on sustainable design. Her work has influenced both national policy and building projects, including the Adaptable Workplace Lab at the U.S. General Services Administration and the Laboratory for Cognition at Electricité de France. As a result of her research, teaching, and professional consulting, Loftness received the 2002 National Educator Honor Award from the American Institute of Architecture Students (AIAS) and a 2003 Sacred Tree Award from the U.S. Green Building Council. In 2005, she was featured as one of 14 design visionaries in *Metropolis* magazine, and one of 25 environmental champions for 2005 by *Environ-Design Journal*. Loftness has Bachelor of Science and Master of Architecture degrees from MIT, is on the national boards of the USGBC, AIACOTE (2005 national chair), TSAC, Turner Construction, and DOE's Federal Energy Management Advisory Council (FEMAC). She is a Fellow of the American Institute of Architects and is a registered architect.

Richard Louv

Richard Louv is the author of seven books, including most recently, *Last Child in the Woods: Saving Our Children from Nature-Deficit Disorder* (Algonquin). Among his other books are *Childhood's Future* (Anchor), *The Web of Life* (Conari), *Fly-Fishing for Sharks: An Angler's Journey Across America* (Simon & Schuster), and *America II* (Houghton Mifflin). He has written for the *New York Times*, the *Washington Post*, and other newspapers and magazines. For 24 years, he was a columnist for the *San Diego Union-Tribune*. He also served as a columnist and member of the editorial advisory board for *Parents* magazine, and was an advisor to the Ford Foundation's Leadership for a Changing World award program. He chairs the Children & Nature Network (www.cnaturenet.org), a nonprofit organization helping to build an international children-and-nature movement, and he is currently working on his eighth book.

Martin Mador

After two immensely satisfying careers involving discrimination litigation in federal courts with the NAACP Legal Defense Fund and medical research computing at Yale Medical School, he treated himself to a life sabbatical to earn a master's degree at Yale School of Forestry and Environmental Studies. His knowledge of water issues, environmental education, and state politics led him to write a prospectus for a world-class museum about the intersection of water and human civilization, to be sited at New Haven Harbor. The museum vision embraces a curriculum covering the historical influences of water and people; the contemporary political, social, and economic issues of quantity and quality; and the manifold biophilic aspects of our attachment to water. It has earned the endorsement of over 60 civic leaders, and continues as a long-term, visionary project. Mador has worked on green building issues in Connecticut, securing funding for a green schools initiative and promoting passage of bills requiring LEED in the state legislature. He currently does biophilic design research at Yale with Steve Kellert. Mador is on the boards of the state Sierra Club, several watershed and river groups, Odyssey of the Mind, and several other civic organizations. He published a book in 2002 on Conservation Commissions in Connecticut. He holds bachelor's and master's degrees from Yale, is an instrument-rated pilot, and is a LEED accredited professional.

Kenneth G. Masden II

Kenneth G. Masden II received his B.Arch. from the University of Kentucky in 1982 and his M.Arch. from Yale University in 2001. While at Yale University, he studied directly with Léon Krier, Fred Koetter, Andrés Duany, and Vincent

Scully. Also during this time, he worked for Peter Eisenman as the project architect on the Memorial to the Murdered Jews of Europe in Berlin and as a project consultant on the Cidade da Cultura de Galicia (Center of Culture) in Santiago, Spain. His work ranges from the design-build of custom homes, to community design work on federal HUD urban renewal and housing projects, to large-scale base relocation and land reclamation projects for the U.S. military totaling nearly $4 billion in projects, which he has designed or managed in Japan, Germany, Spain, Italy, and America. Now an Associate Professor of Architecture at the University of Texas at San Antonio, his research is influenced by his international experience, underpinning his investigations into urban form. His writings and work look specifically at the adaptive and culturally driven urban systems that imbue the built environment with life.

Robin Moore

Robin Moore is a designer and design researcher, specializing in child and family urban environments that support healthy human development, informal play, and nonformal education. His current research interests are focused on landscape design in childcare centers and urban parks sponsored by the National Institutes for Health and the Robert Wood Johnson Foundation. Moore holds degrees in architecture from London University and city and regional planning from MIT. He is professor of Landscape Architecture, adjunct professor of Family and Consumer Sciences, and director of the Natural Learning Initiative, North Carolina State University–Raleigh. As a member of the UNESCO-MOST Growing Up in Cities (GUIC) action research program, he codirected the Buenos Aires project and coordinated the MENA regional program in Amman, Jordan. His publications include "Our Neighbourhood Is like That!" in *Growing Up in an Urbanising World* (2002); "Healing Gardens for Children," in *Healing Gardens* (1999); *Natural Learning* (1997); *Plants for Play* (1993); *Play for All Guidelines* (1987, 1992); and *Childhood's Domain: Play and Place in Child Development* (1986). He is associate editor of the American Journal of Health Promotion and a member of the Editorial Advisory Board for the online journal *Children, Youth and Environments.* Moore is past president of the International Association for the Child's Right to Play (IPA), past chair of the Environmental Design Research Association (EDRA), and a principal in the design and planning firm of Moore Iacofano Goltsman (MIG).

David Orr

David W. Orr is the Paul Sears Distinguished Professor of Environmental Studies and Politics and chair of the Environmental Studies Program at Oberlin College. He is also a James Marsh Professor at Large at the University of Vermont. Born in Des Moines, Iowa, and raised in New Wilmington, Pennsylvania, he holds a B.A. from Westminster College (1965), an M.A. from Michigan State University (1966), and a Ph.D. in International Relations from the University of Pennsylvania (1973). He and his wife have two sons and two grandchildren. David Orr is the author of five books: *The Fifth Revolution: Ecological Design and the Making of the Adam Joseph Lewis Center* (2006); *The Last Refuge: Patriotism, Politics, and the Environment* (2004); *The Nature of Design* (2002); *Earth in Mind* (1994/2004); *Ecological Literacy* (1992); and coeditor of *The Global Predicament* (1979) and *The Campus and Environmental Responsibility* (1992). He has published 150 articles in scientific journals, social science publications, and popular magazines. Orr is contributing editor of *Conservation Biology.* He serves on the Boards of the Rocky Mountain Institute (Colorado), the Center for Ecoliteracy (California), and the Center for Respect of Life and Environment.

Grete Grindal Patil

Grete Grindal Patil (Cand. Agric., Dr. Scient.) is an associate professor at the Department of Plant and Environmental Sciences, Norwegian University of Life Sciences. After completing a Ph.D. in horticulture and plant physiology, she engaged in establishing interdisciplinary research at the university on the benefits of plants in daily life. This involves studies on the potential restorative effects of natural elements in the built environment, as well as interventions with plant activities for people suffering from psychiatric distress. She is currently involved in developing a master's program in public health, in which the nature-and-health theme has a prominent position.

Robert Pyle

Robert Pyle received a B.S. in Nature Perception and Protection and an M.S. in Nature Interpretation from the University of Washington, and a Ph.D. in conservation ecology from the Yale School of Forestry and Environmental Studies. He has worked as butterfly conservation consultant for the government of Papua New Guinea, Northwest Land Steward for The Nature Conservancy, and visiting professor or writer in residence at many colleges and universities, most recently as Kittredge Distinguished Visiting Writer at the University of Montana. A professional writer since 1982, Pyle has written 14 books, which have won the John Burroughs Medal, a Guggenheim Fellowship, and other awards. They include *Wintergreen*, *The Thunder Tree*, *Where Bigfoot Walks*, *Chasing Monarchs*, and *Walking the High Ridge: Life as Field Trip*, as well as several standard butterfly guides. A novel and collections of poems and essays are in progress. His column "The Tangled Bank" appears regularly in *Orion* magazine. Pyle founded the Xerces Society for Invertebrate Conservation, chaired its Monarch project, and received a 1997 Distinguished Service Award from the Society for Conservation Biology. Pyle lives along Gray's River, a tributary of the Lower Columbia River, with botanist and silkscreen artist Thea Linnaea Pyle.

Jonathan Rose

Jonathan Rose is an innovator in bringing together solutions to planning, community development, finance, culture, and land preservation. In 1980, he developed the first live/work community with Internet access in every home. In 1984, he planned the country's first postwar green mixed-income, mixed-use, large-scale transit-oriented development. Most recently, in 2005, he established the first environmentally and socially responsible national real estate acquisition fund. Since then, his projects have consistently modeled new solutions to development, environmental, and community problems. Rose is a leading thinker in the Smart Growth and green building movements, and a frequent speaker on the subjects. His projects range from low-income housing for homeless people with AIDS, seniors, and first-time home buyers, to state-of-the-art academic buildings, performing arts centers, and libraries. His work also includes land preservation, urban infill, inner-city urban industrial, wholesale, artists and telecommunications projects. All of his projects are "green." Rose graduated from Yale University in 1974 with a B.A. in Psychology and received a Master's in Regional Planning from the University of Pennsylvania in 1980. He is married to Diana Rose, dressage rider and president of the Garrison Institute, and has two children.

Nikos Salingaros

Nikos A. Salingaros, M.A., Ph.D., ICTP, ICoH, is the author of *Anti-Architecture and Deconstruction* (2004), *Principles of Urban Structure* (2005), and *A Theory of Architecture* (2006), as well as numerous scientific papers. Both an artist and scientist, he is professor of mathematics at the University of Texas at San Antonio and is also on the architecture faculties of universities in Holland, Italy, and Mexico. His work underpins and helps to link new movements in architecture and urbanism, such as new urbanism, the network city, biophilic design, self-built housing, and sustainable architecture. Salingaros collaborated with Christopher Alexander, helping to edit the four-volume "The Nature of Order" during its 25-year gestation. In 1997, in recognition of his efforts to understand architecture using scientific thinking, he was awarded the first grant ever for research on architecture by the Alfred P. Sloan Foundation. Salingaros is a member of the INTBAU College of Traditional Practitioners and is on the INTBAU Committee of Honor.

Jenifer Seal Cramer

Jenifer Seal Cramer is a leader in research and consulting in high-performance buildings and developments. Her latest work is focused on progressive real estate investment management and fund development. She is the current editor for Urban Land Institute's *Urban Land GreenTech / Sustainable Frontiers* annual magazine and contributing author to Global Green's book *Blueprint for Greening Affordable Housing* (2007). She holds a master's degree in real estate development from Massachusetts Institute of Technology, and a Bachelor's of Architecture and a B.S. in Environmental Design from Ball State University. From 1994 to 2004, she served as a principal in Rocky Mountain Institute's Research and Consulting Group. She is a senior coauthor of RMI's landmark *Green Development: Integrating Ecology and Real Estate* (1998) and *Green Developments CD-ROM* (1998), and a coauthor of Urban Land Institute's *The Green Office Building: A Practical Guide to Development* (2005). Seal Cramer managed and participated in a number of RMI projects such as the Pentagon renovation charrette, California Academy of Sciences LEED Platinum building consultation, Texas Instruments Chip Fab and Office Headquarters charrette, Massachusetts technology collaborative strategic projects, Low-Power Data Centers charrette, Habitat for Humanity charrette, Hypercar charrette, and the Pittsburgh Nine-Mile Run stormwater charrette, as well as a comparative environmental REIT study for Forest City Enterprises. She served as a spokesperson for RMI in press conferences and interviews with CNN, the *Washington Post*, PBS, the *Denver Post*, and other media. Prior to RMI, she worked with William McDonough Architects in New York City. She is the recipient of a national American Institute of Architects Presidential Citation for her work in sustainable architectural education and, with the other members of the RMI GDS team, the 1999 President's Council on Sustainable Development and Renew America Green Building Award.

Megan Snyder

Megan Snyder has a Bachelor of Arts degree from the University of Massachusetts and a Master of Science from Carnegie Mellon University. Her previous research has addressed the environmental performance and life-cycle impacts of green roofs, as well as the relationship between indoor environmental quality and occupant health. She is currently a doctoral student at the Center for Building Performance and Diagnostics in the Department of Architecture at Carnegie Mellon University, with a focus on identifying links between school environments and student health and performance.

Roger Ulrich

Roger Ulrich is Julie and Craig Beale Professor of Health Facilities Design at Texas A&M University and a faculty fellow of the Center for Health Systems and Design, housed jointly in the colleges of architecture and medicine. A behavioral scientist, much of his research focuses on the effects of nature and medical buildings on patient clinical outcomes. He is the most-cited researcher in the area of evidence-based healthcare environmental design, and his work has influenced internationally the architecture and planning of hospitals and other healthcare buildings. He has carried out research in several countries, especially Sweden, where he has worked at the Karolinska Institute of Medicine, Uppsala

University, and Lund University. He has been visiting professor of architecture at the University of Florence, Italy, and the Bartlett School of Architecture, University College–London. He is a recipient of the Japan Society for the Promotion of Science Invitation Research Fellowship in Human-Nature Relations, and he has been senior advisor to the British National Health Service for its program to create scores of new hospitals. A member of the board of directors of the Center for Health Design, California, Ulrich also serves on the Hospital of the Future task force established by the Joint Commission for Accreditation of Healthcare Organizations.

Alex Wilson

Alex Wilson is the president of BuildingGreen, Inc., in Brattleboro, Vermont, and executive editor of *Environmental Building News* and the *GreenSpec® Directory*. A biologist by training, he has written about energy-efficient and environmentally responsible design and construction for more than 25 years. Prior to starting his own company in 1985 (now BuildingGreen, Inc.), he was executive director of the Northeast Sustainable Energy Association for 5 years. Alex is author of *Your Green Home* (New Society Publishers, 2006) and coauthor of the *Consumer Guide to Home Energy Savings* (ACEEE, 9th edition, 2007) and the Rocky Mountain Institute's comprehensive textbook *Green Development: Integrating Ecology and Real Estate* (John Wiley & Sons, 1998). Alex served on the board of directors of the U.S. Green Building Council for five years and he is currently a trustee of The Nature Conservancy–Vermont Chapter.

Edward O. Wilson

Edward O. Wilson was born in Birmingham, Alabama, in 1929. He received his B.S. and M.S. in biology from the University of Alabama and, in 1955, his Ph.D. in biology from Harvard, where he taught for four decades, receiving both of its college-wide teaching awards. He is currently University Research Professor Emeritus and Honorary Curator in Entomology of the Museum of Comparative Zoology at Harvard. He is the author of 25 books, of which 2 won Pulitzer Prizes: *Human Nature* (1978) and *The Ants* (1990, with Bert Hölldobler). He is the recipient of more than 100 other international medals and awards, including the National Medal of Science; the International Prize for Biology from Japan; the Catalonia Prize of Spain; the Presidential Medal of Italy; the Crafoord Prize from the Royal Swedish Academy of Sciences, given in fields of science not covered by the Nobel Prize; and for his conservation efforts, the Gold Medal of the Worldwide Fund for Nature and the Audubon Medal of the National Audubon Society. Six of Wilson's books compose two trilogies. The first, *The Insect Societies*, *Sociobiology*, and *On Human Nature* (1971–78) founded sociobiology and evolutionary psychology. The second, *The Diversity of Life*, *The Future of Life*, and *The Creation* (1992–2006) organized the base of modern biodiversity conservation. Wilson has served on the Boards of Directors of The Nature Conservancy, Conservation International, and the American Museum of Natural History, and gives many lectures throughout the world. His most recent books include *Consilience* (1998), which argues for the uniting of the natural sciences with the humanities. Wilson lives in Lexington, Massachusetts, with his wife, Irene.

Image Credits

1-1: Design: Hopkins Architects, Photographer: Martine Hamilton Knight

1-2: Photo Grant Hildebrand

1-3: Courtesy Kent Bloomer

1-4: Photo: Dramatic Photograph, Ltd.

1-5: Stephen Kellert

1-6: Stephen Kellert

1-7: Stephen Kellert

1-8: Genzyme Center, Photo Anton Grassl

1-9: Stephen Kellert

1-10: Courtesy Javier González-Campaña

1-11: Stephen Kellert

1-12: Grant Hildebrand (© 1976), Visual Resources Collection, College of Architecture and Urban Planning, University of Washington

3-1: Rodd Halstead

3-2: Phil Myers (http://animaldiversity.org)

3-3a: Dyesol

3-3b: Konarka Technologies, Inc. Power Plastic™

3-4a: IStock photos

3-4b: Angel Janer

3-5a: Héctor Landaeta

3-5b: Ernst Haeckel from Kunstformen der Natur (1904)

3-6a: Herman N.L. Hooyschuur

3-6b: Copyright free CD

3-7a: Wilker Group at Purdue University

3-7b: Columbia Forest Products

3-8a: Dr. Barthlott, University of Bonn, Germany

3-8b: Sto

3-8c: Dr. Barthlott, University of Bonn, Germany

3-9: Claudia Himmelreich

3-10a: Daimler-Chrysler3-10b: Daimler-Chrysler

3-10c: Daimler-Chrysler3-11a: Joris Laarman Studio

3-11b: Joris Laarman Studio

3-12: Pat Burke

3-13: Alcatel-Lucent3-14: Autotype

3-15a: Anton Harder

3-15b: Rupert Soar

3-15c: David Brazier

3-16a: Christopher Fay

3-16b: PAX Scientific

3-17a: Lutrus

3-17b: Copyright free CD

3-18: John Todd Ecological Design, Inc.

3-19a: Mary C. Metteer

3-19b: Dr. Kellar Autumn

3-20a: © Mark Hertzberg

3-20b: © Rob W. Sovinski

4-1: © Atelier Dreiseitl

4-2: Douglas Hill

4-3: Design: Perkins+Will, Photography: Craig Dugan, Hedrich Blessing

4-4: Courtesy, National Museum of the American Indian, Smithsonian Institution, Photo by J. Davis

4-5: Bud Weatherby

4-6: Design: Shim-Sutcliffe Architects. Photo credit: James Dow

4-7: Courtesy, National Museum of the American Indian, Smithsonian Institution, Photo by R. A. Whiteside

4-8: Vaughan Landscape Planning and Design Ltd.

4-9: © Photo and Design by Michael Portman

4-10: © Atelier Dreiseitl

4-11: Design: Norwood Oliver Design Associates, Inc. (N.O.D.A.) Photographer: Peter Paige.

4-12: Photograph Copyright Morna Livingston

4-13: Design: Perkins+Will; © James Steinkamp, Steinkamp/Ballogg Photography

4-14: © Atelier Dreiseitl

4-15: © Atelier Dreiseitl

4-16: Design: Kieran Timberlake Associates LLP

4-17: Susan Hebert Imports, Portland, Oregon www.ecobre.com

4-18: ©Atelier Dreiseitl

4-19: ©Atelier Dreiseitl

4-20: Courtesy Ontario Place

6-1: Adapted from Ulrich, R. S., R. F. Simons, B. D. Losito, E. Fiorito, M. A. Miles, and M. Zelson. 1991. "Stress Recovery During Exposure to Natural and Urban Environments. *Journal of Environmental Psychology* 11:201–230.

6-2: Adapted from Ulrich et al. 1991.

6-3: Adapted from Ulrich et al. 1991.

6-4: Architecture by ZGF. Photo by R. Ulrich.

6-5: Photo by R. Ulrich

6-6: Photo by R. Ulrich

6-7: Adapted from Ulrich 1984

6-8: Johnson Design Studio (AIA) and Gretchen Vadnais (LA). Photo by Teresia Hazen

6-9: Architect: Zimmer Gunsul Frasca Architects. Landscape Architect: Walker Macy. Photograph © Eckert & Eckert

8-1: Center for Building Performance and Diagnostics/DOE 1994

8-2: Vivian Loftness, FAIA

8-3: Center for Building Performance and Diagnostics, Carnegie Mellon University

8-4: Center for Building Performance and Diagnostics, Carnegie Mellon University

8-5: Chart created by authors; data from Fjeld et al. 1998

8-6: Vivian Loftness, FAIA

8-7: Vivian Loftness, FAIA

8-8: Vivian Loftness, FAIA

8-9: Architects: ZGF & Robert Murase. Photos: Roger Ulrich, PhD, Texas A&M University

10-1: Photo by Robin Moore

10-2: Photo by Robin Moore

10-3: Photo: Reed Huegerich

10-4: Photo by Robin Moore

10-5: Photo by Nilda Cosco

10-6: Photo by Robin Moore

10-7: Photo by Robin Moore

10-8: Photo by Robin Moore

10-9: Photo by Nilda Cosco

10-10: Photo by Robin Moore

10-11: Photo by Robin Moore

10-12: Photo by Robin Moore

10-13: Photo by Robin Moore

10-14: Photo by Robin Moore

10-15: Photo by Nilda Cosco

10-16: Photo by Nilda Cosco

10-17: Coombes School, Reading UK

10-18: Coombes School, Reading UK

10-19: Coombes School, Reading UK

10-20: Photo by Bruce Cunningham

10-21: Photo by Robin Moore

10-22: Robin Moore, Natural Learning Initiative, NC State University

10-23: Photo by Nilda Cosco

10-24: Photo by Robin Moore

10-25: Photo by Robin Moore

10-26: Photo by Robin Moore

10-27: Photo by Robin Moore

10-28: Photo by Nilda Cosco

10-29: Photo by Robin Moore

10-30: Photo by Robin Moore

10-31: Clare Cooper Marcus

10-32: Clare Cooper Marcus

10-33: Site Plan of St. Francis Square, Marquis and Stoller, Architects

10-34: Clare Cooper Marcus

10-35: Clare Cooper Marcus

10-36: Design by Mogavero Notestine and Associates McCamant & Durrett, Architects

10-37: Clare Cooper Marcus

10-38: Pamella Cavanna on behalf of Roger Cavanna

10-39: Clare Cooper Marcus

10-40: Clare Cooper Marcus

10-41: Clare Cooper Marcus

10-42: Clare Cooper Marcus

10-43: Michael Corbett

10-44: Clare Cooper Marcus

10-45: Clare Cooper Marcus

10-46: Mark Francis, "Children's Use of Open Space in Village Homes," *Children's Environment Quarterly*. 1,4:36–38. 1985.

10-47: Photo by Robin Moore

10-48: Clare Cooper Marcus

10-49: Clare Cooper Marcus

10-50: Clare Cooper Marcus

10-51: Clare Cooper Marcus

10-52: Clare Cooper Marcus

10-53: Photo by Robin Moore

10-54: Photo by Robin Moore

10-55: Kate Herrod, Director, Community Greens—an initiative of Ashoka: Innovators for the Public

10-56: © Eran Ben-Joseph

10-57: Clare Cooper Marcus

13-1: © Mithun, Bert Gregory

13-2: © 2007 JupiterImages. All Rights Reserved.

13-3: © Lara Swimmer Photography

13-4: © 2007 JupiterImages. All Rights Reserved.

13-5: Design by Mithun, © Doug J. Scott / dougscott.com

13-6: © 2007 JupiterImages. All Rights Reserved.

13-7: Design by and © Mithun, Dave Goldberg

13-8: © 2007 JupiterImages. All Rights Reserved.

13-9: Design by Mithun, © Robert Pisano

13-10: © 2007 JupiterImages. All Rights Reserved.

13-11: Design by Mithun, © Robert Pisano

13-12: © 2007 JupiterImages. All Rights Reserved.

13-13: Design by Mithun, © Robert Pisano

13-14: © 2007 JupiterImages. All Rights Reserved.

13-15: Design by and © Mithun, Juan Hernandez

13-16: Photo by Whit Slemmons

13-17: Photo by Whit Slemmons

13-18: © Mithun, Dick Robison

13-19: © Mithun, Dick Robison

13-20: Aerial Photography Copyright the GeoInformation Group 2002

13-21: © Mithun, Dennis Cope

13-22: © Mithun, Yicheng Yang

14-1: Jon Miller of Hedrich Blessing Photographers

14-2: Kieran Timberlake Associates LLP

14-3: Kieran Timberlake Associates LLP

14-4: © 2006 Halkin Architectural Photography

14-5: © 2006 Halkin Architectural Photography

14-6: © 2006 Halkin Architectural Photography

14-7: Photograph © Peter Aaron / Esto. All rights reserved.

14-8: Kieran Timberlake Associates LLP

14-9: Kieran Timberlake Associates LLP

14-10: © 2006 Halkin Architectural Photography

14-11: © 2006 Halkin Architectural Photography

14-12: Kieran Timberlake Associates LLP

14-13: © Walter Striedieck

14-14: © 2006 Halkin Architectural Photography

14-15: Kieran Timberlake Associates LLP

14-16: © 2006 Halkin Architectural Photography

14-17: © 2006 Halkin Architectural Photography

14-18: © 2006 Halkin Architectural Photography

15-1: Photograph by Tim Street-Porter. Richard Neutra, architect: Health House, Philip and Lea Lovell House, Los Angeles, California, 1927–1929.

15-2: Photograph by Julius Shulman. Richard Neutra, architect: Case Study House #20, Pacific Palisades, California, 1948.

15-3: From Claude Bragdon, *The Frozen Fountain* (New York: Alfred A. Knopf, 1932). Claude Bragdon, ornamenter: Window project, early 20th century.

15-4: Photograph by Kent Bloomer. Bay window, The Charles F. Brooker Administration Building, Gaylord Hospital, Wallingford, Connecticut, 1927.

15-5: Photograph by Kent Bloomer. Front-door window, The Charles F. Brooker Administration Building, Gaylord Hospital, Wallingford, Connecticut, 1927.

15-6: Courtesy of Lyndhurst, an historic site of the National Trust for Historic Preservation in the United States. Photographer: James M. Mejuto. A. J. Davis, architect, and L. C. Tiffany, designer: Window, Lyndhurst, Tarrytown, New York, 19th century.

15-7: Photograph by Richard Nickel, Courtesy of the Richard Nickel Committee. Louis Sullivan, architect: Carson Pirie Scott Store, Chicago, Illinois, 1896.

15-8: Photograph by Tim Street-Porter. Frank Lloyd Wright, architect: Hollyhock House, Hollywood, California, 1917.

15-9: Photograph by Tim Street-Porter. Frank Lloyd Wright, architect: Ennis House, Los Angeles, California, 1923.

15-10: Photograph by Tim Street-Porter. Frank Lloyd

Wright, architect: Interior, Hollyhock House, Hollywood, California, 1917.

15-11: Photograph by Bastin and Evrard. Victor Horta, architect: Hotel Tassel, Brussels, Belgium, 1893.

15-12: Photograph by Kent Bloomer. Brent Bowman, architect, Kent Bloomer, ornamenter: Manhattan Public Library, Manhattan, Kansas, 2002.

15-13: From Klaus-Jürgen Sembach, *Art Nouveau* (Köln: Benedikt Taschen, 1991). Antonio Gaudi, architect: Casa Batlló, Barcelona, Spain.

15-14: Photograph by David Lamb. Kent Bloomer, architect and ornamenter: Staircase, Bartels House, Guilford, Connecticut, 2002.

15-15: From François Dumas, *The Temple of Dendura* (Cairo: Ministère de la Culturel Centre de Documentation et D'Étude sur l'Anciene Egypte, 1970). The Temple of Dendura, Thebes 7th century B.C.

15-16: Photograph by Kent Bloomer. Cesar Pelli, architect, Kent Bloomer, ornamenter: Foliated trellis, Ronald Reagan Washington National Airport, Washington, D.C., 1997.

15-17: Photograph by David Lamb. Harold Roth, architect, Kent Bloomer, ornamenter: Interior ceiling, Fair Haven Middle School, New Haven, Connecticut, 2002.

16-1: Photo: Grant Hildebrand

16-2: Photo: Grant Hildebrand

16-3: Photo: Grant Hildebrand

16-4: Photo: Grant Hildebrand

16-5: Photo: Grant Hildebrand

16-6: Photo: Grant Hildebrand

16-7: Photo: Grant Hildebrand

16-8: Photo: Grant Hildebrand

16-9: Photo: Grant Hildebrand

16-10: Photo: Grant Hildebrand

16-11: Photo: Grant Hildebrand

16-12: Photo: Grant Hildebrand

16-13: Photo: Grant Hildebrand

16-14: Photo: Grant Hildebrand

16-15: Photo: Grant Hildebrand

16-16: Photo: Grant Hildebrand

16-17: Photo: Grant Hildebrand

16-18: Photo: Grant Hildebrand

16-19: Photo: Grant Hildebrand

16-20: Photo: Grant Hildebrand

17-1: Photo by Timothy Beatley

17-2: Photo by Timothy Beatley

17-3: Photo by Timothy Beatley

17-4: Photo by Timothy Beatley

17-5: Carlson Architects—Seattle

17-6: Photo by Timothy Beatley

17-7: Photo by Timothy Beatley

17-8: Photo by Timothy Beatley

17-9: Photo by Timothy Beatley

17-10: Photo by Timothy Beatley

17-11: Photo by Timothy Beatley

18-1: Courtesy of Dattner Architects

18-2: Lee Weintraub, Landscape Architect LLC

18-3: Photographer: Scott Dressel-Martin

18-4: Site plan Peter Calthorpe with updates by Klipp Architecture

18-5: Site plan Peter Calthorpe with updates by Klipp Architecture, photo by Jonathan F P Rose

18-6: Phipps Rose Dattner Grimshaw

19-1: © Paul Bardagjy

19-2: © Paul Bardagjy

19-3: © Paul Bardagjy

19-4: Pliny Fisk III

20-1: © Tom Bender

20-2: © Tom Bender

20-3: © Tom Bender

20-4: © Tom Bender

20-5: © Tom Bender

20-6: © Tom Bender

20-7: © Tom Bender

20-8: © Tom Bender

21-1: Photograph The Sky Factory © 2006

21-2: Photograph © Tom Bender

22-1: © Jenifer Seal Cramer

22-2: Photo: Norm Clasen, Rocky Mountain Institute. All Rights Reserved.

22-3: © Mark Hertzberg

22-4: Photographer: Rick Grunbaum, Courtesy HKS, Inc

22-5: © William Browning

22-6: © Jenifer Seal Cramer

22-7: © Jenifer Seal Cramer

22-8: © Huston Eubank

22-9: © William Browning

22-10: Photo: © DesignInc Melbourne; Architects: DesignInc in association with the City of Melbourne

23-1: © Peter Aaron/Esto for Cook+Fox Architects; Design by Richard Cook & Associates

23-2: © Peter Aaron/Esto for Cook+Fox Architects; Design by Richard Cook & Associates

23-3: Copyright © Pat Whalen; Design by BNIM Architects

23-4: © dbox for Cook+Fox Architects

23-5: © dbox for Cook+Fox Architects

23-6: Copyright © Mike Sinclair; Design by BNIM Architects

23-7: Copyright © Paul Kivett; Design by BNIM Architects

23-8: © Bilyana Dimitrova

23-9: Copyright © Bill Truslow, www.truslowphoto.com

23-10: Copyright © Roland Halbe; Design by Steven Holl Architects

23-11: Copyright © Roland Halbe; Design by Steven Holl Architects

23-12: © Seong Kwon for Cook+Fox Architects

23-13: © Whit Slemmons

Index

Aalto, Alvar, 123
Abstract human nature, 66, 67
Action research, 160
Activity cycles, 144–146
ADD, *see* Attention Deficit Disorder
Additive effect variables (preschool play areas), 163
ADHD, *see* Attention Deficit Hyperactivity Disorder
Adolph Coors Company, 302
Advanced Green Builder Demonstration (AGBD), 309
  See also color Figure 19-3
Aesthetic(s):
  and architecture as visual art, 314, 315
  attachment to water, 44–45
  derived from connection to nature, *see* Environmental aesthetic
  epigenetic rules of, 22–24
  genetically-influenced "aesthetic programs," 232
  natural, 228–235
  and needs of children, 174
  sensory, *see* Sensory aesthetics
  and water as life force, 51
Aesthetic judgment, epigenetic rules of, 23
Aesthetic response, neurological basis for, 61–63
Affection for natural world, as design attribute, 14
After-school program professionals, 170
AGBD (Advanced Green Builder Demonstration), 309
  See also color Figure 19-3
Age, as design attribute, 9–10
AIA Journal of Architecture, 300
Air:
  as design attribute, 7
  natural ventilation, 123–125
Air pollution, ix, 218, 220
Aizenberg, Joanna, 35
AKT Development, 210
Albany, California, 191
Alberts, Anton, 342
Aldersey-Williams, Hugh, 33
Alexander, Christopher, 71, 73, 75, 76, 80, 336, 344
Alhambra, Court of Lions, 266
Alienation from nature, ix
Alleys, converted, 192–193
  See also color Figure 10-55
American Electric Power, 218
American Planning Association, 290
American Sports Data, 207
Analogs, natural, 340
Angina treatment, 108–109
Animals:
  in children's play areas, 175
  and child- vs. wildlife-friendly green spaces, 191–192
  as design attribute, 7
  human-animal interactions, 71–73
  loss of wildlife, 215
  resilient relationships among, 233
  sharing homes with, 289
  variation and similarity in, 232
  zoo habitats for, 69
Animal motifs, as design attribute, 8
Anita B. Gorman Discovery Center, Kansas City, Missouri, 348
Appalachia, coal mining in, 217–220
Appleton, J., 234, 235, 265, 266, 269, 338
Arcata, California, community forest, 286
Arches, as design attribute, 8
Arch Coal, 217
Archer, Mike, 289
Architectural transcendence, 74–76
Architecture. *See also* Living architecture
  biophilic strategies/priorities for, 329–330
  processes of building vs. design in, 59–60
  reason for existence of, 244
  survival-advantageous characteristics of, 263–275
  unworkable approaches to, 317
Arkadien Asperg Housing Estate, Stuttgart, Germany, *see* color Figure 4-19
Arlington Row, Bibury, Gloucestershire, United Kingdom, 265
Arnheim, Rudolph, 236, 239–240
Art:
  in healthcare settings, 96–97, 103, 340
  living architecture as, 314, 315
  sacred, 315, 320–321
  water as, 52, 53
Artificial intelligence, 72
Association of Children's Museums, 172
Atlanta Summer Olympic Games asthma rates, 157
Attachment:
  as design attribute, 14
  and real vs. artificial settings, 222
  to universe and beliefs, 74
  values/constructs of, 44–45
Attention, 158, 231
Attention Deficit Disorder (ADD), 157–158, 209, 328
Attention Deficit Hyperactivity Disorder (ADHD), 157–158, 328
Attention restoration theory, 136–137
Attraction to nature, as design attribute, 14
Atwater Commons, Middlebury College, Vermont, 245, 249–251
Audubon Nature Preschool, Edwin Way Teale Learning Center, 165
Augustine, St., 108
Austin, Texas, bats in, 292–293
Australia:
  biophilic spirit in, 293–295
  children's mobility study in, 174
  children's use of school play areas in, 209
  ecosystem mimicry campaign in, 38
  green school projects, 284
Automotive traffic, danger of, 156–157
Autotype, 35
Awe, as design attribute, 14

Backyard Habitat Program (NWF), 220
BAE, 39
Balance, 232, 352–353

Ballard branch, Seattle Library, Seattle, Washington, 278, 279
Bank of America Tower, New York City, 349–350, 354–355
  See also color Figure 23-5
Bank of Astoria, Manzanita, Oregon, *see* color Figure 21-2
Barcelona, Spain, 48
Barnes, Marni, 208
Bastille viaduct, Paris, France, 18
Bat Conservation International, 292
Bateson, Gregory, 308
Baubiologie, 314
Beale, Bob, 289
Beatley, Timothy, 210–211, 298
Beauty. *See also* Aesthetic(s)
  as design attribute, 14
  good design as element of, 27–28
  human preference for, 23
  importance of, 244
  in LEED system, 244
  in living architecture, 315
  in sacred art, 315
Beddington Zero-Energy Development project, 286
Behavioral genetics, 22
Behnisch, Behnisch, and Partner, 17
Bell Labs, 34, 35
Bender, Tom, *see* color Figure 21-2
Ben-Joseph, E., 194
Benyus, Janine, 7
Berry, Wendell, 6, 36
Beverley Minster, Yorkshire, United Kingdom, 264
Bicycles, infrastructure for, 156–157
BIDS, *see* Building Investment Decision Support
Biocultural restoration, 140, 141
Biological anthropology, 22
Biological human nature, 68–73
  environmental complexity/stimulation for children, 69
  human-animal interactions, 71–73
  human-machine interactions, 71–73
  pattern recognition, 70–71
  physiological operation in built environment, 68–69
Biological sciences, 22, 24
Biological wastewater treatment, 55, 56
Biologists, in biomimetic design, 40
Biolytix, 38
Biomimetic innovation credits, 40
Biomimicry, 27–41
  as architectural principle, 28–29
  bio-inspiration gardens, 38–39
  to bring working ecosystems inside, 37–38
  and buildings as chimeras, 30
  for buildings with benevolent presences, 39–40
  colors, 36–37
  daylighting, 34–35
  defined, 28
  as design attribute, 9
  design elements inspired by nature, 30–32
  as design process, 29
  focus on function in, 29
  human comfort with, 40–41
  in landscape design, 38
  organic forms and structures, 32–34
  sounds, 36
  ventilation, 35–36
Biomorphy, as design attribute, 9
Biophilia, 325–326
  benefits of, 326–328, 331
  biomimicry in, 28
  defined, 3, 221
  different interpretations of, 221–222
  and human well-being, 3–4
  nine typologies, 44
  as "weak" biological tendency, 4
Biophilic design, 3–18, 325–333
  as arising from human nature, 65–66
  assumption of positive experiences from, 146
  attributes of, 6–15
  benefits of, 326–328, 331
  branches of, 65
  defined, ix, 3
  dimensions of, 5–6
  elements of, 6–18
  environmental features in, 6–8, 15, 16
  evidence supporting, 107–116
  evolved human-nature relationships in, 13–15, 18
  as expression of inherent need to affiliate with nature, x
  focus of, 206–207
  fundamental questions of, ix
  and genetic dependence on environment, 61–63
  justifying costs of, 331–332
  light and space in, 11, 15, 17
  merging of artificial and natural in, 63–64
  natural patterns and processes in, 9–11, 15, 17
  natural shapes and forms in, 8–9, 15, 16
  and need to affiliate with natural systems/processes, 3–4
  and neurological basis for aesthetic response, 61–63
  and neurological nourishment, 62–65
  next steps for, 332–333
  objectives of, x, 3
  organic or naturalistic dimension of, 5–6
  and other green design priorities, 331
  place-based or vernacular dimension of, 6, 15, 18
  place-based relationships in, 12–13
  positive effects of, 61
  in restorative environmental design, 5, 140
  and sensory/emotional connection to built environment, 61
  steps supporting, 77–80
  strategies and priorities in, 329–330
  in sustainable design, x, 5
Blackberry Creek, Thousand Oaks School, Berkeley, California, 175
Blanchie Carter Discovery Park, Southern Pines Primary School, 159, 168–170
  See also color Figures 10-20, 10-21
Blankenship, Donald, 219
Bloomer, Kent, 16
Body, Memory, and Architecture (Kent Bloomer and Charles Moore), 255
"Bogeyman syndrome," 289
Bold Park, Perth, Australia, 293
Bone chair, 33, 34
Boston Schoolyards Initiative, 168
Botanical motifs, as design attribute, 8
Botta, Mario, 266, 270
"Bottleneck, environmental," 222
Boundaries:
  psychological, around human body, 255
  sensory awareness and lack of, 234
Bounded spaces, as design attribute, 10
Brain. *See also* Neurological science
  "natural," 207
  neocortex, 308–312
  patterns causing arousal of, 23

Brevard Zoo, Melbourne, Florida, 172
Brisbane, Australia, 286, 294
British New Towns, 184
Britton Courts, San Francisco, California, 182–183
Bronchoscopy pain control, 110
Bronson Methodist Hospital, Kalamazoo, Michigan, 327
Brooks, Rodney, 72
Buddha, 113
Buffalo Bayou Partnership, 293
"Buffer effect," 159
Buildings:
  biomimetic, 39–40
  biophilic, 240–241, 339–341
  intermediate fractal patterning in, 233
  sensory aesthetics in, 236–237
  urban design at level of, 278
Building for Life (Stephen Kellert), 206
Building form, landscape features defining, 12
Building Investment Decision Support (BIDS), 125, 337
Building practices:
  to enhance human and natural environments, 339–341
  modern examples of, 341–343
  for productive environments, 336–339
  transformation to biophilic design, 344–345
"Building with Nature" newsletter, 314
Built environment:
  as barrier to children's experience of nature, 155
  dominant paradigm for, ix–x
  extinction of natural experience in, *see* Extinction of natural experience
  as facilitating/impeding contact with nature, 4
  human nature in response to, 59
  informational template for, 63
  intermediate fractal patterning in, 233
  natural environment degraded/depleted by, 5
  need to affiliate with nature in, x
  physiological operation in, 68–69
  prospect and refuge in, 235
  rejuvenating spaces in, *see* color Figure 13-3
  resilience in, 233–234
  restorative/protective/instorative value of, 135
  return to balance in, 352–353
  sense of place and behavior toward, 6
  sensory aesthetics in, 235–236
  sensory/emotional connection to, 61
Burbank Housing Development Corporation, 188
Burch, William, 221
Bushcare groups (Australia), 293, 294
Bush Wardens program, Australia, 284
Butterfield, Deborah, 336
Bystrom, Arne, 270

Calatrava, Santiago, 340
Calthorpe, Peter, 302
Campbell, Craig, 44
Canada, children's use of school play areas in, 209
Carbon, 220, 223
Caring Spaces, Learning Places, (Jim Greenman), 162
Carnegie Mellon University, 126, 337
Cascades, water, 51, 52
Cascadia Region chapter, U.S. Green Building Council, 351
Case-control studies, 111–112
Cathedral of St. Louis, St. Louis, Missouri, 268
Cave Temples of Ellora, India, 321
CDC (Centers for Disease Control and Prevention), 207
Center for Building Performance and Diagnostics, Carnegie Mellon University, 338
  See also color Figure 23-2
Centers for Disease Control and Prevention (CDC), 207
Center for the Built Environment, University of California-Berkeley, 328
Center for Well-Being, Ross Institute, East Hampton, New York, 348
Central focal points, as design attribute, 10
Chains, as design attribute, 10
Change, as design attribute, 9–10, 13
Changefulness, spectacle of, 260
Chawla, Louise, 211
Cherry Hill, Petaluma, California, 188–189
Chicago, birds in, 277
Chicago Landscape Ordinance, 290
Child development centers, 162–164
Childhood spaces, 153–196, 205–212
  action-research design strategy for, 160
  biophilic design for, 209–212
  and biophilic forms of residential neighborhoods, 176–177
  changes in, 214, 215
  clustered housing and shared outdoor space, 177–183
  community nature destinations, 172
  converted back alleys, 192–193
    See also color Figure 10-55
  cul-de-sacs and greenways, 183–192
  and cultural reality of institutionalized childhood, 161
  for early childhood, 162–166
  and ecocommunity concept, 205–206
  and exocommunity concept, 206
  and health of children, 156–161, 163–166, 208–209
  home zones, 194
  LEED Neighborhood Development system, 194–195
  need for biophilic design of, 154–156
  neighborhood parks, 170–171
  for play activities, 174–176
  in residential environments, 173–174
  and restorative environmental design/biophilic design, 206–207
  schoolgrounds as neighborhood parks, 169–170
  school sites, 166–169
  social/cultural obstacles to good design, 207–208
  wild vs. manicured, 221
  woonerf (residential precincts), 193–194
Children:
  art preferences of, 96–97
  contact with nature and development of, 4
  creativity of, 328
  enticement to outdoors, 284–285, 289
  environmental complexity/stimulation for, 69
  lifestyles of, 153, 156–160
  play activities preferred by, 174–176
  school performance of, 328
Children and Nature Design Certification, 211
China, urbanization in, 352
CHRISTUS St. Michael Health Care Center, Texarkana, Texas, 327
Churchill, Winston, 205

CII-Godrej Green Building Center, Hyderabad, India, 351–352
Circadian rhythms, 99, 122
Cité de Refuge, Paris, France, 244
Cities:
 biophilic, *see* Urban design
 and complexity theory, 300
 growth of, 297–298, 352
 sensory aesthetics in, 238–240
City parks, 173
Civano, Tucson, Arizona, 281
Clarke, Peter, 294
Class of 1945 Library, Phillips Exeter Academy, 353–354
Claus, Elvis, 294
Climate change, ix, 216, 222, 298
Clinical epidemiology, 108–110
Clodagh, 341
Clubhouse, Huntington Lakes, Delray Beach, Florida, *see* color Figure 4-5
Clustered housing and shared outdoor space, 177–183
 advantages of, 178
 examples of, 178–182
  See also color Figures 10-35, 10-37, 10-39
 resistance to shared space, 182–183
Coalfields, environmental damage from, 217–220
Cobb, Edith, 328
Cognition, as design attribute, 14
Cognitive development, sedentary lifestyle and, 157
Cognitive neuroscience, 22, 70
Cognitive psychology, 22
Coherence, 21. *See also* Consilience
Cohousing, 180, 181
Color(s):
 biomimicry of, 36–37
 as design attribute, 7
Columbia Forest Products, 31
Columnar supports, as design attribute, 8
Commonwealth Scientific & Industrial Research Organization building, 29
Community gardens, 291
Community Greens: Shared Parks in Urban Blocks, 181
Community nature destinations, 172
Community-supported agriculture (CSA), 287
Complementary contrasts, as design attribute, 10
Complexity:
 connection to, 63
 as design attribute, 13
 for intellectual satisfaction/cognitive prowess, 14
 for neurological nourishment, 64
 in retirement home case study, 274–275
 as survival-advantageous, 264–265
Concrete, 78–79
Congress for New Urbanism, 194
Connecticut, 46, 47
Consilience, 21–22, 24–25
Construction process, human factors in, 79. *See also* Building practices
Contrasts:
 as design attribute, 10
 of light and shadow, 11
Control over nature, as design attribute, 13–14
Converted back alleys, 192–193
 See also color Figure 10-55
Cook+Fox Architects, 353
 See also color Figure 23-8
Cooling, natural, 127
Coombes School, Reading, United Kingdom, 168
Cooper Marcus, Clare, 208
Copenhagen, Denmark, 279
Coping, 134, 135
Corbett, Christopher, 186, 187
Corbett, Judy, 184, 205
Corbett, Lisa, 205
Corbett, Michael, 184, 205–206
Council House 2, Melbourne, Australia, 343
 See also color Figure 22-10
Cowles, Macon, 345
Creating Defensible Space (Oscar Newman), 129
Creativity, access to nature and, 328
Crystal Palace, London, 30
CSA (community-supported agriculture), 287
Cul-de-sacs and greenways, 183–192
 benefits of greenways, 190–191
 citywide greenway networks, 187–190
 conflicts between child- and wildlife-friendly areas, 191
 English Garden City movement, 174
 possible negative unintended consequence of, 191
 urban promenades, 192
 Village Homes, 184–187
Cultural change, living architecture and forces of, 317–320
Cultural connection to place, as design attribute, 12
Cultural evolution, genetic evolution and, 24
Cultural patterns, vulnerability of, 318
Culture, integration of ecology and, 12–13
Curiosity, as design attribute, 13
Cutler-Girdler house, Medina, Washington, 268
Cyberknife Radiosurgery Center of Iowa, Des Moines, Iowa, *see* color Figure 21-1

Dagara tribe, 320
Daimler-Chrysler, 33
Danger (peril):
 thrill of/sensitivity to, 269–271
 from traffic, 156–157, 174
Davenport, Hume, 218
David and Joyce Dinkins Gardens, Harlem, New York City, 301–302
Davis, Alexander Jackson, 257
Day, Chris, 314
Daylighting, 34–35, 127. *See also* Natural light
DC Ranch, Phoenix, Arizona, 187
Debussy, Claude, 46
Demosthenes, 107
Denmark, 164, 175
Density, urban, 298
Denver Business Journal, 299–300
Denver Urban Renewal Authority (DURA), 303, 304
Depression, effects of daylight on, 99–100
Deramus Education Pavilion, Kansas City Zoo, Missouri, 350–351
 See also color Figures 23-6, 23-7
Desealing, 286
Design with Nature (Ian McHarg), 217
Detroit airport, 48
Diaphanous buildings, 127
Diffused light, as design attribute, 11
Direct experience with nature, 5
Discovery of natural processes, as design attribute, 14
Disney Hall, Los Angeles, California, 129
 See also color Figure 8-6

Distraction theory, 93–95, 326
Diurnal changes in nature, 230
Diversity, 11, 220
Doembecher Children's Hospital, Portland, Oregon, 93
Domes, as design attribute, 8
Dominionistic attachment to water, 44
Dreiseitl, Herbert, 284, 342
Dubos, René, 6, 14
Dune (Frank Herbert), 46
DURA, *see* Denver Urban Renewal Authority
Dye-sensitized solar cells, 29, 34
Dynamic balance and tension, as design attribute, 10
Dynamic entanglement or competition, 260

Eastgate office complex, Harare, Zimbabwe, 35–36
  See also color Figures 3-15
Eble, Joachim, 342
E&B (environment and behavior) research, 160
Ecocommunity concept, 205–206
Ecological connection to place, as design attribute, 12
Ecological restoration, 140, 141
Ecological rooftops, 278
Ecology/ecological processes, ix–x, 12–13
The Ecology of Imagination in Childhood (Edith Cobb), 328
Eco-machines, 38
Ecosystems:
  as design attribute, 8
  indoor, 37–38, 51, 63
  See also color Figure 4-5
Eden Center, 30
Edible opportunities, 284
Edible Schoolyard, 167
Edwin Cheney house, Oak Park, Illinois, 266
Efflorescence, as design attribute, 10
Efrati, Amir, 189
Egg forms, as design attribute, 8
Eiffel Tower, 30, 269
Einstein, Albert, 350
Elderly:
  gardens for, 98
  use of shared space by, 178
ELIZA program, 72
Elk Rock Gardens at Bishop's Close, Portland, Oregon, 336
Embassy Suites, 343
Emotional connections, 61, 73–74
Energetic dimensions of place, *see* Living architecture
Energy consumption, ix–x, 298
Energy efficiency, 337, 338
Energy life cycle, 309
English Garden City movement, 184
Engwicht, David, 173
Entelechy II, Sea Island, Georgia, 52
Enterprise, 301
Enticement:
  as design attribute, 13
  in retirement home case study, 274
  as survival-advantageous, 267–269
  in urban design, 284–285
Environment:
  current "bottleneck" in, 222
  genetic dependence on, 61–63
  humane, 227
  mimicking nature, 62
  need for connection to, 59
  and physiological well-being, 62
  for zoo animals, 69
Environmental aesthetic, 243–251
  beauty in, 244
  groundwork for, 244
  integration in, 245–252
  language for, 245
  and reason for separation of man from nature, 244–245
Environmental features (as design element), 6–8, 15, 16
Environmental psychology, 69, 145
Environmental sciences, 22
Environmental stress, 134
Environment and behavior (E&B) research, 160
Epcot Center, 31
Epigenetic rules, 22–24
Erotic aesthetics, 23
Estrogen replacement, 108
Europe:
  gathering spaces, 234, 235
  green neighborhoods, 280
  regional ecological networks, 279
  whole-community approach, 210–211
Eva-Lanxmeer, Culumberg, Netherlands, 280
Evans, Ianto, 314
Evergreen Foundation, 168
Evolution, ix
  attunement to light, 90–91
  epigenetic rules in, 22–24
  increasing complexity in, 62–63
  and need for contact with nature, 3–4
  physiological restoration, 89–91
  preference for environments mimicking, 23
Evolution biology, 22
Evolved human-nature relationships, 13–15, 18
Exocommunity concept, 206
The Experience of Landscape (J. Appleton), 235
Expert knowledge, 70–71
Exploration of natural processes, as design attribute, 14
The Extended Organism (J. Scott Turner), 35
Extinction of natural experience, 213–223
  antidotes to, 217
  and biophilia, 221–223
  and changes in childhood environments, 214, 215, 217
  and climate change, 218
  and creation of sacrifice zones, 222
  defined, 213, 215
  and destruction of environment for energy resources, 217–220
  and diversity within radius of reach, 215–216
  factors in, 214
  and mastery of nature as political, 214
  and New Orleans hurricane damage, 218
  results of, 216
  strategies for green spaces optimization, 220–221

Façade greening, as design attribute, 7
Façade paint, *see* color Figure 3-8a
Fallingwater, 18, 235
False Creek, Vancouver, British Columbia, 182
"Family-Friendly Courtyard Housing," Portland, Oregon, 181–182
Farnsworth House, 245, 248
Fathy, Hassan, 75
Fauna, 50, 289
Fear of nature, as design attribute, 14
Fentress, Curtis W., 300
Fiber optics, 34, 35
Filtered light, as design attribute, 11

Filters, aesthetic, 245
Finland, cul-de-sac and greenway design in, 184
Fire, as design attribute, 8
Fisk, Pliny, 314
Flora, 50, 289
Flowing shapes, as design attribute, 8–9
Focal points, as design attribute, 10
Food production, 283, 314–315
Forest Kindergarten Isarauen, Munich, Germany, *see* color Figure 10-14
Forest kindergartens, 164–166
Forest Service, 114
Forms, *see* Natural shapes and forms
Forman, Richard, 191, 328, 331
Foster, Norman, 30
Fountains, 51–52
Four Seasons Kindergarten, Ringe, Denmark, 165
Fractals:
  as basis of positive responses, 332
  defined, 332
  as design attribute, 10
  and human well-being, 62
  intermediate, preference for, 233
  in natural environment, 232–233
  in Paris, 240
Francis, Mark, 173
Freeness, sense of, 234
Freiburg, Germany, 175
Frei Otto (Munich Olympics), 30
Freshwater, *see* Water
Freud, Sigmund, 23
Freudian theory of incest avoidance, 23
Fuller, Buckminster, 30–32, 351, 356
Function, in biomimicry, 29
Functional site surveys, 40

Gaia, revenge of, 220
Garden Cities of Tomorrow (Ebenezer Howard), 298
Gardens:
  bio-inspiration, 38–39
  choice of plants for, 220
  health benefits of, 326
  in healthcare facilities, 97–98, 102–103, 208
  Japanese, 231
  New York City community gardens, 291
  roof, 50
  urban, 287
  water, 51
  Zen, 299
Gardens:
  indoor, *see* color Figures 4-5, 22-6
Gardner, Howard, 208
Gastric freezing, 108
Gate control theory, 93, 95
Gaudí, Antoni, 78
Gehl, Jan, 154
Gehry, Frank, 129
Gender, restorative design and, 146
Genetic dependence on environment, 61–63
Genetic evolution, cultural evolution and, 24
Genetic heritage:
  epigenetic rules as, 22–24
  transcendence of, ix
Genzyme Building, Cambridge, Massachusetts, 17
Geodesic domes, 31
Geographic connection to place, as design attribute, 12
Geological features, as design attribute, 8
Geometric qualities:
  and human interaction with nature, 80
  of native plants growing wild, 79–80
  for neurological nourishment, 64–65
  preferred by humans, 76
  sacred geometry, 28
Geomorphology, as design attribute, 9
Geothermal springs, 46
Gerbner, George, 207
Germany, 164–165, 279–280
Gewerbehof Prisma, Nuremberg, Germany, 342
Gibson, J. J., 255
Gibson, Larry, 218–219
Gilliam, Frank, 207
Glass, 254. *See also* Viewing nature through glass
Global warming, 222
Going Native (Mike Archer and Bob Beale), 289
Good to Grow initiative, 172
Gratzel cells, 34
Green Builder Program, Austin, Texas, 309
Green building movement, 338
Green Communities Program, 301
Green design:
  focus of, 206
  for schools and school grounds, 167
  success strategies for, 220–221
  for urban areas, *see* Urban design
Green design movement, x
Green exercise, 111
Greening the Building and the Bottom Line (RMI), 337
Greenman, Jim, 162
Green neighborhoods, 279
Green roofs, 50, 278, 353
Greenspaces, 113, 285–287
  See also color Figure 10-37
Green Street program, Portland, Oregon, 175
Green streets, 282–283
Green urbanism, 210–211, 278–280, 297–305
  David and Joyce Dinkins Gardens, 301–302
  Highlands' Garden Village, 302–304
  and importance of connection to nature, 299
  intrinsic implications for, 298
  origin of, 298
  Via Verde, 304
Green Urbanism (Timothy Beatley), 210–211
Greenways, *see* Cul-de-sacs and greenways
Greenwich Millennium Village, London, United Kingdom, 281
Griffin, Corey, 338–339
Grimshaw, Sir Nicholas, 304
Group Zo, 192
Grover, Karan, 351–352
Growing Up in Cities project, 195
Growing Vine Street, Seattle, Washington, 282–283
Growth and development, as design attribute, 10
Guenther, Robin, 331–332

Habitats, as design attribute, 8
Hagen Town Hall, Germany, *see* color Figure 4-14
Hall, Sir James, 245
Halprin, Lawrence, 178
Hammarby Sjöstad, Stockholm, Sweden, 279, 280
Hammill Family Play Zoo, Brookfield, Illinois, 161, 172
Hannover Green Ring, Hannover, Germany, 279–280
Haptic system, 255

Harkness Tower, Yale University, 17
Harlem Congregations for Community Improvement (HCCI), 301
Harman, Jay, 36
Harmonic relationships, sensitivity to, 232
Harmony, spatial, 11
Hartig, Terry, 111, 208
Hattersheim Town Hall, Germany, 284
Havel, Václav, 39
Haworth showroom, Chicago, Illinois, *see* color Figure 4-3
Hawthorne effect, 337–339
Hazards, creating illusion of, 235
HCCI (Harlem Congregations for Community Improvement), 301
Healing spas, 46
Health:
- and access to daylight, 121–123
- and access to fresh air, 123–124
- and access to natural environment, 126
- benefits from contact with nature, 107–116, 208–209
- biophilic urges related to, 4
- childhood lifestyle threats to, 156–160
- coal mining's effects on, 220
- design of childhood spaces for, 163–166
- direct link of nature experiences and, 208–209
- diseases of the spirit, 316
- evidence supporting beliefs about, 109
- impact of low environmental impact design on, 5
- and indoor plants, 128
- and proximity to windows, 120
- and socioeconomic position, 147

Healthcare design, 87–103
- art's effects on health outcomes, 96–97, 340
- biophilic design recommendations for, 102–103
- daylight exposure, 98–100
- environments mimicking nature in, 62, 208
- gardens in, 97–98, 208
- and health outcomes, 88, 138, 299
  - *See also* color Figure 8-8
- for pain reduction, 93–96, 326, 327
- and stress as problem in healthcare, 88–89
- for stress reduction, 89–94, 326–328
- windows in, 130

Health outcomes, 88, 138
Heating, natural, 126–127
Heerwagen, Judith, 69, 267, 325–326, 338, 344
Height, human preference for, 23
Helsinki, Finland, 279
Heraclitean movement, 231
Heraclitus, 45
Herbert, Frank, 46
Herman Miller, 338
Heschong, Lisa, 338
Heschong Mahone Group, 120, 328
Het Groene Dak, Netherlands, 211
Hicks, Edward, 246
Hierarchically organized ratios and scales, as design attribute, 10–11
Highlands' Garden Village, Denver, Colorado, 302–304
- *See also* color Figure 18-3

High Line Canal, Denver, Colorado, 215
High Point, Seattle, Washington, 175
Hildebrand, Grant, 238, 244, 341
Historic connection to place, as design attribute, 12
Holl, Steven, 354
Holocene, 221
"Home," human need for, 6
"Home range," 290
Home zones, 194
Hopkins Architects, 16
Horizon, in imagery of prospect, 235
"Houses That Sing," 314
Houston, Texas, bats in, 293
Howard, Ebenezer, 298
How the Mind Works (Steven Pinker), 228
Hudson River School, 46
Human-animal interactions, 71–73
Humane human environment, 227
Humanistic attachment to water, 44
Humanities/humanistic social sciences, 22–25
Human-machine interactions, 71–73
Human nature, 21–25
- as abstract/mechanistic, 67
- and basis of aesthetic judgment, 23–24
- as biological, 68–73
- biophilic tendencies in, 23
- conceptions/levels of, 66–67
- consilient definition of, 22–23
- in constructing buildings and cities, 59–60
- in relationship/response to built environment, 59
- and sensory/emotional connection to built environment, 61
- and theories of incest avoidance, 22–23
- as transcendent, 73–74

Hume, David, 107
Humphrey, Nicholas, 229, 232
Humphries, Susan, 168
Hundertwasser, Friedrich, 79
HVAC systems, 123–125, 127–128
Hyatt Hotels, 343
Hydromimicry, 54
Hydrotherapy, 46

Ice carnivals, 56
Illich, Ivan, 48
Imaginary elements, 259
Immanuel, Kerry, 216
Immune system health, 159
Implementing biophilic design, 347–356
- by aiming higher, 348–350
- conceptual framework for, 66
- by rebalancing the modern environment, 352–353
- by recognizing new ethic for excellence, 353–355
- by returning to buildings that support life, 350–352

Inca architecture, 315
Indigenous materials, as design attribute, 12
Indirect experience with nature, 5–6
Indoor plumbing, 50
Industrial waste, 48
Information:
- as design attribute, 14
- and human-machine interactions, 71–73

Informational fields, 75
Information richness, as design attribute, 9
Infrastructure, in urban design, 282–285
Inha University Hospital, Korea, 122
Innovation for Conservation program, 40
Inside-outside spaces, as design attribute, 11
Institute of Traffic Engineers, 194
Institutionalized childhood, 161
Instorative environmental design, 135
Integration:
- of culture and ecology, 12–13
- of parts to wholes, 10
- of water and earth, 53

Interconnectedness, 21. *See also* Consilience
Intergovernmental Panel on Climate Change, 216
Interior design, biophilic strategies/priorities for, 330
International Netherlands Group bank, Amsterdam, Netherlands, 342–343
Iridescent color, 37
Isamu Noguchi Sculpture Court, Bloch Building, Nelson-Atkins Museum of Art, *see* color Figures 23-10, 23-11
Ise Shrines, Japan, 321
IslandWood, Bainbridge Island, Washington, *see* color Figures 13-5, 13-7
Ivy, Robert, 236

Jacobson, Max, 341
Japan, decrease in play activities in, 207–208
Japanese gardens, 231
Jenkins, Janet, 293
John Amos coal plant, 218
Johns Hopkins University, 110
Johnson, Nathanael, 243
Johnston, Lindsay, 343
John Todd Ecological Designs, Inc., 37–38
Jonathan Rose Companies, 301
Jones, E. Fay, 236, 237, 355, 356
Joye, Yannick, 62
Jubilee Campus, University of Nottingham, United Kingdom, 16
Jukkasjarvi, Sweden, 56

Kahn, Louis I., 229, 237, 353–354
Kaiser Family Foundation, 207
Kaplan, Rachel, 208, 338
Kaplan, Stephen, 208, 267, 338
Katcher, Aaron, 230–231
Keeney, Brad, 321
Kellert, Stephen, 32, 44, 133–134, 136, 206, 209–210, 244, 325
Kelley, Sun Ray, 314
Kelly, Ellsworth, 247, 248
Keskuspuisto park, Helsinki, Finland, 279
Kids Together Park, Cary, North Carolina, 171
Kieran Timberlake Associates, 245–246
Kinetic sculpture, 53
Kings Park, Perth, Western Australia, 293
Kitazawagawa River Nature Path, Tokyo, Japan, 192
Kitchen, outdoor, *see* color Figure 19-1
Kohler, 50
Konarka, 29, 34
Kronsberg, Hannover, Germany, 280
Kulikauskas, Andrius, 70–71
Kuo, Frances, 110
Ku-ring-gai Council, 294–295
Kurswell, Ray, 308

Laarman, Joris, 33, 34
Landscape design, 38, 329
Landscape ecology, as design attribute, 12
Landscape features defining building form, as design attribute, 12
Landscape orientation, as design attribute, 12
Landscape Park, Duisberg-Nord, Germany, 286–287
Language, aesthetic, 245
Last Child in the Woods (Richard Louv), 153, 217
Las Vegas water entertainment, 48
Lawn, 63
Leadership in Energy and Environmental Design (LEED), x, 5, 211
  beauty ignored in, 244
  biomimetic innovation credits in, 40
  for Neighborhood Development, 194
  rewards for biophilia in, 333, 344
  school sites, 166
Learning Landscapes Alliance, 168
Le Corbusier, 244, 298, 353
Lee, John, 343
LEED, *see* Leadership in Energy and Environmental Design
Legacy Health, Glacier Creek, Washington, 100
Legacy Health, Portland, Oregon, 97
Leonardo da Vinci, 45, 46
Lerner, Kelly, 314
"Letter from Iowa" (Nathanael Johnson), 243
Lewis, C. S., 214
LID (low-impact development), 282
"Life style centers" (shopping), 232
Light, 11, 90–91. *See also* Natural light (daylight/sunlight)
Light, Andrew, 294
Light and space (as design element), 11, 15, 17
Lighting Research Center, Rensselaer Polytechnic Institute, 121–122
Light pools, as design attribute, 11
Linked series, as design attribute, 10
Linz, Austria, 278
Living architecture, 74–76, 313–322
  characteristics of, 313–314
  and current architectural approaches, 317
  deeper essence of beauty in, 315
  drivers toward, 353
  and forces of cultural change, 317–320
  for healing diseases of the spirit, 316
  historical examples of, 315–316
  role and focus of, 314–315
  sacred in, 320–322
  as visual art, 314, 315
Living Building Challenge, 351
Lobell, John, 237
Loblolly House, Taylor's Island, Maryland, 245, 247–248
  See also color Figures 14-10, 14-11
Loftness, Vivian, 209, 332
Loos, Adolph, 266
Los Angeles River, 47, 190–191
Lotus leaves, *see* color Figure 3-8c
Louv, Richard, 153, 217, 222, 289, 326, 328, 350
Love, 73
Lovelock, James, 216, 220
Lovett, Wendell, 266, 268
Low environmental impact design, x, 5
  in restorative environmental design, 139–140
  technologies for, 134
Low-impact development (LID), 282
Lucent Technologies, 34
Lyndhurst, Tarrytown, New York, 257
  See also color Figure 15-6
Lyons, Mohawk Owen, 40

McHarg, Ian, 217
Machu Picchu, 315
McLaren Technology Centre Research Centre, London, United Kingdom, 53
McMaster University, 109
Maeger, Leonard, 326
Malmö, Sweden, green rooftops, 287
MARAG film, 35
Marine sponges, *see* color Figure 3-6b
Market research, 344–345
Marquis, Robert, 178

Marrin, West, 45, 46
Martin Luther King Jr. High School, Berkeley, 167
Mason-Wolf Associates, 283
Massey Energy Inc., 217, 219
Massing, sculptured, 269
Master Gardener Program study, 112–113
Mastery over nature:
  as design attribute, 13–14
  as political, 214
Materials:
  for biophilic design, 78
  as design attribute, 7, 12
  indigenous, 12
    See also color Figure 20-7
  natural, 7
    See also color Figures 20-8, 21-2
  off-the-shelf, 79
Material event cycle, 310
Mattheck, Claus, 33
The Meadows, Berkeley, California, 180–181
  See also color Figure 10-39
"Mean world syndrome," 207
Mechanistic human nature, 66, 67
La Mer (Debussy), 46
Mercedes-Benz, 33
Metamorphosis, as design attribute, 13
Metrovesca, 342
Micro-restorative experiences, 138
Middlebury College, Vermont, 245, 249–251
  See also color Figures 14-17, 14-18
Mies van der Rohe, Ludwig, 245
Migration corridors, 38
Mind, physical basis of, 24
Mineral springs, 46
Minigreenways, 190, 191
Mining, coal, 217–220
Mirrors, 314
Mithen, Steve, 46
Mithun office building, Seattle, Washington, *see* color Figure 13-13
Mixed-use development, 18
  See also color Figure 18-3
Mobility, children's opportunities for, 174
Mogavero Notestine and Associates, 179
Mole Hill, Vancouver, British Columbia, 283
Monet, Claude, 46
Montefiore Hospital, Pittsburgh, Pennsylvania, 122
Montgomery Park, Boston, Massachusetts, 181
Mont-Saint-Michel, France, 18
Moore, Charles, 255, 256
Moore, Ernest, 111
Moore, Robin, 169, 207
Moralistic attachment to water, 45
Morra Park, Drachten, Netherlands, 210–211
Mosweniam, 353
Movement, 230–231
  of people in activity cycles, 144–145
  sense of freeness and options for, 234
  of water, 46, 50
Mud bricks, Egyptian use of, 315
Musée d'Orsay, Paris, France, 265
Music, 46
Mystery:
  preference for, in natural scenes, 267
  theological concept of, 74, 75

National Institute of Environmental Health Sciences, 114
National Institutes of Health (NIH), 114, 332
National Museum of the American Indian, Washington, DC, 51, 52
National Science Foundation (NSF), 332
National Scientific Council on the Developing Child, 208
National Sporting Goods Association (NSGA), 207
National Wildlife Federation (NWF), 220, 331
Natural analogs, 340
Natural Building Colloquia, 314
Natural cooling, 127
Natural heating, 126–127
Natural heritage, transcendence of, ix
Naturalistic attachment to water, 44
Naturalistic dimension of biophilic design, 5–6
Naturalist intelligence, 208
Natural Learning Initiative (NLI), 162, 163, 165
Natural light (daylight/sunlight):
  benefits of, 121–123
  as design attribute, 7, 11
  in healthcare design, 98–100, 102
  human attunement to, 90–91
  and school performance, 328
  and Seasonal Affective Disorder, 230
  variability in, 122
  water interaction with, 50
Natural materials:
  as design attribute, 7
  interaction of water and, 50
  reuse of, 78, 249
Natural patterns and processes, as design element, 9–11, 15, 17
Natural Resources Defense Council, 194
"Natural rights," 223
Natural sciences:
  borderland disciplines in, 22
  common sequence in, 24
  humanities/humanistic social sciences vs., 22–25
Natural shapes and forms, as design element, 8–9, 15, 16
Natural springs, 46
Natural ventilation, 123–125
Nature-deficit disorder, 208, 217, 222, 328
Nature/natural environment:
  adult environmental stewardship and childhood experience of, 155
  basic need for access to, 227
  and childhood health problems, 156–160
  and climate change, 222
  consequences of designing in adversarial relation to, x
  degradation/depletion of, 5
  evolved human-nature relationships, 13–15, 18
  health benefits of contact with, 107–116, 208–209
    See also color Figure 8-8
  human need for contact with, 3–4
  humans' preferred elements in, 23
  liked and disliked features of, 139
  organizing urban life around, 287–289
  real vs. simulated, 332
  sensory richness/variety in, 228–235
  values/constructs of our attachment to, 44–45
\Nature Preschool, Schlitz Audubon Nature Center, 165
Nature preschools, 164–166
Negativistic attachment to water, 44
Neighborhoods, green, 279
Neighborhood parks, 169–171

Nelson-Atkins Museum, Kansas City, Missouri, 354
See also color Figures 23-10, 23-11
Neocortex, 308–312
Netherlands:
children's mobility study, 174
urban green spaces networks, 279
woonerf (residential precincts), 188
Neurological science:
basis for aesthetic response, 61–63
basis for design, 308–312
children's brain architecture, 208
cognitive, 22, 70
monitoring of brain arousal, 23
nourishment, neurological, 62–65, 233
pattern recognition, 70–71
New Housing New York, 304
See also color Figure 18-6
Newland Communities, 210
Newman, Oscar, 129
New Orleans, Louisiana, hurricane damage, 216
New York City, New York, 48, 277
See also color Figure 23-12
New York City green spaces, 291
Nightingale, Florence, 326
NIH, *see* National Institutes of Health
NLI, *see* Natural Learning Initiative
Noise pollution, 36, 221
Noodland Greenway, Stockholm Sweden, 184
North Carolina State University, 162
Northern Life Tower, Seattle, Washington, 269, 270
Northpark, Irvine Ranch, California, 189
Northwestern University, 109
Nourishment, neurological, 62–65, 233
NSF (National Science Foundation), 332
NSGA (National Sporting Goods Association), 207
Nurse logs, *see* color Figure 20-7
NWF, *see* National Wildlife Federation

Oakey, David, 37
Oakland, California, 173
Obesity, 156, 173
Occupation, restorative design and, 147
Oil depletion, 318
Olmstead, Frederick Law, 211
On Adam's House in Paradise (Joseph Rykwert), 244
On Growth and Form (D'Arcy Wentworth Thompson), 32
On Human Nature (Edward Wilson), 65
Ontario Place, Toronto, Ontario, 56
Open spaces:
barbering/manicuring of, 221
high-density development vs., 331
human preference for, 23
Order, as design attribute, 13
Organic dimension of biophilic design, 5–6
Organic forms and structures, mimicking, 32–34
Orians, Gordon, 267, 338
Orientation to landscape, 12
The Origins of Architectural Pleasure (Grant Hildebrand), 341
Ornamentation:
classic function of, 256
interior, 258–260
as natural analog, 340
for neurological nourishment, 64
of picture windows, 256–258
See also color Figures 15-6, 15-8, 15-9, 15-16
serendipity in, 231
Orr, David, ix–x
Ortiz, Iñigo, 342
Ortiz Leon Architects, 342
Outdoor kitchen/breezeway, *see* color Figure 19-1
Outdoor nursery schools, 164–166
Oval forms, as design attribute, 8
Ove Arup & Partners, 35

Paimio Sanatorium, 123
Pain reduction, healthcare design for, 93–96, 100, 326, 327
Pallasmaa, J., 229
Paris, France, 238–240
Parks:
city, 173
neighborhood, 171
pocket, 191
school grounds as, 169–170
urban, standards for, 290
Paseo del Rio River, 48
Passive survivability, 127–128
Pathways to buildings, biophilic features for, 145
See also color Figure 10-48
Patina of time, as design attribute, 9–10
Pattern(s), 80–81
brain-arousing, 23
and complexity theory, 300
as design element, 9–11, 15, 17
fractals, 10, 62
genetic sense of, 232
in human-machine interactions, 72
preferred, 338
from sensory experience, 70–71
similarities in, 232–233
spatial, 341
Patterned wholes, as design attribute, 10
Patterning, rhythmic, 257–258
A Pattern Language (Christopher Alexander), 71, 80, 336, 337
Patterns of Home (Jacobson, Silverstein, and Winslow), 341
Patterson Park neighborhood, Baltimore, Maryland, 192, 193
See also color Figure 10-55
Pavilion and Reflecting Pool, Toronto, Ontario, *see* color Figure 4-6
PAX Scientific, 36
Peaceable Kingdom, 246, 247
Pearce, Mick, 35, 36, 343
Pebble Project, Center for Health Design, 130
Pedestrian spaces, 80
Peggy Notebaert Center, Chicago, IL, 54
Performance:
and access to daylight, 121–122, 328
and positive experience of nature, x
and views of nature, 338
Peril, as survival-advantageous, 269–271
Perry Rose, 302
Peterhouse College, Cambridge, 247
Philadelphia, ants in, 277
Philadelphia Water Department, 191
Philip Merrill Environmental Center, Annapolis, Maryland, 328
Phipps Houses, 304
Photosynthesis, mimicking of, 29
Physical sciences, changes in, 318
Physicians' Health Study, 109
Physiological restoration, 91
Piazza del Duomo, approach to, 268
Picture windows, ornamentation of, 256–258
See also color Figures 15-6, 15-8, 15-9, 15-16
Pike Place Market, Seattle, Washington, 231
Pinker, Steven, 228
Place-based dimension of biophilic design, 6

Place-based relationships, as design element, 12–13
Placelessness, ix, 6, 13
Place of residence, restorative design and, 146–147
Planning and Urban Design Standards (APA), 290
Plants:
accessibility of, 80
as design attribute, 7
human connection with, 64
in children's play areas, *see* color Figures 10-5, 10-10
and sick building syndrome, 128, 129
variation and similarity in, 232
Play activities. *See also* Childhood spaces
adult recollections of, 176
children's views of, 174–176
decrease in, 207–208
UK children's informal play study, 183–184
at Village Homes, 186–187
Pliny the Elder, 46
Pocket parks, 191
Pollution:
air, ix, 218, 220
noise, 221
water, ix, 38, 48, 219
Pools of light, 11
Pools of water, 50–51
Portland, Oregon, ecoroof bonus, 290
Potsdamer Platz, Berlin, Germany, 55, 56
Powerlink, 286
Prisma, Nuremberg, Germany, 54
Productivity:
and access to nature, 126, 328
biophilic urges related to, 4
as driver of green building movement, 338
and efficiency of design, 337, 338
and Hawthorne effect, 337–339
and low environmental impact design, 5
measurement of, 337
with window views, 120, 121
Project PLAE, 170
Promenade Plantée, Paris, France, 18
Proportions, natural, 28, 300
Prospect, 234–235, 341
as design attribute, 13
human preference for, 23
in retirement home case study, 272–274
as survival-advantageous, 265–267
Protection, as design attribute, 13
Protective environmental design, 134–135
Providence, Rhode Island, 48
Providence St. Vincent Hospital, Portland, Oregon, 130
Psychoevolutionary theory, 137
Psychological boundary around bodies, 255
Psychological disorders, benefits of sunlight for, 122
Psychological health, outdoor experiences and, 158
Psychological restoration, 140, 141
Pyle, Robert Michael, 212, 286, 290

Qi energy, 318–319
Quantum nonlocality, 318

Randomized controlled trials, 108–110, 113
Rationality, wholeness vs., 319–320
Ratios:
hierarchical, 10–11
mimicking natural proportions, 28
Recycled materials, *see* color Figure 13-11
Rede Lecture (C. P. Snow), 22
Reflected light, as design attribute, 11
Refuge, 234–235, 341
as design attribute, 13
human preference for, 23
in retirement home case study, 272
as survival-advantageous, 265–267
Regional-level urban design, 279–280
Reid Dennis house, Sun Valley, Idaho, 269–270
REI Denver, *see* color Figure 13-11
REI Seattle, *see* color Figure 13-9
Releasers, 24
Religious architecture, 75–76, 318–319. *See also* Sacred spaces
Religious concept of mystery, 74, 75
Religious uses of water, 46
Relph, Edward, 6
Remington, Charles, 214
Repetition, 259
See also color Figure 13-9
Research on benefits of nature contact:
evidence for benefits, 110–113
funding for, 114
limits to claims, 115–116
need for building evidence base, 113–115
Residential environments:
biophilic forms of, 176–177
children's play areas in, 174
clustered housing and shared outdoor space, 177–183
converted back alleys, 192–193
See also color Figure 10-55
cul-de-sacs and greenways, 183–192
providing for children's needs in, 173–174
street design for, 156, 157
woonerf (residential precincts), 193–194
Resilience, 233–234
See also color Figure 13-11
Resource consumption, ix–x
Reston, Virginia, 187
Restoration theory, 90
Restorative environmental design, x, 5, 133–149
benefits of, 142
biophilic design in, 5
characteristics of, 139–142
and experiences of people in built environment, 145–148
intrinsic implications of, 298
low-impact, environmentally sensitive technologies in, 134
prospects and challenges for, 148–149
protective environmental design vs., 135
protective functions of, 139
restoration perspective in, 134–136
tandem processes in, 139–141
theory/empirical research on, 136–139
time and place matters in, 142–145
Retirement home case study, 271–275
Reuse of materials, 78, 249
Reverence, as design attribute, 14
Rhyming (variation), 232
Rhythm:
appreciation of, 232
in window walls, 257–258
Rhythmization, 260
Richard Dattner Architects, 304
Richness, sensory, 228–230
Rivanna River Greenway, Charlottesville, Virginia, 285
Rivers, 47
RMI, *see* Rocky Mountain Institute
Roberts, Joan, 122
Robert Taylor Homes, Chicago, 110–111
Robots, 72–73, 79
Rock Hill, South Carolina, 190

Rocky Mountain Ditch, 302
Rocky Mountain Institute (RMI Headquarters), Snowmass, Colorado, 336, 337, 338
Ronald Reagan Airport terminal, 16
Roof gardens, 50. *See also* Green roofs
  See also color Figure 18-6
Rose Kennedy Greenway, Boston, Massachusetts, 192
Rose SmartGrowth Fund, 345
Ruskin, John, 7, 238
Rusk Institute of Rehabilitation Medicine, New York City, New York, 97
Rykwert, Joseph, 244, 245
Ryoanji shrine, Kyoto, Japan, 267, 268

Saarinen, Eero, 7, 340
Sacramento Municipal Utility District Call Center, 120, 121
Sacred geometry, 28
Sacred sites, 64–65
Sacred spaces, 75, 320–322
  See also color Figure 20-6
Sacrifice zones, 217, 222
SAD (Seasonal Affective Disorder), 230
St. Francis Square, San Francisco, California, 178–179, 182, 195
  See also color Figure 10-35
Sainte Chapelle, Paris, 236
Saint-Exupéry, Antoine de, 43
Salk Institute, La Jolla, California, 237–238
Salt water, 44
Samara House, West Lafayette, Indiana, 41
  See also color Figures 3-20
San Antonio, Texas, 48
San Antonio River, 48
San Diego Regional Canyonlands Park, 211
San Francisco Museum of Modern Art, 270
Sanitas Corporation headquarters, Madrid, Spain, 341–342
  See also color Figure 22-6
San Raffaele Hospital, Milan Italy, 122
Saratoga Springs, New York, 46
Satisfaction, access to nature and, 328
SBS, *see* Sick building syndrome
Scales:
  for biophilic design, 77–79
  hierarchical, 10–11, 77
  for incorporating water, 49
  of urban design, 278, 279
Scandinavia, urban green spaces networks in, 279
Schools, 166–169. *See also specific schools*
  green design for, 284
  urban design for, 285
School grounds:
  as educational resources, 168
    See also color Figure 10-19
  green design for, 167
  as neighborhood parks, 169–170
Schoolyard Habitats program (NWF), 331
Scientific attachment to water, 45
SC Johnson Administration Building, 340
  See also color Figure 22-3
Scully, Vincent, x
Sculpture, water, 52, 53
Sculptured massing, 269
SEA program, *see* Street Edge Alternatives program
Searles, Harold, 4
Seasonal Affective Disorder (SAD), 230
Seasonal changes, 230
Seattle Library, Ballard branch, 278, 279
Seattle Street Edge Alternatives program, 282–283
Security:
  as design attribute, 13
  and sense of freeness, 234
Semper, Gottfried, 245
Sense of freeness, 234
Sense of place, 6
Sensory aesthetics, 227–241
  in biophilic buildings, 240–241
  in buildings, 236–237
  in cities, 238–240
  in human built environment, 235–326
  movement, 230–231
  prospect and refuge, 234–235
  resilience, 233–234
  sense of freeness, 234
  and sensory richness/variety in nature, 228–235
  serendipity, 231–232
  similarities in patterns, 232–233
  in spaces, 237–238
Sensory connection, to built environment, 61
Sensory experience:
  of built environment, 68–70
  patterns from, 70–71
Sensory variability, as design attribute, 9
SERA Architects, Inc., *see* color Figure 21-2
Serendipity, 231–232, 240
Setagaya Ward, Tokyo, Japan, 192
Seven Sisters Oak, 33, 34
Sexual attraction, 23
Shadow, as design attribute, 11
Shapes, 11. *See also* Natural shapes and forms
Shared outdoor space, *see* Clustered housing and shared outdoor space
Shaw, William, 293
Shayer, Michael, 157, 162
Shells, as design attribute, 8
Shepherd, Benjamin, 338–339
Shonkoff, Jack, 208
Shopping behaviors, 232
Sick building syndrome, 120, 128, 129
Sick building syndrome (SBS), 123–124
Sidwell Friends Middle School, Washington, DC, 54–55, 245–249
  See also color Figures 4-16, 14-4, 14-5, 14-6, 14-7
Silverstein, Murray, 341
Simonds, John Ormsbee, 44
Simulation of natural features, as design attribute, 9
Skin, building, 7, 247
  See also color Figure 14-5
SkyCeilings, 327–328, 331
  See also color Figure 21-1
Sky Factory, 327
Skylights, bio-inspired, 34–35
Snelson, Kenneth, 32
Snow, C. P., 22
Sobel, David, 191
Social ecology of stress and restoration, 143–144
Social health, outdoor experiences and, 158
Social Life of Small Urban Spaces (William Whyte), 231
Socially Responsible Investment (SRI), 345
Social sciences, 22, 24
Socioeconomic position, restorative design and, 147
Soil salinization, 38
Solar cells, photosynthesis-mimicking, 29, 34
Somé, Malidoma, 320
Sonoran Desert Conservation Plan, 293

Soules, Jim, 181
Sounds, 36
complexly ordered, 265
of nature, 230
of water, 46, 51
Southside Park, Sacramento, California, 179–180
See also color Figure 10-37
South Street Seaport Historic District, New York City, New York, 355
Space(s):
as design attribute, 10, 11
as design element, 11, 15, 17
perception of time and size of, 309
prospect and refuge in, 234–235
sense of freeness in, 234
sensory aesthetics in, 237–238
as shape and form, 11
Spaciousness, as design attribute, 11
Spadaro, Jack, 217–218
Spatial harmony, as design attribute, 11
Spatial variability, as design attribute, 11
Spirals, 8, 36
Spirit of place, 6. *See also* Living architecture
in biomimetic buildings, 40
as design attribute, 13
revealed by windows, 130
"Spirit of Place" conferences, 314
Spirituality, 14, 75
SRI (Socially Responsible Investment), 345
Staircases, *see* color Figure 15-14
State Prison of Southern Michigan, 111
Steen, Athena, 314
Steiner, Rudolph, 314, 342
Steiner Schools, 314
Stepner, Mike, 211–212
Stepwell at Chand Baori, Abhaneri, India, 53
Stevens, Wallace, 10
Stoller, Claude, 178
Stormwater management, 54–55, 175, 246
See also color Figures 4-16, 4-19
Straw bale buildings, 314
Strawberry Creek, Berkeley, California, 283
Streets:
green urban design for, 282–283
traffic danger and design of, 156–157
Street Edge Alternatives (SEA) program, 282–283
Stress:
environmental, 134
healthcare design for reduction of, 89–94, 326–328
as problem in healthcare, 88–89
social ecology of, 143–144
Structural color, 37
Structures:
"going against," 258–259
organic, 32–34
Suburban life, contact with nature in, 298
"The Suburbs Under Siege" (Amir Efrati), 189
Sullivan, Louis, 75, 257
Sullivan, William, 110
Sun exposure (for children), 166
Sun Life Plaza, Vancouver, British Columbia, 52
Sunlight, *see* Natural light
Survival-advantageous architectural characteristics, 263–275
complex order, 264–265
enticement, 267–269
peril, 269–271
prospect and refuge, 265–267
retirement home case study, 271–275
Sustainable design, x
biophilic design in, x
focus of, 206
and healthy child development, 160–161
limitation of concept, 206
with low environmental impact design, 5
as new standard practice, 349
Swan, Jim, 314
Sweden:
children's use of school play areas in, 209
cul-de-sac and greenway design in, 184
Swiss Re building, 30
Swiss Re London Headquarters, *see* color Figure 3-6a
Sydney, 16
Symbolic attachment to water, 45
Symbolic experience with nature, 6
Symmetry, 62, 77–78

Tapiola, Helsinki, Finland, 184
Television, impact of, 207
Temple of Hathor, Dendera, Egypt, 315
Tensegrity, 32–33
TERMES project, 36
See also color Figure 3-15
Texas Parks and Wildlife Department, 293
Thomas, Derek, 210
Thompson, D'Arcy Wentworth, 32
Thorncrown Chapel, Eureka Springs, AR, 236, 237, 355–356
Thornton, L. Camille, 299
Thornton, Troy, 299
The Thunder Tree (Robert Michael Pyle), 215, 216
Tidwell, Mike, 216
Time, scales and perceptions of, 308–309
Todd, John, 37–38
Tombs, 259–260
Toronto District School Board, 166
Touch, system of, 255
A Tour Through France (John Ruskin), 238
Toxins, plant, 221
Traffic danger, 156–157, 174
Trance-dancing, 321
Transcendent architecture, 74–76
Transcendent human nature, 67, 73–74
Transitional spaces, as design attribute, 10
Trees, 8, 115, 277
TropWorld Casino, Atlantic City, New Jersey, 53
Trudesland, Denmark, 280
Tsakopoulos, Angelo, 210
Tsakopoulos-Kounalakis, Eleni, 210
Tsui, Eugene, 30
Tubular forms, as design attribute, 8
Turing, Alan, 72
Turing Test, 72
Turner, J. Scott, 35
Twombly, R. C., 238

Uexküll, Ole von, 338–339
Ulrich, Roger, 7, 68–69, 111, 120, 208, 265, 299, 326, 328, 340
Understenshöjden, Stockholm, Sweden, 280–281
Unhealthy indoor environment, ix
United Kingdom:
child-nature gap, 207
children's informal play study, 183–184
children's mobility study, 174
cognitive and conceptual development, 157
cul-de-sac and greenway design, 184
healthcare facilities, 87
home zones, 194

U.S. Green Building Council, x, 194, 344, 348, 351
United States:
  children's use of school play areas, 209
  elimination of school recess, 166
  healthcare spending, 87
  and home zone model, 194
  increased population and building, 335
  nature center preschools, 165
  New Towns and master-planned communities, 187–188
  obesity, 156
  urban green neighborhoods, 281
Unity of knowledge, *see* Consilience
University of California-Santa Barbara, 34
University of Michigan law quadrangle, 16
Urban areas:
  canyons in, 211–212
  children's environments in, 172–174. *See also* Childhood spaces
  child- vs. wildlife-friendly areas in, 191
  parking in, 182
  return to balance in, 352–353
  serendipitous experience in, 232
  stressors in, 147
  sustainability in, 146, 147
  water handling, 55
  water pollution from, 48
  watersheds in, 38
Urban design, 277–295
  "activity friendliness" for children, 155
  biophilic elements, 278
  at building level, 278
  distribution of units/uses, 80
  to entice people outside, 284–285
  food production, 283
  green features, 278
  green neighborhoods, 280–281
  greenspaces, 285–287
  green urbanism, 297–305
  infrastructure, 282–285
  to organize urban life around nature, 287–289
  and reform of urban planning systems, 290–292
  at regional level, 279–280
  scales of, 278, 279
  schools, 285
  streets, 282–283
  vision for biophilic cities, 292–295
  water, 282–284
Urbanisme (Le Corbusier), 298
Urban planning systems, reform of, 289–292
Urban promenade model, 192
Utilitarian attachment to water, 45
Utopian movements, 298
Utzon, Jörn, 7, 16, 266
UV transparency (of glass), 122

Vacant lots, 221
Valley Quest, 191
Variation in nature, 228–230
Vauban, Freiburg, Germany, 280
Vaults, as design attribute, 8
Vegetative façades, as design attribute, 7
Vehicle exhaust, 157
Venice, Italy, 47
Venolia, Carol, 314
Ventilation, 7, 123–125
  See also color Figures 3-6a, 3-15
Vernacular dimension of biophilic design, 6
Vetter, Johanna, 343
Via Verde, New York, 304
Vicarious experience with nature, 6
Vienne, Italy, 297
Views. *See also* Prospect; Windows
  as design attribute, 7
  in healthcare facilities, 102
  scale of, 7
  through glass, *see* Viewing nature through glass
Viewing nature through glass, 253–260. *See also* Windows
  enhancing positive phenomenon of, 254
  and interior ornamentation, 258–260
    See also color Figure 15-10
  lack of touching with, 255, 256
  ornamented picture windows, 256–258
  as passive and only quasi-sensual, 265–266
  and psychological boundary around bodies, 255
Viikki, Helsinki, Finland, 280, 283
Village Homes, Davis, California, 173, 184–187, 190, 205–206, 209
  See also color Figure 10-48
Village of Woodsong, Shallotte, North Carolina, 178
VirtualWindows, 327
Vistas, as design attribute, 7
"Vitamin G," 113
Volme River, Germany, *see* color Figure 4-14

Wageningen, Netherlands, 177
Wales, play activities in, 174–175
Walking School Bus program, 173
Walls, ornamentation on, 258–260
Warm light, as design attribute, 11
Washington Environmental Yard, California, 167, 169–170
Waste generation, ix, 48
Wastewater treatment, biological, 55, 56
  See also color Figures 19-2, 19-3
Water, 43–56
  adjacent to architecture, *see* color Figure 4-14
  animistic traits of, 45
  biophilic aspects of, 44–47
  in children's play areas, 175
    See also color Figure 10-15
  in contemporary Western civilization, 47–49
  as design attribute, 7
  economic value of, 48
  human preference for, 23
  in indoor ecosystem, *see* color Figure 4-5
  in interior reflecting pool, *see* color Figure 4-3
  opportunities to enhance built environment with, 49–56
  religious uses of, 46
  in urban design, 282–284
Water Crater, Westphalen, Germany, 53
Water cycle, 309–310
  See also color Figure 19-2
Waterfalls, 51, 52
Water gardens, 51
  See also color Figure 4-6
Water pollution, ix, 38, 48, 219
Waters, Alice, 167
Waterscapes, 53
Weintraub, Lee, 304
Weizenbaum, Joseph, 72
Well-being:
  and biophilia, 3–4
  contact with nature for, 4
  environmental factors in, 62
  and low environmental impact design, 5
  and neurological nourishment, 69
  and positive experience of nature, x
Wells, Malcolm, 314

West Edmonton, Alberta, 48
Westermarck, Edward, 22
Westermarck effect, 22–23
Western Electric, Hawthorne plant, 337
Western Harbor project, Malmö, Sweden, 281, 288, 290–291
Westin Hotel, Kansas City, Missouri, 339
West Philadelphia Landscape Project, 191
Whewell, William, 21
Wholeness:
as design approach, 317
rationality vs., 319–320
Whyte, William, 231
Wilson, Alex, 127–128
Wilson, Edward O., x, 28, 63, 65, 76, 206, 222, 244, 325
Windows, 119–130. *See also* Viewing nature through glass
for access to views, 119–121
benefits of light from, 121–123
bio-inspired, 34–35
energy, health, and productivity benefits from, 125–126
fresh air and natural ventilation from, 123–125
in healthcare facilities, 102
for natural conditioning, 126–128
openable, 124–125
ornamentation of, 256–258
for passive survivability, 127–128
and spirit of place, 130
for transparency, 129–130
virtual, 327
Windsor, California, 183
Wines, James, 244, 335
Winslow, Barbara, 341
Wise, James, 332, 338
Witherspoon, Bill, 327
Wolch, Jennifer, 211
Woonerf (residential precincts), 188, 193–194
See also color Figure 10-56
Workplace, restorative experiences in, 138, 147
Wright, Frank Lloyd:
on biomimicry of function, 29
Fallingwater, 18, 235
on inspiration from nature, 229
Jones' work with, 236
letting outside in principle of, 355
materials used by, 78, 236–237
mystical perspective of, 75
patterns used by, 341
Prairie-style architecture, 8
refuge and prospect in houses of, 266
Samara House, 41
SC Johnson Administration Building, 340
wall ornamentation placement, 248
window design, 248
See also color Figures 15-8, 15-9, 15-10
The Wright Space (Grant Hildebrand), 341

Yale University Ingalls hockey rink, 7
Yao, Marissa, 339
Yoga Promenade, Tokyo, Japan, 192, 193

Zen gardens, 299
Zoomorphic (Hugh Aldersey-Williams), 33
Zoos, 69, 161, 172

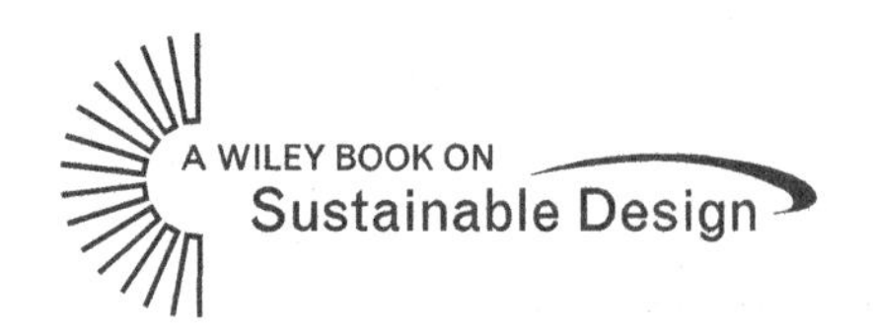

Environmental Benefits Statement

This book is printed with soy-based inks on presses with VOC levels that are lower than the standard for the printing industry. The paper, Rolland Enviro 100, is manufactured by Cascades Fine Paper Group and is made from 100 percent postconsumer, de-inked fiber, without chlorine. According to the manufacturer, the following resources were saved by using Rolland Enviro 100 for this book:

Mature trees	Waterborne waste not created	Water flow saved (in gallons)	Atmospheric emissions eliminated	Energy not consumed	Natural gas saved by using biogas
225	103,500 lbs.	153,000	21,470 lbs.	259 million BTUs	37,170 cubic feet